Instrumentation, Measurements, and Experiments in Fluids

Second Edition

Instrumentation, Measurements, and Experiments in Fluids

Second Edition

Ethirajan Rathakrishnan

Department of Aerospace Engineering
Indian Institute of Technology Kanpur, India

CRC Press
Taylor & Francis Group
Boca Raton London New York

CRC Press is an imprint of the
Taylor & Francis Group, an **informa** business

CRC Press
Taylor & Francis Group
6000 Broken Sound Parkway NW, Suite 300
Boca Raton, FL 33487-2742

First issued in paperback 2020

© 2017 by Taylor & Francis Group, LLC
CRC Press is an imprint of Taylor & Francis Group, an Informa business

No claim to original U.S. Government works

ISBN-13: 978-1-4987-8485-6 (hbk)
ISBN-13: 978-0-367-73670-5 (pbk)

Visit the Taylor & Francis Web site at
http://www.taylorandfrancis.com

and the CRC Press Web site at
http://www.crcpress.com

Dedication

This book is dedicated to my parents,

Mr. Thammanur Shunmugam Ethirajan

and

Mrs. Aandaal Ethirajan

Ethirajan Rathakrishnan

Contents

Preface

The first edition of this book, developed to serve as the text for a course on experiments in fluids at the introductory level for undergraduate courses and advanced level courses at the graduate level, was well received all over the world because of its completeness and easy-to-follow style.

Over the years the feedback received from the faculty and students made the author realize the need for adding exercise problems at the end of different chapters. Also, while using the chapter on wind tunnel in the lectures, students expressed that adding a section on internal balance would be useful. Some faculty conveyed that adding material on PIV would make the chapter on flow visualization more effective. Users of optical visualization for high-speed flows expressed the need for highlighting the possibility of obtaining quantitative information from the schlieren and shadowgraph pictures.

Considering the feedback from faculty and students, the following material is added in this edition.

• A section on internal balance in Chapter 3 on wind tunnels.
• A subsection highlighting the use of schlieren and shadowgraph pictures for qualitative and quantitative analysis of high-speed flows in Chapter 4.
• A detailed section describing the theory and application of particle image velocimetry in Chapter 4.
• A section on water flow channels giving the calibration and use of water flow channels for visualizing vortex formation in Chapter 4.
• The design, fabrication, calibration, and use of Hele-Shaw apparatus has been given in detail, highlighting the capability of this devise to simulate potential flow field, in Example 6.1.
• Some other important aspects, such as the limiting value pressure probe blockage for neglecting the blockage correction, is addressed through a detailed example in Chapter 7 on pressure measurements.
• Some new examples in the chapters on wind tunnel, pressure measurement, temperature measurement, and mass flow measurement.
• A considerable number of exercise problems at the end of many chapters.

I am indebted to Professor Junjiro Iwamoto, Department of Mechanical Engineering, Tokyo Denki University, Japan, for allowing me to take the schlieren and shadowgraph pictures used in Section 4.3.8 in his jet facility.

I thank Dr. Yasumasa Watanabe, Assistant Professor, Department of Aeronautical and Astronautical Engineering Department, University of Tokyo, Japan, for fabricating, calibrating, and testing the Hele-Shaw device during my stay with the Department of Advanced Energy, Graduate School of Frontier Sciences, University of Tokyo, Japan, as Visiting Professor in 2011.

My sincere thanks to Professor Shouiio Iio, Department of Mechanical Engineering, Shinshu University, Nagano, Japan, for providing his PIV facility and working with me in the visualization of the jet, discussed in Section 4.4, during my stay in his lab in the summer of 2014, while on JSPS Fellowship.

Finally, I would like to thank the faculty and students all over the world for adopting this book for their courses.

For instructors, a companion Solutions Manual that contains typed solutions to all the end-of-chapter problems and lecture slides for the complete book are available from the publisher.

Ethirajan Rathakrishnan

About the Book

The first edition was well received. Many faculty wanted exercise problems along with answers, PIV systems of flow visualization, water flow channels for flow visualization, and good pictures with schlieren and shadowgraph, with possible quantitative information extracted from these images. Also, there was a request for adding some more examples with specific details on probe blockage in the case of pressure probes and the theory of internal balances for wind tunnels.

Most of the chapters are provided with exercise problems along with answers. This will enable the users to prepare for the exams in a more effective manner. There is no single book covering the full spectrum of a complete experimental flows course, which should include: wind tunnels, flow visualization, hot-wire anemometers, analogue methods, pressure, temperature, volume/mass and force measurements, data acquisition, and uncertainty. Many books need to be referred to in order to acquire this material. This book covers all the requirements for a full course on experimental fluid mechanics.

About the Author

Ethirajan Rathakrishnan is professor of Aerospace Engineering at the Indian Institute of Technology Kanpur, India. He is well known internationally for his research in the area of high-speed jets. The limit for the passive control of jets, called the *Rathakrishnan Limit*, is his contribution to the field of jet research, and the concept of *breathing blunt nose (BBN)*, which simultaneously reduces the positive pressure at the nose and increases the low pressure at the base, is his contribution to drag reduction at hypersonic speeds. Positioning the twin-vortex Reynolds number at around 5000, by changing the geometry from cylinder, for which the maximum limit for the Reynolds number for positioning the twin-vortex was found to be around 160, by von Karman, to flat plate, is his addition to vortex flow theory. He has published a large number of research articles in many reputed international journals. He is a Fellow of many professional societies including the Royal Aeronautical Society. Rathakrishnan serves as the Editor-in-Chief of the *International Review of Aerospace Engineering* (IREASE) and *International Review of Mechanical Engineering* (IREME) journals. He has authored 11 other books: *Gas Dynamics*, 5th ed. (PHI Learning, New Delhi, 2013); *Fundamentals of Engineering Thermodynamics*, 2nd ed. (PHI Learning, New Delhi, 2005); *Fluid Mechanics: An Introduction*, 3rd ed. (PHI Learning, New Delhi, 2012); *Gas Tables*, 3rd ed. (Universities Press, Hyderabad, India, 2012); *Theory of Compressible Flows* (Maruzen Co., Ltd. Tokyo, Japan, 2008); *Gas Dynamics Work Book*, 2nd ed. (Praise Worthy Prize, Napoli, Italy, 2013); *Applied Gas Dynamics* (John Wiley, New Jersey, USA, 2010); *Elements of Heat Transfer* (CRC Press, Taylor & Francis Group, Boca Raton, Florida, USA, 2012); *Theoretical Aerodynamics* (John Wiley, New Jersey, USA, 2013); *High Enthalpy Gas Dynamics* (John Wiley & Sons Inc., 2015); and *Dynamique Des Gaz* (Praise Worthy Prize, Napoli, Italy, 2015).

In addition to the technical books above, Professor Ethirajan Rathakrishnan has authored the following literary books in Tamil: *Krishna Kaviam* (book on the life of Lord Krishna, from classical Tamil poetry) (Shantha Publishers, Royapettah, Chennai, India, 2014); *Naan Kanda Japan* (The Japan I Saw, in Tamil) (Shantha Publishers, Royapettah, Chennai, India, 2014); *Japanin Munnilai Ragasiam* (The Secrecy of Japanese Success, in Tamil) (Shantha Publishers, Royapettah, Chennai, India, 2015), and *Vallalaar Kapiam* (on St. Ramalinga Swamy, from classical Tamil poetry) (Vanathi Pathippagam, T. Nagar, Chennai, India, 2016).

Chapter 1

Needs and Objectives of Experimental Study

1.1 Introduction

As we know, the theory of potential flow is based on simplifying assumptions such as the fluid is barotropic, inviscid, and so on. Therefore, the potential flow theory cannot account for the profile drag acting on an object present in the flow field and for the boundary layer effects. Also, we know that the mathematical theory of boundary layer motion is highly complex. A considerable amount of theoretical work has already been done on inviscid compressible fluid flow at subsonic and supersonic speeds. But the theory of boundary layer for compressible flow has to develop a lot. Therefore, the theory available for fluid flow analysis is incomplete and needs to be supplemented by experiments. From the design point of view, experiments have two principal objectives:

1. They make it possible to determine the influence of various features of design, and modifications to them, in a safe, quick, direct, and relatively less expensive manner.

2. They provide information of a fundamental nature, usually in conjunction with theoretical work. By this means, the theory is confirmed or extended, thereby laying the foundation for future design improvements of a fundamental character.

The aim of this book is to discuss the fundamental aspects of the *experimentation in fluids*. In other words, our aim here is to acquaint ourselves with the need for experimental study and to gain insight concerning various applications of the available techniques for experimental study of fluid flows.

1.2   Some Fluid Mechanics Measurements

A large variety of measurement techniques are used in the field of experimental fluid mechanics. To have an idea about the different types of measurements, let us see the examples given below:

1.2.1   Wind Tunnel Studies

Wind tunnels are used for numerous investigations ranging from fundamental research to industrial aerodynamics. Many wind tunnel studies aim at the determination of forces on scaled models of aircraft, aircraft components, automobiles, buildings, and so on. Forces such as lift and drag acting on the models being tested are known to obey the following law of similitude

$$F = \frac{1}{2}\rho V^2 S C_N$$

(1.1)

where S is the surface area or cross-sectional area of the model, depending on the application. The force coefficient C_N is known to be a function of several non-dimensional parameters. The prime ones, among such dimensionless parameters used, in aerodynamics are

$$\text{Reynolds number} = \frac{\text{Inertia force}}{\text{Viscous force}} = \frac{\rho V l}{\mu}$$

$$\text{Mach number} = \frac{\text{Inertia force}}{\text{Elastic force}} = \frac{V}{a}$$

where ρ and V are the density and velocity of flow, respectively, μ is the dynamic viscosity coefficient, l is the characteristic length, and a is the speed of sound.

To correlate data, velocity is measured with pitot–static tube or hot–wire anemometer or laser Doppler anemometer, and the temperature and pressure are obtained with appropriate instrumentation. The forces and the moments on a model are usually determined with specially designed balance or surface pressure measurements. The density is usually calculated from the measured pressure and temperature.

1.2.2   Analogue Methods

By analogue methods, problems may be solved by setting up another physical system, such as an electric field, for which the governing equations are of the same form as those for the problem to be solved, with corresponding boundary conditions. The solution of the original problem may be obtained experimentally from measurements on the analogous system. Some of the well-known analogue methods for solving fluid flow problems are the *Hele–Shaw analogy*, the *electrolytic tank*, and the *surface waves in a ripple tank*.

1.2.3 Flow Visualization

Apart from the conventional methods of experimentally investigating flow patterns by means of pressure and velocity surveys, fluid flows lend themselves to numerous visualization techniques. Some of the popularly employed flow visualization methods for fluid flow analysis are flow visualization with *smoke, tuft, chemical coating, interferometer, schlieren,* and *shadowgraph.*

1.3 Measurement Systems

Basically, the main components of a measuring system may be classified into the following three categories.

1. The sensing element.

2. The signal converter.

3. The display.

1.3.1 Sensing Element

A sensing element is also called a *transducer.* For instance, the bulb of a mercury–in–glass thermometer, the diaphragm in a pressure transducer are sensing elements. The transducer is in some way "in contact" with the quantity to be measured and produces some signal which is related to the quantity being measured. A typical sensing process is illustrated schematically in Figure 1.1.

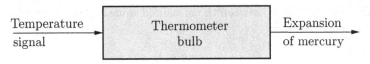

Figure 1.1: Sensing element response

1.3.2 Signal Converter

A signal converter is a device to convert the output from the sensing element to a desired form and feed the same to the display unit. A typical example of a signal converter is the amplifier which receives a small signal from the sensing element and makes it large enough to activate the display.

1.3.3 Display

The display is yet another vital part of a measuring system. It is here that the information from the sensing element, which is converted into a desired form by the signal converter, is read by the experimenter. A typical example of a display system is the combination of a dial and a pointer, as in the case of a dial-type pressure gauge.

1.3.3.1   Performance Terms

There are some commonly used performance terms associated with measurement systems. They are

* Accuracy	* Range
* Error	* Resolution
* Repeatability	* Sensitivity
* Reliability	* Dead space
* Reproducibility	* Threshold
* Lag	* Hysteresis

and so on. Now let us consider these performance terms one by one.

Accuracy

The accuracy of a measuring devise may be defined as the extent to which the reading given by it is close to the exact value. For example, if the accuracy of a mercury manometer is ± 1 mm, it means that when a mercury manometer is used to measure pressure of a flow it can only be stated that the pressure of the flow is lying within ± 1 mm of the manometer reading. Thus, a reading of 500 mm of mercury means that the exact value of the measured pressure is somewhere between 499 and 501 mm of mercury. Accuracy is generally expressed as a percentage of full–scale reading of the instrument.

Error

The error is the difference between the measured value and the true value of the quantity being measured.

Repeatability

The repeatability of an instrument is its ability to display the same reading as long as its sensor element is fed the same signal.

Reliability

The reliability of a measuring system is the probability that it will operate with an agreeable accuracy under the conditions specified for its operation.

Reproducibility

The reproducibility of a measuring device is its ability to display the same reading when it is used to measure the same quantity over a period of time or when that quantity is measured on a number of instants. Reproducibility of a device is also termed as *stability of the device.*

Range

The range of an instrument is the limits between the minimum and the maximum readings measurable by it. For example, the range of a thermometer which is capable of measuring temperatures between $-10°C$ and $110°C$ is $-10°C$ and $110°C$.

Resolution

The resolution of an instrument is the smallest change in the quantity being measured that will produce an observable change in the reading of the instrument.

Sensitivity

The sensitivity of an instrument is its response to any change in the quantity being measured.

Dead space

The dead space of a measuring device is the range of values of the quantity being measured for which it gives no reading.

Threshold

The threshold is the minimum level of the quantity that is being measured, which has to be reached before the instrument responds and gives a detectable reading. In other words, it is just a dead space which occurs when an instrument is used for reading from the minimum limit of its range.

Lag

The lag of an instrument is the time interval between the time of input and the time of display of the reading.

Hysteresis

The hysteresis is that characteristic which makes an instrument give different readings for the same value of measured quantity depending on whether the value has been reached by a continuously increasing change or a continuously decreasing change.

1.4   Some of the Important Quantities Associated with Fluid Flow Measurements

Pressure, temperature, and volume flow/mass flow rate are the prime quantities associated with any fluid flow experimentation. In fact, once these quantities are measured independently, many quantities of practical importance, like acceleration of a fluid stream, density of flow, energy associated with a flow, force acting on an object placed in a flow field, and so on can easily be determined. The units commonly employed for the above quantities are given in Tables 1.1 and 1.2.

Table 1.1 Units

Quantity	Unit
Pressure	$1 \text{ Pa} = 1 \text{ N/m}^2$
	$1 \text{ bar} = 10^5 \text{ Pa}$
	$1 \text{ torr} = 1 \text{ mm of mercury}$
	$1 \text{ atm} = 101.325 \text{ kPa}$
	$= 1.01325 \text{ bar}$
	$= 760 \text{ mm of Hg at } 0°\text{C}$
Specific volume	$1 \text{ m}^3/\text{kg} = 1000 \text{ l/kg}$
	$= 1000 \text{ cm}^3/\text{g}$
Temperature	$T \text{ (K)} = T \text{ (°C)} + 273.15$
	$\Delta T \text{ (K)} = \Delta T \text{ (°C)}$
Volume	$1 \text{ m}^3 = 1000 = 10^6 \text{ cm}^3 \text{(cc)}$

Table 1.2 Conversion factors

Dimension	Unit
Acceleration	$1 \text{ m/s}^2 = 100 \text{ cm/s}^2$
Area	$1 \text{ m}^2 = 10^4 \text{ cm}^2 = 10^6 \text{ mm}^2 = 10^{-6} \text{ km}^2$
Density	$1 \text{ g/cm}^3 = 1 \text{ kg/l} = 1000 \text{ kg/m}^3$
Energy, Heat,	$1 \text{ kJ} = 1000 \text{ J} = 1000 \text{ N.m} = 1 \text{ kPa.m}^2$
Work	$1 \text{ kJ/kg} = 1000 \text{ m}^2/\text{s}^2$
	$1 \text{ kWh} = 3600 \text{ kJ}$
Force	$1 \text{ N} = 1 \text{ kg.m/s}^2$
Length	$1 \text{ m} = 100 \text{ cm} = 1000 \text{ mm}$
	$1 \text{ km} = 1000 \text{ m}$
Mass	$1 \text{ kg} = 1000 \text{ g}$
	$1 \text{ metric ton} = 1000 \text{ kg}$
Power	$1 \text{ W} = 1 \text{ J/s}$
	$1 \text{ kW} = 1000 \text{ W} = 1.341 \text{ hp}$

1.5　Summary

From the discussions on the need and objective of experimental study, it is evident that the experimental studies are important from both fundamental and applied research points of view. The experimental investigations can broadly be classified into

- Direct measurements

- Analogue methods

- Flow visualization

A measurement system basically consists of

- The sensing element

- The signal converter

- The display

In an experimental investigation, an experimenter must have a clear idea about the experimental technique and the measurement system used. In addition to these, the researcher must have a thorough understanding of the problem to be studied, the principles of flow physics associated with the problem being studied, and the principles of operation of the instruments being used. A thorough knowledge of the fundamentals of fluid mechanics is inevitable for any successful experimentalist. Therefore, let us have a quick look at the basic principles underlying the fluid flow processes in Chapter 2, before actually getting into the experimental techniques.

Chapter 2

Fundamentals of Fluid Mechanics

2.1 Introduction

Gases and liquids are generally termed as fluids. Though the physical properties of gases and liquids are different, they are grouped under the same heading since both can be made to flow unlike a solid. Under dynamic conditions, the nature of governing equations is the same for both gases and liquids. Hence, it is possible to treat them under the same heading, namely, fluid dynamics or fluid mechanics. However, certain substances known as viscoelastic materials behave like liquids as well as solids, depending on the rate of application of the force. Pitch and silicone putty are typical examples of viscoelastic material. If the force is applied suddenly the viscoelastic material will behave like a solid, but with gradually applied pressure the material will flow like a liquid. The properties of such materials are not considered in this book. Similarly, non-Newtonian fluids, low-density flows, and two-phase flows such as gas liquid mixtures are also not considered in this book. The experimental techniques described in this book are for well-behaved simple fluids such as air.

2.2 Properties of Fluids

A fluid may be defined *as a substance which will continue to change shape as long as there is a shear stress present, however small it may be.* That is, the basic feature of a fluid is that it can flow, and this is the essence of any definition of it. Examine the effect of shear stress on a solid element and a fluid element, shown in Figure 2.1.

It is seen from this figure that the change in shape of the solid element is characterized by an angle $\Delta\alpha$ when subjected to a shear stress. Whereas, for the fluid element there is no such fixed $\Delta\alpha$ even for an infinitesimal shear stress. A continuous deformation persists as long as shearing stress is applied. The rate of deformation, however, is finite and is determined by the applied shear force and the fluid properties.

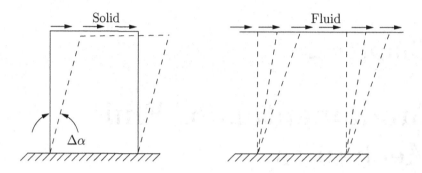

Figure 2.1: Solid and fluid elements under shear stress

2.2.1 Pressure

Pressure may be defined as the *force per unit area which acts normal to the surface of any object which is immersed in a fluid.* For a fluid at rest, at any point the pressure is the same in all directions. The pressure in a stationary fluid varies only in the vertical direction and is constant in any horizontal plane. That is, in stationary fluids the pressure increases linearly with depth. This linear pressure distribution is called *hydrostatic pressure distribution.* The hydrostatic pressure distribution is valid for moving fluids also provided there is no acceleration in the vertical direction. This distribution finds extensive application in manometry.

When a fluid is in motion, the actual pressure exerted by the fluid in the direction normal to the flow is known as the *static pressure.* If there is an infinitely thin pressure transducer which can be placed in a flow field without disturbing the flow, and it can be made to travel with the same speed as that of the flow then it will record the exact static pressure of the flow. From this stringent requirement of the probe for static pressure measurement, it can be inferred that exact measurement of static pressure is impossible. However, there are certain phenomena, like *"the static pressure at the edge of a boundary layer is impressed through the layer,"* which are used for the proper measurement of static pressure. *Total pressure* is that pressure which a fluid will experience if its motion is brought to rest. It is also called *impact pressure.* The total and static pressures are used for computing the flow velocity.

Since pressure is intensity of force, it has the dimensions

$$\frac{\text{Force}}{\text{Area}} = \frac{MLT^{-2}}{L^2} = \left[ML^{-1}T^{-2}\right]$$

and is expressed in the units of newton per square meter (N/m^2) or simply pascal (Pa). At standard sea level condition, the atmospheric pressure is 101325 Pa, which corresponds to 760 mm of mercury column height.

2.2.2 Temperature

In any form of matter the molecules are in motion relative to each other. In gases the molecular motion is a random movement of appreciable amplitude ranging from about 76×10^{-9} m under normal conditions to some tens of millimeters at very low pressures. The distance of free movement of a molecule of a gas is the distance it can travel before colliding with another molecule or the walls of the container. The mean value of this distance for all molecules in a gas is called the molecular *mean free path* length. By virtue of this motion the molecules possess kinetic energy, and this energy is sensed as *temperature* of the solid, liquid, or gas. In the case of a gas in motion it is called the *static temperature*. Temperature has units kelvin (K) or degrees celsius (degC), in SI units. For all calculations in this book, temperature will be expressed in kelvin, i.e., from absolute zero. At standard sea level conditions the atmospheric temperature is 288.15 K.

2.2.3 Density

The total number of molecules in a unit volume is a measure of the density ρ of a substance. It is expressed as mass per unit volume, say kg/m^3. Mass is defined as weight divided by acceleration due to gravity. At standard atmospheric temperature and pressure (288.15 K and 101325 Pa), the density of dry air is 1.225 kg/m^3.

Density of a material is a measure of the amount of material contained in a given volume. In a fluid, density may vary from point to point. Consider the fluid contained within a small spherical region of volume $\delta \mathbb{V}$ centered at some point in the fluid, and let the mass of fluid within this spherical region be δm. Then the density of the fluid at the point on which the sphere is centered can be defined by

$$\rho = \frac{\lim}{\delta \mathbb{V} \to 0} \frac{\delta m}{\delta \mathbb{V}} \tag{2.1}$$

There are practical difficulties in applying the above definition of density to real fluids composed of discrete molecules, since under the limiting condition the sphere may or may not contain any molecule. If it contains a molecule the value obtained for the density will be fictitiously high. If it does not contain a molecule the resultant value of density will be zero. This difficulty can be avoided over the range of temperatures and pressures normally encountered in practice, in the following two ways.

1. The molecular nature of a gas may be ignored and the gas is treated as continuum, i.e., does not consist of discrete particles.

2. The decrease in size of the imaginary sphere may be assumed to be carried
   to a limiting size. This limiting size of the sphere is such that, although
   it is small compared to the dimensions of any physical object present in a
   flow field, e.g., an aircraft, it is large compared to the fluid molecules and,
   therefore, contains a reasonably large number of molecules.

2.2.4 Viscosity

The property which characterizes the resistance that a fluid offers to applied
shear force is termed *viscosity*. This resistance, unlike for solids, does not depend
upon the deformation itself but on the *rate of deformation*. Viscosity is often
regarded as the stickiness of a fluid and its tendency is to resist sliding between
layers. There is very little resistance to the movement of the knife-blade edge-on
through air, but to produce the same motion through a thick oil needs much
more effort. This is because viscosity of oil is higher compared to that of air.

2.2.5 Absolute Coefficient of Viscosity

The absolute coefficient of viscosity is a direct measure of the viscosity of a
fluid. Consider the two parallel plates placed at a distance h apart, as shown in
Figure 2.2.

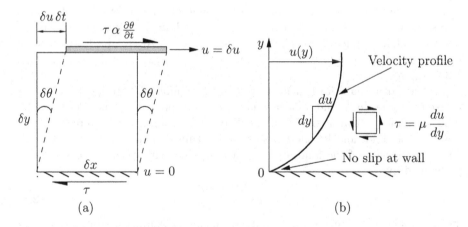

Figure 2.2: Parallel plates with fluid in between

The space between them is filled with a fluid. The bottom plate is fixed and
the other is moved in its own plane at a speed u. The fluid in contact with the
lower plate will be at rest, while that in contact with the upper plate will be
moving with speed u, because of no-slip condition. In the absence of any other
influence, the speed of the fluid between the plates will vary linearly, as shown
in Figure 2.2. As a direct result of viscosity, a force F has to be applied to each
plate to maintain the motion, since the fluid will tend to retard the motion of

the moving plate and will tend to drag the fixed plate in the direction of the moving plate. If the area of each plate in contact with fluid is A, then the shear stress acting on each plate is F/A. The rate of slide of the upper plate over the lower is u/h.

These quantities are connected by Maxwell's equation, which serves to define the absolute coefficient of viscosity μ. The equation is

$$\frac{F}{A} = \mu \left(\frac{u}{h}\right) \tag{2.2}$$

Hence,

$$\left[ML^{-1}T^{-2}\right] = [\mu] \left[LT^{-1}L^{-1}\right] = [\mu] \left[T^{-1}\right]$$

i.e.,

$$[\mu] = \left[ML^{-1}T^{-1}\right]$$

and the unit of μ is therefore kg/(m s). At $0\,\deg$C the absolute coefficient of viscosity of dry air is 1.71×10^{-5} kg/(m s). The absolute coefficient viscosity μ is also called the *dynamic viscosity coefficient*.

Equation (2.2) with μ constant does not apply to all fluids. For a class of fluids, which includes blood, some oils, some paints, and so called thixotropic fluids, μ is not constant but is a function of du/dh. The derivative du/dh is a measure of the rate at which the fluid is shearing. Usually μ is expressed as (N.s)/m^2 or gm/(cm s). One gm/(cm s) is known as a *poise*.

Newton's law of viscosity states that, "*the stresses which oppose the shearing of a fluid are proportional to the rate of shear strain,*" i.e., the shear stress τ is given by

$$\tau = \mu \frac{\partial u}{\partial y} \tag{2.3}$$

where μ is the absolute coefficient of viscosity and $\partial u/\partial y$ is the velocity gradient. The viscosity μ is a property of the fluid. Fluids which obey the above law of viscosity are called *Newtonian fluids*. Some fluids such as silicone oil, viscoelastic fluids, sugar syrup, tar, etc., do not obey the viscosity law given by Equation (2.3) and they are called *non-Newtonian fluids*.

We know that, in incompressible flow, it is possible to separate the calculation of velocity boundary layer from that of thermal boundary layer. But in compressible flow it is not possible, since the velocity and thermal boundary layers interact intimately and therefore, they must be considered simultaneously. This is because, for high-speed flows (compressible flows) heating due to friction as well as temperature changes due to compressibility must be taken into account. Further, it is essential to include the effects of viscosity variation with temperature. Usually large variations of temperature are encountered in high-speed flows.

The relation $\mu(T)$ must be found by experiment. The voluminous data available in literature leads to the conclusion that the fundamental relationship is a complex one and that no single correlation function can be found to apply to all gases. Alternatively, the dependence of viscosity on temperature can be calculated with the aid of the method of statistical mechanics, but as yet no completely satisfactory theory has evolved. Also, these calculations lead to complex expressions for the function $\mu(T)$. Therefore, only semiempirical relations appear to be the means to calculate the viscosity associated with compressible boundary layers. It is important to realize that, even though semiempirical relations are not extremely precise they are reasonably simple relations. For air, it is possible to use an interpolation formula based on D. M. Sutherland's theory of viscosity and express the viscosity coefficient as

$$\frac{\mu}{\mu_0} = \left(\frac{T}{T_0}\right)^{\frac{3}{2}} \frac{T_0 + S}{T + S}$$

where μ_0 denotes the viscosity at the reference temperature, T_0, and S is a constant which assumes the value $S = 110$ K for air.

For air the Sutherland's relation can also be expressed [W.H. Heiser and D.T. Pratt, *Hypersonic air breathing propulsion*, 1994, AIAA Education Series] as

$$\boxed{\mu = 1.46 \times 10^{-6} \left(\frac{T^{3/2}}{T + 111}\right) \frac{\text{N s}}{\text{m}^2}}$$

where T is in kelvin. This equation is valid for the static pressure range of 0.01 to 100 atm, which is commonly encountered in atmospheric flight. The temperature range in which this equation is valid is up to 3000 K. The reasons that the absolute viscosity is a function only of temperature under these conditions are that the air behaves as a perfect gas, in the sense that intermolecular forces are negligible, and that viscosity itself is a momentum transport phenomenon caused by the random molecular motion associated with thermal energy or temperature.

Example 2.1

Determine the absolute viscosity of air at temperatures 0°C, 5°C and 10°C.

Solution

By Sutherland's relation we have

$$\mu = 1.46 \times 10^{-6} \left(\frac{T^{3/2}}{T + 111}\right) \frac{\text{N s}}{\text{m}^2}$$

Substituting the above temperatures in the Sutherland's relation, we get

$$\mu_0 = \boxed{1.71 \times 10^{-5} \frac{\text{N s}}{\text{m}^2}}$$

$$\mu_5 = \boxed{1.74 \times 10^{-5} \, \frac{\text{N s}}{\text{m}^2}}$$

$$\mu_{10} = \boxed{1.76 \times 10^{-5} \, \frac{\text{N s}}{\text{m}^2}}$$

The following program can calculate the viscosity of air at desired temperatures.

PROGRAM
```
------------------------
c    Estimation of viscosity
     real mu
     do it = 0,2000,10
     t=float(it)
     t = t + 273.15
     mu = 1.46E-6 *( t**(1.5)/(t + 111.0))
     print *, it, mu
     enddo
     stop
     end
```

2.2.6 Kinematic Viscosity Coefficient

The kinematic viscosity coefficient is a convenient form of expressing the viscosity of a fluid. It is formed by combining the density ρ and the absolute coefficient of viscosity μ according to the equation

$$\boxed{\nu = \frac{\mu}{\rho}} \qquad (2.4)$$

The kinematic viscosity coefficient ν is expressed as m^2/s and $1 \, \text{cm}^2/\text{s}$ is known as *stoke*.

The kinematic viscosity coefficient is a measure of the relative magnitudes of viscosity and inertia of the fluid. Both dynamic viscosity coefficient μ and kinematic viscosity coefficient ν are functions of temperature. For liquids, μ decreases with increase of temperature, whereas for gases μ increases with increase of temperature. This is one of the fundamental differences between the behavior of gases and liquids. The viscosity is practically unaffected by the pressure.

2.2.7 Thermal Conductivity of Air

At high speeds, heat transfer from vehicles becomes significant. For example, re-entry vehicles encounter an extreme situation where ablative shields are necessary to ensure protection of the vehicle during its passage through the atmosphere. The heat transfer from a vehicle depends on the thermal conductivity K of air. Therefore, a method to evaluate K is also essential. For this case, a relation similar to Sutherland's law for viscosity is found to be useful, and it is

$$K = 1.99 \times 10^{-3} \left(\frac{T^{3/2}}{T + 112} \right) \frac{J}{s \ m \ K}$$

where T is temperature in kelvin. The pressure and temperature ranges in which this equation is applicable are 0.01 to 100 atm and 0 to 2000 K. For the same reason given for viscosity relation, the thermal conductivity also depends only on temperature.

2.2.8 Compressibility

The change in volume of a fluid associated with change in pressure is called compressibility. When a fluid is subjected to pressure it gets compressed and its volume changes. The bulk modulus of elasticity is a measure of how easily the fluid may be compressed, and is defined as the ratio of pressure change to volumetric strain associated with it. The bulk modulus of elasticity, k, is given by

$$k = \frac{\text{Pressure increment}}{\text{Volume strain}} = -\mathbb{V}\frac{dp}{d\mathbb{V}} \tag{2.5}$$

It may also be expressed as

$$k = \frac{\lim}{\Delta v \to 0} \frac{-\Delta p}{\Delta v/v} = \frac{dp}{(d\rho/\rho)} \tag{2.6}$$

where v is specific volume. Since $d\rho/\rho$ represents the relative change in density brought about by the pressure change dp, it is apparent that the bulk modulus of elasticity is the inverse of the compressibility of the substance at a given temperature. For instance, k for water and air are approximately 2 GN/m^2 and 100 kN/m^2, respectively. This implies that, air is about 20,000 times more compressible than water. It can be shown that, $k = a^2/\rho$, where a is the speed of sound. The compressibility plays a dominant role at high speeds. Mach number M (defined as the ratio of local flow velocity to local speed of sound) is a convenient nondimensional parameter used in the study of compressible flows. Based on M the flow is divided into the following regimes. When $M < 1$ the flow is called *subsonic*, when $M \approx 1$ the flow is termed *transonic flow*, M from 1.2 to 5 is called *supersonic regime*, and $M > 5$ is referred to as *hypersonic regime*. When flow Mach number is less than 0.3, the compressibility effects are negligibly small and hence the flow is called *incompressible*. For incompressible flows, density change associated with velocity is neglected and the density is treated as invariant.

2.3 Thermodynamic Properties

We know from thermodynamics that heat is energy in transition. Therefore, heat has the same dimensions as energy, and is measured in units of joule (J).

2.3.1 Specific Heat

The inherent thermal properties of a flowing gas become important when the energies are considered. The specific heat is one such quantity. The specific heat is defined as the amount of heat required to raise the temperature of a unit mass of a medium by one degree. The value of the specific heat depends on the type of process involved in raising the temperature of the unit mass. Usually constant volume process and constant pressure process are used for evaluating specific heat. The specific heats at constant volume and constant pressure processes, respectively, are designated by C_v and C_p. The definitions for these quantities are the following:

$$C_v \equiv \left(\frac{\partial u}{\partial T} \right)_v \qquad (2.7)$$

where u is internal energy per unit mass of the fluid, which is a measure of the potential and more particularly kinetic energy of the molecules comprising the gas. The specific heat C_v is a measure of the energy-carrying capacity of the gas molecules. For dry air at normal temperature, $C_v = 717.5$ J/(kg K).

The specific heat at constant pressure is defined as

$$C_p \equiv \left(\frac{\partial h}{\partial T} \right)_p \qquad (2.8)$$

where $h = u + pv$, the sum of internal energy and flow energy is known as the *enthalpy* or total heat constant per unit mass of fluid. C_p is a measure of the ability of the gas to do external work in addition to possessing internal energy. Therefore, C_p is always greater than C_v. For dry air at normal temperature, $C_p = 1004.5$ J/(kg K).

2.3.2 The Ratio of Specific Heats

The ratio of specific heats given by

$$\boxed{\gamma = \frac{C_p}{C_v}} \qquad (2.9)$$

is an important parameter in the study of compressible flows. This is a measure of *the relative internal complexity of the molecules of the gas*. It has been determined from kinetic theory of gases that the ratio of specific heats can be related to the number of degrees of freedom, n, of the gas molecules by the relation

$$\gamma = \frac{n + 2}{n}$$

At normal temperatures, there are six degrees of freedom, three translational and three rotational, for diatomic gas molecules. For nitrogen, which is a diatomic gas, $n = 5$ since one of the rotational degrees of freedom is small in comparison with the other two. Therefore,

$$\gamma = 7/5 = 1.4$$

A monatomic gas like helium has 3 translational degrees of freedom only, and therefore,

$$\gamma = 5/3 = 1.67$$

This value of 1.67 is the upper limit of the values which γ can take. In general γ varies from 1 to 1.67. i.e.,

$$\boxed{1 \leq \gamma \leq 1.67}$$

The specific heats of a compressible gas are related to the gas constant R. For a perfect gas this relation is

$$\boxed{R = C_p - C_v} \tag{2.10}$$

2.4　Surface Tension

Liquids behave as if their free surfaces were perfectly flexible membranes having a constant tension σ per unit width. This tension is called the *surface tension*. It is important to note that this is neither a force nor a stress but a *force per unit length*. The value of surface tension depends on

- The nature of the fluid.

- The nature of the substance with which it is in contact at the surface.

- The temperature and pressure.

Consider a plane material membrane, possessing the property of constant tension σ per unit width. Let the membrane be a straight edge of length l. The force required to hold the edge stationary is

$$p = \sigma\, l \tag{2.11}$$

Now, suppose that the edge is pulled so that it is displaced normal to itself by a distance x in the plane of the membrane. The work done, F, in stretching the membrane is given by

$$F = \sigma\, l\, x = \sigma\, A \tag{2.12}$$

where A is the area added to the membrane. We see that σ is the free energy of the membrane per unit area. The important point to be noted here is that, if the energy of a surface is proportional to its area, then it will behave exactly as if it were a membrane with a constant tension per unit width and this is totally independent of the mechanism by which the energy is stored. Thus, the existence of surface tension at the boundary between two substances is a manifestation of the fact that the stored energy contains a term proportional to the area of the surface. This energy is attributable to molecular attractions.

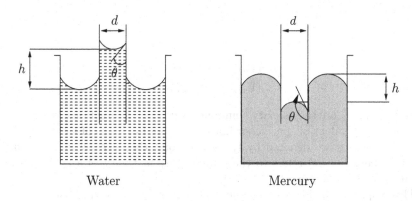

Figure 2.3: Capillary effect of water and mercury

An associated effect of surface tension is the capillary deflection of liquids in small tubes. Examine the level of water and mercury in capillaries, shown in Figure 2.3.

When a glass tube is inserted into a beaker of water, the water will rise in the tube and display a concave meniscus. The deviation of water level h in the tube from that in the beaker can be shown to be

$$h \propto \frac{\sigma}{d} \cos \theta \qquad (2.13)$$

where θ is the angle between the tangent to the water surface and the glass surface. In other words, a liquid like water or alcohol, which wets the glass surface makes an acute angle with the solid, and the level of free surface inside the tube will be higher than that outside. This is termed as capillary action. However, when wetting does not occur, as in the case of mercury in glass, the angle of contact is obtuse and the level of free surface inside the tube is depressed, as shown in Figure 2.3.

Another important effect of surface tension is that, a long cylinder of liquid, at rest or in motion, with a free surface is unstable and breaks up into parts, which then assumes an approximately spherical shape. This is the mechanism of the breakup of liquid jets into drops.

2.5 Analysis of Fluid Flow

Basically two treatments are followed for fluid flow analysis. They are the *Lagrangian* and *Eulerian* descriptions. Lagrangian method describes the motion of each particle of the flow field in a separate and discrete manner. For example, the velocity of the nth particle of an aggregate of particles moving in space can be specified by the scalar equations

$$
\begin{aligned}
(V_x)_n &= f_n(t) \\
(V_y)_n &= g_n(t) \\
(V_z)_n &= h_n(t)
\end{aligned}
\tag{2.14}
$$

where V_x, V_y, V_z are the velocity components in x, y, z directions, respectively. They are independent of space coordinates and are functions of time only. Usually, the particles are denoted by the space point they occupy at some initial time t_0. Thus, $T(x_0, t)$ refers to the temperature at time t of a particle which was at location x_0 at time t_0.

This approach of identifying material points and following them along is also termed the *particle* or *material description*. This approach is usually preferred in the description of low-density flow fields (also called rarefied flows), moving solids, like in describing the motion of a projectile and so on. However, for a deformable system like a continuum fluid, there are infinite numbers of fluid elements whose motion has to be described, and the Lagrangian approach becomes unmanageable. Instead, we can employ spatial coordinates to help to identify particles in a flow. The velocity of all particles in a flow, therefore, can be expressed in the following manner.

$$
\begin{aligned}
V_x &= f(x, y, z, t) \\
V_y &= g(x, y, z, t) \\
V_z &= h(x, y, z, t)
\end{aligned}
\tag{2.15}
$$

This is called the *Eulerian* or *field approach*. If properties and flow characteristics at each position in space remain invariant with time, the flow is called *steady flow*. A time dependent flow is referred to as *unsteady flow*. The steady flow velocity field would then be given as

$$
\begin{aligned}
V_x &= f(x, y, z) \\
V_y &= g(x, y, z) \\
V_z &= h(x, y, z)
\end{aligned}
\tag{2.16}
$$

2.5.1 Relation between Local and Material Rates of Change

The rate of change of a property measured by probes at fixed locations is referred to as *local rate of change*, and the rate of change of properties experienced by a material particle is termed as the *material* or the *substantive rate of change*.

The local rate of change of a property η is denoted by $\partial \eta(x, t)/\partial t$, where it is understood that x is held constant. The material rate of change of property η shall be denoted by $D\eta/Dt$. If η is the velocity V, then DV/Dt is the rate of change of velocity for a fluid particle and thus, is the acceleration that the fluid particle experiences. On the other hand, $\partial V/\partial t$ is just a local rate of change of velocity recorded by a stationary probe. In other words, DV/Dt is the particle or material acceleration and $\partial V/\partial t$ is the local acceleration.

For a fluid flowing with an uniform velocity V_∞, it is possible to write the relation between the local and material rates of change of property η as

$$\frac{\partial \eta}{\partial t} = \frac{D\eta}{Dt} - V_\infty \frac{\partial \eta}{\partial x} \qquad (2.17)$$

Thus, the local rate of change of η is due to the following two effects.

1. Due to the change of property of each particle with time.

2. Due to the combined effect of the spatial gradient of that property and the motion of the fluid.

When a spatial gradient exists, the fluid motion brings different particles with different values of η to the probe, thereby modifying the rate of change observed. This latter effect is termed a *convection effect*. Therefore, $V_\infty(\partial \eta/\partial x)$ is referred to as the convective rate of change of η. Even though Equation (2.17) has been obtained with uniform velocity V_∞, note that, in the limit $\delta t \to 0$ it is only the local velocity V which enters into the analysis and therefore, we have

$$\frac{\partial \eta}{\partial t} = \frac{D\eta}{Dt} - V \frac{\partial \eta}{\partial x} \qquad (2.18)$$

Equation (2.18) can be generalized for a three-dimensional space as

$$\frac{\partial}{\partial t} = \frac{D}{Dt} - (V.\nabla) \qquad (2.19)$$

where ∇ is the gradient operator ($= i\,\partial/\partial x + j\,\partial/\partial y + k\,\partial/\partial z$) and $(V.\nabla)$ is a scalar product ($= V_x\,\partial/\partial x + V_y\,\partial/\partial y + V_z\,\partial/\partial z$). Equation (2.19) is usually written as

$$\frac{D}{Dt} = \frac{\partial}{\partial t} + V.\nabla \qquad (2.20)$$

when η is the velocity of a fluid particle, DV/Dt gives acceleration of the fluid particle and the resultant equation is

$$\boxed{\frac{DV}{Dt} = \frac{\partial V}{\partial t} + (V.\nabla)V} \qquad (2.21)$$

Equation (2.21) is known as *Euler's acceleration formula*.

2.5.2 Graphical Description of Fluid Motion

There are three important concepts for visualizing or describing flow fields. They are

1. Concept of *pathline*.

2. Concept of *streakline*.

3. Concept of *streamline*.

2.5.2.1 Pathline

Pathline may be defined as a line in the flow field describing the trajectory of a given fluid particle. From the Lagrangian viewpoint, namely, a closed system with a fixed identifiable quantity of mass, the independent variables are the initial position with which each particle is identified and the time. Hence, the locus of the same particle over a time period from t_0 to t_n is called the pathline.

2.5.2.2 Streakline

Streakline may be defined as the instantaneous locus of all the fluid elements that have passed the point of injection at some earlier time. Consider a continuous tracer injection at a fixed point Q in space. The connection of all elements passing through the point Q over a period of time is called the streakline.

2.5.2.3 Streamlines

Streamlines are imaginary lines, in a fluid flow, drawn in such a manner that the flow velocity is always tangential to it. Flows are usually depicted graphically with the aid of streamlines. These are *imaginary lines* in the flow field such that the velocity at all points on these lines are always tangential. Streamlines proceeding through the periphery of an infinitesimal area at some instant of time t will form a tube called *streamtube*, which is useful in the study of fluid flow.

From the Eulerian viewpoint, namely, an open system with constant control volume, all flow properties are functions of a fixed point in space and time, if the process is transient. The flow direction of various particles at time t_i forms streamline. The pathline, streamline, and streakline are different in general but coincide in a steady flow.

2.5.2.4 Timelines

In modern fluid flow analysis, yet another graphical representation, namely *timeline*, is used. When a pulse input is periodically imposed on a line of tracer source placed normal to a flow, a change in the flow profile can be observed. The tracer image is generally termed *timeline*. Timelines are often generated in the flow field to aid the understanding of flow behavior such as the velocity and velocity gradient.

From the above-mentioned graphical descriptions, it can be inferred that

- There can be no flow through the lateral surface of the streamtube.

- An infinite number of adjacent streamtubes arranged to form a finite cross-section is often called a bundle of streamtubes.

- Streamtube is a Eulerian (or field) concept.

- Pathline is a Lagrangian (or particle) concept.

- For steady flows, streamlines and streaklines are identical.

2.6 Basic and Subsidiary Laws for Continuous Media

In the range of engineering interest, four basic laws must be satisfied for any continuous medium. They are

- Conservation of matter (continuity equation).
- Newton's second law (momentum equation).
- Conservation of energy (first law of thermodynamics).
- Increase of entropy principle (second law of thermodynamics).

In addition to these primary laws, there are numerous subsidiary laws, sometimes called constitutive relations, that apply to specific types of media or flow processes (e.g., equation of state for perfect gas, Newton's viscosity law for certain viscous fluids, isentropic and adiabatic process relations are some of the commonly used subsidiary equations in flow physics).

2.6.1 Systems and Control Volumes

In employing the basic and subsidiary laws, any one of the following modes of application may be adopted.

- The activities of each and every given element of mass must be such that it satisfies the basic laws and the pertinent subsidiary laws.
- The activities of each and every elemental volume in space must be such that the basic laws and the pertinent subsidiary laws are satisfied.

In the first case, the laws are applied to an identified quantity of matter called the *control mass system*. A control mass system is an identified quantity of matter, which may change shape, position, and thermal condition, with time or space or both, but must always entail the same matter.

For the second case, a definite volume called *control volume* is designated in space, and the boundary of this volume is known as *control surface*. The amount and identity of the matter in the control volume may change with time, but the shape of the control volume is fixed, i.e., the control volume may change its position in time or space or both, but its shape is always preserved.

2.6.2 Integral and Differential Analysis

The analysis where large control volumes are used to obtain the aggregate forces or transfer rates is termed the *integral analysis*. When the analysis is applied to individual points in the flow field, the resulting equations are differential equations, and the method is termed the *differential analysis*.

2.6.3   State Equation

For air at normal temperature and pressure, the density ρ, pressure p, and temperature T are connected by the relation $p = \rho RT$, where R is a constant called gas constant. This is known as the state equation for a perfect gas. At high pressures and low temperatures, the above state equation breaks down. At normal pressures and temperatures the mean distance between molecules and the potential energy arising from their attraction can be neglected. The gas behaves like a perfect gas or ideal gas in such a situation. At this stage, it is essential to understand the difference between the ideal and perfect gases. An *ideal gas is frictionless and incompressible*. The perfect gas has viscosity and can therefore develop shear stresses, and it is compressible according to perfect gas state equation.

Real gases below critical pressure and above the critical temperature tend to obey the perfect-gas law. The perfect-gas law encompasses both Charles' law and Boyle's law. Charles' law states that, *for constant pressure, the volume of a given mass of gas varies directly as its absolute temperature.* Boyle's law (isothermal law) states that, *for constant temperature, the density varies directly as the absolute pressure.*

2.7   Kinematics of Fluid Flow

To simplify the discussions, let us assume the flow to be incompressible, i.e., the density is treated as invariant. The basic governing equations for an incompressible flow are the continuity and momentum equations. The continuity equation is based on the conservation of matter. For steady incompressible flow, the continuity equation in differential form is

$$\boxed{\frac{\partial V_x}{\partial x} + \frac{\partial V_y}{\partial y} + \frac{\partial V_z}{\partial z} = 0} \qquad (2.22)$$

Equation (2.22) may also be expressed as $\nabla . V = 0$, where $V\,(i\,V_x + j\,V_y + k\,V_z)$ is the flow velocity.

The momentum equation which is based on Newton's second law represents the balance between various forces acting on a fluid element, namely,

1. Force due to rate of change of momentum, generally referred to as inertia force.

2. Body forces such as buoyancy force, magnetic force, and electrostatic force.

3. Pressure force.

4. Viscous forces (causing shear stress).

For a fluid element under equilibrium, by Newton's second law, we have the momentum equation as

> Inertia force + Body force + Pressure force + Viscous force = 0

For a gaseous medium, body forces are negligibly small compartialred to other forces and hence can be neglected. For steady incompressible flows, the momentum equation can be written as

$$V_x \frac{\partial V_x}{\partial x} + V_y \frac{\partial V_x}{\partial y} + V_z \frac{\partial V_x}{\partial z} \;=\; -\frac{1}{\rho}\frac{\partial p}{\partial x} + \nu \left(\frac{\partial^2 V_x}{\partial x^2} + \frac{\partial^2 V_x}{\partial y^2} + \frac{\partial^2 V_x}{\partial z^2} \right)$$

$$V_x \frac{\partial V_y}{\partial x} + V_y \frac{\partial V_y}{\partial y} + V_z \frac{\partial V_y}{\partial z} \;=\; -\frac{1}{\rho}\frac{\partial p}{\partial y} + \nu \left(\frac{\partial^2 V_y}{\partial x^2} + \frac{\partial^2 V_y}{\partial y^2} + \frac{\partial^2 V_y}{\partial z^2} \right) \quad (2.23)$$

$$V_x \frac{\partial V_z}{\partial x} + V_y \frac{\partial V_z}{\partial y} + V_z \frac{\partial V_z}{\partial z} \;=\; -\frac{1}{\rho}\frac{\partial p}{\partial z} + \nu \left(\frac{\partial^2 V_z}{\partial x^2} + \frac{\partial V_z}{\partial y^2} + \frac{\partial V_z}{\partial z^2} \right)$$

Equations (2.23) are the x, y, z components of momentum equation, respectively. These equations are generally known as *Navier–Stokes* equations. They are nonlinear partial differential equations and there exists no known analytical method to solve them. This poses a major problem in fluid flow analysis. However, the problem is tackled by making some simplifications to the equation, depending on the type of flow to which it is to be applied. For certain flows, the equation can be reduced to an ordinary differential equation of a simple linear type. For some other type of flows, it can be reduced to a nonlinear ordinary differential equation. For the above types of Navier–Stokes equation governing special category of flows such as potential flow, fully developed flow in a pipe and channel, and boundary layer flows, it is possible to obtain analytical solutions.

It is essential to understand the physics of the flow before reducing the Navier–Stokes equations to any useful form, by making suitable approximations with respect to the flow. For example, let us examine the flow over an aircraft wing, shown in Figure 2.4.

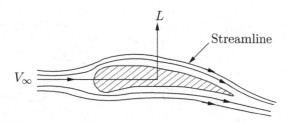

Figure 2.4: Flow past a wing

This kind of problem is commonly encountered in fluid mechanics. Air flow over the wing creates a pressure at the bottom which is larger than that at the top surface. Hence, there is a net resultant force component normal to the freestream flow direction called *lift*, L. The velocity varies along the wing chord as well as in the direction normal to its surface. The former variation is due to the shape of the aerofoil and the latter is due to the no-slip condition at the wall. In the direction normal to wing surface the velocity gradients are very large near the wall and the flow reaches asymptotically a constant velocity in a short distance. This thin region adjacent to the wall where the velocity increases from zero to freestream value is known as the *boundary layer*. Inside the boundary layer the viscous forces are predominant. Further, it so happens that the static pressure outside the boundary layer, acting in the direction normal to the surface, is transmitted through the boundary layer without appreciable change. In other words, the pressure gradient across the boundary layer is zero. Neglecting the interlayer friction between the streamlines, in the region outside the boundary layer, it is possible to treat the flow as inviscid. Inviscid flow is also called potential flow, and for this case the Navier–Stokes equation can be made linear. It is possible to obtain the pressures in the field outside the boundary layer and treat this pressure to be invariant across the boundary layer, i.e., the pressure in the freestream is impressed through the boundary layer. For low-viscous fluids such as air, we can assume with a high degree of accuracy that, the flow is frictionless over the entire flow field except for a thin region near solid surfaces. In the vicinity of solid surface, owing to high velocity gradients the frictional effects become significant. Such regions near solid boundaries where the viscous effects are predominant are termed *boundary layers*.

In general, for streamlined bodies these boundary layers are extremely thin. There may be laminar and turbulent flow within the boundary layer, and its thickness and profile may change along the direction of the flow. Different zones of boundary layer over a flat plate are shown in Figure 2.5. The *laminar sublayer* is that zone near the boundary where the turbulence is suppressed to such a degree that only the laminar effects predominate. The various regions shown in Figure 2.5 are not sharp demarcations of different zones. There is actually a smooth variation from a region where certain effect predominates to another region where some other effect is predominant.

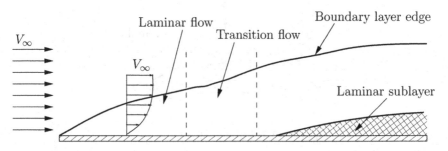

Figure 2.5: Flow over a flat plate

Although the boundary layer is thin, it plays a vital role in fluid dynamics. The drag on ships and missiles, the efficiency of compressors and turbines for jet engines, the effectiveness of ram and turbojets, and the efficiencies of numerous other engineering devices are all influenced by the boundary layer to a significant extent. The performance of a device depends on the behavior of boundary layer and its effect on the main flow. The following are some of the important parameters associated with boundary layers.

2.7.1 Boundary Layer Thickness

Boundary layer thickness, δ, may be defined as the distance from the wall in the direction normal to the wall surface, where the fluid velocity is within 1 percent of the local main stream velocity. It may also be defined as the distance, δ normal to the surface, in which the flow velocity rises from zero to some specified value (e.g., 99%) of its local main stream flow. The boundary layer thickness may be shown schematically as in Figure 2.6.

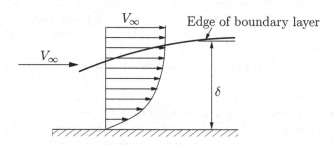

Figure 2.6: Illustration of boundary layer thickness

2.7.2 Displacement Thickness

Displacement thickness, δ^*, may be defined as the distance by which the boundary would have to be displaced if the entire flow were imagined to be frictionless and the same mass flow maintained at any section.

Consider unit width in the flow across an infinite flat plate at zero angle of attack, and let x-component of velocity to be V_x and y-component of velocity be V_y. The volume flow rate Δq through this boundary layer segment of unit width is given by

$$\Delta q = \int_0^\infty (V_m - V_x)\, dy$$

where V_m is the main stream frictionless velocity component and V_x is the actual local velocity component. To maintain the same volume flow rate, q, for the frictionless case as in the actual case, the boundary must be shifted out by a distance δ^* so as to cut off the amount Δq of flow.

Thus,

$$V_m \delta^* = \Delta q = \int_0^\infty (V_m - V_x)dy$$

$$\delta^* = \int_0^\infty \left(1 - \frac{V_x}{V_m}\right) dy \qquad (2.24)$$

The displacement thickness is illustrated in Figure 2.7.

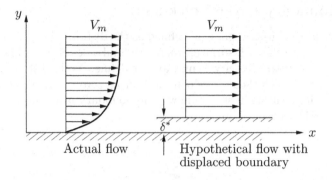

Figure 2.7: Displacement thickness

The main idea of this postulation is to permit the use of a displaced body in place of the actual body such that, the frictionless mass flow around the displaced body is the same as the actual mass flow around the real body. The displacement thickness concept is made use of in the design of wind tunnels, air intakes for jet engines, etc.

There are other (thickness) measures pertaining to the thickness of boundary layer, such as *momentum thickness*, θ, and *energy thickness*, δ_e. They are defined mathematically as follows.

$$\theta = \int_0^\infty \left(1 - \frac{V_x}{V_m}\right) \frac{\rho V_x}{\rho_m V_m} dy \qquad (2.25)$$

$$\delta_e = \int_0^\infty \left(1 - \frac{V_x^2}{V_m^2}\right) \frac{\rho V_x}{\rho_m V_m} dy \qquad (2.26)$$

where V_m and ρ_m are the velocity and density at the edge of the boundary layer and V_x and ρ are the velocity and density at any y location normal to the body surface. In addition to boundary layer thickness, displacement thickness, momentum thickness, and energy thickness, we can define the transition point and separation point also with the help of boundary layer.

2.7.3 Transition Point

Transition point may be defined as the end of the region at which the flow in the boundary layer on the surface ceases to be laminar and becomes turbulent.

2.7.4   Separation Point

Separation point is the position at which the boundary layer leaves the surface of a solid body. If the separation takes place while the boundary layer is still laminar, the phenomenon is termed *laminar separation*. If it takes place for a turbulent boundary layer it is called *turbulent separation*.

The boundary layer theory makes use of Navier–Stokes equation (Equation (2.23)) with the viscous terms in it but in a simplified form. On the basis of many assumptions, such as boundary layer thickness being small compared to the body length and similarity between velocity profiles in a laminar flow, the Navier–Stokes equation can be reduced to a nonlinear ordinary differential equation, for which special solutions exist. Some such problems for which Navier–Stokes equations can be reduced to boundary layer equations and closed form solutions can be obtained are flow past a flat plate or Blassius problem, Hagen–Poiseuille flow through pipes, Couette flow, and flow between rotating cylinders.

2.7.5   Rotational and Irrotational Motion

When a fluid element is subjected to a shearing force, a velocity gradient is produced perpendicular to the direction of shear, i.e., a relative motion occurs between two layers. To achieve this relative motion the fluid elements have to undergo rotation. A typical example of this type of motion is the motion between two roller chains rubbing each other, but moving at different velocities. It is convenient to use an abstract quantity called *circulation*, Γ, defined as the line integral of velocity vector between any two points (to define rotation of the fluid element). By definition, the circulation is given as

$$\Gamma = \oint_c V.\, dl \qquad (2.27)$$

where dl is an elemental length. Circulation per unit area is known as *vorticity* ζ,

$$\zeta = \Gamma/A \qquad (2.28)$$

In vector form ζ becomes

$$\zeta = \nabla \times V = \operatorname{curl} V \qquad (2.29)$$

For a two-dimensional flow in xy plane, ζ becomes

$$\zeta_z = \frac{\partial V_y}{\partial x} - \frac{\partial V_x}{\partial y} \qquad (2.30)$$

where ζ_z is the vorticity about the z-direction, which is normal to the flow field.

Likewise, the other components of vorticity about x and y axes are

$$\zeta_x = \frac{\partial V_z}{\partial y} - \frac{\partial V_y}{\partial z}$$

$$\zeta_y = \frac{\partial V_x}{\partial z} - \frac{\partial V_z}{\partial x}$$

If $\zeta = 0$, the flow is known as *irrotational flow*. Inviscid flows are basically irrotational flows.

2.8   Streamlines

These are imaginary lines in the flow field such that the velocity at all points on these lines is always tangential to them. Flows are usually depicted graphically with the aid of streamlines. Streamlines proceeding through the periphery of an infinitesimal area at some time t form a tube called a *stream tube*, which is useful for the study of fluid flow phenomena. From the definition of streamlines, it can be inferred that

- Flow cannot cross a streamline, and the mass flow between two streamlines is confined.

- Based on the streamline concept, a function ψ called *stream function* can be defined. The velocity components of a flow field can be obtained by differentiating the stream function.

In terms of stream function, ψ, the velocity components of a two-dimensional incompressible flow are given as

$$V_x = \frac{\partial \psi}{\partial y}, \quad V_y = -\frac{\partial \psi}{\partial x} \qquad (2.31)$$

If the flow is compressible the velocity components become

$$\boxed{V_x = \frac{1}{\rho}\frac{\partial \psi}{\partial y}, \quad V_y = -\frac{1}{\rho}\frac{\partial \psi}{\partial x}} \qquad (2.32)$$

It is important to note that the stream function is defined only for two-dimensional flows, and the definition does not exist for three-dimensional flows. Even though some books define ψ for axisymmetric flows, they again prove to be equivalent to two-dimensional flow. We must realize that the definition of ψ does not exist for three-dimensional flows. This is because such a definition demands a *single tangent* at any point on a streamline, which is possible only in two-dimensional flows.

2.8.1 Relationship between Stream Function and Velocity Potential

For irrotational flows (the fluid elements in the field are free of angular motion), there exists a function ϕ called *velocity potential* or *potential function*. For two-dimensional flows, ϕ must be a function of x, y, and t. The velocity components are given by

$$V_x = \frac{\partial \phi}{\partial x}, \qquad V_y = \frac{\partial \phi}{\partial y} \tag{2.33}$$

From Equations (2.31) and (2.33), we can write

$$\boxed{\frac{\partial \psi}{\partial y} = \frac{\partial \phi}{\partial x}, \qquad \frac{\partial \psi}{\partial x} = -\frac{\partial \phi}{\partial y}} \tag{2.34}$$

These relations between stream function and potential function given by Equation (2.34) are the famous *Cauchy–Riemann equations* of complex-variable theory. It can be shown that the lines of constant ϕ or potential lines form a family of curves which intersect the streamlines in such a manner as to have the tangents of the respective curves always at right angles at the point of intersection. Hence, the two sets of curves given by $\psi =$ constant and $\phi =$ constant form an orthogonal grid system or flow net.

Unlike stream function, potential function exists for three-dimensional flows also. This is because there is no condition such as the local flow velocity must be tangential to the potential lines imposed in the definition of ϕ. The only requirement for the existence of ϕ is that the flow must be potential.

2.9 Potential Flow

Potential flow is based on the concept that the flow field can be represented by a potential function ϕ such that,

$$\boxed{\nabla^2 \phi = 0} \tag{2.35}$$

This linear partial differential equation is popularly known as the *Laplace equation*. Derivatives of ϕ give velocities, as given in Equation (2.33), for a two-dimensional flow. Unlike the stream function ψ, the potential function can exist only if the flow is *irrotational*, that is, when viscous effects are absent. All inviscid flows must satisfy the irrotationality condition, namely,

$$\boxed{\nabla \times V = 0} \tag{2.36}$$

For two-dimensional potential flows, by Equation (2.30), we have the vorticity ζ as

$$\zeta_z = \frac{\partial V_y}{\partial x} - \frac{\partial V_x}{\partial y} = 0$$

Equation (2.36) can be rewritten, using Equation (2.33), as

$$\frac{\partial^2 \phi}{\partial x \partial y} - \frac{\partial^2 \phi}{\partial x \partial y} = 0$$

This shows that the flow is irrotational. For two-dimensional flows, the continuity equation given by Equation (2.22) becomes

$$\frac{\partial V_x}{\partial x} + \frac{\partial V_y}{\partial y} = 0$$

Using Equation (2.33) this equation can be expressed as

$$\frac{\partial^2 \phi}{\partial x^2} + \frac{\partial^2 \phi}{\partial y^2} = 0$$

i.e.,

$$\boxed{\nabla^2 \phi = 0} \tag{2.37}$$

For flows with finite vorticity the potential function ϕ does not exist, and the linear equation $\nabla^2 \phi = 0$ cannot be obtained.

For potential flows, the Navier–Stokes equations (2.23) reduce to the form

$$V_x \frac{\partial V_x}{\partial x} + V_y \frac{\partial V_x}{\partial y} + V_z \frac{\partial V_x}{\partial z} = -\frac{1}{\rho} \frac{\partial p}{\partial x}$$

$$V_x \frac{\partial V_y}{\partial x} + V_y \frac{\partial V_y}{\partial y} + V_z \frac{\partial V_y}{\partial z} = -\frac{1}{\rho} \frac{\partial p}{\partial y} \tag{2.38}$$

$$V_x \frac{\partial V_z}{\partial x} + V_y \frac{\partial V_z}{\partial y} + V_z \frac{\partial V_z}{\partial z} = -\frac{1}{\rho} \frac{\partial p}{\partial z}$$

Equations (2.38) are known as *Euler equations*.

2.9.1   Two-Dimensional Source and Sink

A type of flow in which the fluid emanates from origin and spreads radially outwards to infinity, as shown in Figure 2.8, is called a *source*. The volume flow rate q crossing a circular surface of radius r and unit depth is given by

$$q = 2\pi r V_r \tag{2.39}$$

where V_r is the radial component of velocity. For a source, the radial lines are the streamlines. Therefore, the potential lines must be concentric circles, represented by

$$\phi = A \ln(r)$$

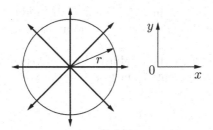

Figure 2.8: A two-dimensional source

where A is a constant. The radial velocity component $V_r = \partial\phi/\partial r = A/r$. Substituting this into Equation (2.39), we get

$$\frac{2\pi r A}{r} = q$$

or

$$A = \frac{q}{2\pi}$$

Thus, the velocity potential for a two-dimensional source of strength q becomes

$$\boxed{\phi = \frac{q}{2\pi}\ln(r)} \tag{2.40}$$

In a similar manner as above, the stream function for a source of strength q can be obtained as

$$\boxed{\psi = \frac{q}{2\pi}\theta} \tag{2.41}$$

where θ stands for the location of the streamline in θ-direction. Similarly, for a sink, which is *a type of flow in which the fluid at infinity flows radially towards the origin*, we can show that the potential and stream functions are given by

$$\boxed{\phi = -\frac{q}{2\pi}\ln(r)}$$

and

$$\boxed{\psi = -\frac{q}{2\pi}\theta}$$

where q is the strength of the sink. Note that the volume flow rate is termed as the strength of source and sink. Also, for both source and sink the origin is a singular point.

2.9.2 Simple Vortex

A *simple* or *free vortex* is a flow in which the fluid elements simply move along concentric circles, without spinning about their own axes. The fluid elements have only translatory motion in a free vortex. In addition to moving along concentric paths, if the fluid elements spin about their own axes, the flow is termed *forced vortex*.

A *simple* or *free vortex* can be established by selecting the stream function, ψ, of the source to be the potential function ϕ of the vortex. Thus, for a simple vortex

$$\boxed{\phi = \frac{q}{2\pi}\theta} \tag{2.42}$$

It can be easily shown from Equation (2.42) that the stream function for a simple vortex is

$$\boxed{\psi = -\frac{q}{2\pi}\ln(r)} \tag{2.43}$$

It follows from Equations (2.42) and (2.43) that, the velocity components of the simple vortex, shown in Figure 2.9, are

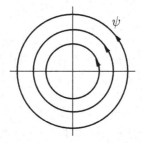

Figure 2.9: Simple vortex flow

$$V_\theta = \frac{q}{2\pi r}, \qquad V_r = 0 \tag{2.44}$$

Here again the origin is a singular point, where the tangential velocity approaches infinity, as seen from Equation (2.44). The flow in a simple or free vortex resembles part of the common whirlpool found while paddling a boat or while emptying water from a bathtub. An approximate profile of a whirlpool is as shown in Figure 2.10.

For the whirlpool shown in Figure 2.10, the circulation along any path about the origin is given by,

$$\Gamma = \oint V.dl$$

$$= \int_0^{2\pi} V_\theta\, r\, d\theta$$

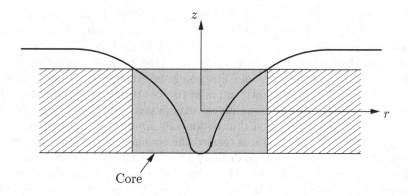

Figure 2.10: A whirlpool flow field

By Equation (2.44), $V_\theta = \dfrac{q}{2\pi r}$, therefore, the circulation becomes

$$\Gamma = \int_0^{2\pi} \frac{q}{2\pi r} r d\theta = q$$

Since there are no other singularities for the whirlpool shown in Figure 2.10, this must be the circulation for all paths about the origin. Consequently, q in the case of vortex is the measure of circulation about the origin and is also referred to as the *strength of the vortex*.

2.9.3 Source-Sink Pair

This is a combination of a source and sink of equal strength, situated (located) at a distance apart. The stream function due to this combination is obtained simply by adding the stream functions of source and sink. When the distance between the source and sink is made negligibly small, in the limiting case, the combination results in a *doublet*.

2.10 Viscous Flows

In the previous sections of this chapter, we have seen many interesting concepts of fluid flows. With this background, let us observe some of the important aspects of fluid flow from a practical or application point of view.

We are familiar with the fact that viscosity produces shear force which tends to retard the fluid motion. It works against inertia force. The ratio of these two forces governs (dictates) many properties of the flow, and the ratio expressed in the form of a nondimensional parameter known as the famous *Reynolds number*, Re_L.

$$\boxed{Re_L = \frac{\rho V L}{\mu}} \qquad (2.45)$$

where V and ρ are the velocity and density of the flow, respectively, μ is the dynamic viscosity coefficient of the fluid, and L is a characteristic dimension. The Reynolds number plays a dominant role in fluid flow analysis. This is one of the fundamental dimensionless parameters which must be matched for similarity considerations in most of the fluid flow analysis. At high Reynolds numbers, the inertia force is predominant compared to viscous forces. At low Reynolds numbers the viscous effects predominate everywhere, whereas at high Re the viscous effects confine only to a thin region just adjacent to the surface of the object present in the flow, and it is termed the *boundary layer*. Since the length and velocity scales are chosen according to a particular flow, when comparing the flow properties at two different Reynolds numbers, only flows with geometric similarity should be considered. In other words, flow over a circular cylinder should be compared only with flow of another circular cylinder whose dimensions can be different, but not the shape. Flow in pipes with different velocities and diameters, flow over aerofoils of the same kind, and the boundary layer flows are also some geometrically similar flows. From the above-mentioned similarity consideration, we can infer that geometric similarity is a prerequisite for dynamic similarity. That is, *dynamically similar flows must be geometrically similar,* but the converse need not be true. Only similar flows can be compared, that is, when comparing the effect of viscosity, the changes in flow pattern due to body shape should not interfere with the problem.

For calculating Reynolds number, different velocity and length scales are used. Some of the length scales we often encounter in fluid flow studies are given below.

Cylinder

$Re_D = \frac{\rho_\infty V_\infty D}{\mu_\infty}$ $\qquad\qquad$ D is cylinder diameter

Aerofoil

$Re_C = \frac{\rho_\infty V_\infty c}{\mu_\infty}$ $\qquad\qquad$ c is aerofoil chord

Pipe flow (fully developed)

$Re_D = \frac{\rho \overline{V} D}{\mu}$ $\qquad\qquad$ $\overline{V}$ is average velocity

$\qquad\qquad\qquad\qquad\qquad\qquad$ D is pipe diameter

Channel flow (two-dimensional and fully developed)

$Re_h = \frac{\rho \overline{V} h}{\mu}$ $\qquad\qquad$ $\overline{V}$ is the average velocity

$\qquad\qquad\qquad\qquad\qquad\qquad$ h is the height of the channel

Flow over a grid

$Re_M = \frac{\rho V M}{\mu}$ $\qquad\qquad$ V is the velocity upstream or

$\qquad\qquad\qquad\qquad\qquad\qquad$ downstream of the grid

$\qquad\qquad\qquad\qquad\qquad\qquad$ M is the mesh size

Boundary layer

$Re_\delta = \frac{\rho V \delta}{\mu}$ $\qquad\qquad$ V is the outer velocity

$\qquad\qquad\qquad\qquad\qquad\qquad$ δ is the boundary layer thickness

$Re_\theta = \frac{\rho V \theta}{\mu}$ $\qquad\qquad$ θ is the momentum thickness

$Re_x = \frac{\rho V x}{\mu}$ $\qquad\qquad$ x is the distance from the leading edge

---

In the above descriptions of Reynolds number, the quantities with subscript ∞

are at the freestream and quantities without subscript are the local properties. Reynolds number is basically a similarity parameter. It is used to determine the laminar and turbulent nature of flow. Below certain Reynolds number, the flow cannot be turbulent. Under such condition any disturbance introduced into the flow will be dissipated out by viscosity. It is known as the *lower critical Reynolds number*. Some of the well-known critical Reynolds number are listed below:

Pipe flow $-$ Re_D $=$ 2300 : based on mean velocity and diameter.

Channel flow $-$ Re_h $=$ 1000 (two-dimensional) : based on height and mean velocity.

Boundary layer flow $-$ Re_θ $=$ 350 : based on freestream velocity and momentum thickness.

Circular cylinder $-$ Re_w $=$ 200 (turbulent wake) : based on wake width and wake defect.

Flat plate $-$ Re_x $=$ 5 $\times 10^5$: based on length from the leading edge.

Circular cylinder $-$ Re_d $=$ 1.66 $\times 10^5$: based on cylinder diameter.

The *lower critical Reynolds number* is that Reynolds number below which the entire flow is laminar. The Reynolds number above which the entire flow is turbulent is termed *upper critical Reynolds number*. *Critical Reynolds number* is that at which the flow field is a mixture of laminar and turbulent flows.

Note: It is important to note that when Re is low due to large μ, the flow is termed *stratified flow*, e.g., flow of tar, honey, etc., are stratified flows. When Re is low because of low density, the flow is termed *rarefied flow*. For instance, flows in space and very high altitudes are rarefied flows.

2.10.1   Drag of Bodies

When a body moves in a fluid, it experiences forces and moments due to the relative fluid flow which is taking place around it. If the body has an arbitrary shape and orientation, the flow will exert forces and moments about all three coordinate axes, as shown in Figure 2.11. The force on the body along the flow direction is called *drag*.

The drag is essentially a force opposing the motion of the body. Viscosity is responsible for a part of the drag force, and the body shape generally determines the overall drag. In the design of transport vehicles, shapes experiencing minimum drag are considered to keep the power consumption at a minimum. Low drag shapes are called *streamlined bodies* and high drag shapes are termed *bluff bodies*.

Drag arises due to (a) the difference in pressure between the front and back regions and (b) the friction between the body surface and the fluid. Drag force (a) is known as *pressure drag* and (b) is known as *skin friction drag* or *shear drag*.

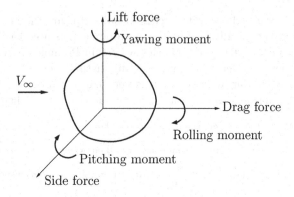

Figure 2.11: Forces acting on an arbitrary body

2.10.1.1 Pressure Drag

The pressure drag arises due to the separation of boundary layer, whenever it encounters an adverse pressure gradient. The phenomenon of separation and how it causes the pressure drag can be explained better by considering flow around a body like a circular cylinder. If there is no viscosity, and hence no boundary layer, the flow would have gone around the cylinder like a true potential flow, as shown in Figure 2.12. For this flow the pressure distribution will be the same on the front and back sides, and the net force along the freestream direction would be zero. That is, there would not be any drag acting on the cylinder.

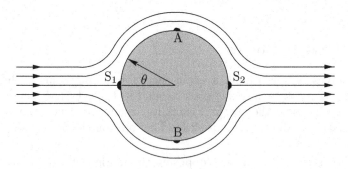

Figure 2.12: Potential flow past a circular cylinder

But in real flow, because of viscosity, a boundary layer is formed on the surface of the cylinder. The flow experiences a favorable pressure gradient from forward stagnation point S_1 to the topmost point A on the cylinder at $\theta = 90°$, as shown in Figure 2.12. Therefore, the flow accelerates from $\theta = 0°$ to $90°$. However, beyond $\theta = 90°$ the flow is subjected to an adverse pressure gradient and hence decelerates. Under this condition the pressure is acting against the fluid flow. In a boundary layer, the velocity near the surface is small and hence the force due to its momentum is unable to counteract the pressure force. The boundary layer flow gets retarded and the velocity near the wall region reduces to zero and then the flow is pushed back in the opposite direction, as illustrated in Figure 2.13. This phenomenon is called *separation*.

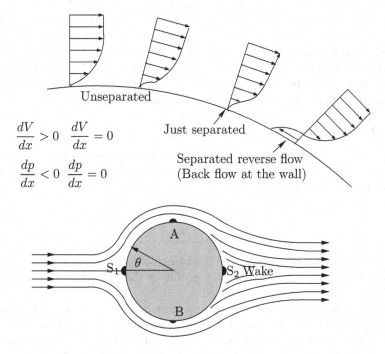

Figure 2.13: Illustration of separation process

Across the separated region, the pressure is nearly constant and lower than what it would have been if the flow did not separate. The pressure does not recover completely as in the case of potential flow. Thus, on account of the incomplete recovery of pressure due to separation, a net drag force opposing the body motion is generated. We can easily see that the pressure drag will be less if the separation had taken place later, that is, the area over which the pressure unrecovered is less. To minimize pressure drag, the separation point should be as far as possible from the leading edge. This is true for any shape. Streamlined bodies are designed on this basis and the adverse pressure gradient is kept as small as possible by keeping the curvature very small.

The separation of boundary layer depends not only on the strength of the adverse pressure gradient but also on the nature of the boundary layer, namely, laminar or turbulent. Laminar flow has the tendency to separate earlier than a turbulent flow. This is because the laminar velocity profiles in a boundary layer have less momentum near the wall. This is conspicuous in the case of flow over a circular cylinder. Laminar boundary layer separates nearly at $\theta = 90°$ whereas, for a highly turbulent boundary layer, the separation is delayed and the attached flow continues up to as far as $\theta = 150°$ on the cylinder. The reduction of pressure drag when the boundary layer changes from laminar to turbulent is of the order of 5 times for bluff bodies. The flow behind a separated region is called the *wake*. For low drag, the wake width should be small.

Although separation is shown to take place at well-defined locations on the body, in the illustration in Figure 2.13, it actually takes place over a zone on the surface which cannot be identified easily. Therefore, theoretical estimation of separation especially for a turbulent boundary layer is difficult and hence the pressure drag cannot be easily calculated. Some approximate methods exist but they can serve only as guide lines for the estimation of pressure drag.

2.10.1.2 Skin Friction Drag

The friction between the surface of the body and the fluid causes viscous shear stress and this force is known as *skin friction drag*. Wall shear stress τ at the surface of a body is given by

$$\tau = \mu \frac{\partial V_x}{\partial y} \tag{2.46}$$

where μ is the dynamic viscosity coefficient and $\partial V_x / \partial y$ is the velocity gradient at body surface $y = 0$. If the velocity profile in the boundary layer is known, then the shear stress can be calculated.

For streamlined bodies, the separated zone being small, the major portion of the drag is from skin friction. Turbulent boundary layer results in more skin friction than a laminar one. Examine the friction coefficient C_f variation with Reynolds number, for a flat plate kept at zero angle of attack in an uniform stream, plotted in Figure 2.14.

The characteristic length for Reynolds number is the plate length from its leading edge. It can be seen from Figure 2.14 that, the C_f is more for a turbulent flow than laminar flow. The friction coefficient is defined as

$$C_f = \frac{\text{Total frictional force}}{\frac{1}{2}\rho V_\infty^2 S} \tag{2.47}$$

where V_∞ and ρ are the freestream velocity and density, respectively, and S is the wetted surface area of the flat plate.

For bluff bodies the pressure drag is many times larger than the skin friction drag and for streamlined bodies the condition is the reverse. In the case of streamlined bodies like aerofoil the designer aims at keeping the skin friction drag as low as possible. Maintaining laminar boundary layer conditions all along the surface is the most suitable arrangement. Though such aerofoils have been designed, they have many limitations. Even small surface roughness and disturbances make the flow turbulent, which spoils the purpose. In addition, there is a

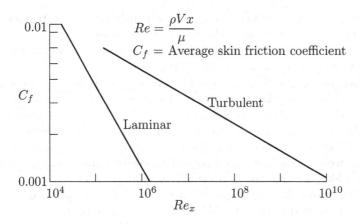

Figure 2.14: C_f variation with Reynolds number

tendency for the flow to separate even at small angles of attack, which severely restricts the use of such aerofoils.

2.10.1.3 Comparison of Drag of Various Bodies

In low-speed flow past geometrically similar bodies with identical orientation and relative roughness, the drag coefficient should be a function of the Reynolds number only.

$$C_D = f(Re) \tag{2.48}$$

The Reynolds number is based upon freestream velocity V and a characteristic length L of the body. The drag coefficient C_D could be based upon L^2, but it is customary to use a characteristic area of the body instead of L^2. Thus, the drag coefficient becomes

$$C_D = \frac{\text{Drag}}{\frac{1}{2}\rho V^2 S} \tag{2.49}$$

The factor 1/2 is our traditional tribute to Euler and Bernoulli. The area S is usually one of the following three types.

1. Frontal area of the body as seen from the stream. This is suitable for thick stubby bodies, such as spheres, cylinders, cars, missiles, projectiles, and torpedos.

2. Planform area of the body area as seen from above. This is suitable for wide flat bodies such as wings and hydrofoils.

3. Wetted area. This is appropriate for surface ships and barges.

While using drag or other fluid force data, it is important to note what length and area are being used to scale the measured coefficients.

Table 2.1 gives a few data on drag, based on frontal area, of two-dimensional bodies of various cross section, at $R_e \geq 10^4$.

The sharp-edged bodies, which tend to cause flow separation regardless of the nature of boundary layer, are insensitive to Reynolds number. The elliptic cylinders, being smoothly rounded, have the laminar turbulent transition effect and are therefore quite sensitive to the nature of the boundary layer.

Table 2.2 lists drag coefficients of some three-dimensional bodies. For these bodies also we can conclude that sharp edges always cause flow separation and high drag which is insensitive to Reynolds number.

Rounded bodies like the ellipsoid have drag which depends upon the point of separation, so that both Reynolds number and the nature of boundary layer are important. Increase of body length will generally decrease pressure drag by making the body relatively more slender, but sooner or later the skin friction drag will catch up. For the flat-faced cylinder in Table 2.2, pressure drag decreases with L/d but skin friction increases, so that minimum drag occurs at about $L/d = 2$.

2.10.2   Turbulence

Turbulent flow is described as a flow with irregular fluctuations. In nature, most of the flows are turbulent. Turbulent flows have characteristics which are appreciably different from those of laminar flows. Likewise, we have to explain all the characteristics of turbulent flow to completely describe it. Incorporating all the important characteristics the turbulence may be described as *a three-dimensional, random phenomenon, exhibiting multiplicity of scales, possessing vorticity, and showing very high dissipation.* Turbulence is described as a three-dimensional phenomenon. It means that, even in a one-dimensional flow field the turbulent fluctuations are always three-dimensional. In other words, the mean flow may be one- or two- or three-dimensional, but the turbulence is always three-dimensional. From the above discussions, it is evident that *turbulence can only be described and cannot be defined.*

A complete theoretical approach to turbulent flow paralleling that of laminar flow is impossible because of the complexity and apparently random nature of the velocity fluctuations in turbulent flow. Nevertheless, semitheoretical analysis aided by limited experimental data can be carried out for turbulent flows, with instruments which have the capacity to detect high-frequency fluctuations. For flows at very low speeds, say around 20 m/s, the frequencies encountered will be 2 to 500 Hz. Hot-wire anemometer is well suited for measurements in such flows. A typical hot wire velocity trace of a turbulent flow is shown in Figure 2.15.

Turbulent fluctuations are random in amplitude, phase, and frequency. If an instrument such as a pitot-static tube, which has a low-frequency response of the order of 30 seconds, is used for the measurement of velocity, the manometer will read only a steady value, ignoring the fluctuations. This means that, the

Table 2.1: Drag of two-dimensional bodies at $R_e \geq 10^4$

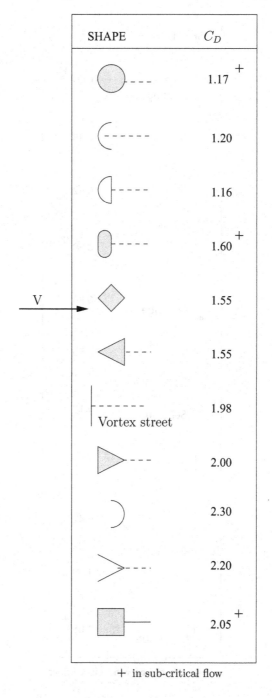

Table 2.2: Drag of three-dimensional bodies at $R_e \geq 10^4$

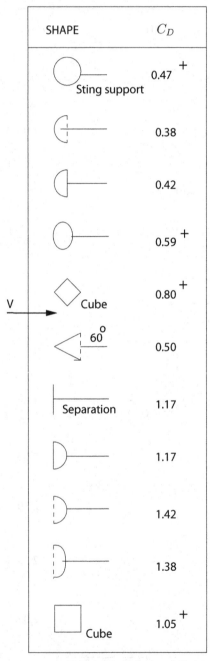

+   Tested on wind tunnel floor

Note: From *Fluid Dynamic Drag*, Hoerner, 1975.

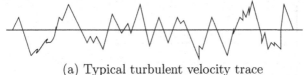

(a) Typical turbulent velocity trace

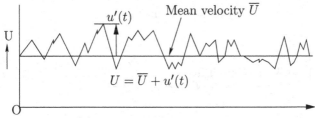

(b) Mean and fluctuating velocities

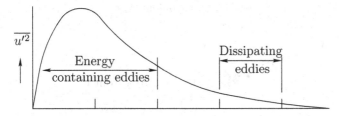

(c) Distribution of turbulent kinetic energy at various frequencies

Figure 2.15: Hot-wire trace of a turbulent flow

turbulent flow consists of a steady velocity component which is independent of time, over which the fluctuations are superimposed, as shown in Figure 2.16(b). That is,

$$U(t) = \overline{U} + u'(t) \qquad (2.50)$$

where U(t) is the instantaneous velocity, $\overline{U}$ is the time averaged velocity, and $u'(t)$ is the turbulent fluctuation around the mean velocity. Since $\overline{U}$ is independent of time, the time average of $u'(t)$ should be equal to zero. That is,

$$\frac{1}{t} \int_0^t u'(t) = 0 \ ; \ \overline{u'} = 0$$

provided the time t is sufficiently large. In most of the laboratory flows, averaging over a few seconds is sufficient if the main flow is kept steady.

In the beginning of this section, we saw that the turbulence is always three-dimensional in nature even if the main flow is one-dimensional. For example, in a fully developed pipe or channel flow, as far as the mean velocity is concerned only the x-component of velocity $\overline{U}$ exists, whereas all three components of turbulent

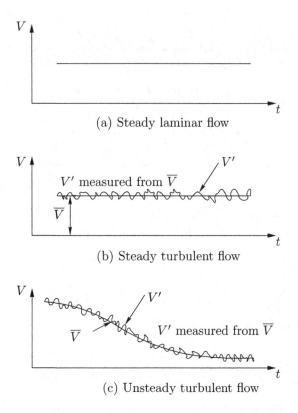

(a) Steady laminar flow

(b) Steady turbulent flow

(c) Unsteady turbulent flow

Figure 2.16: Variation of flow velocity with time

fluctuations u', v', and w' are always present. The intensity of the turbulent velocity fluctuations is expressed in the form of its root mean square value. That is, the velocity fluctuations are instantaneously squared, then averaged over a certain period and finally the square root is taken. The root mean square (RMS) value is useful in estimating the kinetic energy of fluctuations. The turbulence level for any given flow with a mean velocity $\overline{U}$ is expressed as a *turbulence number*, n, defined as

$$n = 100 \frac{\sqrt{\overline{u'^2} + \overline{v'^2} + \overline{w'^2}}}{3\overline{U}} \tag{2.51}$$

In the laboratory, turbulence can be generated in many ways. A wire-mesh placed across an air stream produces turbulence. This turbulence is known as *grid turbulence*. If the incoming air stream as well as the mesh size are uniform then the turbulent fluctuations behind the grid are isotropic in nature, i.e., u', v', and w' are equal in magnitude. In addition to this, the mean velocity is the same across any cross-section perpendicular to the flow direction. That is, no shear stress exists. As the flow moves downstream the fluctuations diedown due to viscous effects. Turbulence is produced in jets and wakes also. The mean velocity in these flows varies and they are known as *free shear flows*. Fluctuations exist up to some distance and then slowly decay. Another type of turbulent flow often encountered in practice is the *turbulent boundary layer*. It is a shear flow with zero velocity at the wall. These flows maintain the turbulence level even at large distances, unlike grid or free shear flows. In wall shear flows or boundary layer type flows, turbulence is produced periodically to counteract the decay.

A turbulent flow may be visualized as a flow made up of eddies of varies sizes. Large eddies are first formed, taking energy from the mean flow. They then break up into smaller ones in a sequential manner until they become very small. At this stage the kinetic energy gets dissipated into heat due to viscosity. Mathematically it is difficult to define an eddy in a precise manner. It represents, in a way, the frequencies involved in the fluctuations. Large eddy means low-frequency fluctuations and smallest eddy means highest-frequency fluctuations encountered in the flow. The kinetic energy distribution at various frequencies can be represented by an energy spectrum as shown in Figure 2.15(c).

The problem of turbulence is yet to be solved completely. Different kinds of approach are employed to solve these problems. The well-known method is to write the Navier–Stokes equations for the fluctuating quantities and then average them over a period of time, substituting the following in Navier–Stokes equations (Equation (2.23)).

$$V_x = \overline{V}_x + u' \qquad p = \overline{p} + p'$$

$$V_y = \overline{V}_y + v' \qquad V_z = \overline{V}_z + w' \tag{2.52}$$

where V_x, u', V_y, v', V_z, and w' are the mean and fluctuational velocity components along x, y, and z-directions, respectively. Bar denotes the mean values, that is, time averaged quantities.

Let us now consider the x-momentum equation (Equation 2.23) for a two-dimensional flow.

$$V_x \frac{\partial V_x}{\partial x} + V_y \frac{\partial V_x}{\partial y} = -\frac{1}{\rho} \frac{\partial p}{\partial x} + \nu \left(\frac{\partial^2 V_x}{\partial x^2} + \frac{\partial^2 V_x}{\partial y^2} \right) \tag{2.53}$$

In Equation (2.53), ν is the *kinetic viscosity*, given by

$$\nu = \mu/\rho$$

Substituting Equation (2.52) into Equation (2.53), we get

$$(\overline{V_x} + u') \frac{\partial(\overline{V_x} + u')}{\partial x} + (\overline{V_y} + v') \frac{\partial(\overline{V_x} + u')}{\partial y}$$

$$= -\frac{1}{\rho} \frac{\partial(\overline{p} + p')}{\partial x} + \nu \frac{\partial^2(\overline{V_x} + u')}{\partial x^2} + \nu \frac{\partial^2(\overline{V_x} + u')}{\partial y^2} \tag{2.54}$$

Expanding Equation (2.54), we obtain

$$\overline{V}_x \frac{\partial \overline{V}_x}{\partial x} + \overline{V}_x \frac{\partial u'}{\partial x} + u' \frac{\partial \overline{V}_x}{\partial x} + u' \frac{\partial u'}{\partial x} + \overline{V}_y \frac{\partial V_x}{\partial y} + \overline{V}_y \frac{\partial u'}{\partial y} + u' \frac{\partial \overline{V}_x}{\partial y} + v' \frac{\partial \overline{V}_x}{\partial y}$$

$$= -\frac{1}{\rho} \frac{\partial \overline{p}}{\partial x} - \frac{1}{\rho} \frac{\partial p'}{\partial x} + \nu \frac{\partial^2 \overline{V}_x}{\partial x^2} + \nu \frac{\partial^2 u'}{\partial x^2} + \nu \frac{\partial^2 \overline{V}_x}{\partial y^2} + \nu \frac{\partial^2 u'}{\partial y^2} \qquad (2.55)$$

In this equation, time averaging of the individual fluctuations is zero. Their products or square quantities are not zero. Taking time average of Equation (2.55), we get

$$\overline{V}_x \frac{\partial \overline{V}_x}{\partial x} + \overline{u' \frac{\partial u'}{\partial x}} + \overline{V}_y \frac{\partial V_x}{\partial y} + \overline{v' \frac{\partial u'}{\partial y}}$$

$$= -\frac{1}{\rho} \frac{\partial \overline{p}}{\partial x} + \nu \frac{\partial^2 \overline{V}_x}{\partial x^2} + \nu \frac{\partial^2 \overline{V}_x}{\partial y^2} \qquad (2.56)$$

Equation (2.56) is slightly different from the laminar Navier–Stokes equation (2.53).

The continuity equation for the two-dimensional flow under consideration is

$$\frac{\partial (\overline{V}_x + u')}{\partial x} + \frac{\partial (\overline{V}_y + v')}{\partial y} = 0$$

This can be expanded to result in

$$\frac{\partial \overline{V}_x}{\partial x} + \frac{\partial \overline{V}_y}{\partial y} = 0 \quad \text{and} \quad \frac{\partial u'}{\partial x} + \frac{\partial v'}{\partial y} = 0 \qquad (2.57)$$

The terms involving turbulent fluctuational velocities u' and v' on the left-hand side of Equation (2.56) can be written as

$$\overline{u' \frac{\partial u'}{\partial x}} + \overline{v' \frac{\partial u'}{\partial y}} = \frac{\partial}{\partial x} \overline{(u'^2)} - \overline{u' \frac{\partial u'}{\partial x}} + \overline{v' \frac{\partial u'}{\partial y}}$$

Using Equation (2.57) the above equation can be expressed as

$$\overline{u' \frac{\partial u'}{\partial x}} + \overline{v' \frac{\partial u'}{\partial y}} = \frac{\partial}{\partial x} \overline{(u'^2)} + \frac{\partial}{\partial y} \overline{(u'v')} \qquad (2.58)$$

Combination of Equations (2.56) and (2.58) results in

$$\rho \overline{V}_x \frac{\partial \overline{V}_x}{\partial x} + \rho \overline{V}_y \frac{\partial \overline{V}_x}{\partial y} = -\frac{\partial \overline{p}}{\partial x} + \frac{\partial}{\partial x} \left(\mu \frac{\partial \overline{V}_x}{\partial y} - \rho \overline{u'^2} \right) + \frac{\partial}{\partial y} \left(\mu \frac{\partial \overline{V}_x}{\partial y} - \rho \overline{u'v'} \right)$$

$$\qquad (2.59)$$

The terms $-\rho \overline{u'^2}$ and $-\rho \overline{u'v'}$ in Equation (2.59) are due to turbulence. They are popularly known as *Reynolds* or *turbulent stresses*. For a three-dimensional flow the turbulent stress terms are $\overline{\rho u'^2}, \overline{\rho v'^2}, \overline{\rho w'^2}, \overline{\rho u'v'}, \overline{\rho v'w'}$ and $\overline{\rho u'w'}$. Solutions of Equation (2.59) is rather cumbersome. Assumptions like eddy viscosity, mixing length are made to find a solution for this equation.

At this stage, it is important to have proper clarity about the laminar and turbulent flows. The laminar flow may be described as *a well-ordered pattern where fluid layers are assumed to slide over one another*, i.e., in laminar flow the fluid moves in layers, or laminas, one layer gliding over an adjacent layer with interchange of momentum only at molecular level. Any tendencies toward instability and turbulence are damped out by viscous shear forces that resist the relative motion of adjacent fluid layers. In other words, *laminar flow is an orderly flow in which the fluid elements move in an orderly manner such that the transverse exchange of momentum is insignificant*. Whereas, *the turbulent flow is a three-dimensional random phenomenon, exhibiting multiplicity of scales, possessing vorticity, and showing very high dissipation*. Turbulent flow is basically an *irregular flow*. Turbulent flow has very erratic motion of fluid particles, with a violent transverse exchange of momentum.

The laminar flow, while having irregular molecular motions, is macroscopically a well-ordered flow. But in the case of turbulent flow there is the effect of a small but macroscopic fluctuating velocity superimposed on a well-ordered flow. A graph of velocity versus time at a given position in a pipe flow would appear as shown in Figure 2.16(a), for laminar flow, and as shown in Figure 2.16(b), for turbulent flow. In Figure 2.16(b) for turbulent flow, an average velocity denoted as $\overline{V}$ has been indicated. Because this average is constant with time, the flow has been designated as steady. An unsteady turbulent flow may prevail when the average velocity field changes with time, as shown in Figure 2.16(c).

2.10.3 Flow through Pipes

Fluid flow through pipes with circular and noncircular cross-sections is one of the commonly encountered problems in many practical systems. Flow through pipes is driven mostly by pressure or gravity or both.

Consider the flow in a long duct, shown in Figure 2.17. This flow is constrained by the boundary walls. At the entrance, the flow (assumed to be inviscid) converges and enters the tube.
Because of the viscous friction between the fluid and pipe wall, viscous boundary layer grows downstream of the entrance. The boundary layer growth makes the effective area of the pipe decrease progressively downstream, thereby making the flow along the pipe accelerate. This process continues up to the point where the boundary layer from the wall grows and meets at the pipe centerline, i.e., fills the pipe.

The zone upstream of the boundary layer merging point is called the *entrance* or *flow development length* and the zone downstream of the merging point is termed *fully developed region*. In the fully developed region, the velocity profile remains unchanged. Dimensional analysis shows that Reynolds number is the only parameter influencing the entrance length. In the functional form,

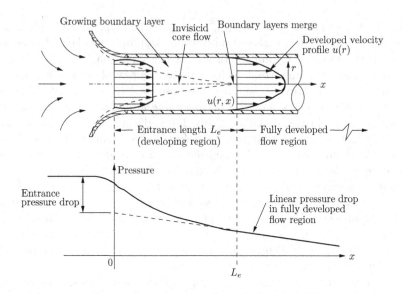

Figure 2.17: Flow development in a long duct

the entrance length can be expressed as

$$L_e \;=\; f(\rho, V, d, \mu)$$

$$\frac{L_e}{d} \;=\; f_1\left(\frac{\rho V d}{\mu}\right) = f_1(Re)$$

where ρ, V, and μ are the flow density, velocity, and viscosity, respectively, and d is the pipe diameter.

For laminar flow, the accepted correlation is

$$\boxed{\frac{L_e}{d} \approx 0.06\, Re_d}$$

At the critical Reynolds number $Re_c = 2300$, for pipe flow, $L_e = 138d$, which is the maximum development length possible.

For turbulent flow the boundary layer grows faster, and L_e is given by the approximate relation

$$\boxed{\frac{L_e}{d} \approx 4.4\, (Re_d)^{\frac{1}{6}}} \tag{2.60}$$

Now, examine the flow through an inclined pipe, shown in Figure 2.18, considering the control volume between sections 1 and 2.

Treating the flow to be incompressible, by volume conservation, we have

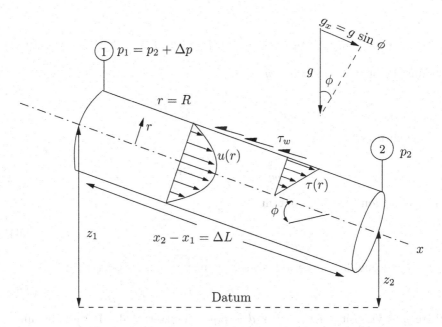

Figure 2.18: Fully developed flow in an inclined pipe

$$Q_1 = Q_2 = \text{constant}$$

$$V_1 = \frac{Q_1}{A_1} = V_2 = \frac{Q_2}{A_2}$$

where Q_1 and Q_2, respectively, are the volume flow rates and A_1, A_2, V_1, and V_2 are the local areas and velocities at states 1 and 2. The velocities V_1 and V_2 are equal, since the flow is fully developed and also $A_1 = A_2$.

By incompressible Bernoulli equation, we have

$$\frac{p_1}{\rho} + \frac{1}{2}V_1^2 + gz_1 = \frac{p_2}{\rho} + \frac{1}{2}V_2^2 + gz_2 \tag{2.61}$$

Since $V_1 = V_2$, we can write from Equation (2.61) the head loss due to friction as

$$h_f = \left(z_1 + \frac{p_1}{\rho g}\right) - \left(z_2 + \frac{p_2}{\rho g}\right) = \Delta z + \frac{\Delta p}{\rho g} \tag{2.62}$$

that is, the head loss in the pipe, due to friction, is equal to the sum of the change in gravity head and pressure head.

By momentum balance, we have

$$\Delta p \pi R^2 + \rho g(\pi R^2)\Delta L \sin\theta - \tau_w(2\pi R)\Delta L = \dot{m}(V_1 - V_2) = 0 \qquad (2.63)$$

Dividing throughout by $(\pi R^2)\rho g$, we get

$$\frac{\Delta p}{\rho g} + \Delta L \sin\theta = \frac{2\tau_w}{\rho g}\frac{\Delta L}{R}$$

But $\Delta L \sin\theta = \Delta z$. Thus,

$$\frac{\Delta p}{\rho g} + \Delta z = \frac{2\tau_w}{\rho g}\frac{\Delta L}{R}$$

Using Equation (2.62), we obtain

$$\frac{\Delta p}{\rho g} + \Delta z = h_f = \frac{2\tau_w}{\rho g}\frac{\Delta L}{R} \qquad (2.64)$$

In the functional form, the wall shear, τ_w, may be expressed as

$$\tau_w = F(\rho, V, \mu, d, \epsilon) \qquad (2.65)$$

where μ is viscosity of the fluid, d is pipe diameter, and ϵ is wall roughness height. By dimensional analysis, Equation (2.65) may be expressed as

$$\frac{8\tau_w}{\rho V^2} = f = F\left(Re_d, \frac{\epsilon}{d}\right) \qquad (2.66)$$

where f is called the *Darcy friction factor*, which is a dimensionless parameter. Combining Equations (2.64) and (2.66), we obtain the pipe head loss as

$$\boxed{h_f = f\frac{L}{d}\frac{V^2}{2g}} \qquad (2.67)$$

This is called the *Darcy–Weisbach* equation, valid for flow through ducts of any cross-section. Further, in the derivation of the above relation, there was no mention about whether the flow was laminar or turbulent and hence Equation (2.67) is valid for both laminar and turbulent flow. The value of friction factor f for any given pipe (i.e., for any surface roughness ϵ and d) at a given Reynolds number can be read from the Moody chart (which is a plot of f as a function of Re_d and ϵ/d).

2.11   Gas Dynamics

In the preceding sections of this chapter, the discussions were for fluid motions where the density can be regarded as constant, i.e., incompressible. But in many engineering applications, such as designing buildings to withstand winds, the design of engines and of vehicles of all kinds - cars, yachts, trains, aeroplanes, missiles, and launch vehicles - require a study of the flow with velocities at which the gas cannot be treated as incompressible. Indeed, the flow becomes compressible. Study of such *flows where the changes in both density and temperature associated with pressure change become appreciable is called gas dynamics.* In other words, *gas dynamics is the science of fluid flows where the density and temperature changes become important.* The essence of the subject of gas dynamics is that the entire flow field is dominated by Mach waves, expansion waves, and shock waves, when the flow speed is supersonic. It is through these waves that the change of flow properties from one state to another takes place. In the theory of gas dynamics, change of state in flow properties is achieved by three means: (a) with area change, treating the fluid to be inviscid and passage to be frictionless, (b) with friction, treating the heat transfer between the surrounding system to be negligible, and (c) with heat transfer, assuming the fluid to be inviscid. These three types of flows are called *isentropic flow, frictional* or *Fanno-type flow*, and *Rayleigh-type flow*, respectively.

All problems in gas dynamics can be classified under the three flow processes described above, of course with the assumptions mentioned. Although it is impossible to have a flow process which is purely isentropic or Fanno type or Rayleigh type, in practice, it is justified in assuming so, since the results obtained with these treatments prove to be accurate enough for most practical problems in gas dynamics. Even though it is possible to solve problems with mathematical equations and working formulae associated with these processes, it is found to be extremely useful and time saving if the working formulae are available in the form of tables with Mach number which is the dominant parameter in compressible flow analysis.

2.11.1 Perfect Gas

In principle, it is possible to do gas dynamic calculations with the general equation of state relations, for fluids. But in practice most elementary treatments are confined to perfect gases with constant specific heats. For most problems in gas dynamics, the assumption of the perfect gas law is sufficiently in accord with the properties of actual gases, hence it is acceptable.

For perfect gases, the pressure-density-temperature relation or the *thermal equation of state*, is given by

$$\boxed{p = \rho R T} \tag{2.68}$$

where R is the gas constant and T is absolute temperature. All gases obeying the thermal state equation are called *thermally perfect gases*. A perfect gas must obey at least two calorical state equations, in addition to the thermal state equation. The C_p, C_v relations given below are two well-known *calorical state equations*.

$$C_p = \frac{\partial h}{\partial T}$$

$$C_v = \frac{\partial u}{\partial T}$$

where h is specific enthalpy and u is specific internal energy, respectively. Further, for perfect gases with constant specific heats, we have

$$R = C_p - C_v$$

$$\gamma = \boxed{\frac{C_p}{C_v}} \qquad (2.69)$$

where C_p and C_v are the specific heats at constant pressure and constant volume, respectively, and γ is the isentropic index. For all real gases C_p, C_v, and γ vary with temperature, but only moderately. For example, C_p of air increases about 30 percent as temperature increases from 0 to 3000°C. Since we rarely deal with such large temperature changes, it is reasonable to assume specific heats to be constants in our studies.

2.11.2   Velocity of Sound

In the beginning of this section, it was stated that gas dynamics deals with flows in which both compressibility and temperature changes are important. The term compressibility implies variation in density. In many cases, the variation in density is mainly due to pressure change. The rate of change of density with respect to pressure is closely connected with the velocity of propagation of small pressure disturbances, i.e., with the velocity of sound "a."

The velocity of sound may be expressed as

$$a^2 = \left(\frac{\partial p}{\partial \rho}\right)_s \qquad (2.70)$$

In Equation (2.70), the ratio $dp/d\rho$ is written as partial derivative at constant entropy because the variations in pressure and temperature are negligibly small, and consequently, the process is nearly reversible. Moreover, the rapidity with which the process takes place, together with the negligibly small magnitude of the total temperature variation, makes the process nearly adiabatic. In the limit, for waves with infinitesimally small thickness, the process may be considered both reversible and adiabatic, and thus, isentropic.

It can be shown that, for an isentropic process of a perfect gas, the velocity of sound can be expressed as

$$\boxed{a = \sqrt{\gamma R T}} \qquad (2.71)$$

where T is absolute static temperature.

2.11.3   Mach Number

Mach number M is a dimensionless parameter, expressed as the ratio between the magnitudes of local flow velocity and local velocity of sound, i.e.,

$$M = \frac{\text{Local flow velocity}}{\text{Local velocity of sound}} = \frac{V}{a} \tag{2.72}$$

Mach number plays a dominant role in the field of gas dynamics.

2.11.4 Flow with Area Change

If the flow is assumed to be isentropic for a channel flow, all states along the channel or stream tube lie on a line of constant entropy and have the same stagnation temperature. The state of zero velocity is called the *isentropic stagnation state*, and the state with $M = 1$ is called the *critical state*.

2.11.4.1 Isentropic Relations

The relations between pressure, temperature, and density for an isentropic process of a perfect gas are

$$\frac{p}{p_0} = \left(\frac{\rho}{\rho_0}\right)^\gamma \quad \text{and} \quad \frac{T}{T_0} = \left(\frac{p}{p_0}\right)^{\frac{\gamma-1}{\gamma}} \tag{2.73}$$

Also, the pressure-temperature density relation of a perfect gas is

$$\frac{p}{\rho T} = \frac{p_0}{\rho_0 T_0} = R \tag{2.74}$$

The temperature, pressure, and density ratios as functions of Mach number are

$$\frac{T_0}{T} = \left(1 + \frac{\gamma - 1}{2}M^2\right) \tag{2.75}$$

$$\frac{p_0}{p} = \left(1 + \frac{\gamma - 1}{2}M^2\right)^{\frac{\gamma}{\gamma-1}} \tag{2.76}$$

$$\frac{\rho_0}{\rho} = \left(1 + \frac{\gamma - 1}{2}M^2\right)^{\frac{1}{\gamma-1}} \tag{2.77}$$

where T_0, p_0, and ρ_0 are the temperature, pressure, and density, respectively, at the stagnation state. The particular value of temperature, pressure, and density ratios at the critical state (i.e., at the choked location in a flow passage) are found by setting $M = 1$ in Equations (2.75)–(2.77). For $\gamma = 1.4$, the following are the temperature, pressure, and density ratio at the critical state.

$$\frac{T^*}{T_0} = \frac{a^{*2}}{a_0^2} = \frac{2}{\gamma+1} = 0.8333 \tag{2.78}$$

$$\frac{p^*}{p_0} = \left(\frac{2}{\gamma+1}\right)^{\frac{\gamma}{\gamma-1}} = 0.5283 \tag{2.79}$$

$$\frac{\rho^*}{\rho_0} = \left(\frac{2}{\gamma+1}\right)^{\frac{1}{\gamma-1}} = 0.6339 \tag{2.80}$$

where, T^*, p^*, and ρ^* are the temperature, pressure, and density, respectively, at the critical state.

The critical pressure ratio p^*/p_0 is of the same order of magnitude for all gases. It varies almost linearly with γ from 0.6065, for $\gamma = 1$, to 0.4867, for $\gamma = 1.67$.

The *dimensionless velocity* M^* is one of the most useful parameters in gas dynamics. Generally it is defined as

$$M^* \equiv \frac{V}{a^*} \equiv \frac{V}{V^*} \tag{2.81}$$

where a^* is the critical speed of sound. This dimensionless velocity can also be expressed in terms of Mach number as

$$M^{*2} = \frac{(\gamma+1)M^2}{(\gamma-1)M^2+2} \tag{2.82}$$

2.11.4.2　Area-Mach Number Relation

For an isentropic flow of a perfect gas through a duct, the Area-Mach number relation may be expressed, assuming one-dimensional flow, as

$$\left(\frac{A}{A^*}\right)^2 = \frac{1}{M^2}\left[\frac{2}{\gamma+1}\left(1+\frac{\gamma-1}{2}M^2\right)\right]^{\frac{\gamma+1}{\gamma-1}} \tag{2.83}$$

where A^* is called the sonic or critical throat area.

2.11.4.3　Prandtl–Meyer Function

Prandtl–Meyer function ν is an important parameter to solve supersonic flow problems involving isentropic expansion or isentropic compression. Basically Prandtl–Meyer function is a *similarity parameter*. The Prandtl–Meyer function can be expressed in terms of M as

$$\boxed{\nu = \frac{\gamma+1}{\gamma-1}\ \text{arc}\tan\sqrt{\frac{\gamma-1}{\gamma+1}(M^2-1)} - \text{arc}\tan\sqrt{M^2-1}} \tag{2.84}$$

From Equation (2.84) it is seen that, for a given M, ν is fixed.

2.11.5 Normal Shock Relations

The shock may be described as a compression front, in a supersonic flow field, across which the flow properties jump. The thickness of the shock is comparable to the mean free path of the gas molecules in the flow field. When the shock is normal to the flow direction it is called *normal shock*, and when it is inclined at an angle to the flow it is termed *oblique shock*. For a perfect gas, it is known that all the flow property ratios across a normal shock are unique functions of specific heats ratio, γ, and the upstream Mach number, M_1.

Considering the normal shock shown in Figure 2.19, the following normal shock relations, assuming the flow to be one-dimensional can be obtained:

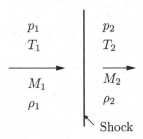

Figure 2.19: Flow through a normal shock

$$M_2^2 = \frac{2 + (\gamma - 1)\, M_1^2}{2\gamma M_1^2 - (\gamma - 1)} \tag{2.85}$$

$$\frac{p_2}{p_1} = 1 + \frac{2\gamma}{\gamma + 1}(M_1^2 - 1) \tag{2.86}$$

$$\frac{\rho_2}{\rho_1} = \frac{V_1}{V_2} = \frac{(\gamma + 1)M_1^2}{(\gamma - 1)M_1^2 + 2} \tag{2.87}$$

$$\frac{T_2}{T_1} = \frac{h_2}{h_1} = \frac{a_2^2}{a_1^2} = 1 + \frac{2(\gamma - 1)}{(\gamma + 1)^2} \frac{(\gamma M_1^2 + 1)}{M_1^2}(M_1^2 - 1) \tag{2.88}$$

In Equation (2.88), h_1 and h_2 are the static enthalpies upstream and downstream of the shock, respectively.

The stagnation pressure ratio across a normal shock, in terms of the upstream Mach number is

$$\frac{p_{02}}{p_{01}} = \left[1 + \frac{2\gamma}{\gamma + 1}(M_1^2 - 1)\right]^{\frac{-1}{\gamma - 1}} \left[\frac{(\gamma + 1)M_1^2}{(\gamma - 1)M_1^2 + 2}\right]^{\frac{\gamma}{\gamma - 1}} \tag{2.89}$$

The change in entropy across the normal shock is given by

$$s_2 - s_1 = R \ln \frac{p_{01}}{p_{02}} \tag{2.90}$$

2.11.6   Oblique Shock Relations

Consider the flow through an oblique shock wave, as shown in Figure 2.20.

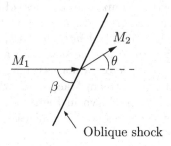

Figure 2.20: Flow through an oblique shock

The component of M_1 normal to the shock wave is

$$M_{n1} = M_1 \sin \beta \tag{2.91}$$

where β is the shock angle. The shock in Figure 2.20 can be visualized as a normal shock with upstream Mach number $M_1 \sin \beta$. Thus, replacement of M_1 in the normal shock relations, Equations (2.80) to (2.84), by $M_1 \sin \beta$, results in the corresponding relations for the oblique shock.

$$M_{n2}^2 = \frac{M_1^2 \sin^2 \beta + \dfrac{2}{\gamma - 1}}{\dfrac{2\gamma}{\gamma - 1} M_1^2 \sin^2 \beta - 1} \tag{2.92}$$

$$\frac{p_2}{p_1} = 1 + \frac{2\gamma}{\gamma + 1}(M_1^2 \sin^2 \beta - 1) \tag{2.93}$$

$$\frac{\rho_2}{\rho_1} = \frac{(\gamma + 1)M_1^2 \sin^2 \beta}{(\gamma - 1)M_1^2 \sin^2 \beta + 2} \tag{2.94}$$

$$\frac{T_2}{T_1} = \frac{a_2^2}{a_1^2} = 1 + \frac{2(\gamma - 1)}{(\gamma + 1)^2} \frac{\gamma M_1^2 \sin^2 \beta + 1}{M_1^2 \sin^2 \beta}\left(M_1^2 \sin^2 \beta - 1\right) \tag{2.95}$$

$$\frac{p_{02}}{p_{01}} = \left[1 + \frac{2\gamma}{\gamma + 1}\left(M_1^2 \sin^2 \beta - 1\right)\right]^{\frac{-1}{\gamma - 1}} \left[\frac{(\gamma + 1)M_1^2 \sin^2 \beta}{(\gamma - 1)M_1^2 \sin^2 \beta + 2}\right]^{\frac{\gamma}{\gamma - 1}} \tag{2.96}$$

The entropy change across the oblique shock is given by

$$s_2 - s_1 = R \ln \frac{p_{01}}{p_{02}}$$

Equation (2.92) gives only the normal component of Mach number M_{n2} behind the shock. But the Mach number of interest is M_2. It can be obtained from Equation (2.92) as follows:

From the geometry of the oblique shock flow field in Figure 2.20, it is seen that M_2 is related to M_{n2} by

$$M_2 = \frac{M_{n2}}{\sin(\beta - \theta)} \qquad (2.97)$$

where θ is the flow turning angle across the shock. Combining Equations (2.92) and (2.97), the Mach number M_2 after the shock can be obtained.

2.11.7 Flow with Friction

In Section 2.11.3, on flow with area change, it was assumed that the changes in flow properties for compressible flow of gases in ducts were brought about solely by area change. That is, the effects of viscosity and heat transfer have been neglected. But, in practical flow situations like stationary power plants, aircraft engines, high vacuum technology, transport of natural gas in long pipe lines, transport of fluids in chemical process plants, and various types of flow systems, the high-speed flow travels through passages of considerable length and hence the effects of viscosity (friction) cannot be neglected for such flows. In many practical flow situations, friction can even have a decisive effect on the resultant flow characteristics.

Consider one-dimensional steady flow of a perfect gas with constant specific heats through a constant area duct. Assume that, there is neither external heat exchange nor external shaft work and the difference in elevation produces negligible changes in flow properties compared to frictional effects. The flow with the above said conditions, namely, adiabatic flow with no external work, is called *Fanno line flow*. For Fanno line flow, the wall friction (due to viscosity) is the chief factor bringing about changes in flow properties.

Working Formulae for Fanno-Type Flow

Consider the flow of a perfect gas through a constant area duct shown in Figure 2.21. Choosing an infinitesimal control volume as shown in the figure, the relation between Mach number M and friction factor f can be written as

$$\int_0^{L_{\max}} 4f \frac{dx}{D} = \int_{M^2}^1 \frac{1 - M^2}{\gamma M^4 \left(1 + \dfrac{\gamma - 1}{2} M^2\right)} dM^2 \qquad (2.98)$$

In this relation, the integration limits are taken as (1) the section where the Mach number is M, and the length x is arbitrarily set equal to zero, and (2) the section where Mach number is unity and x is the maximum possible length of duct, $L_{\max}$ and D is the *hydraulic diameter*, defined as

$$D \cong \frac{4 \, (\text{Cross-section area})}{\text{Wetted perimeter}}$$

On integration, Equation (2.98) yields

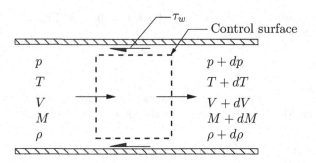

Figure 2.21: Control volume for Fanno flow

$$4\bar{f}\frac{L_{\max}}{D} = \frac{1 - M^2}{\gamma M^2} + \frac{\gamma + 1}{2\gamma}\ln\left(\frac{(\gamma + 1)M^2}{2(1 + \frac{\gamma - 1}{2}M^2)}\right) \tag{2.99}$$

where $\bar{f}$ is the mean friction coefficient with respect to duct length, defined by

$$\bar{f} = \frac{1}{L_{\max}}\int_0^{L_{\max}} f\,dx$$

Likewise, the local flow properties can be found in terms of local Mach number. Indicating the properties at $M = 1$ with superscripted with "asterisk," and integrating between the duct sections with $M = M$ and $M = 1$, the following relations can be obtained(*Gas Dynamics*, Rathakrishnan, 1995).

$$\frac{p}{p^*} = \frac{1}{M}\left[\frac{\gamma + 1}{2\left(1 + \frac{\gamma + 1}{2}M^2\right)}\right]^{\frac{1}{2}} \tag{2.100}$$

$$\frac{V}{V^*} = M\left[\frac{\gamma + 1}{2\left(1 + \frac{\gamma - 1}{2}M^2\right)}\right]^{\frac{1}{2}} \tag{2.101}$$

$$\frac{T}{T^*} = \frac{a^2}{a^{*2}} = \frac{\gamma + 1}{2\left(1 + \frac{\gamma - 1}{2}M^2\right)} \tag{2.102}$$

$$\frac{\rho}{\rho^*} = \frac{V^*}{V} = \frac{1}{M}\left[\frac{2\left(1+\frac{\gamma-1}{2}M^2\right)}{\gamma+1}\right]^{\frac{1}{2}} \tag{2.103}$$

$$\frac{p_0}{p_0^*} = \frac{1}{M}\left[\frac{2\left(1+\frac{\gamma-1}{2}M^2\right)}{\gamma+1}\right]^{\frac{\gamma+1}{2(\gamma-1)}} \tag{2.104}$$

$$\frac{F}{F^*} = \frac{1+\gamma M^2}{M\left(2(\gamma+1)\left(1+\frac{\gamma-1}{2}M^2\right)\right)^{\frac{1}{2}}} \tag{2.105}$$

In Equation (2.105), the parameter F is called *Impulse Function*, defined as

$$F \cong pA + \rho A V^2 = pA(a + \gamma M^2)$$

2.11.8   Flow with Simple T_0-Change

In the section on flow with area change, the process was considered to be isentropic with the assumption that the frictional and energy effects were absent. In Fanno line flow, only the effect of wall friction was taken into account in the absence of area change and energy effects. In the present section, the processes involving change in the stagnation temperature or the stagnation enthalpy of a gas stream, which flows in a frictionless constant area duct are considered. From one-dimensional point of view, this is yet another effect producing continuous changes in the state of a flowing stream and this factor is called the energy effect, such as external heat exchange, combustion, or moisture condensation. Though a process involving simple stagnation temperature (T_0) change is difficult to achieve in practice, many useful conclusions of practical significance may be drawn by analyzing the process of simple T_0-change. This kind of flow involving only T_0-change is called *Rayleigh-type flow*.

Working Formulae for Rayleigh-Type Flow

Consider the flow of a perfect gas through a constant-area duct without friction, shown in Figure 2.22.

Considering a control volume, as in Figure 2.22, the normalized expressions (working formulae) for the flow process involving only heat transfer can be obtained as (*Gas Dynamics*, Rathakrishnan, 1995):

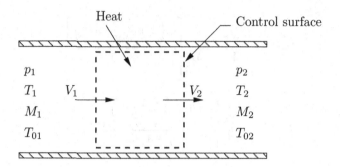

Figure 2.22: Control volume for Rayleigh flow

$$\frac{p}{p^*} = \frac{\gamma+1}{1+\gamma M^2} \tag{2.106}$$

$$\frac{\rho}{\rho^*} = \frac{V^*}{V} = \frac{1+\gamma M^2}{(\gamma+1)M^2} \tag{2.107}$$

$$\frac{T}{T^*} = \frac{(\gamma+1)^2 M^2}{(1+\gamma M^2)^2} \tag{2.108}$$

$$\frac{T_0}{T_0^*} = \frac{2(\gamma+1)M^2\left(1+\dfrac{\gamma-1}{2}M^2\right)}{(1+\gamma M^2)^2} \tag{2.109}$$

$$\frac{p_0}{p_0^*} = \frac{\gamma+1}{1+\gamma M^2}\left[\frac{2\left(1+\dfrac{\gamma-1}{2}M^2\right)}{\gamma+1}\right]^{\frac{\gamma}{\gamma-1}} \tag{2.110}$$

2.12   Summary

Fluid may be defined as *a substance which will continue to change shape as long as there is a shear stress present, however small it may be.*

Pressure is the *force per unit area which acts normal to the surface of any object which is immersed in a fluid.* For a fluid at rest, the pressure at any point is the same in all directions. The pressure in a stationary fluid varies only in the vertical direction and is constant in any horizontal plane. When a fluid is in motion, the actual pressure exerted by the fluid is known as the *static pressure.*

Total pressure is that pressure which a fluid will experience if its motion is brought to rest. It is also called *impact pressure.* The total and static pressures are used for computing the flow velocity. The unit of pressure is N/m^2 or simply pascal (Pa). At standard sea level condition, the atmospheric pressure is 101325 Pa, which corresponds to 760 mm of mercury column height.

By virtue of motion the molecules possess kinetic energy, and this energy is sensed as *temperature* of the solid, liquid, or gas. In the case of a gas in motion it is called the *static temperature*. The unit for temperature is kelvin (K) or degree celsius (°C), in SI units. For all calculations in this book, temperature will be expressed in kelvin, i.e., from absolute zero. At standard sea level conditions the atmospheric temperature is 288.15 K.

The total number of molecules in a unit volume is a measure of the density, ρ, of the substance. It is expressed as mass per unit volume, say kg/m^3. Mass is defined as weight divided by acceleration due to gravity. At standard atmospheric temperature and pressure (288.15 K and 101325 Pa), the density of dry air is 1.225 kg/m^3.

The property which characterizes the resistance that a fluid offers to applied shear force is termed *viscosity*. The absolute coefficient of viscosity is a direct measure of the viscosity of a fluid. For air the Sutherland's relation for viscosity is

$$\boxed{\mu = 1.46 \times 10^{-6} \left(\frac{T^{3/2}}{T + 111} \right) \frac{\text{N s}}{\text{m}^2}}$$

where T is in kelvin.

At high speeds, heat transfer from vehicles becomes significant. The heat transfer from a vehicle depends on the thermal conductivity, K, of air. For this case, a relation similar to Sutherland's law for viscosity is found to be useful, and it is

$$K = 1.99 \times 10^{-3} \left(\frac{T^{3/2}}{T + 112} \right) \frac{\text{J}}{\text{s m K}}$$

where T is temperature in kelvin. The pressure and temperature ranges in which this equation is applicable are 0.01 to 100 atm and 0 to 2000 K, respectively.

The change in volume of a fluid associated with change in pressure is called *compressibility*. When a fluid is subjected to pressure it gets compressed and its volume changes. The bulk modulus of elasticity is a measure of how easily the fluid may be compressed, and is defined as the ratio of pressure change to volumetric strain associated with it.

The specific heats at constant volume and constant pressure processes, respectively, are designated by C_v and C_p.

$$C_v \equiv \left(\frac{\partial u}{\partial T} \right)_v$$

C_v is a measure of the energy-carrying capacity of the gas molecules.

$$C_p \equiv \left(\frac{\partial h}{\partial T} \right)_p$$

where $h = u + pv$, the sum of internal energy and flow energy is known as the *enthalpy* or total heat constant per unit mass of fluid and C_p is a measure of the ability of the gas to do external work in addition to possessing internal energy.

The specific heats ratio, γ, given by

$$\boxed{\gamma = \frac{C_p}{C_v}}$$

is an important parameter in the study of compressible flows. This is a measure of *the relative internal complexity of the molecules of the gas.* It has been determined from kinetic theory of gases that the ratio of specific heats can be related to the number of degrees of freedom n of the gas molecules by the relation

$$\gamma = \frac{n+2}{n}$$

At normal temperatures, there are six degrees of freedom, three translational and three rotational, for diatomic gas molecules. For nitrogen, which is a diatomic gas, $n = 5$, since one of the rotational degrees of freedom is small in comparison with the other two. Therefore,

$$\gamma = 7/5 = 1.4$$

A monatomic gas like helium has 3 translational degrees of freedom only, and therefore,

$$\gamma = 5/3 = 1.67$$

This value of 1.67 is the upper limit of the values which γ can take. In general γ varies from 1 to 1.67. i.e.,

$$\boxed{1 \leq \gamma \leq 1.67}$$

Basically two treatments are followed for fluid flow analysis. They are Lagrangian and Eulerian descriptions.

The rate of change of a property measured by probes at fixed locations are referred to as *local rates of change* and the rate of change of properties experienced by a material particle is termed the *material* or the *substantive rates of change.*

For a fluid flowing with an uniform velocity V_∞, it is possible to write the relation between the local and material rates of change of property η as

$$\frac{\partial \eta}{\partial t} = \frac{D\eta}{Dt} - V_\infty \frac{\partial \eta}{\partial x}$$

Pathline may be defined as a line in the flow field describing the trajectory of a given fluid particle.

Streakline may be defined as the instantaneous locus of all the fluid elements that have passed the point of injection at some earlier time.

Streamline is an imaginary line, in a fluid flow, drawn in such a manner that the flow velocity is always tangential to it. Flows are usually depicted graphically with the aid of streamlines. These are *imaginary lines* in the flow field such that the velocity at all points on these lines are always tangential.

Streamlines proceeding through the periphery of an infinitesimal area at some instant of time t will form a tube called *streamtube*, which is useful in the study of fluid flow. In modern fluid flow analysis, yet another graphical representation, namely *timeline* is used. When a pulse input is periodically imposed on a line of tracer source placed normal to a flow, a change in the flow profile can be observed. The tracer image is generally termed a timeline.

The four basic laws that must be satisfied by any continuous medium are

- Conservation of matter (continuity equation).

- Newton's second law (momentum equation).

- Conservation of energy (first law of thermodynamics).

- Increase of entropy principle (second law of thermodynamics).

In addition to these primary laws, there are numerous subsidiary laws, sometimes called constitutive relations, that apply to specific types of media or flow processes (e.g., equation of state for perfect gas, Newton's viscosity law for certain viscous fluids, isentropic process, adiabatic process).

An identified quantity of matter is called the *control mass system*. A control mass system may change its shape, position, and thermal condition, with time or space or both but must always entail the same matter.

Control volume is designated in space, and the boundary of this volume is known as *control surface*. The amount and identity of the matter in the control volume may change with time, but the shape of the control volume is fixed, i.e., the control volume may change its position in time or space or both, but its shape is always preserved.

The analysis where large control volumes are used to obtain aggregate forces or transfer rates is termed the *integral analysis*. When the analysis is applied to individual points in the flow field, the resulting equations are differential equations and the method is termed the *differential analysis*.

For steady incompressible flow, the continuity equation in differential form is

$$\boxed{\frac{\partial V_x}{\partial x} + \frac{\partial V_y}{\partial y} + \frac{\partial V_z}{\partial z} = 0}$$

This equation may also be expressed as $\nabla . V = 0$, where V is the velocity vector.

For gaseous medium, body forces are negligibly small compared to other forces and hence can be neglected. For steady incompressible flows, the momentum equation can be written as

$$V_x \frac{\partial V_x}{\partial x} + V_y \frac{\partial V_x}{\partial y} + V_z \frac{\partial V_x}{\partial z} = -\frac{1}{\rho} \frac{\partial p}{\partial x} + \nu \left(\frac{\partial^2 V_x}{\partial x^2} + \frac{\partial^2 V_x}{\partial y^2} + \frac{\partial^2 V_x}{\partial z^2} \right)$$

$$V_x \frac{\partial V_y}{\partial x} + V_y \frac{\partial V_y}{\partial y} + V_z \frac{\partial V_y}{\partial z} = -\frac{1}{\rho} \frac{\partial p}{\partial y} + \nu \left(\frac{\partial^2 V_y}{\partial x^2} + \frac{\partial^2 V_y}{\partial y^2} + \frac{\partial^2 V_y}{\partial z^2} \right)$$

$$V_x \frac{\partial V_z}{\partial x} + V_y \frac{\partial V_z}{\partial y} + V_z \frac{\partial V_z}{\partial z} = -\frac{1}{\rho} \frac{\partial p}{\partial z} + \nu \left(\frac{\partial^2 V_z}{\partial x^2} + \frac{\partial V_z}{\partial y^2} + \frac{\partial V_z}{\partial z^2} \right)$$

These equations are the x, y, z components of momentum equation, respectively. These equations are popularly known as *Navier–Stokes equations.*

Boundary layer thickness δ may be defined as the distance from the wall out to where the fluid velocity is within 1 percent of the local main stream velocity. It may also be defined as the distance δ normal to the surface, in which the flow velocity rises from zero to some specified value (e.g., 99%) of its local main stream flow.

Displacement thickness δ^* may be defined as the distance by which the boundary would have to be displaced if the entire flow were imagined to be frictionless and the same mass flow is maintained at any section.

$$\delta^* = \int_0^\infty \left(1 - \frac{V_x}{V_m}\right) dy$$

Other (thickness) measures pertaining to the thickness of boundary layer are *momentum thickness, θ,* and *energy thickness, δ_e,* expressed as

$$\theta = \int_0^\infty \left(1 - \frac{V_x}{V_m}\right) \frac{\rho V_x}{\rho_m V_m} dy$$

$$\delta_e = \int_0^\infty \left(1 - \frac{V_x^2}{V_m^2}\right) \frac{\rho V_x}{\rho_m V_m} dy$$

where V_m and ρ_m are the velocity and density at the edge of the boundary layer and V_x and ρ are the velocity and density at any y.

Transition point may be defined as the end of the region at which the flow in the boundary layer on a surface ceases to be laminar and becomes turbulent.

Separation point is the position at which the boundary layer leaves the surface of a solid body.

Circulation Γ is defined as the line integral of velocity vector between any two points, i.e.,

$$\Gamma = \oint_c V. \, dl$$

where dl is an elemental length. Circulation per unit area is known as *vorticity* ζ,

$$\zeta = \Gamma/A$$

In vector form ζ becomes

$$\zeta = \nabla \times V = \operatorname{curl} V$$

In terms of stream function ψ, the velocity components of a two-dimensional incompressible flow are given by

$$V_x = \frac{\partial \psi}{\partial y}, \quad V_y = -\frac{\partial \psi}{\partial x}$$

For compressible flow, the velocity components become

$$V_x = \frac{1}{\rho}\frac{\partial \psi}{\partial y}, \quad V_y = -\frac{1}{\rho}\frac{\partial \psi}{\partial x}$$

For irrotational flows (the fluid elements in the field are free of angular motion), there exists a function ϕ called *velocity potential or potential function*. For two-dimensional flows ϕ must be a function of x, y, and t. The velocity components are given by

$$V_x = \frac{\partial \phi}{\partial x}, \quad V_y = \frac{\partial \phi}{\partial y}$$

We can write the relation between ϕ and ψ as

$$\frac{\partial \psi}{\partial y} = \frac{\partial \phi}{\partial x}, \quad \frac{\partial \psi}{\partial x} = -\frac{\partial \phi}{\partial y}$$

These relations are the famous *Cauchy–Riemann equations* of complex-variable theory.

Potential flow is based on the concept that the flow field can be represented by a potential function ϕ such that,

$$\boxed{\nabla^2 \phi = 0}$$

This linear partial differential equation is popularly known as *Laplace equation*. Unlike the stream function ψ, the potential function can exist only if the flow is *irrotational*, that is, when viscous effects are absent.

Source is a flow in which the fluid emanates from origin and spreads radially outwards to infinity. The velocity potential for a two-dimensional source of strength q is

$$\boxed{\Phi = \frac{q}{2\pi}\ln(r)}$$

The stream function for a source of strength q is

$$\boxed{\psi = \frac{q}{2\pi}\theta}$$

Sink, is a type of flow in which the fluid at infinity flows radially towards the origin. The stream functions for a sink is

$$\boxed{\phi = -\frac{q}{2\pi}\ln(r)}$$

Simple vortex or *free vortex* is a flow in which the fluid elements translate along circular paths, without spinning about their own axes. For a simple vortex, the potential and stream functions are

$$\boxed{\phi = \frac{q}{2\pi}\theta}$$

$$\psi = -\frac{q}{2\pi}\ln(r)$$

Forced vortex is a flow in which the fluid elements, in addition to moving along circular paths, spin about their own axes.

Doublet is a combination of a source and a sink placed with a negligibly small distance between them.

Viscosity produces shear force which tends to retard the fluid motion. It works against inertia force. The ratio of these two forces governs (dictates) many properties of the flow and the ratio expressed in the form of a nondimensional parameter known as the famous *Reynolds number* Re_L.

$$Re_L = \frac{\rho V L}{\mu}$$

where V is flow velocity, ρ is flow density, μ is dynamic viscosity of the fluid, and L is a characteristic dimension.

Lower critical Reynolds number is that Reynolds number below which the entire flow is laminar. The Reynolds number above which the entire flow is turbulent is termed as *upper critical Reynolds number*. *Critical Reynolds number* is that at which the flow field is a mixture of laminar and turbulent flows.

Drag is essentially a force opposing the motion of the body. Viscosity is responsible for a part of the drag force, and the body shape generally determines the overall drag. Low drag shapes are called *streamlined bodies* and high drag shapes are termed *bluff bodies*.

Drag due to the difference in pressure between the front and back regions of a body is known as *pressure drag* and drag due to the friction between the body surface and the fluid is known as *skin friction drag* or *shear drag*. For bluff bodies the pressure drag is many times larger than the skin friction drag and for streamlined bodies the condition is the reverse.

Turbulence may be described as *a three-dimensional, random phenomenon, exhibiting multiplicity of scales, possessing vorticity and showing very high dissipation*. The turbulence level for any given flow with a mean velocity $\overline{U}$ is expressed as a turbulence number n, defined as

$$n = 100\frac{\sqrt{\overline{u'^2} + \overline{v'^2} + \overline{w'^2}}}{3\overline{U}}$$

Laminar flow is an orderly flow in which the fluid elements move in an orderly manner such that the transverse exchange of momentum is insignificant.

The head loss in a pipe can be expressed as

$$h_f = f\frac{L}{d}\frac{V^2}{2g}$$

Gas dynamics is the science of fluid flow where both density and temperature changes become important. The essence of the subject of gas dynamics is that the entire flow field is dominated by Mach waves, expansion waves, and shock waves, when the flow speed is supersonic. It is through these waves, the change of flow properties from one state to another takes place. In the theory of gas dynamics, change of state in flow properties is achieved by three means; (a) with area change, treating the fluid to be inviscid, passage to be frictionless, and the process to be adiabatic, (b) with friction, treating the heat transfer between the surroundings and system to be negligible, and (c) with heat transfer, assuming the fluid to be inviscid. These three types of flows are called *isentropic flow, frictional or Fanno-type flow and Rayleigh-type flow*, respectively.

A perfect gas is that which is thermally and calorically perfect. That is, thermal perfectness is a prerequisite for calorical perfectness.

The rate of change of density with respect to pressure is closely connected with the velocity of propagation of small pressure disturbances, i.e., with the velocity of sound "a." The velocity of sound may be expressed as

$$a^2 = \left(\frac{\partial p}{\partial \rho}\right)_s$$

For a perfect gas,

$$\boxed{a = \sqrt{\gamma RT}}$$

where T is absolute static temperature.

Mach number M is a dimensionless parameter, expressed as the ratio between the magnitudes of local flow velocity and local velocity of sound. i.e.,

$$M = \frac{\text{Local flow velocity}}{\text{Local velocity of sound}} = \frac{V}{a}$$

If the flow is assumed to be isentropic for a channel flow, all states along the channel or stream tube lie on a line of constant entropy and have the same stagnation temperature. The state of zero velocity is called the *isentropic stagnation state*, and the state with $M = 1$ is called the *critical state*.

The temperature, pressure, and density ratios as functions of Mach number are

$$\frac{T_0}{T} = \left(1 + \frac{\gamma - 1}{2} M^2\right)$$

$$\frac{p_0}{p} = \left(1 + \frac{\gamma - 1}{2} M^2\right)^{\frac{\gamma}{\gamma - 1}}$$

$$\frac{\rho_0}{\rho} = \left(1 + \frac{\gamma - 1}{2} M^2\right)^{\frac{1}{\gamma - 1}}$$

where T_0, p_0, and ρ_0 are the temperature, pressure, and density, respectively, at the stagnation state.

For an isentropic flow of a perfect gas through a duct, the area-Mach number relation may be expressed, assuming one-dimensional flow, as

$$\left(\frac{A}{A^*}\right)^2 = \frac{1}{M^2}\left[\frac{2}{\gamma+1}\left(1+\frac{\gamma-1}{2}M^2\right)\right]^{\frac{\gamma+1}{\gamma-1}}$$

where A^* is called the sonic or critical throat area.

Prandtl–Meyer function ν is an important parameter to solve supersonic flow problems involving isentropic expansion or isentropic compression. Basically the Prandtl–Meyer function is a *similarity parameter*. The Prandtl–Meyer function can be expressed in terms of M as

$$\nu = \frac{\gamma+1}{\gamma-1}\ \mathrm{arc\ tan}\sqrt{\frac{\gamma-1}{\gamma+1}(M^2-1)} - \mathrm{arc\ tan}\sqrt{M^2-1}$$

A *shock* may be described as a compression front in a supersonic flow field and flow process across which results in an abrupt change in fluid properties. When the shock is normal to the flow direction it is called *normal shock* and when it is inclined at an angle to the flow it is termed *oblique shock*. For a perfect gas, it is known that all the flow property ratios across a normal shock are unique functions of μ and the upstream Mach number. The normal shock relations are,

$$M_2^2 = \frac{1+\dfrac{\gamma-1}{2}M_1^2}{\gamma M_1^2 - \dfrac{\gamma-1}{2}}$$

$$\frac{p_2}{p_1} = 1 + \frac{2\gamma}{\gamma+1}(M_1^2-1)$$

$$\frac{\rho_2}{\rho_1} = \frac{V_1}{V_2} = \frac{(\gamma+1)M_1^2}{(\gamma-1)M_1^2+2}$$

$$\frac{T_2}{T_1} = \frac{h_2}{h_1} = \frac{a_2^2}{a_1^2} = 1 + \frac{2(\gamma-1)}{(\gamma+1)^2}\frac{(\gamma M_1^2+1)}{M_1^2}(M_1^2-1)$$

$$\frac{p_{02}}{p_{01}} = \left[1 + \frac{2\gamma}{\gamma+1}(M_1^2-1)\right]^{\frac{-1}{\gamma-1}}\left[\frac{(\gamma+1)M_1^2}{(\gamma-1)M_1^2+2}\right]^{\frac{\gamma}{\gamma-1}}$$

The change in entropy across the normal shock is given by

$$s_2 - s_1 = R\ln\frac{p_{01}}{p_{02}}$$

The component of M_1 normal to the shock wave is

$$M_{n1} = M_1\sin\beta$$

where β is shock angle.

The oblique shock can be visualized as a normal shock with upstream Mach number $M_1 \sin \beta$. Thus, replacement of M_1 in normal shock relations by $M_1 \sin \beta$ results in the corresponding relations for the oblique shock.

Fanno flow is an adiabatic flow with no external work. For Fanno line flow, the wall friction (due to viscosity) is the chief factor bringing about changes in flow properties.

For flow of a perfect gas through a constant area duct the relation between Mach number M and friction factor f can be written as

$$4\bar{f}\frac{L_{max}}{D} = \frac{1-M^2}{\gamma M^2} + \frac{\gamma+1}{2\gamma} \ln\left(\frac{(\gamma+1)M^2}{2(1+\frac{\gamma-1}{2}M^2)}\right)$$

where $\bar{f}$ is the mean friction coefficient with respect to duct length, defined by

$$\bar{f} = \frac{1}{L_{max}} \int_0^{L_{max}} f \, dx$$

Rayleigh-type flow is that involving only T_0-change. In Rayleigh flow, the process is considered to involve only change in the stagnation temperature or the stagnation enthalpy of a gas stream.

In our discussions on measurement devices for flow properties and their working principles, different aspects of flow and fluid properties scanned in this chapter, namely, the viscosity, compressibility, friction, heat transfer, isentropic and adiabatic nature, Mach and shock waves, and so on will find decisive effects on the performance of the devices. Therefore, it will be beneficial to go through this chapter before taking up our discussions on measuring devices in the chapters to follow.

Exercise Problems

2.1 The turbulence number of a uniform horizontal flow of velocity 25 m/s is 6. If the turbulence is isotropic, determine the mean square values of the fluctuations.

[Answer: 6.75 m^2/s^2]

2.2 Flow through the convergent nozzle shown in Figure P2.2 is approximated as one-dimensional. If the flow is steady, will there be any fluid acceleration? If there is acceleration, obtain an expression for it in terms of volumetric flow rate $\dot{Q}$, if the area of cross-section is given by $A(x) = e^{-x}$.

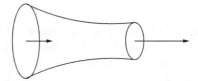

Figure P2.2: Flow through a convergent nozzle

$$\left[\text{Answer: } \left(\frac{\dot{Q}}{e^{-x}}\right)^2\right]$$

2.3 Atmospheric air is cooled by a desert cooler by 18°C and sent into a room. The cooled air then flows through the room and picks up heat from the room at a rate of 0.15°C/s. The air speed in the room is 0.72 m/s. After some time from switching on, the temperature gradient assumes a value of 0.9°C/m in the room. Determine $\partial T/\partial t$ at a point 3 m away from the cooler.

[Answer: $-0.498\,$°C/s]

2.4 For proper functioning, an electronic instrument onboard a balloon should not experience temperature change more than $\pm0.006\,\text{K/s}$. The atmospheric temperature is given by

$$T = \left(288 - 6.5 \times 10^{-3}\, z\right)\left(2 - e^{-0.02t}\right)\,\text{K}$$

where z is the height in meters above the ground and t is the time in hours after sunrise. Determine the maximum allowable rate of ascent, without affecting the instrument performance, when the balloon is on the ground at $t = 2$ hr.

[Answer: 1.12 m/s]

2.5 Flow through a tube has a velocity given by

$$u = u_{\text{max}} \left(1 - \frac{r^2}{R^2}\right)$$

where R is the tube radius and u_{max} is the maximum velocity, which occurs at the tube centerline. (a) Find a general expression for the volume flow rate and the average velocity through the tube, (b) compute the volume flow rate, if $R = 25$ mm and $u_{\text{max}} = 10$ m/s, and (c) compute the mass flow rate, if $\rho = 1000$ kg/m^3.

$$\left[\text{Answer: (a) } \frac{1}{2}u_{\text{max}}\,\pi R^2,\ \frac{1}{2}u_{\text{max}},\ \text{(b) } 0.00982\,\text{m}^3/\text{s},\ \text{(c) } 9.82\,\text{kg/s}\right]$$

2.6 A two-dimensional velocity field is given by

$$V = \left(x - y^2\right) i + \left(xy + 2y\right) j$$

in arbitrary units. Compute (a) the acceleration components a_x and a_y, (b) the velocity component in the direction $\theta = 30°$, and (c) the magnitude and direction of maximum velocity and maximum acceleration, at the point (2, 1).

[Answer: (a) $a_x = -7$ units, $a_y = 17$ units, (b) 2.87 units, (c) $V = 4.123$ units at 75.96° from x-axis, $a = 18.385$ units at 292.38° from x-axis]

2.7 A tank is placed on an elevator which starts moving upwards at time $t = 0$ with a constant acceleration a. A stationary hose discharges water into the tank at a constant rate as shown in Figure P2.7. Determine the time required to fill the tank if it is empty at $t = 0$.

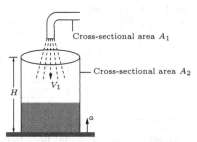

Figure P2.7: A tank on an elevator moving up

$$\left[\text{Answer: } \frac{-V_1 \pm \sqrt{V_1^2 + 2\dfrac{A_2}{A_1}aH}}{a}\right]$$

2.8 Develop the differential form of continuity equation for the cylindrical polar coordinates shown by taking an infinitesimal control volume, as shown in Figure P2.8.

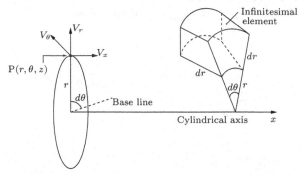

Figure P2.8: Cylindrical polar coordinates

$$\left[\text{Answer: } \frac{\partial \rho}{\partial t} + \frac{1}{r}\frac{\partial(\rho r V_r)}{\partial r} + \frac{1}{r}\frac{\partial(\rho V_\theta)}{\partial \theta} + \frac{\partial(\rho V_z)}{\partial z} = 0\right]$$

2.9 For the flow field is given by

$$V = 3x\,i + 4y\,j - 5t\,k$$

(a) find the velocity at position $(10, 6)$ at $t = 3$ s, (b) determine the slope of the streamlines for this flow at $t = 0$ s?, (c) find the equation of the streamlines at $t = 0$ up to an arbitrary constant and (d) sketch the streamlines at $t = 0$.

[Answer: (a) $V = (30\,i + 24\,j - 15\,k)$ m/s, (b) $4y/3x$, (c) $y = \frac{4}{5}x + c$, where c is an arbitrary constant, (d) at $t = 0$, the streamlines are straight lines at an angle of $38.66°$ to the x-axis]

2.10 For the fully developed two-dimensional flow of water between two impervious flat plates, shown in Figure P2.10. Also, show that $V_y = 0$ everywhere.

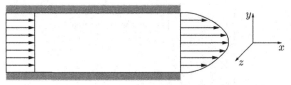

Figure P2.10: Fully developed two-dimensional flow between two impervious flat plates

2.11 Water flows through an elbow, entering section 1 at 200 N/s and leaving section 2 at $30°$ angle, as shown in Figure P2.11. At section 1 the flow has the laminar velocity profile $u = u_{m1}\left(1 - \dfrac{r^2}{R^2}\right)$, while at section 2 the flow has the turbulent profile $u = u_{m2}\left(1 - \dfrac{r}{R}\right)^{1/7}$. If the flow is steady and incompressible (water), find the maximum velocities u_{m1} and u_{m2}, at sections 1 and 2, respectively, in m/s. Assume $u_{av} = 0.5\,u_m$, for laminar flow, and $u_{av} = 0.82\,u_m$, for turbulent flow.

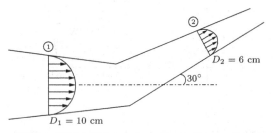

Figure P2.11: Water flow through an elbow

[Answer: 5.2 m/s, 8.79 m/s]

2.12 Consider a jet of fluid directed at the inclined plate shown in Figure P2.12. Obtain the force necessary to hold the plate in equilibrium against the jet pressure. Also, obtain the volume flow rates $\dot{Q}_1$ and $\dot{Q}_2$ in terms of the incoming flow rate $\dot{Q}_0$. Assume that, $V_0 = V_1 = V_2$ and the fluid is inviscid.

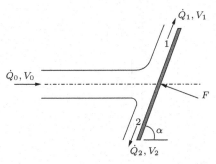

Figure P2.12: Jet impinging on an inclined plate

[Answer: $\rho V_0 \dot{Q}_0 \sin \alpha$, $\dot{Q}_1 = \dfrac{\dot{Q}_0}{2}(1 + \cos \alpha)$, $\dot{Q}_2 = \dfrac{\dot{Q}_0}{2}(1 - \cos \alpha)$]

2.13 Consider a laminar fully developed flow, without body forces, through a long straight pipe of circular cross-section (Poiseuille flow), shown in Figure P2.13. Apply the momentum equation and show that

$$\tau_{rz} = \frac{p_1 - p_2}{l}\frac{r}{2}$$

Assuming $(p_1 - p_2)/l = $ constant, obtain the velocity profile using the relation

$$\tau_{rz} = -\mu\left(\frac{dV_z}{dr}\right)$$

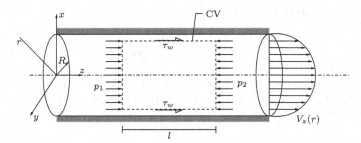

Figure P2.13: Fully developed flow through a pipe

$$\left[\text{Answer: } V_z = \left(\frac{p_1 - p_2}{l}\right)\frac{1}{4\mu}\left(R^2 - r^2\right)\right]$$

2.14 A liquid of density ρ and viscosity μ flows down a stationary wall, under the influence of gravity, forming a thin film of constant thickness h, as shown in Figure P2.14. An upward flow of air next to the film exerts a constant shear stress τ, in the direction upwards, on the surface of the liquid layer, as shown in the figure. If the pressure in the film is uniform, derive expressions for (a) the film velocity V_y, as a function of y, ρ, μ, h, and τ, and (b) the shear stress τ that would result in a zero net volume flow rate in the film.

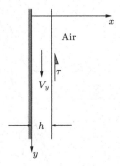

Figure P2.14: Liquid flow over a vertical wall

$$\left[\text{Answer: (a) } V_y = \frac{\rho g \left(hx - \dfrac{x^2}{2}\right) - \tau x}{\mu}, \text{ (b) } \tau = \frac{2}{3}\rho g h\right]$$

2.15 Show that the head loss for laminar, fully developed flow in a straight circular pipe is given by

$$h_l = \frac{64}{\text{Re}} \frac{L}{D} \frac{V_{av}^2}{2g}$$

where Re is the Reynolds number defined as $(\rho V_{av} D)/\mu$.

2.16 A horizontal pipe of length L and diameter D conveys air. Assuming the air to expand according to the law $p/\rho = $ constant and that the acceleration effects are small, prove that,

$$\frac{p_1}{\rho_1}\left(\rho_1^2 - \rho_2^2\right) = \frac{16 f L \dot{m}^2}{\pi^2 D^5}$$

where $\dot{m}$ is the mass flow rate of air through the pipe, f is the average friction coefficient, and subscripts 1 and 2 refer to the inlet and discharge ends of the pipe, respectively.

2.17 In the boundary layer over the upper surface of an airplane wing, at a point A near the leading edge, the flow velocity just outside the boundary layer is 250 km/hour. At another point B, which is downstream of A, the velocity outside the boundary layer is 470 km/hour. If the temperature at A is 288 K, calculate the temperature and Mach number at point B.

[Answer: 281.9 K, 0.388]

2.18 A long right circular cylinder of diameter a meter is set horizontally in a steady stream of velocity u m/s and made to rotate at an angular velocity of ω radians/second. Obtain an expression for the ratio of pressure difference between the top and bottom of the cylinder to the dynamic pressure of the stream, in terms of ω and u.

$$\left[\text{Answer: } -\frac{8a\omega}{u}\right]$$

2.19 The velocity and temperature fields of a flow are given by

$$V = x\,i + \left(3y + 3t^2 y\right) j + 12\,k$$
$$T = x + y^2 z + 5t$$

Find the rate of change of temperature that would be recorded by a floating probe (thermocouple), when it is at $3\,i + 5\,j + 2\,k$, at time $t = 2$ units.

[Answer: 1808]

2.20 If a parachute of 10 m diameter, when carrying a load W, descends at a constant velocity of 5.5 m/s, in atmospheric air at 18°C and 10^5 Pa, determine the load W if the drag coefficient for the parachute is 1.4.

[Answer: 1.991 kN]

Chapter 3

Wind Tunnels

3.1 Introduction

Wind tunnels are devices which provide air streams flowing under controlled conditions so that models of interest can be tested using them. From an operational point of view, wind tunnels are generally classified as *low-speed*, *high-speed*, and *special purpose tunnels*.

3.1.1 Low–Speed Wind Tunnels

Low-speed tunnels are those with test-section speed less than 650 kmph. Depending upon the test-section size they are referred to as small size or full scale tunnels. They are further classified into the following categories: *open-circuit tunnels*, having no guided return of air, as shown in Figure 3.1.

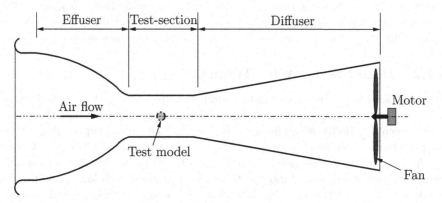

Figure 3.1: Open circuit wind tunnel

After leaving the diffuser, the air circulates by devious paths back to the intake.

If the tunnel draws air directly from the atmosphere, all the time entirely fresh air flows through the tunnel.

The second type, called *closed-circuit* or *return-flow tunnel*, has a continuous path for the air, as shown in Figure 3.2.

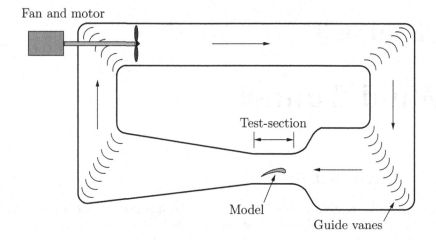

Figure 3.2: Typical closed-circuit wind tunnel

Both open-circuit and closed-circuit tunnels can operate with either *open-jet* or *closed-jet* test-sections. Open-jet is that test-section without side walls and closed-jet test-section is that with side walls. The cross-section of the test-section can have different shapes such as, rectangular, circular, elliptical, octagonal, etc.

In low-speed tunnels, the predominant factors influencing the tunnel performance are inertia and viscosity. The effect of compressibility is negligible for these tunnels. Thus, if the Reynolds number of the experimental model and full scale prototype are equal, any difference in viscosity becomes unimportant.

3.1.2   High-Speed Wind Tunnels

Tunnels with test-section speed more than 650 kmph are called high-speed tunnels. The predominant aspect in high-speed tunnel operation is that the influence of compressibility is significant. This means that, in high-speed flows, it is essential to consider *Mach number* as a more appropriate parameter than velocity. A lower limit of *high-speed* might be considered to be the flow with Mach number approximately 0.5 (about 650 kmph) at standard sea level conditions.

Based on the test-section Mach number, M, range, the high-speed tunnels are classified as follows.

- $0.8 < M < 1.2$ *Transonic tunnel*

- $1.2 < M < 5$ *Supersonic tunnel*

- $M > 5$ *Hypersonic tunnel*

Like low-speed tunnels, high-speed tunnels are also classified as *intermittent* or *open-circuit tunnels* and *continuous return-circuit tunnels*, based on the type of operation. The power to drive a low-speed wind tunnel varies as the cube of the test-section velocity. Although this rule does not hold in the high-speed regime, the implication of rapidly increasing power requirements with increasing test-section speed holds for high-speed tunnels also. Because of the power requirements, high-speed wind tunnels are often of the *intermittent type*, in which energy is stored in the form of pressure or vacuum or both and is allowed to drive the tunnel only a few seconds out of each pumping hour.

3.1.2.1 General Features

All modern wind tunnels have four important components; the *effuser*, the *working* or *test-section*, the *diffuser*, and the *driving unit*.

3.1.2.2 The Effuser

This is a converging passage located upstream of the test-section. In this passage fluid gets accelerated from rest (or from very low speed) at the upstream end of it to the required conditions at the test-section. In general, the effuser contains honeycomb and wire gauze screens to reduce the turbulence and produce a uniform air stream at the exit. The effuser is usually referred to as *contraction cone*.

3.1.2.3 Test-Section

The model to be tested is placed here in the air-stream, leaving the downstream end of the effuser, and the required measurements and observations are made. If the test-section is bounded by rigid walls, the tunnel is called a *closed-throat tunnel*. If it is bounded by air at different velocity (usually at rest), the tunnel is called *open-jet tunnel*. The test-section is also referred to as *working-section*.

3.1.2.4 Diffuser

The diffuser is used to re-convert the kinetic energy of the air-stream leaving the working-section into pressure energy, as efficiently as possible. Essentially it is a passage in which the flow decelerates.

3.1.2.5 Driving Unit

If there were no losses, steady flow through the test-section could continue forever, once it is established, without the supply of energy from an external agency. But in practice, losses do occur, and kinetic energy is being dissipated as heat in vorticity, eddying motion, and turbulence. Moreover, as the expansion of the diffuser cannot continue to infinity, there is rejection of some amount of kinetic energy at the diffuser exit. This energy is also converted to heat in mixing with the surrounding air. To compensate for these losses, energy from an external agency becomes essential for wind tunnel operation. Since power must be supplied continuously to maintain the flow, the fourth essential component, namely, some form of driving unit, is essential for wind tunnel operation. In low-speed tunnels this usually takes the form of a fan or propeller. Thus, for a low-speed tunnel, the simplest layout is similar to that shown in Figure 3.1.

The overall length of the wind tunnel may be shortened, and the rejection of kinetic energy at the diffuser exit eliminated, by the construction of some form of return circuit. Even then the driving unit is necessary to overcome the losses occurring due to vorticity, eddying motion, and turbulence. The skin friction at the walls and other surfaces will be large since the velocity at all points in the circuit will be large (of the same order as the test-section velocity). Also, a construction ahead of the test-section, as shown in Figure 3.2, is necessary if the turbulence at the test-section has to be low, and particularly if the velocity distribution has to be uniform. To achieve this, usually guide vanes are placed in the corners.

3.1.3   Special Purpose Tunnels

These are tunnels with layout totally different from that of low-speed and high-speed tunnels. Some of the popular special purpose tunnels are: *spinning tunnels*, *free-flight tunnels*, *stability tunnels*, and *low-density tunnels*. The discussions on these types of tunnels are not considered in this book.

3.2   Low–Speed Wind Tunnels

A general utility low-speed tunnel has four important components, namely, the effuser, the test-section, the diffuser, and the driving unit.

3.2.1   Effuser

This is basically a contraction cone, as shown in Figure 3.1. Its application is to bring down the level of turbulence and increase the velocity of flow. The contraction ratio n of an effuser is defined as

$$n = \frac{\text{Area at entry to convergent cone}}{\text{Area at exit of convergent cone}}$$

The contraction ratio usually varies from 4 to 20 for conventional low-speed tunnels.

3.2.2   Test-Section

The portion of the tunnel with constant flow characteristics across its entire section is termed the test- or working-section. Since boundary layer is formed along the test-section walls, the walls are given a suitable divergence so that the net cross-sectional area of the uniform flow is constant along the length of the test-section.

3.2.3 Diffuser

The purpose of the diffuser is to convert the kinetic energy of the flow coming out of the test-section to pressure energy, before it leaves the diffuser, as efficiently as possible. Generally, the smaller the diffuser divergence angle, the more efficient is the diffuser. Near the exit, its cross-section should be circular to accommodate the fan.

3.2.4 Driving Unit

Generally the driving unit consists of a motor and a propeller or fan combination. The fan is used to increase the static pressure of the stream leaving the diffuser.

The wind tunnel fan, looking similar to the propeller of an airplane, operates under peculiar conditions that put it in a class by itself. Since the thrust of the fan and the drag of the various tunnel components vary with the square of the fan *rpm*, it would appear that to maintain a uniform velocity distribution in the test-section, speed adjustments should be made by varying the fan *rpm* rather than fan pitch. Although this conclusion is justified in short tunnels of low contraction ratio, for large tunnels it is not true. Indeed, many large tunnels which are equipped with both *rpm* and pitch change mechanisms use only the latter, being quick and simpler.

3.3   Power Losses in a Wind Tunnel

The total power loss in a wind tunnel may be split into the following components.

- Losses in cylindrical parts.
- Losses in guide vanes at the corners (in closed circuit tunnels).
- Losses in diffuser.
- Losses in contraction cone.
- Losses in honeycomb, screens etc.
- Losses in test-section (jet losses in case of open jet).
- Losses in exit in case of open-circuit tunnel.

3.3.1   Calculation of Percentage Energy Loss in the Various Parts of Wind Tunnel

The loss of energy is expressed in terms of static pressure drop Δp, in the dimensionless form, called pressure drop coefficient K, as follows

$$K = \frac{\Delta p}{q} \tag{3.1}$$

where q is the dynamic pressure of the flow. In terms of local velocity V, Equation (3.1) becomes

$$K = \frac{\Delta p}{\frac{1}{2}\rho V^2}$$

where ρ is the density of flow.

It will be convenient to refer the local losses at different parts of the wind tunnel to the jet or test-section dynamic pressure, defining the coefficient of loss as

$$K_0 = \frac{\Delta p}{q_0} = \frac{\Delta p}{q}\frac{q}{q_0} = K\frac{q}{q_0}$$

where q_0 is the test-section dynamic pressure. But $q \propto V^2$ and $V \propto \frac{1}{A} \propto \frac{1}{D^2}$, therefore, the above equation can be rewritten as

$$K_0 = K\left(\frac{D_0}{D}\right)^4 \tag{3.2}$$

where D_0 is the test-section diameter and D is the local tunnel diameter.

Using the above definitions, the section energy loss in the wind tunnel, ΔE, may be expressed as,

$$\begin{aligned} \Delta E &= \Delta p A V \\ &= K q A V \end{aligned}$$

or

$$\Delta E = K\frac{1}{2}\rho A V^3 \tag{3.3}$$

But $Kq = K_0 q_0$, therefore, Equation (3.3) becomes

$$\Delta E = K_0\, q_0\, A_0\, V_0 = K_0\, \frac{1}{2}\, \rho\, A_0 V_0^3 \tag{3.4}$$

where A_0 is the test-section area and A is the local cross-sectional area.

Note that, in the above discussions on energy loss, the density of the flow is treated as invariant, even though it is valid only for incompressible flows. Such an assumption is usually made in the study of low-speed wind tunnels; since the Mach number involved is always less than 0.5, the compressibility effect associated with the flow will be only marginal and hence assuming the density, ρ, as an invariant will not introduce any significant error to the measured values.

3.3.1.1 Energy Ratio

The ratio of the energy of air-stream at the test-section to the input energy to the driving unit is a measure of the efficiency of a wind tunnel. It is nearly always greater than unity, indicating that the amount of stored energy in the wind stream is capable of doing work at a higher rate than what it is doing in a wind tunnel, before being brought to rest. The energy ratio ER is in the range of 3 to 7, for most closed-throat wind tunnels. The energy ratio is defined as

$$ER = \frac{\text{Kinetic energy of jet}}{\text{Energy loss}}$$

$$= \frac{\frac{1}{2}\rho A_0 V_0^3}{\sum K_0 \frac{1}{2}\rho A_0 V_0^3}$$

that is,

$$ER = \frac{1}{\sum K_0} \tag{3.5}$$

The definition of the energy ratio given by Equation (3.5) excludes the fan and motor efficiency.

The magnitudes of the losses in the various components of a wind tunnel of circular cross-section may be calculated as follows.

3.3.1.2 Losses in Cylindrical Section

We know that the pressure drop in a cylindrical section of length L can be expressed as

$$\Delta p = f \frac{L}{D} \frac{\rho}{2} V^2 \tag{3.6}$$

where f is friction coefficient and D is diameter of the cylindrical section. Combining Equations (3.1) and (3.6), we get

$$K = \frac{\Delta p}{q} = \frac{fL}{D}$$

Substitution of the above equation into Equation (3.2), results in

$$K_0 = \frac{fL}{D} \left(\frac{D_0}{D}\right)^4 \tag{3.7}$$

The value of the friction coefficient, f, may be computed from the Von Karman formula (White, 1986), which gives

$$\frac{1}{\sqrt{f}} = 2\log_{10}\left(Re_D\sqrt{f}\right) - 0.8$$

for smooth pipes at high Reynolds numbers.

Some numerical values of f, at specified Reynolds numbers are listed below.

It is seen that f drops by only a factor of 5 over 10,000-fold increase in Reynolds number. The above equation for f in terms of Re_D is usually used for calculating f. There are many alternate approximations in the literature from which f can be computed explicitly from Re_D. However, for wind tunnel applications the above relation is good enough.

Re_D	400	10^4	10^5	10^6	10^7	10^8
f	0.0899	0.0309	0.0180	0.0116	0.0081	0.0059

The friction coefficient variation with Reynolds number is shown as a plot in Figure 3.3.

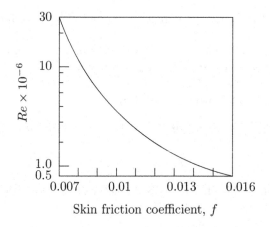

Figure 3.3: Friction coefficient variation with Reynolds number

Equation (3.7) can be used to compute K_0 for (i) test-section and (ii) return passage (where there is no divergence in the section). For cross-sections other than circular, an equivalent diameter has to be used for the calculation of Re, f, etc.

3.3.1.3 Losses in Convergent Cone

Consider the contraction cone shown in Figure 3.4, with diameters D_1, and D_0 at its entrance and exit, respectively. The loss in convergent section is mainly due to friction. This loss may be expressed in terms of pressure loss as

$$\Delta p = f \frac{L}{D_1} \frac{\rho}{2} \left(V_0^2 - V_1^2 \right) \tag{3.8}$$

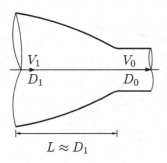

Figure 3.4: Contraction cone

where V_1 and V_0 are the velocities at the inlet and exit of the convergent section, respectively.

Usually, for contraction cones $L \approx D_1$, therefore, Equation (3.8) becomes

$$\Delta p = f \left[1 - \left(\frac{D_0}{D_1} \right)^4 \right] \frac{\rho}{2} V_0^2 \qquad (3.9)$$

i.e.,

$$\frac{\Delta p}{\frac{1}{2}\rho V_0^2} = K_0 = f \left[1 - \left(\frac{D_0}{D_1} \right)^4 \right]$$

For contraction cones of good shapes with smooth walls, experimental results give $f = 0.005$; therefore,

$$K_0 = 0.005 \left[1 - \left(\frac{D_0}{D_1} \right)^4 \right]$$

3.3.1.4 Losses in Diffuser

The loss of energy in the diffuser is due to (i) skin friction and (ii) expansion. Consider the divergent section shown in Figure 3.5, with V_1, D_1 and V_2, D_2 as the velocity and diameter at its entrance and exit, respectively.

Taking f_n as the average value of the friction coefficient, the pressure loss in a divergent section may be expressed as

$$\Delta p = f_n \frac{\rho}{2} \int_0^{L_n} \frac{V_n^2}{D_n} \, dL_n$$

For an incompressible flow through the diffuser, by continuity, we have

$$D_n^2 V_n = D_1^2 V_1$$

Therefore,

$$V_n^2 = \frac{V_1^2 D_1^4}{D_n^4}$$

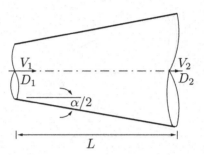

Figure 3.5: Diffuser section

In the above equations the subscript n stands for conditions at the exit of the nth section of the diffuser. This implies that, the diffuser may have n sections with different divergence angles α between the opposite walls. In the present case, the diffuser shown in Figure 3.5 has only one portion with divergence angle α, therefore, $D_n = D_2$, $V_n = V_2$.

Substituting of the above expression for V_n^2 into the Δp equation above, we get

$$\Delta p = f_n \frac{\rho}{2} \int_0^{L_n} V_1^2 \, D_1^4 \, \frac{dL_n}{D_n^5} \qquad (3.10)$$

From the geometry of the diffuser shown in Figure 3.5, we have

$$D_n = D_1 + 2 L_n \tan\left(\frac{\alpha}{2}\right)$$

In the differential form, this becomes

$$dD_n = 2 \tan\left(\frac{\alpha}{2}\right) dL_n$$

Therefore,

$$dL_n = \frac{dD_n}{2 \tan\left(\frac{\alpha}{2}\right)}$$

Substitution of this into Equation (3.10) results in

$$\Delta p = f_n \frac{\rho}{2} V_1^2 \, D_1^4 \frac{1}{2 \tan\left(\frac{\alpha}{2}\right)} \int_{D_1}^{D_2} \frac{dD_n}{D_n^5}$$

On integration, this yields

$$\Delta p = \frac{f_n}{8 \tan\left(\frac{\alpha}{2}\right)} \left[1 - \left(\frac{D_1}{D_2}\right)^4\right] \frac{\rho}{2} V_1^2 \qquad (3.11)$$

Therefore,

$$K = \frac{\Delta p}{q}$$

$$= \frac{f_n}{8 \tan\left(\frac{\alpha}{2}\right)} \left[1 - \left(\frac{D_1}{D_2}\right)^4\right]$$

In terms of test-section dynamic pressure, the above equation becomes

$$K_{01} = \frac{f_n}{8 \tan\left(\frac{\alpha}{2}\right)} \left[1 - \left(\frac{D_1}{D_2}\right)^4\right] \left(\frac{D_0}{D_1}\right)^4 \tag{3.12}$$

The expansion losses may be calculated from *Fleigners formula*, which expresses the pressure loss (except for tapered water pipe) as

$$\Delta p = \frac{\rho}{2} (\sin \alpha) (V_1 - V_2)^2 \tag{3.13}$$

Therefore,

$$K = \sin \alpha \left[1 - \left(\frac{D_1}{D_2}\right)^2\right]^2$$

In terms of test-section dynamic pressure, the above K becomes

$$K_{02} = \sin \alpha \left[1 - \left(\frac{D_1}{D_2}\right)^2\right]^2 \left(\frac{D_0}{D_1}\right)^4 \tag{3.14}$$

The total loss coefficient for the diffuser is the sum of frictional and expansion loss coefficients. Thus,

$$K_0 = K_{01} + K_{02} \tag{3.15}$$

The variation of pressure loss coefficients K_{01}, K_{02}, and K_0 with diffuser divergence angle α is shown in Figure 3.6.

It is seen from Figure 3.6 that, divergence angle α between 6 and 8 degrees proves to be the optimum for diffusers, resulting in minimum pressure loss.

3.3.1.5   Honeycombs

Wind tunnels have honeycombs in the settling chamber, in order to improve the flow quality in the test-section. Usually, the honeycombs are made of octagonal or hexagonal or square or circular cells with their length 5 to 10 times their width (diameter). Some typical honeycombs used in wind tunnels are shown in Figure 3.7. The value of loss coefficient K, shown in Figure 3.7, are for honeycombs with a (length/diameter) $= 6.0$, and equal tube areas. The loss in the honeycombs is usually less than 5 percent of the total loss of a tunnel.

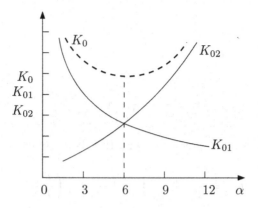

Figure 3.6: Losses in diffuser

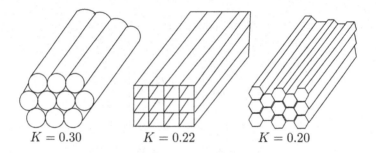

Figure 3.7: Honeycombs

3.3.1.6   Guide Vanes

In wind tunnel design, it is not practical to make the corners of the return passage so gradual that the air can follow the corner walls, with very small pressure loss. Such corners would require more space and more construction cost. Abrupt corners are therefore used, and their losses are kept to a minimum by means of corner or guide vanes. The losses in guide vanes are due to

- the skin friction of the vanes (approximately 33 percent of the total corner loss).

- rotational component due to change of flow direction (rest of the corner loss).

Abrupt corners without guide vanes may show a loss of even 100 percent cent of velocity head. Well-designed corners with guide vanes can reduce the loss by 15 to 20 percent. Here, basically the corner is divided into many vanes of high aspect ratio, defined as the ratio of vane gap G to height h, as shown in Figure 3.8.

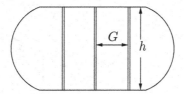

Figure 3.8: Corner vane geometry

Commonly employed corner vanes have an aspect ratio of 6. In general, this criterion defines the vane gap since the height is known. Some typical vane profiles are shown in Figure 3.9, with the loss experienced under test conditions at Reynolds number around 40,000.

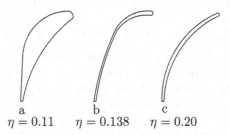

a
$\eta = 0.11$

b
$\eta = 0.138$

c
$\eta = 0.20$

Figure 3.9: Corner vanes

In the corners, friction in the guide vanes accounts for about one-third of the total corner loss, and rotation losses account for the other two-thirds. For guide vanes of the type shown in Figure 3.9, the pressure loss coefficient K_0 may be calculated by the following formula, provided $\Delta p/q \approx 0.15$ (based on Re = 500,000, Pope and Goin, 1965).

$$K_0 = \left[0.10 + \frac{4.55}{(\log_{10} R_e)^{2.58}} \right] \left(\frac{D_0}{D} \right)^4 \qquad (3.16)$$

3.3.1.7   Losses due to Open Jet Test-Section

Consider the open jet test-section, as shown in Figure 3.10.
If the jet is assumed to be a closed jet, as shown by the dashed lines, the losses will become very small. For example, let the length of the jet be 1.5 times the diameter D_0. For a smooth wall the friction coefficient is very low and is of the order 0.008. Treating the jet as a cylindrical portion, the loss becomes, by Equation (3.7),

$$K_0 = 0.008 \times 1.5 \times 100 = 1.2 \, \text{percent}$$

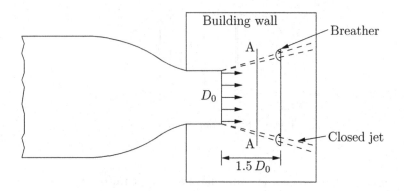

Figure 3.10: Open jet test-section

Instead, when the jet is open the coefficient of friction is approximately 0.08 and therefore,

$$K_{\text{open jet}} = 0.08 \times 1.5 \times 100 = 12 \, \text{percent}$$

That is, the jet loss for a open test-section is approximately 10 times that for a closed test-section. Further, for open test-section operation, due consideration must be given to the possibility of pulsations similar to the vibrations in an organ pipe, arising at the jet boundaries. This phenomenon, believed to be a function of jet length, can be quite serious. The simplest solution generally applied to overcome this problem is to provide vents in the diffuser, as shown in Figure 3.10, which connects it to atmosphere. Such an arrangement is called a *breather*.

3.3.1.8   Screens (wire gauze)

The prime uses of wire gauze screens in wind tunnels are the following.

(1a) To provide uniformity of velocity distribution in a duct. Consider the wire gauze screen shown in Figure 3.11. The wire gauze offers higher resistance to a jet of a higher velocity and vice versa. Therefore, by having a number of screens, it is possible to achieve uniform velocity distribution. The resistance offered by the wire gauze to the flow is proportional to its velocity. In Figure 3.11, let the velocities at two locations in the flow field upstream of the screen be V_1 and V_2. Downstream of the screen, the velocities become

$$\boxed{V_1 - \Delta V_1 \approx V_2 - \Delta V_2}$$

since,
$$\Delta V_1 \propto V_1 \quad \text{and} \quad \Delta V_2 \propto V_2$$

(1b) In a diffusing duct, the gauze penetrates upstream, reducing the causes of nonuniformity as well as the nonuniformity itself.

For case 1a, the porosity of the screen should be carefully chosen. Wherever high value of K_{screen} is required, a large number of screens with small

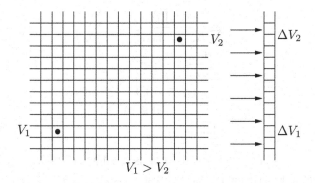

Figure 3.11: Wire gauze screen

individual K values should be used instead of fewer screens with high individual K values. Screens should be placed at locations where the velocity should be small.

(2) To reduce the turbulence of the stream. For a general purpose wind tunnel, the turbulence level should be less than 5 percent (commonly acceptable limit). Fine gauzes are used in low turbulence wind tunnels. The components of turbulence are modified in the contraction on entering the gauze and if the porosity is suitably chosen the overall turbulence downstream of the screen can be reduced.

(3) To introduce high turbulence artificially, in a low-turbulence wind tunnel, to study the flow transition, etc., with varying turbulence.

(4) To introduce known pressure drops when required in experiments.

Example 3.1

An open circuit subsonic wind tunnel of test-section 1.2 m × 0.9 m is run by a 110 kW motor. If the test-section speed is 90 m/s, calculate the energy ratio of the tunnel. Also, find the total loss in the tunnel in terms of test-section kinetic energy. Take the air density as the standard sea level value.

Solution

The energy ratio ER is defined as

$$\text{ER} = \frac{\text{Kinetic energy of air stream in the test-section}}{\text{Input energy}}$$

$$= \frac{\frac{1}{2} \rho A_0 V_0^3}{\sum K_0 \frac{1}{2} \rho A_0 V_0^3}$$

$$= \frac{1}{\sum K_0}$$

$$= \frac{\frac{1}{2} \times 1.225 \times 0.9 \times 1.2 \times 90^3}{110 \times 10^3}$$

$$= \boxed{4.38}$$

The total loss $= \sum K_0 = 1/\text{ER} = 1/4.38 = 0.2283$. That is, total loss is $\boxed{22.83 \text{ percent}}$ of kinetic energy at the test-section.

Example 3.2

An open jet test-section of a subsonic wind tunnel expands freely into a still environment. The test-section length is 1.5 times the diameter of the contraction cone exit. The friction coefficient for the free jet is 10 times that of the closed throat with smooth wall. If the friction coefficient of the smooth wall is 0.008, determine the increase of loss when the jet is open, treating the jet as a cylindrical duct.

Solution

Given that the length-to-diameter ratio of the jet is $L/D_0 = 1.5$ and the friction coefficient of the closed jet wall, $f = 0.008$.

For cylindrical ducts, the pressure loss can be expressed as

$$\Delta p = \frac{\rho V^2}{2} \frac{L}{D} f$$

Thus, the loss coefficient becomes

$$K_0 = \frac{\Delta p}{\frac{\rho V^2}{2}} = \frac{L}{D_0} f$$

where V_0 is the test-section velocity. Thus,

$$K_0 = 1.5 \times 0.008 = \boxed{0.012}$$

For open jet, $f = 0.08$, therefore,

$$K_0 = 1.5 \times 0.08 = \boxed{0.12}$$

Thus, increase of loss for the open jet is $(0.12 - 0.012) = 0.108$, i.e., $\boxed{10.8 \text{ percent}}$

Example 3.3

A subsonic open-circuit wind tunnel runs with a test-section speed of 40 m/s. The temperature of the lab environment is 16°C. If a turbulent sphere measures the turbulence factor, TF (defined as the ratio of theoretical critical Reynolds number for the sphere to the actual critical Reynolds number) of the tunnel as 1.2, determine the sphere diameter. Assume the test-section pressure as the standard sea level pressure.

Solution

Density of air in the test-section can be expressed as

$$\rho = \frac{p}{RT}$$

where p is sea level pressure, and R is the gas constant for air. The test-section temperature, T, can be determines as follows.

By energy equation, we have

$$h_0 = h + \frac{V^2}{2}$$

For perfect gas, $h = C_p T$, thus,

$$C_p T_0 = C_p T + \frac{V^2}{2}$$

$$T = T_0 - \frac{V^2}{2C_p}$$

where T_0 is the test-section stagnation temperature, which is the same as the lab temperature, and C_p is the specific heat at constant pressure. Thus,

$$T = 289.15 - \frac{40^2}{2 \times 1004.5}$$

$$= 288.35 \, \text{K}$$

The flow density at the test-section becomes

$$\rho = \frac{101325}{287 \times 288.35} = 1.224 \, \text{kg/m}^3$$

The test-section Reynolds number is

$$Re = \frac{\rho V d}{\mu}$$

where d is the sphere diameter and μ is dynamic viscosity coefficient at 288.35 K. The viscosity coefficient at 288.35 K is

$$\mu = 1.46 \times 10^{-6} \times \frac{288.35^{\frac{3}{2}}}{288.35 + 111}$$

$$= 1.79 \times 10^{-5} \, \text{kg/(m s)}$$

The critical Reynolds number becomes

$$Re_c = \frac{1.224 \times 40 \times d}{1.79 \times 10^{-5}}$$

The turbulence factor, TF, is given by

$$TF = \frac{385000}{Re_c}$$

$$Re_c = \frac{385000}{1.2} = 320833.3$$

Thus,

$$d = \frac{Re_c \times 1.79 \times 10^{-5}}{1.224 \times 40}$$

$$= \frac{320833.3 \times 1.79 \times 10^{-5}}{1.224 \times 40} = 0.1173 \, \text{m}$$

$$= \boxed{11.73 \, \text{cm}}$$

Example 3.4

A closed-return type wind tunnel of large contraction ratio has air at standard sea-level conditions in the settling chamber upstream of the contraction to the test-section. Assuming isentropic compressible flow in the tunnel, estimate the speed and the kinetic energy per unit area in the working section when the Mach number is 0.75.

Solution

The given sea-level pressure and temperature upstream of the test-section can be taken as the stagnation values, i.e., $p_0 = 101325$ Pa, and $T_0 = 288.15$ K. Also, the p_0 and T_0 are invariants, since the flow is isentropic.

By isentropic relation, we have

$$T = T_0 \left(1 + \frac{\gamma - 1}{2} M^2\right)^{-1}$$

$$= 288.15(1 + 0.2 \times 0.75^2)^{-1}$$

$$= 259 \, \text{K}$$

Also, the pressure in the test-section is given by

$$p = p_0 \left(1 + \frac{\gamma - 1}{2} M^2 \right)^{-\frac{\gamma}{\gamma-1}}$$
$$= \frac{101325}{(1 + 0.2 \times 0.75^2)^{3.5}}$$
$$= 69769.65 \, \text{Pa}$$

Therefore, the flow density in the test-section is

$$\rho = \frac{p}{RT} = \frac{69769.65}{287 \times 259}$$
$$= 0.9386 \, \text{kg/m}^3$$

Thus, the speed and the kinetic energy in the test-section becomes

$$V = M a = 0.75 \times \sqrt{\gamma RT}$$
$$= 0.75 \times \sqrt{1.4 \times 287 \times 259}$$
$$= \boxed{242 \, \text{m/s}}$$
$$ke = \frac{1}{2}\rho V^3 = 0.5 \times 0.9386 \times 242^3$$
$$= \boxed{6.65 \, \text{MW/m}^2}$$

3.4 High–Speed Wind Tunnels

High-speed tunnels are those with test-section speed more than 650 kmph (Pope and Goin, 1965). The power to drive a low-speed wind tunnel varies as the cube of the test-section velocity. Although this rule is not valid for the high-speed regime, the implication of rapidly increasing power requirement with increasing test-section speed is true for high-speed tunnels also. Because of the power requirements, high-speed wind tunnels are often of the *intermittent* type in which energy is stored in the form of pressure or vacuum or both and is allowed to drive the tunnel only a few seconds out of each pumping hour.

High-speed tunnels are generally grouped into intermittent and continuous operation tunnels, based on the type of operation. The intermittent tunnels are further divided into blowdown tunnels and induction tunnels, based on type of the operational procedure.

Even though the flow in the Mach number range from 0.5 to 5.0 is usually termed as high-speed flow, the tunnels with test–section Mach number less than 0.9 are generally grouped and treated under subsonic wind tunnels. Wind tunnels with Mach numbers from 1.5 to 5.0 are classified as supersonic tunnels and those with Mach number more than 5 are termed hypersonic tunnels. The wind tunnels in the Mach number range from 0.9 to 1.5 are called transonic tunnels.

The intermittent blowdown and induction tunnels are normally used for Mach numbers from 0.5 to about 5.0, and the intermittent pressure-vacuum tunnels are normally used for higher Mach numbers. The continuous tunnel is used throughout the speed range. Both intermittent and continuous tunnels have their own advantages and disadvantages.

3.4.1   Blowdown–Type Wind Tunnels

Essential features of the intermittent blowdown wind tunnels are schematically shown in Figure 3.12.

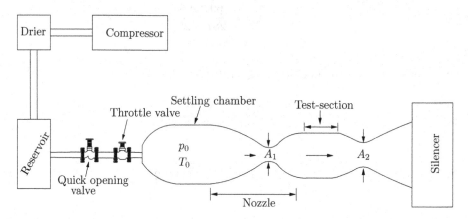

Figure 3.12: Schematic layout of intermittent blowdown tunnel

3.4.1.1   Advantages

The main advantages of blowdown-type wind tunnels are the following.

- They are the simplest among the high-speed tunnel types and most economical to build.

- Large size test-sections and high Mach numbers (up to $M = 4$) can be obtained.

- Constant blowing pressure can be maintained and running time of considerable duration can be achieved.

These are the primary advantages of intermittent blowdown tunnels. In addition to these, there are many additional advantages for this type of tunnel, like a single drive may easily run several tunnels of different capabilities, failure of a model usually will not result in tunnel damage. Extra power is available to start the tunnel and so on.

3.4.1.2   Disadvantages

The major disadvantages of blowdown tunnels are the following.

- Charging time to running time ratio will be very high for large size tunnels.

- Stagnation temperature in the reservoir drops during tunnel run, thus changing the Reynolds number of the flow in the test-section.

- Adjustable (automatic) throttling valve between the reservoir and settling chamber is necessary for constant stagnation pressure (temperature varying) operation.

- Starting load is high (no control possible).

- Reynolds number of flow is low due to low static pressure in the test-section.

The commonly employed reservoir pressure range is from 100 to 300 psi for blowdown tunnel operations. As large as 2000 psi is also used where space limitations require.

3.4.2   Induction–Type Tunnels

In this type of tunnel, a vacuum created at the downstream end of the tunnel is used to establish the flow in the test-section. A typical induction tunnel circuit is shown schematically in Figure 3.13.

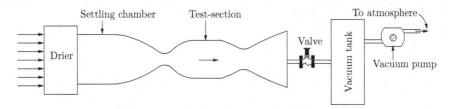

Figure 3.13: Schematic diagram of induction tunnel

3.4.2.1   Advantages

The *advantages* of induction tunnels are the following.

- Stagnation pressure and stagnation temperature are constants.

- No oil contamination in air, since the pump is at the downstream end.

- Starting and shutdown operations are simple.

3.4.2.2   Disadvantages

The *disadvantages* of induction-type supersonic tunnels are the following.

- Size of the air drier required is very large, since it has to handle a large mass flow in a short duration.

- Vacuum tank size required is also very large.

- High Mach numbers ($M > 2$) are not possible because of large suction requirements for such Mach numbers.

- Reynolds number is very low, since the stagnation pressure is atmospheric.

The above-mentioned blowdown and induction principles can also be employed together for supersonic tunnel operation to derive the benefits of both types.

3.4.3   Continuous Supersonic Wind Tunnels

The essential features of a continuous flow supersonic wind tunnel are shown in Figure 3.14.

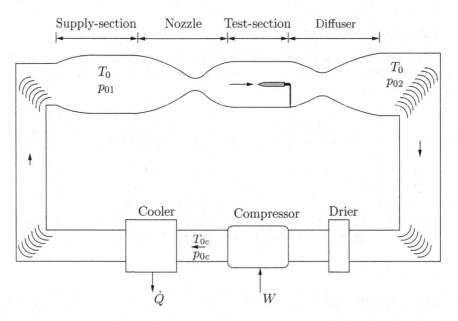

Figure 3.14: Schematic of closed-circuit supersonic wind tunnel

Like intermittent tunnels, the continuous tunnels also have some advantages and disadvantages.

The main *advantages* of continuous supersonic wind tunnels are the following.

- Better control over the Reynolds number possible, since the shell is pressurized.

- Only a small capacity drier is required.

- Testing conditions can be held the same over a long period of time.

- The test-section can be designed for high Mach numbers ($M > 4$) and large size models.

- Starting load can be reduced by starting at low pressure in the tunnel shell.

The major *disadvantages* of continuous supersonic tunnels are the following.

- Power required is very high.

- Temperature stabilization requires large size cooler.

- Compressor drive to be designed to match the tunnel characteristics.

- Tunnel design and operation are more complicated.

It is seen from the foregoing discussions that, both *intermittent* and *continuous* tunnels have certain specific advantages and disadvantages. Before going into the specific details about supersonic tunnel operation, it will be useful to note the following details about supersonic tunnels.

- Axial flow compressor is better suited for large pressure ratio and mass flow.

- Diffuser design is critical since increasing diffuser efficiency will lower the power requirement considerably. Supersonic diffuser portion (geometry) must be carefully designed to decrease the Mach number of the flow to be as low as possible, before shock formation. Subsonic portion of the diffuser must have an optimum angle, to minimize the frictional and separation losses.

- Proper nozzle geometry is very important to obtain good distribution of Mach number and freedom from flow angularity in the test-section. Theoretical calculation to high accuracy and boundary layer compensation, etc., have to be carefully worked out for large test-sections. Fixed nozzle blocks for different Mach numbers are simple but very expensive and quite laborious for change over in the case of large size test-sections. The flexible wall type nozzle is complicated and expensive from a design point of view and Mach number range is limited (usually $1.5 < M < 3.0$).

- Model size is determined from the test-rhombus, shown in Figure 3.15.

The model must be accommodated inside the rhombus formed by the incident and reflected shocks for proper measurements.

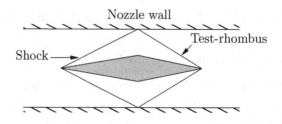

Figure 3.15: Test-rhombus

3.4.4 Losses in Supersonic Tunnels

The total power loss in a continuous supersonic wind tunnel may be split into the following components.

1. Frictional losses (in the return circuit).

2. Expansion losses (in diffuser).

3. Losses in contraction cone and test-section.

4. Losses in guide vanes.

5. Losses in cooling system.

6. Losses due to shock wave (in diffuser supersonic part).

7. Losses due to model and support system drag.

The first five components of losses represent the usual low-speed tunnel losses. All the five components together constitute only about 10 percent of the total loss. Components 6 and 7 are additional losses in a supersonic wind tunnel and usually amount to approximately 90 percent of the total loss, with shock wave losses alone accounting for nearly 80 percent and model and support system drag constituting nearly 10 percent of the total loss. Therefore, it is customary in estimating the power requirements to determine the pressure ratio required for supersonic tunnel operation so that the pressure ratio across the diffuser alone is considered and a correction factor is applied to take care of the rest of the losses.

The pressure ratio across the diffuser multiplied by the correction factor must therefore be equal to the pressure ratio required across the compressor to run the tunnel continuously. The relationship between these two vital pressure ratios, namely the diffuser pressure ratio, p_{01}/p_{02}, and compressor pressure ratio, p_{0c}/p_{03}, may be related as follows.

$$\text{Compressor pressure ratio } \frac{p_{0c}}{p_{03}} = \frac{p_{01}}{p_{02}} \frac{1}{\eta} \tag{3.17}$$

where

p_{0c} = stagnation pressure at compressor exit,

p_{03} = stagnation pressure at compressor inlet,

p_{01} = stagnation pressure at diffuser inlet,

p_{02} = stagnation pressure at diffuser exit,

and

$$\eta = \frac{\text{Diffuser losses}}{\text{Total loss}}$$

is the correction factor.

The value of η varies from 0.6 to 0.85, depending on the kind of shock pattern through which the pressure recovery is achieved in the diffuser. The variation of compressor pressure ratio, p_{0c}/p_{03}, with the test-section Mach number, M, is shown in Figure 3.16.

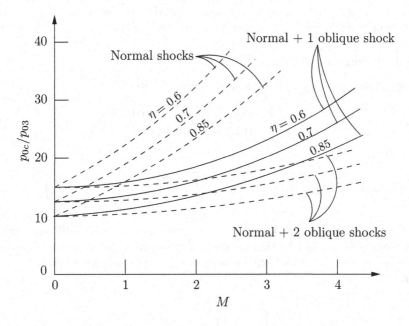

Figure 3.16: Compressor pressure ratio variation with Mach number

3.4.5 Supersonic Wind Tunnel Diffusers

Basically a diffuser is a device to convert the kinetic energy of a flow to pressure energy. The diffuser efficiency may be defined, in two ways, as

1. Polytropic efficiency η_d.

2. Isentropic efficiency η_σ.

3.4.5.1 Polytropic Efficiency

It is known that, at any point in a diffuser, a small change of kinetic energy of unit mass of fluid results in an increase of pressure energy as per the equation

$$\Delta p = \eta_d\, d\left(\frac{V^2}{2}\right) \tag{3.18}$$

and the pressure ratio is given by

$$\frac{p_{02}}{p_1} = \left(\frac{T_{02}}{T_1}\right)^{\frac{\gamma}{\gamma-1}\eta_d} = \left(1 + \frac{\gamma-1}{2}M_1^2\right)^{\frac{\gamma}{\gamma-1}\eta_d} \tag{3.19}$$

where p_1 and p_{02} are the static and stagnation pressures upstream and downstream of the point under consideration, respectively, and η_d is the polytropic efficiency. M_1, T_1, and T_{02}, respectively, are the Mach number, static and stagnation temperatures at the appropriate locations.

3.4.5.2 Isentropic Efficiency

The isentropic efficiency of a diffuser may be defined as

$$\eta_\sigma = \frac{\text{Ideal KE required for observed power}}{\text{Actual KE transferred}}$$

and

$$\text{Ideal } KE \text{ from } p_1 \text{ to } p_{02} \text{ (without loss)} = \int_{p_1}^{p_{02}} \frac{dp}{\rho} = \frac{\gamma}{\gamma-1}\frac{p_1}{\rho_1}\left[\left(\frac{p_{02}}{p_1}\right)^{\frac{\gamma-1}{\gamma}} - 1\right] \tag{3.20}$$

Note that, in Equations (3.19) and (3.20), the velocity at the diffuser outlet is assumed to be negligible; that is why the pressure at location 2 is taken as p_{02}, the stagnation pressure. With Equation (3.20) the isentropic efficiency, η_σ, becomes

$$\eta_\sigma = \frac{\dfrac{\gamma}{\gamma-1}\dfrac{p_1}{\rho_1}\left[\left(\dfrac{p_{02}}{p_1}\right)^{\frac{\gamma-1}{\gamma}} - 1\right]}{\frac{1}{2}V_1^2} = \frac{2}{\gamma-1}\frac{1}{M_1^2}\left[\left(\frac{p_{02}}{p_1}\right)^{\frac{\gamma-1}{\gamma}} - 1\right]$$

From the above equation, the pressure ratio p_{02}/p_1 becomes

$$\frac{p_{02}}{p_1} = \left(1 + \frac{\gamma-1}{2}M_1^2\eta_\sigma\right)^{\frac{\gamma}{\gamma-1}} \tag{3.21}$$

From Equations (3.19) and (3.21), we get

$$\left(1 + \frac{\gamma-1}{2}M_1^2\right)^{\eta_d} = \left(1 + \frac{\gamma-1}{2}M_1^2\eta_\sigma\right) \tag{3.22}$$

Let H be total pressure (total head) upstream of the test-section, and p_1 be the static pressure there; then we have by isentropic relation,

$$\frac{H}{p_1} = \left(1 + \frac{\gamma - 1}{2}M^2\right)^{\frac{\gamma}{\gamma - 1}} \tag{3.23}$$

Therefore, the overall pressure ratio, H/p_{02}, for the tunnel becomes

$$\frac{H}{p_{02}} = \frac{H}{p_1}\frac{p_1}{p_{02}}$$

But this is also *the compressor pressure ratio required to run the tunnel*. Hence, using Equations (3.21) and (3.23), the compressor pressure ratio, p_σ, can be expressed as

$$p_\sigma = \frac{H}{p_{02}} = \left[\frac{1 + \dfrac{\gamma - 1}{2}M_1^2}{1 + \dfrac{\gamma - 1}{2}M_1^2\eta_\sigma}\right]^{\frac{\gamma}{\gamma - 1}} \tag{3.24}$$

For continuous and intermittent supersonic wind tunnels, the energy ratio, ER, may be defined as follows.

1. For continuous tunnel

$$ER = \frac{KE \text{ at the test-section}}{\text{Work done in isentropic compression per unit time}}$$

Using Equation (3.24), this may be expressed

$$ER = \frac{1}{\left(p_\sigma^{\frac{\gamma - 1}{\gamma}} - 1\right)\left(\dfrac{2}{(\gamma - 1)M_1^2} + 1\right)} \tag{3.25}$$

2. For intermittent tunnel

$$ER = \frac{(KE \text{ in test-section})(\text{Time of tunnel run})}{\text{Energy required for charging the reservoir}} \tag{3.26}$$

From the above discussions we can infer that,

- For $M < 1.7$, induced flow tunnels are more efficient than blowdown tunnels.

- In spite of this advantage, most of the supersonic tunnels even over this Mach number range are operated as blowdown tunnels and not as induced flow tunnels. This is because vacuum tanks are more expensive than compressed air storage tanks.

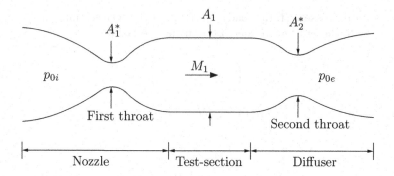

Figure 3.17: Schematic of supersonic wind tunnel with second throat

3.4.6   Effects of Second Throat

A typical supersonic tunnel with second throat is shown schematically in Figure 3.17.

The second throat, shown in Figure 3.17, is used to provide isentropic deceleration and highly efficient pressure recovery after the test-section. Neglecting frictional and boundary layer effects, a wind tunnel can be run at design conditions indefinitely, with no pressure difference requirement to maintain the flow, once started. But this is an ideal situation which cannot be encountered in practice. Even under the assumptions of this ideal situation, during start-up a pressure difference must be maintained across the entire system, shown in Figure 3.17, to establish the flow. For the supersonic tunnel sketched in Figure 3.17, the following observation may be made.

- As the pressure ratio p_{0e}/p_{0i} is decreased below 1.0, the flow situation is the same as that in a convergent-divergent nozzle, where p_{0i} and p_{0e} are the stagnation pressures at the nozzle inlet and the diffuser exit, respectively.

- Now, any further decrease in p_{0e}/p_{0i} would cause a shock to appear downstream of the nozzle throat.

- Further decrease in p_{0e}/p_{0i} moves the shock downstream, towards the nozzle exit.

- With a shock in the diverging portion of the nozzle, there is a severe stagnation pressure loss in the system.

- To pass the flow after the shock, the second throat must be at least of an area A_2^*.

- The worst case causing maximum loss of stagnation pressure is that with a normal shock in the test-section. For this case, the second throat area must be at least A_2^*.

- If the second throat area is less than this, it cannot pass the required flow and the shock can never reach the test-section, and will remain in the divergent part of the nozzle.

- Under these conditions, supersonic flow can never be established in the test-section.

- As p_{0e}/p_{0i} is further lowered, the shock jumps to an area in the divergent portion of the diffuser which is greater than the test-section area, i.e., the shock is swallowed by the diffuser.

- To maximize the pressure recovery in the diffuser, p_{0e}/p_{0i} can now be increased, which makes the shock move upstream to the diffuser throat, and the shock can be positioned at the location where the shock strength is at a minimum.

From the above observations, it is evident that the second throat area must be large enough to accommodate the mass flow, when a normal shock is present in the test-section. Assuming the flow to be one-dimensional in the tunnel sketched in Figure 3.17, it can be shown from the continuity equation that,

$$\rho_1^* a_1^* A_1^* = \rho_2^* a_2^* A_2^*$$

The flow process across a normal shock is adiabatic and therefore,

$$T_1^* = T_2^*$$

and

$$\rho_1^*/\rho_2^* = p_1^*/p_2^* = p_{01}/p_{02}$$

Also,

$$a_2^* = a_1^*$$

since $T_1^* = T_2^*$. Therefore, the minimum area of the second throat required for starting the tunnel becomes

$$\boxed{\frac{A_2^*}{A_1^*} = \frac{p_{01}}{p_{02}}} \tag{3.27}$$

where p_{01} and p_{02} are the stagnation pressures upstream and downstream, respectively, of the normal shock just ahead of the second throat. The pressures p_{01} and p_{02} are identically equal to p_{0i} and p_{0e}, respectively. Instead of the ratio of the throats area, it is convenient to deal with the ratio of test-section area A_1 to diffuser throat area A_2^*. This is called the *diffuser contraction ratio*, ψ. Thus, the maximum permissible contraction ratio for starting the tunnel is given by

$$\psi_{\max} = \frac{A_1}{A_2^*} = \frac{A_1}{A_1^*} \times \frac{A_1^*}{A_2^*} = \frac{A_1}{A_1^*} \times \frac{p_{02}}{p_{01}} = f(M_1) \tag{3.28}$$

when the second throat area is larger than the minimum required for any given condition, the shock wave is able to "jump" from the test-section to the downstream side of the diffuser throat. This is termed *shock swallowing*. The complete test-section has supersonic flow, which is the required state for a supersonic wind tunnel test-section. However, the second throat and part of the diffuser also have supersonic flow. Apparently we have only shifted the shock from the test-section to the diffuser. This again will result in considerable loss. In principle, it is possible to bring down the loss to a very low level by reducing the area of the second throat, after starting the tunnel. As A_2^* is reduced, the shock becomes weaker (as seen from Equation (3.27)) and moves upstream towards the second throat. When $A_2^* = A_1^*$, the shock just reaches the second throat, and its strength becomes vanishingly small. This is the ideal situation, resulting in supersonic flow in the test-section and isentropic flow in the diffuser.

At this stage, we should realize that the above model is based on the assumption that the flow is one-dimensional and inviscid, with a normal shock in the test-section. A more realistic model might have to take into account the nonstationary effects of the shock, the possibility of oblique shocks, and the role of boundary layer development. Further, reduction of A_2^* to A_1^*, which is the ideal value, is not possible in practice. However, some contraction after starting is possible, up to a limiting value at which the boundary layer effects prevent the maintenance of sufficient mass flow for maintaining a supersonic test-section, and beyond that the flow *breaks down*.

Experimental studies confirm, in a general way, the theoretical considerations outlined above, although there are modifications owing to viscous effects.

The skin friction at the wall of course causes some additional loss of stagnation pressure. Some of the diffuser problems outlined here may be avoided to a large extent by

- Using a variable-geometry diffuser.

- Using a variable-geometry diffuser in conjunction with a variable-geometry nozzle.

- Driving the shock through the diffuser throat by means of a large-amplitude pressure pulse.

- Taking advantage of effects which are not one-dimensional.

3.4.7 Compressor Tunnel Matching

Usually the design of a continuous supersonic wind tunnel has either of the following two objectives.

1. Choose a compressor for specified test-section size, Mach number, and pressure level.

2. Determine the best utilization of an already available compressor.

In the first case, wind tunnel characteristics govern the selection of compressor and in the second case it is the other way about. In either case the characteristics to be matched are the *overall pressure ratio* and *mass flow*.

The compressor characteristics are usually given in terms of the volumetric flow $\mathbb{V}$ rather than mass flow. Therefore, it is convenient to give the wind tunnel characteristics also in terms of $\mathbb{V}$. We know that, the volume can be expressed as

$$\mathbb{V} = \frac{m}{\rho}$$

since the density ρ varies in the tunnel circuit, the volumetric flow also varies for a given constant mass flow m. For the compressor, we specify the intake flow as

$$\mathbb{V}_i = \frac{m}{\rho_i} \tag{3.29}$$

which is essentially the same as the volume flow at the diffuser exit.

On the other hand, the volume flow at the supply section (wind tunnel settling chamber) is

$$\mathbb{V}_0 = \frac{m}{\rho_0} \tag{3.30}$$

Using throat as the reference section, the mass flow can be expressed as

$$m = \rho^* a^* A^* = \left(\frac{2}{\gamma+1}\right)^{\frac{\gamma+1}{2(\gamma-1)}} \rho_0 a_0 A^* \tag{3.31}$$

where a^* and a_0 are the sonic speeds at the throat and stagnation state, respectively.

With Equation (3.31), Equation (3.30) can be rewritten as

$$
\begin{aligned}
\mathbb{V}_0 &= \left(\frac{2}{\gamma+1}\right)^{\frac{\gamma+1}{2(\gamma-1)}} a_0 A^* \\
&= \left(\frac{2}{\gamma+1}\right)^{\frac{\gamma+1}{2(\gamma-1)}} \sqrt{\gamma R T_0} \frac{A^*}{A} A \\
&= \text{constant } \sqrt{T_0} A \left(\frac{A^*}{A}\right)
\end{aligned}
$$

From this equation it is seen that, the volume flow rate $\mathbb{V}_0$ depends on the stagnation temperature, test-section area, and test-section Mach number (since A/A^* is a function of M).

The compressor intake flow and the supply section (settling chamber) flow may easily be related, using Equations (3.29) and (3.30), to result in

$$\frac{\mathbb{V}_i}{\mathbb{V}_0} = \frac{\rho_0}{\rho_i} = \frac{p_0}{p_i} \times \frac{T_i}{T_0} = \Lambda \tag{3.32}$$

since $T_i = T_0$, Λ is simply the pressure ratio at which the tunnel is actually operating. This pressure ratio Λ must always be more than the minimum pressure ratio required for supersonic operation at any desired Mach number.

Equation (3.32) gives the relation between the operating pressure ratio, Λ, and the compressor intake volume, $\mathbb{V}_0$, as

$$\Lambda = \left(\frac{1}{\mathbb{V}_0}\right)\mathbb{V}_i$$

The plot of Λ verses $\mathbb{V}_i$ is a straight line through the origin, with slope $1/\mathbb{V}_0$, as shown in Figure 3.18. (Liepmann and Roshko, 1956).

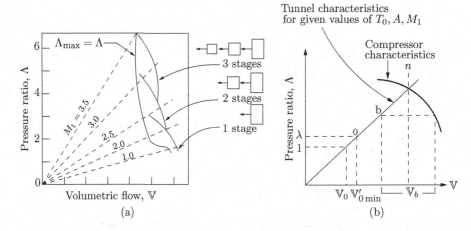

Figure 3.18: Wind tunnel and compressor characteristics: (a) operation over a range of M, using multistage compressor; (b) matching of wind tunnel compressor characteristics (one test-section condition): n, matching point; b, matching point with bypass; 0, match point at minimum operating pressure ratio

The power requirement for a multistage compressor is given by

$$HP = \left(\frac{1}{746}\right)\left(\frac{N\gamma}{\gamma - 1}\right)\dot{m}RT_s\left[\left(\frac{p_{0c}}{p_{03}}\right)^{\frac{\gamma - 1}{\gamma N}} - 1\right] \qquad (3.33)$$

where $\dot{m}$ is the mass flow rate of air in kg/s, p_{03} and p_{0c} are the total pressures at the inlet and outlet of the compressor, respectively, N is the number of stages, and T_s is the stagnation temperature.

Example 3.5

Determine the minimum possible diffuser contraction ratio and the power required for a two-stage compressor to run a closed-circuit supersonic tunnel at $M = 2.2$. The efficiency of the compressor is 85 percent, $p_{01} = 4$ atm, $T_0 = 330$ K and $A_{TS} = 0.04$ m^2.

Solution

Compressor pressure ratio is

$$\frac{p_{0c}}{p_{03}} = \frac{p_{01}}{p_{02}} \frac{1}{\eta}$$

Given that, $M = 2.2$, $\eta = 0.85$, $N = 2$, $T_0 = 330\,\text{K}$, $p_{01} = 4\,\text{atm}$, $A_{TS} = 0.04\,\text{m}^2$.

For $M_1 = 2.2$, $\dfrac{p_{02}}{p_{01}} = 0.6281$, from normal shock table, and $\dfrac{A_1}{A_1^*} = 2.005$, from isentropic table.

Therefore, the maximum possible contraction ratio becomes

$$\begin{aligned}
\psi_{\text{max}} &= \frac{A_1}{A_1^*} \times \frac{p_{02}}{p_{01}} \\[2mm]
&= 2.005 \times 0.6281 \\[2mm]
&= \boxed{1.26}
\end{aligned}$$

The mass flow rate is given by

$$\begin{aligned}
\dot{m} &= \frac{0.6847}{\sqrt{RT_0}} p_0 A^* \\[2mm]
&= \frac{0.6847 \times 4 \times 101325}{\sqrt{287 \times 330}} \times \frac{0.04}{2.005} \\[2mm]
&= 17.99\,\text{kg/s}
\end{aligned}$$

The power required to run the tunnel is

$$\text{Power} = \frac{1}{746} \times \frac{2 \times 1.4}{0.4} \times 17.99 \times 287 \times 330 \left[\left(\frac{1/0.6281}{0.85} \right)^{0.2857/2} - 1 \right]$$

$$= \boxed{1499.50\,\text{hp}}$$

3.4.7.1 Basic Formulas for Supersonic Wind Tunnel Calculations

From our discussions so far, it is easy to identify that the following are the important relations required for supersonic tunnel calculations.

$$\frac{p_1}{p_2} = \left(\frac{\rho_1}{\rho_2}\right)^{\gamma} = \left(\frac{T_1}{T_2}\right)^{\frac{\gamma}{\gamma-1}}$$

$$a = \sqrt{\gamma RT} = 20.04\sqrt{T} \text{ m/s}$$

$$\frac{\gamma}{\gamma-1}\frac{p}{\rho} + \frac{V^2}{2} = \text{constant} = \frac{\gamma}{\gamma-1}\frac{p_t}{\rho_t}$$

where p_t and ρ_t are the stagnation pressure and density, respectively.

$$\frac{p_2}{p_1} = \left(\frac{1+\dfrac{\gamma-1}{2}M_1^2}{1+\dfrac{\gamma-1}{2}M_2^2}\right)^{\frac{\gamma}{\gamma-1}}$$

$$\frac{p_t}{p} = \left(1+\frac{\gamma-1}{2}M^2\right)^{\frac{\gamma}{\gamma-1}}$$

$$\frac{\rho_t}{\rho} = \left(1+\frac{\gamma-1}{2}M^2\right)^{\frac{1}{\gamma-1}}$$

$$\frac{T_t}{T} = \left(1+\frac{\gamma-1}{2}M^2\right)$$

where p, ρ, and T are the local pressure, density, and temperature, respectively, and p_1 and p_2 are the pressures upstream and downstream of a normal shock.

3.4.8   The Mass Flow

Mass flow rate is one of the primary considerations in sizing a wind tunnel test-section and the associated equipment, such as compressor and diffuser. The mass flow rate is given by

$$\dot{m} = \rho AV$$

From isentropic relations, for air with $\gamma = 1.4$, we have

$$\rho = \rho_t(1+0.2M^2)^{-\frac{5}{2}}$$

where ρ_t is the total or stagnation density. By the perfect gas state equation, we have

$$\rho_t = \frac{p_t}{RT_t}$$

Therefore,

$$\rho = \left(\frac{p_t}{RT_t}\right)(1+0.2M^2)^{-\frac{5}{2}}$$

where, $R = 287 \, \text{m}^2/(\text{s}^2 \, \text{K})$ is the gas constant for air, p_t is the total pressure in pascal, and T_t is the total temperature in kelvin.

Also, the local temperature and velocity are given by

$$
\begin{aligned}
T &= T_t(1 + 0.2M^2)^{-1} \\
V &= M(1.4\,R\,T)^{\frac{1}{2}}
\end{aligned}
$$

Substituting the above expression for T into V expression, we get

$$
V = M \left[\frac{1.4\,R\,T_t}{(1 + 0.2M^2)} \right]^{\frac{1}{2}}
$$

Using the above expressions for V and ρ in the $\dot{m}$ equation, we get the mass flow rate as

$$
\boxed{\dot{m} = \left(\frac{1.4}{R\,T_t} \right)^{\frac{1}{2}} \frac{M\,p_t\,A}{(1 + 0.2M^2)^3}} \tag{3.34}
$$

This equation is valid for both subsonic and supersonic flows. When the mass flow rate being calculated is for subsonic Mach number, Equation (3.34) is evaluated using test-section Mach number in conjunction with the total temperature and pressure. For supersonic flows, it is usually convenient to make the calculations at the nozzle throat, where the Mach number is 1.0. Further, it should be noted that blowdown tunnels are usually operated at a constant pressure during run. The main objective of constant pressure run is to obtain a steady flow while data is being recorded. Thus, the total pressures to be used in the evaluation of Equation (3.34) are the minimum allowable (or required) operating pressures.

Example 3.6

A continuous wind tunnel operates at Mach 2.5 at test-section, with static conditions corresponding to 10,000 m altitude. The test-section is 150 mm × 150 mm in cross-section, with a supersonic diffuser downstream of the test-section. Determine the power requirements of the compressor during start-up and during steady-state operation. Assume the compressor inlet temperature to be the same as the test-section stagnation temperature.

Solution

At the test-section, $M = 2.5$. At 10,000 m altitude, from atmospheric table, we have

$$
p = 26.452 \, \text{kPa}, \qquad T = 223.15 \, \text{K}
$$

These are the pressure and temperature at the test-section.
    From isentropic table, for $M = 2.5$, we have

$$
\frac{p}{p_0} = 0.058528 \qquad \frac{T}{T_0} = 0.4444
$$

Therefore, the stagnation pressure and temperature at the test-section are,

$$p_0 = \frac{26.452}{0.058528} = 451.95\,\text{kPa}$$

$$T_0 = \frac{223.15}{0.44444} = 502.1\,\text{K}$$

During steady-state operation, the mass flow rate through the test-section is

$$\dot{m} = \rho A V$$

$$= \frac{p}{RT} A M \sqrt{\gamma R T}$$

$$= \frac{26452}{287 \times 223.15}(0.15 \times 0.15)(2.5)\sqrt{1.4 \times 287 \times 223.15}$$

$$= 6.96\,\text{kg/s}$$

From isentropic table, for $M = 2.5$, we have

$$\frac{A}{A^*} = 2.63671$$

Therefore,

$$A^* = \frac{0.15 \times 0.15}{2.63671}$$

$$= 0.00853\,\text{m}^2$$

This is the area of the first throat.

During start-up, a shock wave is formed when the flow becomes supersonic. The pressure loss due to this shock is maximum when it is at the test-section.

For $M = 2.5$, from normal-shock table, we have

$$\frac{p_{02}}{p_{01}} = 0.499$$

Also, we know that,

$$\frac{p_{02}}{p_{01}} = \frac{A_1^*}{A_2^*}$$

Therefore,

$$\frac{A_1^*}{A_2^*} = 0.499$$

$$A_2^* = \frac{A_1^*}{0.499} = \frac{0.00853}{0.499}$$

Thus,

$$\frac{A}{A_2^*} = \frac{0.15 \times 0.15}{0.00853} \times 0.499$$

$$= 1.316$$

For this area ratio, from isentropic table, we get $M = 1.68$. This is the Mach number ahead of the shock when the shock is at the second throat.

For $M = 1.68$, from normal shock table, we have

$$\frac{p_{02}}{p_{01}} = 0.86394$$

This pressure loss must be compensated by the compressor. The power input required for the compressor to compensate for this loss is

$$\text{Power} = h_o - h_i = C_p \left(T_o - T_i \right)$$

where the subscripts o and i, respectively, refer to compressor outlet and inlet conditions. For an isentropic compressor,

$$\frac{T_o}{T_i} = \left(\frac{p_o}{p_i} \right)^{\frac{\gamma-1}{\gamma}}$$

$$T_o - T_i = T_i \left(\left(\frac{p_o}{p_i} \right)^{\frac{\gamma-1}{\gamma}} - 1 \right)$$

$$= 502.1 \left(\left(\frac{1}{0.86394} \right)^{0.286} - 1 \right)$$

$$= 21.45 \, \text{K}$$

Thus, the power input required becomes

$$\text{Power} = 1004.5 \times 21.45 = 21546.52 \, \text{J/kg}$$

The horse power required for the compressor is

$$\text{Power} = \frac{\dot{m} \, W}{746}$$

$$= \frac{6.96 \times 21546.52}{746}$$

$$= \boxed{201 \, \text{hp}}$$

This is the running horse power required for the compressor.

During start-up, $M_1 = 2.5$, the corresponding $p_{02}/p_{01} = 0.499$, from normal shock table.

The isentropic work required for the compressor during start-up is

$$W = \left[\left(\frac{1}{0.499} \right)^{0.286} - 1 \right] 502.1 \, C_p$$

$$= 110.4 \, C_p$$

$$= 1004.5 \times 110.4$$

$$= 110896.8 \, \text{J/kg}$$

Thus, the power required is

$$\text{Power} = \frac{6.96 \times 110896.8}{746}$$

$$= \boxed{1034.7 \, \text{hp}}$$

Example 3.7

Estimate the settling chamber pressure and temperature and the area ratio required to operate a Mach 2 tunnel under standard sea-level conditions. Assume the flow to be one-dimensional and the tunnel is operating with correct expansion.

Solution

The tunnel is operating with correct expansion. Therefore, the sea-level pressure and temperature become the pressure and temperature in the test-section (i.e., at the nozzle exit). Thus, $p_e = 101.325$ kPa and $T_e = 15°C$.

This problem can be solved by using the appropriate equations or by using gas tables. Let us solve the problem by both methods.

Solution Using Equations

Let subscripts e and 0 refer to nozzle exit and stagnation states, respectively.

From isentropic relations, we have the temperature and pressure ratio as

$$\frac{T_0}{T_e} = 1 + \frac{\gamma - 1}{2} M_e^2$$

$$= 1 + \frac{1.4 - 1}{2} \times 2^2 = 1.8$$

$$T_0 = 1.8 \, T_e = 1.8 \times 288.15$$

$$= \boxed{518.67 \, \text{K}}$$

$$\frac{p_0}{p_e} = \left(1 + \frac{\gamma - 1}{2} M_e^2\right)^{\frac{\gamma}{\gamma - 1}}$$

$$= 1.8^{3.5} = 7.824$$

$$p_0 = 7.824 \, p_e$$

$$= \boxed{792.77 \, \text{kPa}}$$

From isentropic relations we have the area ratio as

$$\left(\frac{A_e}{A_{\text{th}}}\right)^2 = \frac{1}{M_e^2}\left[\frac{2}{\gamma + 1}\left(1 + \frac{\gamma - 1}{2} M_e^2\right)\right]^{\frac{\gamma + 1}{\gamma - 1}}$$

$$= \frac{1}{2^2}\left[\frac{2}{2.4} \times 1.8\right]^6 = 2.8476$$

$$\frac{A_e}{A_{\text{th}}} = \boxed{1.687}$$

Solution Using Gas Tables

From the isentropic table, for $M_e = 2$, we have

$$\frac{p_e}{p_0} = 0.1278, \quad \frac{T_e}{T_0} = 0.55556, \quad \frac{A_e}{A_{\text{th}}} = \boxed{1.6875}$$

Thus,

$$p_0 = \frac{p_e}{0.1278}$$

$$= \frac{101325}{0.1278} = \boxed{792.84\,\text{kPa}}$$

$$T_0 = \frac{T_e}{0.55556}$$

$$= \frac{288.15}{0.55556} = \boxed{518.67\,\text{K}}$$

3.4.9   Blowdown Tunnel Operation

In a blowdown tunnel circuit, the pressure and temperature of air in the compressed air reservoir (also called storage tank) change during operation. This change of reservoir pressure causes the following effects.

- The tunnel stagnation and settling chamber pressures fall correspondingly.

- The tunnel is subjected to dynamic condition.

- Dynamic pressure in the test-section falls and hence, the forces acting on the model change during test.

- Reynolds number of the flow changes during tunnel run.

Usually three methods of operation are adopted for blowdown tunnel operation. They are

- Constant Reynolds number operation.

- Constant pressure operation.

- Constant throttle operation.

The ratio between the settling chamber initial pressure p_{bi} and reservoir initial pressure p_{0i} is an important parameter influencing the test-section Reynolds number. Let

$$\frac{p_{bi}}{p_{0i}} = \frac{\text{Settling chamber initial pressure}}{\text{Reservoir initial pressure}} = \alpha$$

The variation of Reynolds number with tunnel running time t, as a function of α will be as shown in Figure 3.19.

As seen from Figure 3.19, the Reynolds number increases with running time for constant pressure operation, and decreases with running time for constant throttle operation. The change in Reynolds number results in the change of boundary layer thickness, which in turn causes area and Mach number change in the test-section. Usually Mach number variation due to the above causes is small.

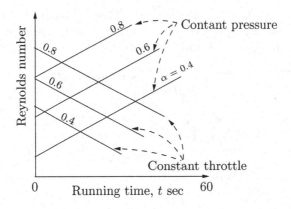

Figure 3.19: Reynolds number variation with tunnel running time

3.4.9.1 Reynolds Number Control

By definition, Reynolds number is the ratio between the inertia and viscous forces.

$$Re = \frac{\text{Inertia force}}{\text{Viscous force}}$$

It can be shown that

$$Re = \frac{\rho V L}{\mu}$$

where ρ, V, and μ are the density, velocity, and viscosity, respectively, and L is a characteristic dimension of the model being tested. The above equation may be expressed as

$$\frac{Re}{L} = \frac{\rho V}{\mu} \tag{3.35}$$

Also, the viscosity coefficient may be expressed as

$$\mu = C T^m V = C T^m M \sqrt{\gamma R T} = C_1 f(M) \sqrt{T}$$

In the above expression, C_1 and C are constants, m is viscosity index, γ is isentropic index, and R is gas constant.

Let p_b and p_{bi} be the instantaneous and initial pressures in the settling chamber, respectively, and T_b and T_{bi} be the corresponding temperatures. With the above relations for μ, Equation (3.35) can be expressed as

$$\frac{R_e}{L} = g_1 f(M, m) \left[\frac{p_b/p_{bi}}{(T_b/T_{bi})^{(m+\frac{1}{2})}} \right] \tag{3.36}$$

where g_1 is a function of initial (starting) conditions (p_{bi}, p_{ti}). From Equation (3.36) it is seen that the Reynolds number during tunnel run is influenced only by the quantities within the square brackets. These quantities can easily be held constant by suitable manipulation of a throttling valve located between reservoir and settling chamber, as shown in Figure 3.20.

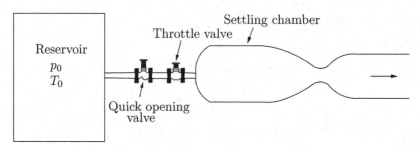

Figure 3.20: Blowdown tunnel layout

The throttling process may be expressed by the following equation:

$$p_{bi} = \alpha \, p_0^{\beta} \qquad (3.37)$$

where p_{bi} and p_0 are the total pressures after (stagnation pressure in the settling chamber) and before (stagnation pressure in the reservoir) throttling, respectively, and α and β are constants.

The function g in Equation (3.36), at settling chamber conditions, is

$$g_i = \left(\frac{\gamma p_{bi}}{a_{bo} \mu_{bo}} \right)$$

where a_{bo} is the proportionality constant and μ_{bo} is the viscosity coefficient of air in the settling chamber.

The function $f(M, m)$, from isentropic relations, is

$$f(M, m) = \frac{M}{\left[1 + \dfrac{\gamma - 1}{2} M^2 \right]^{\left(\frac{\gamma+1}{2(\gamma-1)} - m \right)}}$$

Now, applying the polytropic law for the expansion of gas in the storage tank, we can write

$$\frac{p_0}{p_{0i}} = \left(\frac{T_0}{T_{0i}} \right)^{\frac{n}{n-1}}$$

where subscripts 0 and 0i refer to instantaneous and initial conditions in the reservoir and n is the polytropic index.

Also, from Equation (3.37), we have

$$\frac{p_b}{p_{bi}} = \left(\frac{p_0}{p_{0i}} \right)^{\beta}$$

Therefore, with the above relations, Equation (3.36) can be expressed as,

$$\frac{Re}{L} = g_1 f(M, m) \left[\left(\frac{p_0}{p_{0i}} \right)^{\left(\beta - \frac{(2m+1)(n-1)}{2n} \right)} \right] \tag{3.38}$$

This is the general relation between test-section Reynolds number and reservoir pressure. From this equation, the following observations can be made.

- For R_e = constant; $\quad \beta = \dfrac{(2m+1)(n-1)}{2n}$

- For constant "p_b" operation, $p_b = \alpha p_0^\beta$ = constant, and $\beta = 0$. Thus, Equation (3.38) simplifies to

$$\frac{R_e}{L} = K_3 \left(\frac{p_{0i}}{p_0} \right)^{\frac{(2m+1)(n-1)}{2n}}$$

  where K_3 is a constant. This implies that, R_e increases with time t, since p_0 decreases with t.

- For constant throttle operation, $\beta = 1$ and

$$p_b = \alpha p_0^\beta = \alpha p_0$$

Therefore,

$$\frac{Re}{L} = K_3 \left(\frac{p_0}{p_{0i}} \right)^{\left[1 - \frac{(2m+1)(n-1)}{2n} \right]} \qquad 0 < \frac{(2m+1)(n-1)}{2n} < 1$$

This implies that, R_e decreases with t for constant throttle operation.

From the above observations it can be inferred that, for a given settling chamber pressure and temperature, the running time is

- The shortest for constant throttle operation.

- The longest for constant Reynolds number operation.

- In between the above two for constant pressure operation.

3.4.10 Optimum Conditions

For optimum performance of a tunnel in terms of running time "t", the drop in reservoir pressure should be as slow as possible. To achieve this slow rate of fall in reservoir pressure, the pressure regulating valve should be adjusted after the tunnel has been started, in such a manner that the pressure in the settling chamber is the minimum pressure, p_{bmin}, required for the run.

The performance of the tunnel, the test-section Mach number M versus the tunnel run time t, for different methods of control mentioned above, should be evaluated for the entire range of operation. These performances can be recorded in the form of graphs for convenient reference. From such graphs, the best suited method of operation for any particular test and the required settings of the throttle valve (α, β, etc.) can be chosen. A typical performance chart will look like the one shown in Figure 3.21.

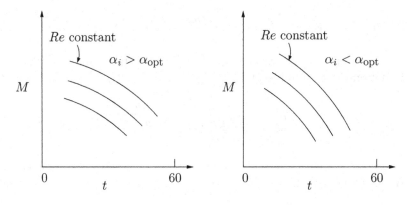

Figure 3.21: Wind tunnel performance chart

For a given test-section Mach number M there is a p_b minimum in the settling chamber, given by the pressure ratio relation. The Reynolds number in the test-section depends on this p_b value and constant Reynolds number operation is possible only if

- p_{bi} value is so chosen that as "t" proceeds (increases) so both p_0 and p_b reach p_{bmin} value simultaneously (to result in an optimum constant Reynolds number).

- $p_{bi} > p_{\text{bopt}}$ the reservoir pressure will become equal to p_b at some instant and then subsequently constant Reynolds number operation is not possible.

- $p_{bi} < p_{\text{bopt}}$ the $p_b = p_{\text{bmin}}$ state will be reached at time "t" when $p_0 > p_b$ and supersonic operation will not be possible thereafter.

3.4.11   Running Time of Blowdown Wind Tunnels

Blowdown supersonic wind tunnels are usually operated with either constant dynamic pressure (q) or constant mass flow rate ($\dot{m}$).

For constant q operation, the only control necessary is a pressure regulating valve (PRV) that holds the stagnation pressure in the settling chamber at a constant value. The stagnation pressure in the storage tank falls according to the polytropic process; with the polytropic index $n = 1.4$ for short duration runs, with high mass flow, approaching $n = 1.0$ for long duration runs with thermal mass[1] in the tank.

For constant mass flow run, the stagnation temperature and pressure in the settling chamber must be held constant. For this, either a heater or a thermal mass external to the storage tank is essential. The addition of heat energy to the pressure energy in the storage tank results in longer running time of the tunnel. Another important consequence of this heat addition is that the constant settling chamber temperature of the constant-mass run keeps the test-section Reynolds number at a constant value.

For calculating the running time of a tunnel, let us make the following assumptions.

- Expansion of the gas in the storage tank is polytropic.

- Gas temperature in the storage tank is held constant with a heater.

- Gas pressure in the settling chamber is kept constant with a pressure regulating valve.

- No heat is lost in the pipe lines from the storage tank to the test-section.

- Expansion of the gas from the settling chamber to test-section is isentropic.

- Test-section speed is supersonic.

The mass flow rate $\dot{m}$ through the tunnel, as given by Equation (3.34), is

$$\dot{m} = \left(\frac{1.4}{R\,T_t}\right)^{1/2} \frac{M\,p_t\,A}{(1 + 0.2M^2)^3}$$

where M is the test-section Mach number, p_t and T_t, respectively, are the pressure and temperature in the settling chamber.

We know that, for supersonic flows it is convenient to calculate the mass flow rate with nozzle throat conditions. At the throat, $M = 1.0$ and Equation (3.34) becomes

$$\dot{m} = 0.0404 \frac{p_t A^*}{\sqrt{T_t}} \tag{3.39}$$

The value of the gas constant used in the above equation is $R = 287 \text{ m}^2/(\text{s}^2\,\text{K})$, which is the gas constant for air.

The product of mass flow rate and run time gives the change of mass in the storage tank. Therefore,

$$\dot{m}t = (\rho_i - \rho_f)\,\mathbb{V}_t \tag{3.40}$$

where $\mathbb{V}_t$ is the tank volume and ρ_i and ρ_f are the initial and final densities in the tank, respectively.

---

[1]Thermal mass is a material which has a high value of thermal capacity.

From Equation (3.40), the running time t is obtained as

$$t = \frac{(\rho_i - \rho_f)}{\dot{m}} V_t$$

Substituting for $\dot{m}$ from Equation (3.39) and arranging the above equation, we get

$$t = 24.728 \frac{\sqrt{T_t}}{p_t} \frac{V_t}{A^*} \rho_i \left(1 - \frac{\rho_f}{\rho_i}\right) \tag{3.41}$$

For polytropic expansion of air in the storage tank, we can write

$$\frac{\rho_f}{\rho_i} = \left(\frac{p_f}{p_i}\right)^{\frac{1}{n}}; \quad \rho_i = \frac{p_i}{RT_i}$$

where subscripts i and f denote the initial and final conditions in the tank, respectively.

Substitution of the above relations into Equation (3.41) results in

$$t = 0.086 \frac{V_t}{A^*} \frac{\sqrt{T_t}}{T_i} \frac{p_i}{p_t} \left[1 - \left(\frac{p_f}{p_i}\right)^{\frac{1}{n}}\right] \tag{3.42}$$

with V_t in m^3, this gives the run time in seconds for the general case of blowdown tunnel operation with constant mass flow rate condition.

From Equation (3.42) it is obvious that, for $t_{\max}$, the condition required is p_t minimum. At this stage we should realize that the above equation for running time has to be approached from a practical point of view and not from a purely mathematical point of view. Realizing this, it can be seen that the tunnel run does not continue until the tank pressure drops to the settling chamber stagnation pressure p_t, but stops when the storage pressure reaches a value which is appreciably higher than p_t, i.e., when $p_f = p_t + \Delta p$. This Δp is required to overcome the frictional and other losses in the piping system between the storage tank and the settling chamber. Value of Δp varies from about $0.1p_t$ for very-small-mass flow runs to somewhere around $1.0p_t$ for high-mass flow runs.

The proper value of the polytropic index n in Equation (3.42) depends on the rate at which the stored high-pressure air is used, total amount of air used, and the shape of the storage tank. The value of n tends towards 1.4 as the storage tank shape approaches spherical shape. With heat storage material in the tank (i.e., for isothermal condition), the index n approaches unity. Equation (3.42) may also be used with reasonable accuracy for constant-pressure runs in which the change in total temperature is small, since these runs approach the constant-mass flow rate situation.

Example 3.8

Determine the running time for a Mach 2 blowdown wind tunnel with test-section cross-section of 300 mm × 300 mm. The storage tank volume is 20 m^3

and the pressure and temperature of air in the tank are 20 atm and $25^{\deg}$C, respectively. The tank is provided with a heat sink material inside. Take the starting pressure ratio required for Mach 2.0 to be 3.0, the loss in pressure regulating valve (PRV) to be 50 percent, and the polytropic index $n = 1.0$.

Solution

Given that the settling chamber pressure required to start the tunnel is $p_t = 3.0 \times 101.3$ kPa. The pressure loss in the PRV is 50 percent, therefore,

$$p_f = 1.5 \times 303.9 = 455.85 \text{ kPa.}$$

From isentropic tables, for $M = 2.0$, we have $A^*/A = 0.593$. Therefore,

$$A^* = 0.593 \times 0.09 = 0.0534 \text{ m}^2$$

By Equation (3.42), the running time, t, is given by

$$t = 0.086 \left(\frac{20}{0.0534}\right)\left(\frac{\sqrt{298}}{298}\right)\left(\frac{2026}{303.9}\right)\left[1 - \left(\frac{455.85}{2026}\right)\right]$$

$$= \boxed{9.64 \text{ s}}$$

Example 3.9

A two-dimensional, symmetric wedge of nose-angle 10° is placed, with its axis coinciding with the tunnel axis, in a Mach 2.2 tunnel. (a) If the pressure coefficient over the upper surface of the wedge is -0.0347, find the flow angularity in the test-section, with respect to the tunnel axis. (b) If there is no flow angularity, what will be the pressure coefficient over the upper surface of the wedge? Also, find the Mach number of the flow over the upper surface.

Solution

If the flow in the test-section is uniform and parallel to the tunnel axis, the flow past the given wedge will be as illustrated in Figure E3.9(a). It is seen that the flow turning caused by the oblique shock at the wedge nose is just 5°. Therefore, the shock at the nose can be treated as weak and the entire flow can be solved treating the flow traversed by the weak oblique shock as isentropic. However, it is essential to ensure that the pressure loss caused by the shock is less than 5 percent, to justify the assumption of weak shock.

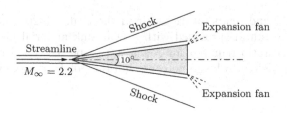

Figure E3.9a: Waves over the wedge, with its axis parallel to the freestream flow

Let us first calculate the pressure loss associated with the shock.

For $M_\infty = 2.2$ and $\theta = 5°$, from an oblique shock table or chart, we have the shock angle as

$$\beta = 31.10°$$

Thus, the normal component of flow Mach number ahead of the shock is

$$M_{1n} = M_\infty \sin \beta$$

$$= 2.2 \times \sin 31.10$$

$$= 1.14$$

From a normal shock table, for $M_{1n} = 1.14$, we have

$$\frac{p_{02}}{p_{01}} = 0.9973, \quad \frac{p_2}{p_1} = 1.3495$$

where subscripts 1 and 2 refer to state ahead of and behind the shock, respectively. The pressure loss is

$$\frac{p_{01} - p_{02}}{p_{01}} = 1 - \frac{p_{02}}{p_{01}}$$

$$= 1 - 0.9973$$

$$= 0.0027$$

$$= 0.27 \text{ percent}$$

It is seen that the pressure loss is just 0.27 percent, hence the entire flow can be treated as isentropic.

(a) Given, $M_\infty = 2.2$, $C_{pu} = -0.0347$.

The pressure coefficient, C_{pu}, over the upper surface of the wedge is given as negative. Therefore, the pressure over the upper surface, p_u, has to be less than the freestream pressure, p_∞. Thus, the flow at the nose of the wedge should pass through an expansion fan, as illustrated in Figure E3.9(b).

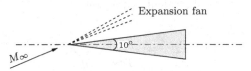

Figure E3.9b: Expansion fan at the nose of the wedge, turning the flow parallel to the upper surface

From isentropic table, for $M_\infty = 2.2$, $\dfrac{p_\infty}{p_{0\infty}} = 0.0935$, $\nu_\infty = 31.732°$.

By definition, the pressure coefficient, C_p, is

$$C_p = \frac{p - p_\infty}{\frac{1}{2}\rho_\infty V_\infty^2}$$

where p is the local static pressure and p_∞, ρ_∞, and V_∞ are the static pressure, static density, and velocity of the freestream flow, respectively.

The dynamic pressure, $\frac{1}{2}\rho_\infty V_\infty^2$, can be expressed in terms of freestream Mach number, M_∞, as

$$\frac{1}{2}\rho_\infty V_\infty^2 = \frac{1}{2}\left(\frac{p_\infty}{RT_\infty}\right)V_\infty^2$$

$$= \frac{1}{2}\left(\frac{\gamma p_\infty}{\gamma RT_\infty}\right)V_\infty^2$$

$$= \frac{\gamma p_\infty}{2}\frac{V_\infty^2}{a_\infty^2}$$

$$= \frac{\gamma p_\infty}{2}M_\infty^2$$

Substituting this, we have the C_p as

$$C_p = \frac{p - p_\infty}{\frac{1}{2}\gamma p_\infty M_\infty^2}$$

$$= \frac{2\left(\frac{p}{p_\infty} - 1\right)}{\gamma M_\infty^2}$$

where p is the local static pressure. Therefore, the pressure coefficient, C_{pu}, over the upper surface of the wedge is

$$C_{pu} = \frac{2\left(\frac{p_u}{p_\infty} - 1\right)}{\gamma M_\infty^2} \qquad (i)$$

where subscript u refers to the upper surface of the wedge.

The pressure ratio $\dfrac{p_u}{p_\infty}$ can be calculated as follows.

$$\frac{p_u}{p_\infty} = \frac{p_u}{p_{0\infty}} \times \frac{p_{0\infty}}{p_\infty}$$

$$\frac{p_u}{p_{0\infty}} = \frac{p_u}{p_\infty} \times \frac{p_\infty}{p_{0\infty}}$$

From Equation (i), we have

$$\frac{p_u}{p_\infty} = 1 + \frac{\gamma M_\infty^2 C_{pu}}{2}$$

Therefore,

$$\begin{aligned}
\frac{p_u}{p_{0\infty}} &= \left(1 + \frac{\gamma M_\infty^2 C_{pu}}{2}\right) \frac{p_\infty}{p_{0\infty}} \\
&= \left(1 + \frac{1.4 \times 2.2^2 \times (-0.0347)}{2}\right) \times 0.0935 \\
&= 0.0825
\end{aligned}$$

From isentropic table, for $\frac{p_u}{p_{0\infty}} = 0.0825$, the Prandtl–Meyer function, ν_u, is

$$\nu_u = 33.78^\circ$$

Also,

$$\begin{aligned}
\nu_u &= \nu_\infty + \theta \\
&= 31.732 + \theta \\
\theta &= 33.78 - 31.732 \\
&= 2.05^\circ
\end{aligned}$$

Therefore, the flow turns by $5 + 2.05 = 7.05^\circ$ and flows over the upper surface of the wedge. Thus, the flow angularity is $\boxed{7.05^\circ}$ (from below), with reference to the tunnel axis.

(b) If there is no flow angularity, the flow should turn only by 5°, to flow parallel to the surface of the wedge, as illustrated in Figure E3.9(a). Thus,

$$\nu_u = \nu_\infty - \theta$$

$$= 31.732 - 5$$

$$= 26.732°$$

From isentropic table, for $\nu_u = 26.732°$,

$$M_u \approx \boxed{2.0}$$

For $M_u = 2$, from isentropic table, $\frac{p_u}{p_{0u}} = 0.1278$. Also, because of the isentropic assumption, $p_{0u} = p_{0\infty}$.

The pressure coefficient over the upper surface is

$$C_{pu} = \frac{2\left(\frac{p_u}{p_\infty} - 1\right)}{\gamma M_\infty^2}$$

$$= \frac{2\left(\frac{p_u/p_{0u}}{p_\infty/p_{0\infty}} - 1\right)}{\gamma M_\infty^2}$$

$$= \frac{2 \times (0.1278/0.0935 - 1)}{1.4 \times 2.2^2}$$

$$= \frac{2 \times (1.367 - 1)}{1.4 \times 2.2^2}$$

$$= \boxed{0.1083}$$

Note: The angularity of 7.5° is very large for any practical tunnel. This problem should be taken only as an exercise meant to practice the concept.

3.5   Hypersonic Tunnels

Hypersonic tunnels operate with test-section Mach numbers above 5. Generally they operate with stagnation pressures in the range from 10 to 100 atmosphere and stagnation temperatures in the range from 50°C to 2000°C. Contoured nozzles which are more often axially symmetric are used in hypersonic tunnels.

Models that can be tested in hypersonic tunnels are usually larger than those meant for test in supersonic tunnels. The model frontal area can go up to 10 percent of the test-section cross-sectional area. Model size will probably be restricted by the wake behind it, which takes too much flow area in the diffuser and blocks it during tunnel starting.

Use of dry and heated air is necessary for hypersonic operation to avoid condensation effects and liquefaction during expansion to the high Mach number and corresponding low-temperatures. The requirement of heated air is the major factor making hypersonic tunnel operation more complicated than supersonic tunnel operation. To get a feel about the drastic changes in the flow properties at hypersonic speeds, let us examine the parameters listed in Table 3.1, for isentropic index $\gamma = 1.4$.

Table 3.1 Flow Parameters

M	A/A^*	p_0/p	T_0/T
5	25.00 E 00	529.10 E 00	6.000
10	53.59 E 01	424.39 E 03	21.000
15	37.55 E 02	660.15 E 03	46.000
20	15.38 E 03	478.29 E 04	80.998
25	46.31 E 03	224.54 E 05	126.000

Now, we should note that the above table is based on isentropic relations. As we know, in isentropic relations the index γ is treated as a constant. However, we are familiar with the fact that γ is constant only for gases which are thermally as well as calorically perfect; simply termed *perfect gases* (*Gas Dynamics*, Rathakrishnan, 1995). Therefore, in hypersonic flows if the test-section flow temperatures are to be at room temperature levels, the storage temperature has to be increased to very high values, which will pose metallurgical problems. Because of these considerations the temperatures in the test-section of hypersonic tunnels are usually quite low, in spite of the fact that the storage temperature is kept appreciably above the ambient temperature.

Example 3.10

Find the test-section temperature for a hypersonic stream of air at Mach 7 with stagnation temperature at 700 K.

Solution

From the isentropic table ($\gamma = 1.4$), for $M = 7.0$, we have

$$\frac{T}{T_0} = 0.09592$$

Therefore, the temperature of air in the test-section is

$$T = (0.09592)(700) = 67.144 \, \text{K}$$

$$= \boxed{-205.856^\circ \text{C}}$$

It is seen from Table 3.1 that the pressure ratios involved in hypersonic tunnel flow process are very high. In order to achieve these pressure ratios, it is customary to employ a combination of high pressure and vacuum together, in hypersonic tunnel operations. A typical hypersonic tunnel circuit is schematically shown in Figure 3.22.

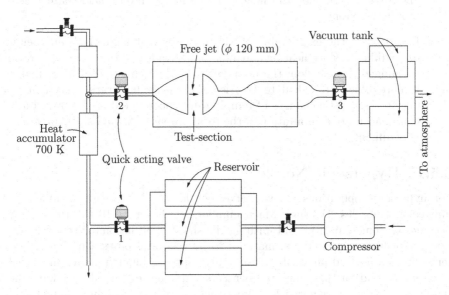

Figure 3.22: Hypersonic tunnel circuit

From the above discussions on hypersonic tunnel operation, the following observations can be made.

- The pressure ratio, area ratio, and temperature ratio for $M > 5$ increase very steeply with increase of M. Usually both the high pressure tank at the nozzle inlet and the vacuum tank at the diffuser end are necessary for hypersonic operations.

- The very low temperatures encountered in the test-section results in liquefaction of air and hence preheating of air to 700 K to 1000 K is common in hypersonic tunnel operation. For air, up to Mach 8 preheating to about 1300 K is satisfactory. For $M > 10$, gas like helium is better suited.

- Because of very low density in the test-section, optical flow visualization of viscous shock waves, etc., becomes more difficult.

- Shock wave angle (e.g., on wedge, cone, etc.,) changes appreciably with moisture content of air and hence, measurements have to be done with extra care.

- The heating of air introduces additional problems, like material require- ments for settling chamber, nozzle, test-section glass window, and distri- bution of parts (the tunnel structure, test-section walls, etc.) to stand high temperatures.

- Because of low pressure and temperature, the flow at the test-section has low Reynolds number and hence, the boundary layer thickness increases to a large extent.

- Determining the exact value of Mach numbers at high Mach number is quite difficult, since heating expands tunnel walls and therefore, area ratio is changed. In addition, the boundary layer (which is quite thick) makes it more difficult to calculate M. Also, the specific heat ratio γ is changing due to the drastic changes of temperature encountered in the tunnel and hence, accurate computation of the total pressure, p_0, and static pressure, p, is difficult.

3.5.1   Hypersonic Nozzle

For hypersonic operations axisymmetric nozzles are better suited than two-dimensional nozzles. For high Mach numbers of the order 10 the throat size becomes extremely narrow and forming the shape itself becomes very difficult. Because of high temperatures, material to be used also poses a problem. Ma- terial liners scale and pit easily at these high speeds. Though porcelain coated nozzles are good for these high temperatures and speeds, they do not have the smoothness which is required for hypersonic speeds. A minimum diameter of around 12 mm for the nozzle throat is arrived at from practical considerations. A typical shape of hypersonic nozzles is schematically shown in Figure 3.23.

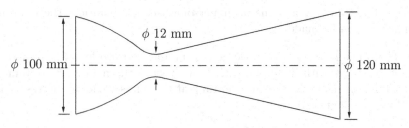

Figure 3.23: Hypersonic nozzle

3.6 Instrumentation and Calibration of Wind Tunnels

Calibration of wind tunnel test-section to ensure uniform flow characteristics everywhere in the test-section is an essential requirement in wind tunnel operation. In this section, let us see the calibration required and the associated instrumentation to study the flow characteristics in the test-sections of subsonic and supersonic wind tunnels. The instruments used for calibration will be only briefly touched upon in this section. The detailed study of these measuring devices will be done in the chapters to follow.

3.6.1 Low-Speed Wind Tunnels

We know that a low-speed air stream is characterized by the distribution of its dynamic pressure, static pressure, total pressure, temperature, flow direction, and turbulence level. From these details, the flow velocity and Reynolds number for any specific model can be computed. In other words, the instrumentation and calibration of low-speed tunnels involve the determination of the following.

1. Speed setting; calibration of true air speed in the test-section.

2. Flow direction; determining the flow angularity (pitch and yaw) in the test-section.

3. Turbulence level.

4. Velocity distribution; determination of flow quality.

5. Wake survey; determination of flow field in the wake of any model.

3.6.2 Speed Setting

Consider the flow in a subsonic wind tunnel, schematically shown in Figure 3.24.

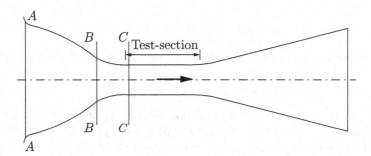

Figure 3.24: Open circuit tunnel

Measure the static pressure at the entry and exit of the contraction cone, at stations AA and BB. Applying the incompressible Bernoulli equation, we can write

$$p_A + q_A = p_B + q_B - k_1 q_B$$

where, p_A and p_B are the static pressures at sections AA and BB and q_a and q_B are the corresponding dynamics pressures, respectively, and k_1 is the pressure drop coefficient due to frictional loss between stations AA and BB. Therefore,

$$\Delta p = p_A - p_B = q_B - q_A - k_1 q_B \qquad (3.43)$$

From incompressible flow continuity equation, we have

$$A_A V_A = A_B V_B$$

Squaring and multiplying both sides by $\frac{\rho}{2}$, we get

$$\frac{\rho}{2} A_A^2 V_A^2 = \frac{\rho}{2} A_B^2 V_B^2$$

But $\frac{\rho}{2} V^2 = q$ is the dynamics pressure, thus,

$$q_A A_A^2 = q_B A_B^2$$

Let

$$\frac{A_B^2}{A_A^2} = k_2$$

Therefore,

$$q_A = k_2 q_B$$

Using this relation in Equation (3.43), we obtain

$$\Delta p = q_B - k_1 q_B - k_2 q_B = (1 - k_1 - k_2) q_B \qquad (3.44)$$

Applying Bernoulli equation between stations BB and CC, we get

$$q_B A_B^2 = q_t A_t^2$$

$$q_B = \left(\frac{A_t}{A_B} \right)^2 q_t = k_3 q_t$$

where subscript t refers to test-section. Using this, Equation (3.44) can be rewritten as

$$\boxed{\Delta p = p_A - p_B = (1 - k_1 - k_2) k_3 q_t} \qquad (3.45)$$

The pressure drop, Δp, and the test-section velocity, V_t, in Equation (3.45) can be measured independently, by wall static pressure taps and a standard pitot-static tube placed in the empty test-section, respectively. The typical variations of test-section velocity, V_t and dynamic pressure, q_t with static pressure drop is given in Figure 3.25. From this figure, the advantage of using q_t instead of V_t is obvious from its linear variation with Δp.

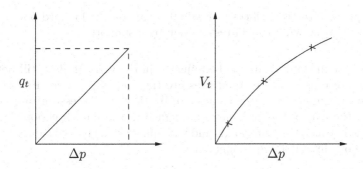

Figure 3.25: Test-section dynamic pressure and velocity variation with static pressure drop in the contraction cone

3.6.3 Flow Direction

The flow in the test-section has to be uniform throughout its cross-section, with all streamlines parallel to the tunnel axis, when the tunnel is run without any model in the test-section. Therefore, ensuring the flow direction along the tunnel axis is essential before putting the tunnel to use. A yaw head probe can be used to measure the angularity of the flow. The yaw head probes used for measuring the flow angularity in a wind tunnel test-section are usually sphere type or claw type.

3.6.3.1 Yaw Sphere

Schematic diagram of a typical sphere type yaw meter is shown in Figure 3.26.

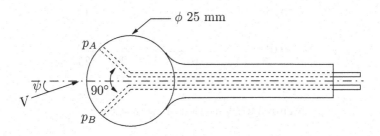

Figure 3.26: Yaw sphere

For measuring the angularity of planar or two-dimensional flows we need a two-hole yaw sphere, whereas, yaw meter with four-holes is necessary for three-dimensional flows, since the yaw and pitch of the stream need to be measured simultaneously. The procedure followed for angularity measurement is the following.

- A sphere with two orifices located 90° apart on the forward port is to be placed in the wind tunnel test-section by a support.

- The instrument axis has to be adjusted in the vertical plane till both the pressure taps measure equal pressure (i.e., $p_A = p_B$, refer Figure 3.26). The measurement of equal pressure by these two pressure taps implies that the axis of the yaw meter is aligned to the flow direction. Now the angle between the tunnel axis and yaw sphere axis gives the angle between the flow direction and tunnel axis.

- Align the instrument axis parallel to the tunnel axis and note the pressure difference $(p_A - p_B)$. The instrument may be calibrated by standard experiments for yaw head, defined as the ratio between ΔK and $\Delta\psi$, where $\Delta K = \Delta p/q$ and ψ is the yaw angle.

Theoretical and actual values of $\Delta p/q$ for a spherical yaw head are compared in Figure 3.27.

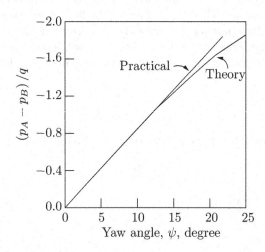

Figure 3.27: Variation of $\Delta p/q$ with ψ

The yaw head ($\Delta K/\Delta\psi$) varies from 0.04 to 0.07/degree. Therefore, for each instrument its yaw head constant, $\Delta K/\Delta\psi$, should be determined experimentally.

In addition to the yaw holes, a total-head orifice may also be provided at the front of the yaw head sphere. From Figure 3.27, it can be noted that the total pressure head orifice reading will be correct only for small flow deflection angles. Indeed, at 5° yaw the total head reading is down by 1.2 percent.

3.6.3.2 Claw Yaw Meter

The principle and the functioning of claw-type yaw meters is similar to that of the spherical yaw meter. A typical claw-type yaw meter is shown schematically in Figure 3.28.

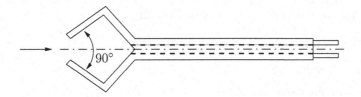

Figure 3.28: Claw yaw meter

Because of less interfering geometrical construction, it is generally used to measure the direction, rotation, and so on of the flow field at any point near model surfaces.

3.6.4 Turbulence

From our discussions on turbulence in Chapter 2, it is evident that basically turbulence is a measure of chaos or disorderliness in the flow. Further, generation of turbulent flow is easier compared to laminar flow. In fact, generation of laminar flow in the laboratory is a cumbersome task. In nature, most of the flows (air flow over plains, sea breeze, water flow in rivers, streams and so on) are turbulent. Therefore, it becomes imperative to simulate the level of turbulence in the actual flow field, where a prototype is going to operate, in the wind tunnel test-section, where a scale model of the prototype is tested, for establishing dynamically similar flow conditions between the prototype and model flow fields. If there is any disagreement between tests made in different wind tunnels at the same Reynolds number and between tests made in wind tunnels and in the actual field, then correction has to be applied on the effect of the turbulence produced in the wind tunnel by the propeller, guide vanes, screens and wire meshes, and vibration of the tunnel walls. From experience it is known that turbulence causes the flow pattern in the tunnel to be similar to the flow pattern in free air at a higher Reynolds number. Hence, the wind tunnel test Reynolds number could be said to have a higher *effective Reynolds number* compared to free flight in the actual flow field. In other words, a tunnel with a certain level of turbulence exhibits a flow pattern at a test-section Reynolds number Re, which will be identical to a free field flow at a Reynolds number which is much higher than the test-section Reynolds number Re. That is, the actual Reynolds number of the test-section is equivalent to a much higher free field Reynolds number. This increase is called the *turbulence factor*, TF, defined as

$$\boxed{Re_e = TF \times Re_c} \tag{3.46}$$

where the subscript e stands for effective Reynolds number, and Re_c is the measured critical Reynolds number in the tunnel test-section. Now it is clear that, measurement of turbulence in the test-section is essential for determining the Re_e. The turbulence may be measured with

1. Turbulence sphere

2. Pressure sphere

3. Hot-wire anemometer

3.6.4.1　Turbulence Sphere

The drag coefficient, C_D, of a sphere is greatly influenced by changes in flow velocity. The C_D for a sphere decreases with increasing air speed, since the earlier transition to turbulent flow results in reduced wake behind the sphere. This action decreases the form or pressure drag, resulting in lower total drag coefficient. The decrease of drag coefficient is rapid in a range of speed in which both the drag coefficient and drag go down. The Reynolds number at which the transition occurs at a given point on the sphere is a function of the turbulence already present in the air stream, and hence the drag coefficient of the sphere can be used to measure turbulence of the air stream. The procedure usually adopted for this measurement is the following.

- Measure the drag, D, for a small sphere of about 150 mm diameter, at many speeds of the test-section.

- After subtracting the buoyancy, the drag coefficient may be calculated from the relation,

$$C_D = \frac{D}{\frac{1}{2}\rho V^2 \left(\frac{\pi d^2}{4}\right)}$$

where d is the sphere diameter.

- The sphere drag coefficient is plotted against Reynolds number Re, as shown in Figure 3.29.

- The Re at which $C_D = 0.30$ is noted and termed the critical Reynolds number, Re_c.

For the drag sphere placed in an undisturbed flow, the theoretical value of $C_D = 0.3$ occurs at the theoretical critical Reynolds number of $Re = 385,000$. Therefore,

$$\boxed{TF = \frac{385,000}{Re_c}} \tag{3.47}$$

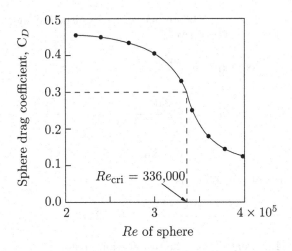

Figure 3.29: C_D versus Re for drag sphere

where Re_c is the actual critical Reynolds number for the tunnel. Using this equation and Equation (3.46), the effective Reynolds number can be calculated.

Example 3.11

A subsonic wind tunnel of square test-section runs at 30 m/s, with pressure 97.325 kPa and temperature 22°C, in the test-section. A turbulence sphere with theoretical surface finish offering 4 percent blockage experiences critical Reynolds number at this state. Determine the test-section height.

Solution

The flow density in the test-section is

$$\rho = \frac{p}{RT} = \frac{97.325 \times 10^3}{287 \times 295} = 1.15\,\text{kg/m}^3$$

The viscosity of air at 22°C is

$$
\begin{aligned}
\mu &= 1.46 \times 10^{-6}\,\frac{T^{3/2}}{T + 111} \\[2mm]
&= 1.46 \times 10^{-6}\,\frac{(295)^{3/2}}{295 + 111} \\[2mm]
&= 1.822 \times 10^{-5}\,\text{kg/(m s)}
\end{aligned}
$$

For a turbulence sphere with theoretical surface finish, the theoretical critical Reynolds number is 385,000. Therefore,

$$385000 = \frac{\rho V d}{\mu}$$

$$d = \frac{385000 \times (1.822 \times 10^{-5})}{1.15 \times 30}$$

$$= 0.203 \,\mathrm{m}$$

The projected area of the sphere is

$$A = \frac{\pi \times 0.203^2}{4} = 0.0324 \,\mathrm{m}^2$$

The blockage is 4 percent. Thus, the test-section area is

$$A_{\mathrm{TS}} = \frac{0.0324}{0.04} = 0.81 \,\mathrm{m}^2$$

This gives the test-section height as $\boxed{0.9 \,\mathrm{m}}$.

3.6.4.2 Pressure Sphere

In this method, no force measurement is necessary. Therefore, the difficulty of finding the support drag is eliminated. The pressure sphere has an orifice at the front stagnation point and four equally spaced interconnected orifices at $22\frac{1}{2}$ degrees from the theoretical rear stagnation point. A lead from the front orifice is connected across a manometer to the lead from the four rear orifices. After the pressure difference due to the static longitudinal pressure gradient is subtracted, the resultant pressure difference Δp for each Re at which measurements were made is divided by the dynamic pressure for the appropriate Re. The resulting dimensionless pressure difference, also known as the pressure coefficient, C_p ($C_p = \Delta p/q$), is plotted against Re, as shown in Figure 3.30.

It has been proved that, $\Delta p/q = 1.22$ corresponds to $C_D = 0.30$ and hence this value of $\Delta p/q$ determines the critical Reynolds number Re_c. The turbulence factor may now be found from Equation (3.46).

At this stage, it is proper to question the accuracy of the turbulence measurement with *turbulence sphere* or *pressure probe*, since we know that turbulence is first of all a *random phenomenon* and further the flow transition from laminar to turbulent nature takes place over a range of Reynolds numbers and not at a particular Reynolds number. The answer to this question is that, in all probability, the turbulence factor will change slightly with tunnel speed. This variation of turbulence factor may be obtained by finding it with spheres of several different diameters.

The turbulence will also vary slightly across the jet of the test-section. Particularly high turbulence is usually noted at the center of the jet of double-return

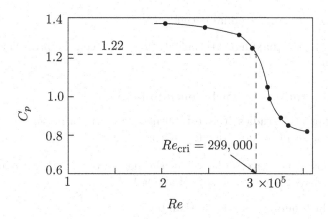

Figure 3.30: C_p variation with Re

tunnels, because this air has scraped over the walls of the return passage (Pope and Goin, 1965).

Turbulence factor varies from 1.0 to about 3.0. Values above 1.4 possibly indicate that (Pope and Goin, 1965) the air has very high turbulence for good testing results. Turbulence should be very low for research on low-drag airfoils. The variation of the turbulence factor with the degree of turbulence is shown in Figure 3.31.

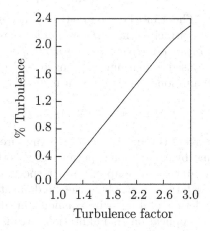

Figure 3.31: Turbulence factor variation with turbulence level

The turbulence in a tunnel may be kept at low level by the following means.

- Using the maximum number of fan blades.

- With anti-swirl vanes.

- With a very long, gradual nacelle.

- Providing the maximum possible distance between the propeller and test-section.

3.6.4.3   Limitations of Turbulence Sphere

From experimental studies it is established that the accuracy of turbulence sphere measurements is not adequate when

- The turbulence level is less than the degree of turbulence which corresponds to a turbulence factor of about 1.05.

- The Mach number is greater than 0.35.

Under these circumstances, we have to resort to devices which are capable of handling these situations. One such popular device is the hot-wire anemometer.

3.6.5   Hot-Wire Anemometer

The hot-wire anemometer can be used to measure freestream turbulence directly. It can be used for the measurement of turbulence at very low speeds also. It measures the instantaneous values of speed accurately and hence the turbulence, defined as

$$\text{Turbulence} = \frac{\text{Deviation from mean speed}}{\text{Mean speed}}$$

can be found directly. The time lag is practically negligible. The hot-wire anemometer works on the principle that *"the rate of heat loss from a wire heated electrically and placed in an air stream is proportional to the stream velocity"* (Hinze, 1987). The wire used is platinum or tungsten wire of about 0.015 mm diameter and about 10 mm long. The rate of heat dissipation H is given by

$$H = I^2 R = \left(A + B\sqrt{V} \right) (T - T_a) \tag{3.48}$$

where I is the electric current through the wire, R is the wire resistance, A and B are constants to be found by calibration experiments, V is the flow velocity, and T and T_a are the wire and room temperatures, respectively. Equation (3.48) is known as the King's formula, which is the fundamental relation for hot-wire anemometry. The details of a hot-wire anemometer, its working principle, measurement procedure, application, and limitations are discussed in Chapter 5.

3.6.6   Rakes

For the simultaneous measurement of a large number of total-head readings, a bank of total-head tubes, generally termed *rake* is commonly employed. Usually, brass or stainless steel tubes of diameter about 1.5 mm, arranged with a lateral spacing of around 3 mm on a single unit support, as shown in Figure 3.32, is used as a rake for measurements in wake behind a body and so on. It is particularly important that the lateral spacing be exact, since even a slight misalignment of a tube can cause considerable error in wake survey measurement. The total head tubes should be cut off to have square-edged entry, and their length is immaterial.

Even though the static pressure in the wakes rapidly reaches the freestream static pressure, it is advisable to make frequent checks of this. Hence, it is customary that several static tubes are kept along with the total-head tubes, as shown in Figure 3.32.

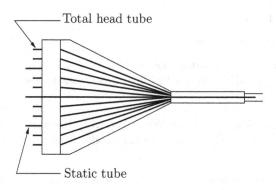

Figure 3.32: Total pressure rake

3.6.7 Surging

One of the most troublesome problems associated with wind tunnels is *tunnel surging*. It is *a low-frequency vibration in velocity that may run as high as 5 percent of the dynamic pressure "q"*. In practice, large numbers of tunnels suffer from this defect at one time or other; some live with it; some find a cure.

Surging is mainly due to the separation and reattachment of flow in the diffuser, and usually it can be cured by providing a substantial number of tripper strips in the diffuser. After a cure has been found, we may successively cut back on tripper numbers and size to save power. Another means to overcome surge is by employing some sort of boundary-layer control which corrects the surge rather than hides its effects.

The following are the difficulties caused by surging.

- Surging makes the wind tunnel balances try to keep up.

- Surging disturbs the reference pressure in pressure measurements.

- Surging makes the validity of assigning a Reynolds number to the test doubtful.

- It usually makes the dynamic testings impossible.

3.7　Wind Tunnel Balance

Basically, wind tunnel balance is a device to measure the actual forces and moments acting on a model placed in the test-section stream. Based on the constructional details, the wind tunnel balances are broadly classified into,

- Wire-type balance.
- Strut-type balance.
- Platform-type balance.
- Yoke-type balance.
- Strain gauge-type balance.

Irrespective of its type of construction, a wind tunnel balance should have certain basic features and characteristics for proper measurements. They are the following.

1. The balance should be capable of measuring the various loads (1, 2, 3, 4, 5, or 6 components) acting on the model with a very high degree of accuracy.

2. The interaction between the different load components should be kept small.

3. The balance should have provision to vary the angle of incidence, pitch, yaw, roll, etc., of the model which is mounted on the balance, within the normal range.

4. The balance and the supporting structures should be designed for very high rigidly, so that the deflection of the parts under the influence of maximum load is negligible.

5. Damping devices should be incorporated in the measuring system.

6. Use of bearings (ball, roller type, etc.) should as far as possible be avoided, since they cause large hysteresis and zero error.

The range of loads and accuracy for the balance are determined from the tunnel size, test-section velocity, model size, configuration, etc.

3.7.1　Wire Balances

In wire-type wind tunnel balances only wires are used to support the model. All the load components are transmitted to the measuring device by these wires. Wire-type balances are probably the simplest and easiest to build. But they have several disadvantages due to the use of too many bearings and bell crank systems, friction of the wires, and high damping requirements. Further, it is extremely bulky since the support system should be very rigid. The disadvantages of wire balances are the following.

- Large tare drag because of the exposed wires. It is difficult to streamline the wires and hence it is difficult to determine the magnitude of tare drag accurately.

- Bearings and linkages cause zero error.

- Wires have the tendency to crystallize and break.

- Space occupied is very large.

In spite of these disadvantages, the wire balances are still in use because of their simplicity and low cost. The usual arrangements for a wire-type balance are the following.

1. Models are tested in inverted position to avoid unloading of wires.

2. Since the wires can take only tension, it is essential to arrange the wires such that, for all types of loading (± lift, ± drag, etc.) the necessary wires are in tension. The number of wires required, in view of the above requirements, is considerable for securing a model.

3. Conventional three-point attachment is provided on the model. Pulleys, ball bearings, lever arms, etc., are freely used in the transfer of loads to the measuring devices.

4. The support structure is usually very heavy and bulky to give the necessary rigidity.

A typical six-component wire balance, supporting an airfoil model, is shown in Figure 3.33.

The balance shown in Figure 3.33 mounts the models at the quarter chord. The forces and moments acting on the wing section mounted on the balance are

$$\text{Lift} \quad = \quad L_W + M_F$$

$$\text{Drag} \quad = \quad D$$

$$\text{Side force} \quad = \quad C + E + F$$

$$\text{Pitching moment} \quad = \quad -M_F \times l$$

$$\text{Rolling moment} \quad = \quad (C - F)(h/2)$$

$$\text{Yawing moment} \quad = \quad -E \times l$$

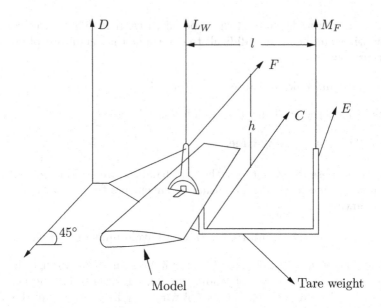

Figure 3.33: A wire balance

3.7.2   Strut-Type Balances

The wire-type balances are seldom used in large tunnels because of the disadvantages associated with them. The strut-type balances are proved to be suitable for such tunnels. Based on the structural construction, strut-type balances are broadly classified into

1. Yoke-type.

2. Platform-type.

3. Pyramid-type.

As indicated by the name, struts are used to support the model and transmit the loads to the measuring devices. Conventionally, models are mounted on the balance with three-point mounting (two on the wings and one on the rear fuselage for aircraft models, for instance). A typical strut-type balance with an aircraft model mounted on it is shown schematically in Figure 3.34.

Advantages of strut-type balances are the following.

1. The struts being rigid, their deflection can be kept at very small value.

2. The arrangement enables the struts to be faired so that the tare and interference drag are minimized.

3. By using cross spring pivot instead of knife-edge ball bearings, etc., in the beam, the zero error can be minimized.

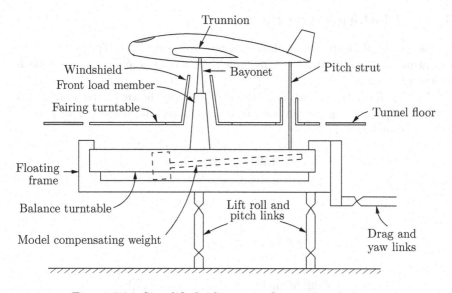

Figure 3.34: Simplified schematic of strut-type balance

4. By choosing proper linkages, interaction loads can be reduced to a small value.

5. Weight of the support structure can be kept very low, since efficient (tubular, for instance) members can be easily adapted.

Schematic of the cross spring pivot used in strut-type balances is shown in Figure 3.35.

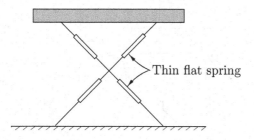

Figure 3.35: Cross spring pivot

The principle of load measurement for yoke-, platform-, and pyramid-type are the same, and only the method of model suspension is different in them.

3.7.3   Platform-Type Balance

Consider the platform-type balance shown in Figure 3.36. There is only one mounting point on the fuselage, by which the model is mounted; thus, the interference is minimized.

The platform is supported at three points, as shown in the Figure 3.36. Model is supported on the platform by a single streamlined strut. The loads are calculated as follows.

$$\text{Lift} \;=\; Z_a + Z_b + Z_c$$

$$\text{Drag} \;=\; X_d + X_c$$

$$\text{Side force} \;=\; Y_f$$

$$\text{Rolling moment} \;=\; M_x = (Z_a - Z_b)\,\frac{l}{2}$$

$$\text{Pitching moment} \;=\; M_y = Z_c \times m$$

$$\text{Yawing moment} \;=\; M_z = (X_e - X_d)\,\frac{l}{2}$$

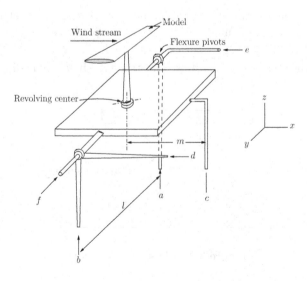

Figure 3.36: Platform-type balance

3.7.4 Yoke-Type Balance

Consider the yoke-type balance shown in Figure 3.37. Measurements of Z_A, Z_B, X_C, X_D, X_E, and Y_F are made for finding the six components of the load acting on the model. The model is mounted on yoke, as shown in Figure 3.37.

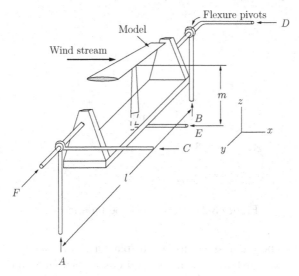

Figure 3.37: Yoke-type balance

The loads are calculated as follows.

$$\text{Lift} = Z_A + Z_B$$

$$\text{Drag} = X_C + X_D + X_E$$

$$\text{Side force} = Y_F$$

$$\text{Rolling moment} = (Z_A - Z_B)\frac{l}{2}$$

$$\text{Pitching moment} = X_E \times m$$

$$\text{Yawing moment} = (X_C - X_D)\frac{l}{2}$$

3.7.5 Pyramid-Type Balance

In pyramid-type balances, the model mounting is about one point connecting the model to the platform (by four arms), as shown in Figure 3.38.

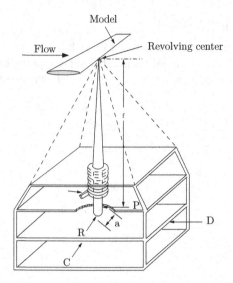

Figure 3.38: Pyramid-type balance

The pyramid-type balance, compared to platform- or yoke-type, has a specific advantage. In this balance, the forces and moments acting on the model are measured with respect to a single point (attachment point), and hence by locating this point at an advantageous position like aerodynamic center of model, the readings obtained are free from error associated with the conversion of a measured quantity to the required quantity. An accuracy of the order of ±0.1 percent, at full load, can easily be achieved with a pyramid-type balance.

3.7.6   Strain Gauge Balance

Balances with strain gauges as load-sensing elements are termed *strain gauge balances*. Based on strain gauge fixing on the model, the strain gauge balances are generally classified as

1. Internal balance.

2. Semi-internal balance.

3. External balance.

In an internal balance, all the measuring elements are located inside the model; in a semi-internal balance, the measuring elements are located partially inside and partially outside; and in an external balance, all the measuring elements are located outside the model. The types of strain gauges commonly used in wind tunnel balances are shown in Figure 3.39.

The gauges consist of a grid of very fine wire (10 to 30 microns in diameter) or very thin foil (thickness less than 30 microns) embedded on a sheet of bakelite

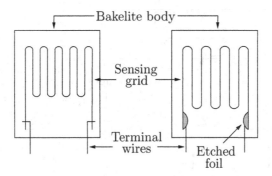

Figure 3.39: Strain gauge balance

having a thickness comparable to that of thick paper. The grid material is usually constantan, nichrome, or nichrome with small additions of iron and aluminum. The grid length varies from about 1 mm to several centimeters.

3.7.6.1 Strain Gauge Operation Theory

When the bakelite body of the gauge is intimately connected to the surface of a structure, it will stretch or contract as the outer fibers of the structure to which it is attached. The grid wires embedded in the bakelite will stretch or contract with the bakelite body and thus with the outer fibers of the structure. As the grid wires are stretched, their cross-sectional area decreases, causing an increase in electrical resistance. Similarly, as the grid wires are compressed, their cross-sectional area increases, causing a decrease in electrical resistance. In both events the changes in resistance are actually greater than the changes that the area would indicate, because of the change in the length. From experience it has been found that, the change in resistance of the types of strain gauges normally used in wind tunnel balances is directly proportional to the stress in the outer fibers of the structure to which it is attached.

From the above discussion, it is evident that great care must be exercised in the installation of the strain gauge. Strain gauges in wind tunnel balances are normally located on a member in which the desired component of loading is a bending moment. A typical strain gauge installation is schematically shown in Figure 3.40.

Two gauges are placed side by side on the compression as well as the tension surface of the member. The four gauges are wired together into a bridge circuit, as shown in Figure 3.41.

The supply voltage is generally between 5 and 10 volts and the current may be either direct or alternating.

The signal voltage from a strain gauge bridge can be calculated using Ohm's law, as follows. The current flow through gauges 1 and 3 and through gauges 2 and 4 of Figure 3.41 are

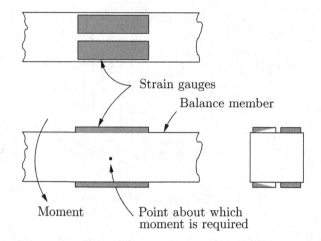

Figure 3.40: Strain gauge installation on a balance member

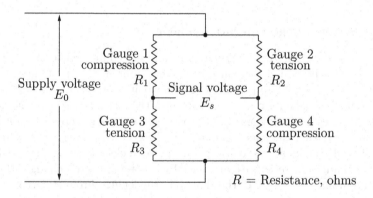

Figure 3.41: Strain gauges in a bridge circuit

$$I_{13} = \frac{E_0}{R_1 + R_3}$$

$$I_{24} = \frac{E_0}{R_2 + R_4}$$

The drops in the voltage across gauges 1 and 2 are

$$\Delta E_1 = I_{13} R_1 = E_0 \frac{R_1}{R_1 + R_3}$$

$$\Delta E_2 = I_{24} R_2 = E_0 \frac{R_2}{R_2 + R_4}$$

The signal voltage is equal to

$$E_s = (E_0 - \Delta E_1) - (E_0 - \Delta E_2)$$

This can be reduced to

$$\frac{E_s}{E_0} = \frac{R_2 R_3 - R_1 R_4}{(R_1 + R_3)(R_2 + R_4)}$$

with the matched gauges on a symmetrical section of the balance having both axial and bending loads, the resistances of the individual gauges are

$$R_1 = R_4 = R_0 - \Delta R_b + \Delta R_a$$

$$R_2 = R_3 = R_0 + \Delta R_b + \Delta R_a$$

where R_0 is the initial resistance of the gauge, ΔR_a and ΔR_b are the increments in the gauge resistance due to the axial and bending stresses, respectively. Using these, we get

$$\boxed{\frac{E_s}{E_0} = \frac{\Delta R_b}{R_0 + \Delta R_a}} \tag{3.49}$$

Since ΔR_a is normally very small compared to R_0, it can be neglected in the above equation.

3.7.6.2 Basic Equations of Strain Gauge Transducer

The fundamental equations associated with strain gauges are

$$GF = \frac{\Delta R / R}{\Delta L / L} \tag{3.50}$$

$$e = \frac{\sigma}{E} = \frac{\Delta L}{L} \tag{3.51}$$

where GF is the gauge factor (strain gauge), R is the resistance of strain gauge, L is the length of the strain gauge, ΔR and ΔL are the changes in resistance and length, respectively, due to load, E is Young's modulus, σ is stress, and e is strain.

From Equations (3.56) and (3.57), the stress can be expressed as

$$\sigma = \frac{\Delta R}{R} \frac{E}{GF} \tag{3.52}$$

From this equation it is seen that, for computing the load on a member an accurate measurement of ΔR only is necessary. Wheatstone-type bridge circuit (modified) is used for this measurement.

The Dynamometer Block

This unit is designed as a system of transducers to convert the six components of load: the normal force, tangential force, side force, rolling moment, pitching moment, and yawing moment, into corresponding electrical quantities.

3.7.6.3 Strain Gauge Signal–Measuring Devices

The transducer, namely the strain gauge used here is a device to convert the quantity to be measured (force, etc.) into an electrical quantity. For a detailed description of devices for measuring strain gauge signals the reader is encouraged to refer to books devoted to this topic. However, a brief description of the principle involved is given in the following subsection. The measurement principle is as shown below.

$$\boxed{\text{Dynamometer block}} \rightarrow \boxed{\text{Balancing bridge}} \rightarrow \boxed{\text{Recorder}}$$

3.7.6.4 Balancing Bridge

A typical balancing bridge circuit is shown in Figure 3.42.

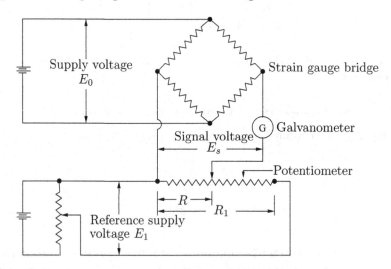

Figure 3.42: Balancing bridge circuit

The principle involved here is the comparison of strain gauge signal voltage with a known reference voltage which is varied until the reference and the signal voltages are equal. A voltage E_1 is applied across the resistance of a potentiometer. This voltage is a small fraction of the voltage, E_0, applied to the strain gauge bridge, but larger than the strain gauge signal voltage, E_s. The voltage E_1 is divided to provide a potential $E_{ref} = E(R/R_1)$ between one end of the resistor and the movable contact of the potentiometer. One of the strain gauge signal leads is attached to the end of the resistor and the other to the movable contact of the potentiometer, through a galvanometer. The movable contact of the potentiometer is moved until the galvanometer indicates no current flow. At this point of zero current flow, the reference potential E_{ref} equals the strain gauge signal voltage E_s. Knowing the voltage E_1 and the variation of potentiometer resistance with movable contact position, the signal variation from contact position and the relation $E_s = E_{ref} = E_1(R/R_1)$ can be determined.

In wind tunnel operations, with a system of the type described above, the current flow that operates the galvanometer is amplified and drives a motor. The motor drives a movable contact of the potentiometer to a null position. Additional circuitry is provided to eliminate the necessity of reversing signal leads when the signal voltage from the strain gauge changes polarity. Though the modern measuring systems are highly sophisticated compared to the one described, the basic principle underlying them is the same as that briefed above.

3.7.7 Balance Calibration

All balances (wire, strut, or strain gauge type) are required to be calibrated after assembly, and checked periodically. It is customary to check the calibration before running the wind tunnel for a model test. The calibration of wind tunnel balances consists of the following procedure.

1. Applying known loads, stepwise, in fixed directions for each component and noting the balance readings. The load is usually applied by means of a wire pulley arrangement or a pull rod with flexer pivots at the end depending on the size of the unit. Alignment of the load applying unit (wire or pull rod) should be adjusted very accurately. Dead weights (standardized units 10 kg $\pm$ 5 g, 5 kg $\pm$ 1 g etc.) are used for varying the magnitude of the load.

2. For each component of the load applied, the readings of all the indicators are noted after balancing the system.

3. In a similar manner, the experiments for other load components are carried out.

4. Combined loads are applied in discrete steps (3 or 6 components) and the various readings are noted.

5. From the test data, the interaction and percentage error of the different load components are computed and plotted in the form of graphs.

6. The deflection of the system (balance, support structure) should be measured under different loading conditions and the correction factor determined.

3.7.8   Wind Tunnel Boundary Correction

We know that the conditions under which a model is tested in a wind tunnel are not the same as those in free air. Even though there is practically no difference between having the model stationary and the air moving instead of vice versa, there is longitudinal static pressure gradient usually present in the test-section, and the open or closed jet boundaries in most cases produce extraneous forces that must be subtracted from the measured values.

The variation of static pressure along the test-section produces a drag force known as *horizontal buoyancy*. It is usually small in closed test-section, and is negligible in open jets.

The presence of the lateral boundaries around the test-section (the walls of the test-section) produces the following.

- A lateral constraint to the flow pattern around a body is known as *solid blocking*. For closed throats solid blocking is the same as an increase in dynamic pressure, increasing all forces and moments at a given angle of attack. For open test-sections it is usually negligible, since the air stream is free to expand.

- A lateral constraint to flow pattern about the wake is known as *wake blocking*. This effect increases with increase of wake size. For closed test-section wake blocking increases the drag of the model. Wake blocking is usually negligible for open test-sections.

- An alteration to the local angle of attack along the span.

3.8   Internal Balance

In aerospace applications, wind tunnels are used to test models of aircraft and engine components. During a test, the model is placed in the *test-section* of the tunnel and air, at desired Mach numbers, is made to flow past the model, to measure the pressure distribution over the model and the aerodynamic forces, such as lift, drag, and side force, acting on the model, and moments, such as rolling, pitching, and yawing moments. The instrument used for measuring the pressure distribution is a manometer or pressure transducer. For measuring the forces and moments, usually wind tunnel balance is used. We must measure six components, three forces (lift, drag, and side force) and three moments (pitching, rolling, and yawing moments), to completely describe the conditions on the model when subjected to the flight condition.

The wind tunnel balances can broadly be classed as *external balance* and *internal balance*, depending on their location. The balance located external to the model and the test-section is referred to as external balance and that placed inside the model is called internal balance. The location of the balance affects the choice of mounting system for the model and the data reduction necessary to determine the aerodynamic forces.

We saw some of the commonly used external balances in Section 3.7. Now let us have a look at the internal balance. Due to the limited space available inside the model itself, internal balances have to be relatively small in comparison to external balances. There are two main types of internal balances used in wind tunnel measurements. They are the *monolithic type*, in which the balance body consists of a single piece of material, designed such that certain areas are primarily stressed by the applied loads. The other internal balance type uses small transducers which are oriented with their sensing axes in the direction of the applied loads. Such a balance is combined into a solid structure. A balance measures the total model loads and therefore is placed at the center of gravity of the model and is generally constructed from one solid piece of material.

The word *load* in wind tunnel measurement refers to both the applied forces and moments. The task of a balance is to measure the aerodynamic loads, which act on the model or on the components of the model. In total there are six different components of aerodynamic loads, three forces in the direction of the coordinate axes and the three moments around these axes. These force and moment components are measured in a certain coordinate system which can be either fixed to the model or to the wind tunnel. For the measurement of loads on aircraft model parts such as rudders, flaps and missiles, normally fewer than six components are required.

3.8.1 Coordinate Systems

One possible coordinate system is fixed to the wind tunnel-the *wind axis system*-and is aligned to the main flow direction. The *lift force* is generally defined as the force on the model acting vertical to the main flow direction, whereas the *drag force* is defined as the force acting in the main flow direction. This definition is common all over the world. However, the definition of the positive direction of the forces is not universal. In the USA, lift (normal force, NF) and drag (axial force, AF) are defined positive, as illustrated in Figure 3.43.

But in Europe, weight and thrust are defined as positive in the model-fixed axis system, as shown in Figure 3.44.

To form a right-hand axis system, the side force in the USA (Figure 3.43) has to be positive in the starboard direction, as shown in Figure 3.44. The definitions of the positive moments do not follow the sign rules of the right-hand system. The pitching moment is defined as positive turning right around the y-axis, but yawing and rolling moments are defined positive turning left around their corresponding axes. This makes this system inconsistent in a mathematical sense.

The European axis system is consistent with the right-hand system. A balance which always stays fixed in the tunnel, and relative to the wind axis system, always gives the pure aerodynamic loads on the model.

In the case of the model-fixed axis system, the balance does not measure the aerodynamic loads directly. The loads acting on the model are given by the balance and the pure aerodynamic loads must then be calculated from these components by using the correct yaw and pitch angles. The difference between

American and European definitions of the positive direction remains the same in this case.

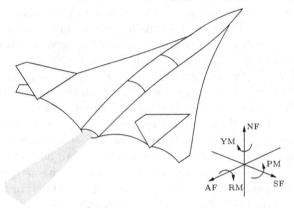

Figure 3.43: Illustration of wind axis system; AF-aerodynamic force, NF-normal force, SF-side force, PM-pitching moment, RM-rolling moment, YM-yawing moment

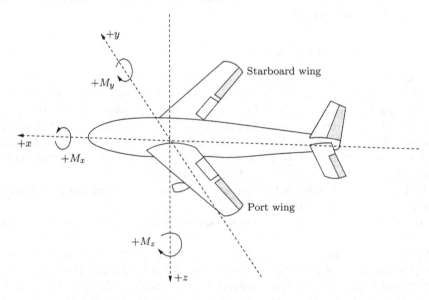

Figure 3.44: Illustration of model-fixed system

3.8.2 Specification of Balance Load Ranges

For designing a balance, the specifications of the load ranges and the space available for placing the balance are required. The maximum combined loads specify the load ranges for the balance design. The maximum design loads of a balance are defined in various manners. For example, if several loads act simultaneously, then the load range must be specified as the *maximum combined load*. If the maximum load acts alone, the load range is defined then as the *maximum single load*. Usually such single loads do not exist in wind tunnel tests and combined loads must be expected. Such combined loads stress the balance in a much more complicated manner and therefore deserve very careful attention. The stress analysis of the balance has to be taken into account in this situation. Furthermore, the combination of two loads usually requires that the balance carries higher loads. To determine which balance can be used for a given setup, the balance manufacturer provides the test engineer with loading diagrams which define the maximum load combinations for various available models.

An aircraft model mounted on a 3-component sting balance is shown in Figure 3.45.

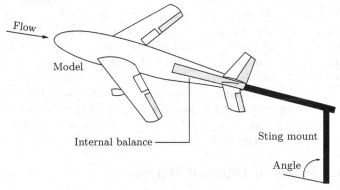

(a) An aircraft model with internal balance.

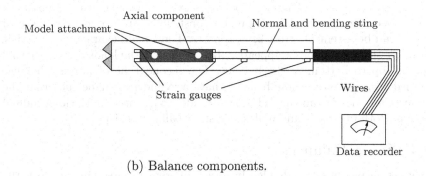

(b) Balance components.

Figure 3.45: An aircraft model mounted on a 3-component sting balance

As shown in Figure 3.45(a), a plane model is attached to a sting and placed in the test-section of a wind tunnel. Figure 3.45(b) shows the details of the model attachment to the sting. The model is actually attached to a three-component balance system and the balance system is attached to the sting. The three-component balance can detect the axial and perpendicular, or normal, forces and the bending along an axis perpendicular to the axial and normal directions. From these measurements one can derive the lift, drag, and pitch of the model, but cannot determine the side force, roll, or yaw. Forces on the model are detected by strain gauges located on the balance.

Each strain gauge measures a force by the stretching of an electrical element in the gauge. The stretching changes the resistance of the gauge which changes the measured current through the gauge according to Ohm's law. Wires carry electricity to the gauges through the hollow sting and carry the resulting signal back through the sting to recording devices in the control room.

Multiple strain gauges are arranged on the balance to account for temperature changes on the model during the test. A Wheatstone bridge electrical circuit is used to provide temperature compensation. Because this example uses only a three-component balance, the model must be aligned with the flow in the tunnel to eliminate the side force, roll, and yawing moments. We can only vary the angle of attack of the model, as shown at the left bottom of Figure 3.45(a). With an internal balance, the forces are measured in a coordinate system attached to the model. The resulting measurements must be corrected before determining the lift, drag, and pitching moment in the tunnel coordinate system. When using a sting with an internal balance, the aft geometry of the model is often modified for proper attachment of the sting.

3.8.3   Types of Internal Balances

We saw that balance placed inside the model is referred to as internal balance. It is evident that the internal balances need to be relatively smaller than the external balances, because only a limited space is available inside the model for placing the balance.

Two general groups of internal balances exist. The first group consists of the so-called *box balance*. These can be constructed from one solid piece of metal or can be assembled from several parts. Their main characteristic is that their outer shape most often appears cubic, such that the loads are transferred from the top to bottom of the balance. The second type of internal balance is termed *sting balance*. These balances have a cylindrical shape such that the loads are transferred from one end of the cylinder to the other in the longitudinal direction. Both one-piece and multipiece sting balances exist.

3.8.4   Sting Balances

Internal sting balances are divided into two different groups. One group is the force-balance type and the other group is the *moment-balance type*. If the bridge output is directly proportional to one load component then these balances are termed *direct read balances*. Typically for all groups, the axial force and rolling moment are directly measured with one bridge. The measurements of lift force and pitching moment or for side force and yawing moment are done in different ways characterizing each group.

Moment-type balances and force-type balances have one main feature in common, being the lack of a direct output proportional to lift/pitch and side force/yaw. The signals which are proportional to each of these loads should be calculated by summing or subtracting the signal from one another, before being fed into the data reduction process. The advantage is that the associated concentrated wiring on each section is much less sensitive to temperature effects.

3.8.4.1 Force Balances

This type of balance uses two measurement sections placed in the forward and the aft section of the balance. In these measurement sections forward and aft forces are measured most often through tension and compression transducers. These forward and aft force components are used to calculate the resulting force in the plane as well as the moment around the axis (perpendicular to the measurement plane). Schematic diagram of a typical force balance is shown in Figure 3.46.

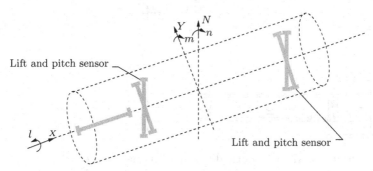

Figure 3.46: Schematic diagram of force balance with tension transducers in forward and aft sections; X-axial force, Y-side force, N-normal force, l-rolling moment, m-pitching moment, n-yawing moment

3.8.4.2 Moment-Type Balances

Moment-type balances have a bending moment measuring section in the front as well as in the aft regions of the balance (S_1 and S_2 in Figure 3.47). The measurement of the two bending moments (S_1 and S_2) is used to obtain a signal which is proportional to the force in the measurement plane and a second one which is proportional to the moment around the axis (perpendicular to the measurement plane). The stress distribution shows that the moment M_y (M_z) is proportional to the sum of S_1 and S_2. However, the force F_y (F_z) is proportional to the difference in the signals S_1 and S_2.

To measure the rolling moment (M_x) one bending section must be applied with shear stress gauges to detect the shear stress τ. The most complicated part of the balance is the axial force section which consists of flexures and a bending

beam to detect axial force. These flexures enable axial movement while carrying
the other loads.

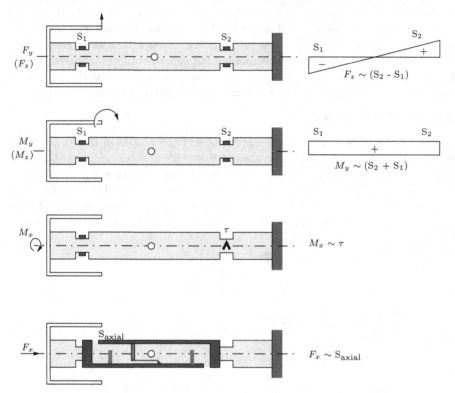

Figure 3.47: Schematic diagram of moment-type balance

3.8.4.3   Direct-Read Balances

A direct-read balance can be categorized as either a force-balance type or as
a moment-balance type. Instead of measuring a force or a bending moment at
each section separately, half bridges on every section are directly wired to a
moment bridge while the other set of half bridges are directly wired to a force
bridge. Thus the difference between direct-read balances and the other types
is only in the wiring of the bridges. The disadvantage of such a wiring is the
length of the wires from the front to the aft ends. Temperature changes inside
these wires cause errors in the output signals.

3.8.4.4   Box Balances

The main difference between the box and sting balances are the model and sting
attachment area. The load transfer in such balances is from top to bottom along
the vertical-axis. Therefore these balances use a central sting arrangement, as
shown in Figure 3.48(a), for the case of an airplane configuration. The monopiece

balance is constructed from one single piece of material. The advantages of this relatively expensive manufacturing process are the low hysteresis and good creep behavior, which are basic requirements for good balance repeatability. Multipiece box balances are built from several parts which can in turn be manufactured separately. The load transducers can either be integrated in the structure or separate load cells can be used instead (as shown in Figure 3.48(b)). This enables a parallel manufacturing process with a final assembly at the end, in turn making the whole process quicker than that for a monopiece balance. Box-type balances are considered internal balances, but have actually more in common with semi-span balances. In particular their temperature-sensitivity behavior is similar to that of semi-span balances.

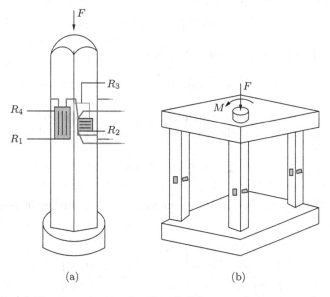

(a)   (b)

Figure 3.48: (a) Tension-type load cell and (b) cage of tension and compression columns

Semi-span models are used to increase the effective Reynolds number of the tests by increasing the geometry of the model. Besides the higher Reynolds number, the larger model size makes the models with variable flaps and slats much easier to construct, thus semi-span models are often used for the testing of takeoff and landing configurations.

3.8.5   Shear–Type Load Cells

The maximum shear stress on a cantilever beam, shown in Figure 3.49, produced by force (F), appears at the center line of the beam at $45 \pm \delta\theta$ degree, where the tension of compression stresses is nearly zero. In such a case the maximum strain also appears at $45 \pm \delta\theta$ degree. Therefore to get the maximum output the gauges must be bonded as shown in Figure 3.49. The dimensions of the beam can be calculated by determining the stresses expressed as

$$\tau_{\max} = \frac{Fc}{A} \ \text{newton/mm}^2$$

where F is the applied force, $A = bh$ mm^2 is the beam cross-section, and c is a form factor which depends on the shape of the beam cross-section. For a rectangular cross-section $b \le h/2$ and $c = 3/2$, whereas for a circular cross-section $c = 4/3$. These formulas are approximations for the centerline. The strain gauge covers a certain area around the center line such that the integrated value measured by the strain gage will be smaller. The signal of the transducer can be determined using the equation

$$\Delta U/U = k\epsilon_{45^\circ}$$

for the full bridge output as well as the equation

$$\epsilon_{45^\circ} = \frac{\tau_{\max}}{2G}$$

for the strain under the gauge, where G is the shear modulus.

For a torque transducer the shear stress must be calculated using

$$\tau_{\max} = \frac{M_t}{W_p} \ \text{newton/mm}^2$$

where M_t (newton-meter) is the torque moment and W_p mm^3 is the polar section modulus.

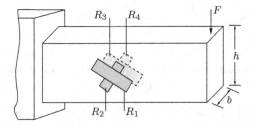

Figure 3.49: Shear-type load cell on a cantilever beam

3.8.6   Tension and Compression Load Cells

The main difference between shear- and bending-stress transducers and load cells using tension and compression stress measurements is that in the case of the latter there is no positive and negative stress of equal value. If the tension stress $\sigma_{\text{tension}} = F/A$ is defined as positive, the negative compression stress becomes $\sigma_{\text{comp}} = \nu F/A$, termed the Poisson stress. For metals $\nu = 0.3$ while the compression stress is only about $1/3$ of the tension stress. Using a Wheatstone bridge with four gauges applied, the output of such a transducer related to the force (F) will be nonlinear. This is not a big disadvantage as long as these nonlinearities are taken into account through calibration.

For standard load cells a nonlinear characteristic is not common, however for applications in wind tunnel testing they are sometimes advantageous since they do not require much space. Another advantage of such a transducer is that the strain gauges are placed very close to each other, thus remaining at the same temperature level. This minimizes the zero-drift and temperature-gradient sensitivity of the transducer. In some wind-tunnel balances such tension and compression columns can be arranged in a cage. In such a way it is possible to measure tension and compression as well as the moments acting on the transducer.

3.8.7 Loads

Both the applied forces and moment are referred to as *loads* in balance measurement. The task of a balance is to measure the aerodynamic loads, which act on the model or on components of the model itself. In total there are six different components of aerodynamic loads, three forces in the direction of the coordinate axes and the moments around these axes themselves. These components are measured in a certain coordinate system which can be either fixed to the model or to the wind tunnel. For the measurement of loads on model parts such as rudders, flaps, and missiles, normally less than six components are required.

3.9 Calibration of Supersonic Wind Tunnels

Supersonic tunnels operate in the Mach number range of about 1.4 to 5.0. They usually have operating total pressures from about atmospheric to 2 MPa ($\approx 300\,\text{psi}$) and operating total temperatures of about ambient to 100°C. Maximum model cross-section area (projected area of the model, normal to the test-section axis) of the order of 4 percent of the test-section area is quite common for supersonic tunnels. *Model size is limited by tunnel chocking and wave reflection considerations.* When proper consideration is given to choking and wave reflection while deciding on the size of a model, there will be no effects of the wall on the flow over the models (unlike low-speed tunnels), since the reflected disturbances will propagate only downstream of the model. However, there will be a buoyancy effect if there is a pressure gradient in the tunnel. Luckily, typical pressure gradients associated with properly designed tunnels are small, and the buoyancy effects in such tunnels are usually negligible. The Mach number in a supersonic tunnel with solid walls cannot be adjusted, because it is set by the geometry of the nozzle. Small increases in Mach number usually accompany large increases in operating pressure (the stagnation pressure in the settling chamber in the case of constant back pressure or the nozzle pressure ratio in the case of blowdown indraft combination), in that the boundary layer thickness is reduced and consequently the effective area ratio is increased.

During calibration as well as testing, the condensation of moisture in the test

gas must be avoided. To ensure that condensation will not be present in significant amounts, the air dewpoint in the tunnel should be continuously monitored
during tunnel operation. The amount of moisture that can be held by a cubic
meter of air increases with increasing temperature, but is independent of the
pressure. The moist atmospheric air cools as it expands isentropically through
a wind tunnel. The air may become supercooled (cooled to a temperature below the dewpoint temperature) and the moisture will then condense out. If the
moisture content is sufficiently high, it will appear as a dense fog in the tunnel.

Condensation causes changes in local Mach number and other flow characteristics such that the data taken in a wind tunnel test may be erroneous. The
flow changes depend on the amount of heat released through condensation. The
functional dependence of Mach number and pressure on the heat released may
be expressed as

$$\frac{dM^2}{M^2} = \frac{1 + \gamma M^2}{(1 - M^2)} \left(\frac{dQ}{H} - \frac{dA}{A} \right)$$

$$\frac{dp}{p} = -\frac{\gamma M^2}{(1 - M^2)} \left(\frac{dQ}{H} - \frac{dA}{A} \right)$$

where M is the Mach number, γ is the specific heats ratio, dQ is the heat added
through condensation, H is the enthalpy, A is the duct cross-sectional area, and
p is the static pressure.

From the above relations it is seen that, at subsonic speeds the Mach number
decreases and the pressure increases with condensation, whereas at supersonic
speeds the reverse is taking place. It is important to note that the presence of
water vapor without condensation is of no significance as for as the temperature
ratio, pressure ratio, and Mach number determination with the isentropic relations are concerned. The condensation depends on the amount of moisture in
the air stream, the static temperature of the stream, the static pressure of the
stream, and the time duration during which the stream is at a low temperature.

The amount of moisture that may be contained in normal atmospheric air
usually varies in the range of 0.004 to 0.023 kg/(kg of dry air). The air temperature at the test-section of a supersonic tunnel is usually quite low. For example,
let us assume that the total temperature of the stored air is 40°C. From isentropic table (for $\gamma = 1.4$), we can see that for a test-section Mach number of
1.0, the static temperature is −12°C, for $M = 1.5$, $T = -57$°C, and will go
to a low value of −220°C at $M = 5.0$. The static temperatures reached during
expansion of air at a stagnation temperature of 40°C to Mach number above 1
are considerably below the dewpoint temperatures, normally found in the atmosphere. Hence, the static temperature in a supersonic tunnel can easily be low
enough to condense out normal atmospheric water vapor. The static pressure
in a wind tunnel drops more rapidly than the static temperature, with Mach
number increase.

The condensation of moisture in an air stream is the result of molecules
colliding and combining and eventually building up into droplet size. The likelihood of condensation in a supersonic wind tunnel with supercooling (cooling of

the air below the dewpoint) of less than $-12°C$ is negligible. Since condensation is a result of a gradual buildup from molecular to droplet size due to molecular collisions, it is obviously a time-dependent process.

There are two ways of solving the problem of condensation in supersonic tunnels. The first is to heat the air so that upon expansion to the desired Mach number, its static temperature will be above the temperature corresponding to $-12°C$ of supercooling. But this approach requires heating of air to high temperatures, and hence is impractical. For example, with a $3°C$ dewpoint, $-12°C$ of supercooling would correspond to a static air temperature of $-9°C$. If this occurs at Mach 2.0, the total temperature is required to be $202°C$. The total temperature required would increase rapidly with increasing Mach number.

The second method is to dry the air before storing it in the pressure tanks. Being simpler, this is the commonly used approach. Equipment for drying air to dewpoints in the neighborhood of $-20°C$ is commercially available and not expensive.

3.9.1 Calibration

The calibration of a supersonic wind tunnel includes determining the test-section flow Mach number throughout the range of operating pressure of each nozzle, determining flow angularity, and determining an indication of turbulence level effects.

3.9.2 Mach Number Determination

The following methods may be employed for determining the test-section Mach number of supersonic wind tunnels.

- Mach numbers from close to the speed of sound to 1.6 are usually obtained by measuring the static pressure (p) in the test-section and the total pressure (p_{01}) in the settling chamber and using the isentropic relation

$$\frac{p_{01}}{p} = \left(1 + \frac{\gamma - 1}{2} M^2\right)^{\frac{\gamma}{\gamma - 1}}$$

- For Mach numbers above 1.6, it is more accurate to use the pitot pressure in the test-section (p_{02}) with the total head in the settling chamber (p_{01}) and the normal shock relation.

$$\frac{p_{02}}{p_{01}} = \left[1 + \frac{2\gamma}{\gamma + 1}\left(M_1^2 - 1\right)\right]^{-\frac{1}{\gamma - 1}} \left[\frac{(\gamma + 1)\, M_1^2}{(\gamma - 1)\, M_1^2 + 2}\right]^{\frac{\gamma}{\gamma - 1}}$$

- Measurement of static pressure, p_1, using a wall pressure tap in the test-section and measurement of pitot pressure, p_{02}, at the test-section axis, above the static tap can be used through the Rayleigh pitot formula,

$$\frac{p_1}{p_{02}} = \frac{\left(\dfrac{2\gamma}{\gamma+1} M_1^2 - \dfrac{\gamma-1}{\gamma+1} \right)^{\frac{1}{\gamma-1}}}{\left(\dfrac{\gamma+1}{2} M_1^2 \right)^{\frac{\gamma}{\gamma-1}}}$$

for accurate determination of the Mach number.

- Measurement of shock wave angle, β, from schlieren and shadowgraph photograph of flow past a wedge or cone of angle θ can be used to obtain the Mach number through $\theta - \beta - M$ relation,

$$\tan \theta = 2 \cot \beta \left[\frac{M_1^2 \sin^2 \beta - 1}{M_1^2 (\gamma + \cos 2\beta) + 2} \right]$$

- The Mach angle, μ, measured from a schlieren photograph of a clean test-section can also be used for determining the Mach number with the relation

$$\sin \mu = \frac{1}{M_1}$$

For this the schlieren system used must be powerful enough to capture the Mach waves in the test-section.

- Mach number can also be obtained by measuring pressures on the surface of cones or two-dimensional wedges, although this is rarely done in calibration.

3.9.3   Pitot Pressure Measurement

Pitot pressures are measured by using a pitot probe. The pitot probe is simply a tube with a blunt end facing into the air stream. The tube will normally have an inside to outside diameter ratio of 0.5 to 0.75, and a length aligned with the air stream of 15 to 20 times the tube diameter. The inside diameter of the tube forms the pressure orifice. For test-section calibration, a rake consisting of a number of pitot probes is usually employed. The pitot tube is simple to construct and accurate to use. It should always have a squared-off entry and the largest practical ratio of hole (inside) diameter to outside diameter.

At this stage, it is important to note that an open-ended tube facing into the air stream always measures the stagnation pressure (a term identical in meaning to the "total head") it sees. For flows with Mach number greater than 1, a bow shock wave will be formed ahead of the pitot tube nose. Therefore, the flow reaching the probe nose is not the actual freestream flow, but the flow traversed by the bow shock at the nose. Thus, what the pitot probe measures is not the actual static pressure but the total pressure behind a normal shock (the portion of the bow shock at the nose hole can be approximated to a normal shock). This new value is called *pitot pressure* and in modern terminology refers to pressure measured by a pitot probe in a supersonic stream.

It will be seen in Chapter 7 that, the pressure measured by a pitot probe is significantly influenced by very low Reynolds numbers based on probe diameter. However, this effect is seldom a problem in supersonic tunnels, since a reasonable-size probe will usually have a Reynolds number well above 500. Reynolds numbers below this are the troublesome range for pitot pressure measurements.

3.9.4 Static Pressure Measurement

Supersonic flow static pressure measurements are much more difficult than the measurement of pitot and static pressures in a subsonic flow. The primary problem in the use of static pressure probes at supersonic speeds is that the probe will have a shock wave (either attached or detached shock) at its nose, causing a rise in static pressure. The flow passing through the oblique shock at the nose will be decelerated. However, the flow will continue to be supersonic, since all naturally occurring oblique shocks are weak shocks with supersonic flow on either side of them. The supersonic flow of reduced Mach number will get decelerated further, while passing over the nose-cone of the probe, since decrease of streamtube area would decelerate a supersonic stream. This progressively decelerating flow over the nose-cone would be expanded by the expansion fan at the nose-cone shoulder junction of the probe. Therefore, the distance over the shoulder is sufficient for the flow to get accelerated to the level of the undisturbed freestream static pressure, in order to measure the correct static pressure of the flow. The static pressure hole should be located at the point where the flow comes to the level of freestream Mach number. Here, it is essential to note that the flow deceleration process through the oblique shock at the probe nose and over the nose-cone portion can be made to be approximately isentropic, if the flow turning angles through these compression waves are kept less than 5° (*Gas Dynamics*, Rathakrishnan, 1995). The design and measurement procedure of static pressure probe are discussed in detail in Chapter 7.

Static pressures on the walls of supersonic tunnels are often used for rough estimation of test-section Mach numbers. However, it should be noted that the wall pressures will not correspond to the pressures on the tunnel center line if compression or expansion waves are present between the wall and the center line. When Mach number is to be determined from static pressure measurements, the total pressure of the stream is measured in the settling chamber simultaneously with the test-section static pressure. Mach number is then calculated using isentropic relation.

3.9.5 Determination of Flow Angularity

The flow angularity in a supersonic tunnel is usually determined by using either cone or wedge yaw meters. Sensitivities of these yaw meters are maximum when the wedge or cone angles are maximum. They work below Mach numbers for which wave detachment occurs, and are so used. The cone yaw meter is more extensively used than the wedge yaw meter, since it is easier to fabricate.

3.9.6 Determination of Turbulence Level

Measurements with a hot-wire anemometer demonstrate that there are high-frequency fluctuations in the air stream of supersonic tunnels that do not occur in free air. These fluctuations, broadly grouped under the heading of "turbulence," consist of small oscillations in velocity, stream temperature (entropy), and static pressure (sound). Some typical values of these fluctuations are given in Table 3.2.

Table 3.2 Turbulence level in the settling chamber and test-section of a supersonic tunnel

Parameter	Settling chamber	Test-section	
M	all	2.2	4.5
Sound, $\frac{\Delta p}{p}$	< 0.1 %	0.2 %	1 %
Entropy, $\frac{\Delta T}{T}$	< 0.1 %	< 0.1 %	< 1 %
Velocity, $\frac{\Delta V}{V}$	0.5 to 1 %	< 0.1 %	< 1 %

The pressure regulating valve, drive system, after cooler, and test-section boundary layer are the major causes for the fluctuations. Velocity fluctuations due to upstream causes may be reduced at low and moderate Mach numbers by the addition of screens in the settling chambers. At high Mach numbers, upstream pressure and velocity effects are usually less, since the large nozzle contraction ratios damp them out. Temperature fluctuations are unaffected by contraction ratio.

3.9.7 Determination of Test-Section Noise

The test-section noise is defined as pressure fluctuations. Noise may result from unsteady settling chamber pressure fluctuations due to upstream flow conditions. It may also be due to weak unsteady shocks originating in a turbulent boundary layer on the tunnel wall. Noise in the test-section is very likely to influence the point of boundary layer transition on a model. Also, it is probable that the noise will influence the other test results as well.

Test-section noise can be detected by either hot-wire anemometry measurements or by high-response pitot pressure measurements. It is a usual practice to make measurements in both the test-section and the settling chamber of the tunnel to determine whether the noise is coming from the test-section boundary layer. It is then possible to determine whether fluctuations in the two places are related. The *test-section noise usually increases with increasing tunnel operating pressure, and test-section noise originating in the settling chamber usually decreases as tunnel Mach number increases.*

3.9.8 The Use of Calibration Results

The Mach number in the vicinity of a model during a test is assumed to be equal to an average of those obtained in the same portion of the test-section during calibrations. With this Mach number and the total pressure (p_{01}) measured in the settling chamber, it is possible to define the dynamic pressure, q, as

$$\frac{q}{p_{01}} = \frac{\gamma}{2} M^2 \left(1 + \frac{\gamma - 1}{2} M^2 \right)^{-\frac{\gamma}{\gamma - 1}}$$

for use in data reduction. If the total temperature is also measured in the settling chamber, all properties of the flow in the test-section can be obtained using isentropic relations. The flow angularities measured during calibration are used to adjust model angles set with respect to the tunnel axis to a mean flow direction reference. The transition point and noise measurements made during the calibration may be used to decrease the tunnel turbulence and noise level.

3.9.9 Starting of Supersonic Tunnels

Supersonic tunnels are usually started by operating a quick-operating valve, which causes air to flow through the tunnel. In continuous operation tunnels, the compressors are normally brought up to the desired operating speed with air passing through a bypass line. When operating speed is reached, a valve in the bypass line is closed, which forces the air through the tunnel. In blowdown tunnels a valve between the pressure storage tanks and the tunnel is opened.

Quick starting is desirable for supersonic tunnels, since the model is subjected to high loads during the starting process. Also, the quick start of the blowdown tunnel conserves air. To determine when the tunnel is started, the pressure at an orifice in the test-section wall near the model nose is usually observed. When this pressure suddenly drops to a value close to the static pressure for the design Mach number, the tunnel is started. If the model is blocking the tunnel, the pressure will not drop. We can easily identify the starting of the tunnel from the sound it makes.

Some tunnels are provided with variable second-throat diffusers, designed to decrease the pressure ratio required for tunnel operation. These diffusers are designed to allow the setting of a cross-sectional area large enough for starting the tunnel and to allow the setting of a less cross-sectional area for more efficient tunnel operation. When used as designed, the variable diffuser throat area is reduced to a predetermined area as soon as the tunnel starts.

3.9.10   Starting Loads

Whenever a supersonic tunnel is being started or stopped, a normal shock passes through the test-section and large forces are imposed on the model. The model oscillates violently at the natural frequency of the model support system and normal force loads of about 5 times those which the model would experience during steady flow in the same tunnel at an angle of attack of 10 degrees are not uncommon. The magnitudes of starting loads on a given model in a given tunnel are quite random and exactly what causes the large loads is not yet understood.

Starting loads pose a serious problem in the design of balances for wind tunnel models. If the balances are designed to be strong enough to withstand these severe starting loads, it is difficult to obtain sensitivities adequate for resolving the much smaller aerodynamic loads during tests. A number of methods have been used for alleviating this problem. Among them the more commonly used methods are

- Starting at a reduced total pressure in continuous tunnels.

- Changing the model during starting.

- Shielding the model with retractable protective shoes, at start.

- Injecting the model into the air stream after the tunnel is started.

3.9.11   Reynolds Number Effects

The primary effects of Reynolds number in supersonic wind tunnel testing are on drag measurements. The aerodynamic drag of a model is usually made up of the following four parts.

1. The skin friction drag, which is equal to the momentum loss of air in the boundary layer.

2. The pressure drag, which is equal to the integration of pressure loads in the axial direction, over all surfaces of the model ahead of the base.

3. The base drag, which is equal to the product of base pressure differential and base area.

4. The drag due to lift, which is equal to the component of normal force in the flight direction.

The pressure drag and drag due to lift are essentially independent of model scale or Reynolds number, and can be evaluated from wind tunnel tests of small models. But the skin friction and base drags are influenced by Reynolds number. In the supersonic regime, the skin friction is only a small portion of the total drag due to the increased pressure drag over the fore body of the model. However, it is still quite significant and needs to be accounted for. Although the probability of downstream disturbances affecting the base pressure and hence the base drag is reduced because of the inability of downstream disturbances to move upstream in supersonic flow, enough changes make their way through the subsonic wake to cause significant base interference effects.

3.9.12 Model Mounting-Sting Effects

Any sting extending downstream from the base of a model will have an effect on the flow and therefore, is likely to affect model base pressure. For actual tests the sting must be considerably larger than that to withstand the tunnel starting loads and to allow testing to the maximum steady load condition, with a reasonable model deflection. Sting diameters of 1/4 to 3/4 model base diameters are typical in the wind tunnel tests. The effects on the base pressure of typical sting diameters are significant, but represent less than 1 percent of the dynamic pressure and therefore, a small amount of the total drag of most of the models.

3.10 Calibration and Use of Hypersonic Tunnels

Hypersonic tunnels operate in the Mach number range of 5 to 10 or more. The stagnation pressure varies from 1 MPa to 10 MPa, and the stagnation temperature varies from 60°C to 2000°C. They mostly have solid-walled test-sections and require contoured nozzles which are most frequently axially symmetric instead of two-dimensional. Models which are larger than those that can be tested in supersonic tunnels can be tested in hypersonic tunnels. Hypersonic tunnel models sometimes have frontal area as high as 10 percent of the test-section area. Model size will probably be limited by the large model wake, which takes up too much flow area in the diffuser and blocks it during tunnel starting. The tunnel wall effect is unlikely to affect the flow over the model.

The air used in a hypersonic tunnel is heated to avoid liquefaction during expansion to the high Mach number and the corresponding low temperatures, and to facilitate heat transfer studies. In fact, the use of heated air is the major factor that makes the hypersonic tunnel operation more complicated than supersonic tunnel operation. The air in hypersonic tunnels must also be dry to avoid condensation effects due to the expansion of the air to high Mach numbers and the associated low temperatures. However, this problem is less serious here than in supersonic tunnels, because in the process of compressing the air to the necessary high pressures for hypersonic flow, most of the natural water will be simply squeezed out.

3.10.1 Calibration of Hypersonic Tunnels

The calibration procedure for hypersonic tunnel test-section is generally the same as that of a supersonic tunnel. However, in hypersonic tunnels it is much more important to calibrate over the complete range of conditions through which the tunnel will operate. The boundary layers at the nozzle wall are much thicker and subject to larger changes in thickness than in supersonic tunnels, due to operating pressure and temperature. Also, the real gas effects make the test-section

Mach number very sensitive to total temperature. Further, a significant axial temperature gradient may exist in the settling chamber, with the temperature decaying as the nozzle throat is approached. In addition to axial gradients, serious lateral temperature gradient is also present in the settling chamber. These must be eliminated before uniform flow can be achieved in the test-section.

3.10.2   Mach Number Determination

As in the supersonic tunnels, Mach numbers in hypersonic tunnels are usually obtained by using pitot pressure measurements, which differ from those in supersonic tunnels in pressure and Reynolds number range. Pitot pressure in hypersonic tunnels will usually be lower. It should be ensured that the Reynolds number based on probe diameter is above 500 (or preferably 1000) since inaccurate measurements are likely if it is lower.

The determination of Mach number from the measured pitot and total pressures becomes highly complicated if the air temperature is 500°C or above, because of the real gas effects. The procedure for determining the Mach number from the measured pitot and total pressures and a measured total temperature is as follows.

1. From the measured p_{02}/p_{01}, determine the corresponding Mach number from the normal shock table for perfect gas.

2. From the chart of $(p_{02}/p_{01})_{\text{thermperf}} / (p_{02}/p_{01})_{\text{perf}}$ versus M, determine this ratio of pressure ratios at the above M and measured T_0.

3. Divide the experimental pressure ratio by the ratio determined above to obtain the corresponding value of $(p_{02}/p_{01})_{\text{perf}}$.

4. For the new $(p_{02}/p_{01})_{\text{perf}}$, determine the corresponding M from perfect gas normal shock table.

5. If the Mach number obtained above is not equal to that used in step 2, enter step 2 with the Mach number from step 4 and repeat. When the two Mach numbers agree closely, the interpolation is complete.

Note that, for accurate determination of M, we need pressure-Mach number charts for high temperature ($> 500°C$) regime. For this we should refer to books specializing on high-temperature gases. This method of determining Mach number from measured pressure ratio is cumbersome and inaccurate. A high-speed computer may be used for this purpose.

The following facts about hypersonic tunnel operation will be of high value for experimentation with hypersonic tunnels. If the air in the settling chamber is at room temperature, we can achieve a test-section Mach number of 5 without liquefaction of air. But to avoid liquefaction of air as it expands to the test-section conditions where the Mach number is 10, the stagnation temperature (i.e., the temperature in the settling chamber) should be approximately 1060

K. Because of the limitation of heater capacity, the maximum Mach number for a continuous-flow wind tunnel using air as the test gas is approximately 10. However, using better insulation and heater buildup methods, there are tunnels operating at Mach numbers of 12 and even 14 with air as the test gas.

One of the major concerns in hypersonic tunnel operations is the combination of maximum Mach number and minimum stagnation temperature for which condensation-free flow can be generated. Daum and Gyarmathy (1968) showed, based on their experimental study of air condensation that, in a rapidly expanding nozzle flow at low stream pressures (less than about 0.05 mm Hg), significant supercooling of the air can be achieved, since the onset of condensation was due to the spontaneous condensation of nitrogen. At these low pressures, an approximately constant experimental supercooling value of about 22 K was obtained.

To achieve higher Mach number flows, the stagnation temperature must be in excess of 1060 K. The high-pressure–high-temperature condition required for hypersonic tunnels can be generated in many ways. For example, use of flow conditions downstream of the reflected shock wave in a shock tube as the reservoir conditions for a shock tunnel is one such means. But the tunnel run time for such facilities is very short. The short run time reduces the energy requirements and alleviates tunnel and model thermal-structural interactions.

Based on run time, hypersonic wind tunnels are classified into the following three categories.

- Impulse facilities, which have run times of about 1 second or less.

- Intermittent tunnels (blowdown or indraft), which have run times of a few seconds to several minutes.

- Continuous tunnels, which can operate for hours (this is only of theoretical interest, since continuous hypersonic tunnels are extremely expensive to build and operate).

The facilities with the shortest run times have the higher stagnation temperatures. Arc discharge or reflected shock waves in a shock tube are used to generate the short-duration, high-temperature stagnation conditions.

The flow quality (including uniformity, noise, cleanliness, and steadiness) in the test-section can affect the results obtained in a ground-test program. Disturbance modes in the hypersonic tunnels include vorticity (turbulence fluctuations), entropy fluctuations (temperature spottiness) which are traceable to the settling chamber, and pressure fluctuations (radiated aerodynamic noise). These disturbances can affect the results of boundary layer transition studies conducted in hypersonic wind tunnels.

It is known that even the high-enthalpy, short-duration facilities operate on the borderline between perfect gas and real gas flow. Let us look at the reservoir temperature required to maintain perfect air at a test-section temperature of 50 K and Mach number M_1. Table 3.3 gives these values for different M_1.

When the test-section Mach number is 8.5, the stagnation temperature for perfect air must be 772 K. However, it is well established that the vibrational state is excited beginning approximately at 770 K. Thus, for the high Mach number facilities, vibrational excitation occurs in the settling chamber, followed by vibrational freezing downstream of the throat, and subsequent rapid relaxation in the downstream section of the nozzle. This kind of improper characteristic of hypersonic flow field manifests itself as an error in the Mach number. Hypersonic tunnels heated by conventional clean air heaters have exhibited a Mach number error of as much as 1.5 percent compared to that predicted by isentropic flow using the ratio of freestream pitot pressure to reservoir pressure. Even these relatively small errors in Mach number can result in significant errors in the nondimensionalized data, if they are ignored.

Table 3.3 Stagnation temperature (T_{01}) variation with M_1 for $T_1 = 50$ K, based on perfect gas assumption

M_1	5	6	7	8	9	10	11	12	13	14	15
T_{01} (K)	300	410	540	690	860	1050	1260	1490	1740	2010	2300

It is important to note that usually hypersonic wind tunnels operate such that the static temperature in the test-section approaches the liquefaction limit. Thus, hypersonic Mach numbers are achieved with relatively low freestream velocities (which is related to the kinetic energy), because the speed of sound (which is related to the static temperature) is relatively low.

3.10.3 Determination of Flow Angularity

Flow angularity in hypersonic tunnels is usually determined by cones with included angle in the range of 20 to 90 degrees. The shock waves on cones with higher angles are detached throughout the hypersonic Mach number range and the surface pressure variation with angle of attack cannot be easily calculated.

3.10.4 Determination of Turbulence Level

The large contraction ratios of hypersonic tunnels have a tendency to reduce the turbulence percentage level in the test-section to insignificant values. Therefore, there is no need to determine the turbulence level of hypersonic tunnels.

3.10.4.1 Blockage Tests

Blockage tests are made during the calibration phase, to determine the size of the models that may be tested in the tunnel, and to find the effect of model size on the starting and operating pressure ratios of the tunnel.

3.10.4.2 Starting Loads

There is no published data available on starting loads in hypersonic tunnels. But from experience it is found that, at Mach 7, the starting loads are not so severe as indicated for supersonic tunnels.

3.10.5   Reynolds Number Effects

At hypersonic speeds the boundary layers are relatively thicker and prone to separate in the presence of unfavorable (adverse) pressure gradients than at supersonic speeds. Also, there is likely to be an intense interaction between the shock waves and boundary layers. For example, on a wedge or cone leading edge, the shock at hypersonic speeds will be very close to the surface. The boundary layer on the wedge or cone will be an important part of the distance between the surface and the shock. Under these conditions, loads on the model can no longer be considered simply as those due to an inviscid flow field which exerts pressure through the boundary layer and onto the model surface. Since the boundary layer primarily depends on the Reynolds number, we can say that the complete flow field around a vehicle at hypersonic speeds is dependent to a significant extent on Reynolds number. Thus, force and moment coefficients in addition to drag are likely to be influenced significantly by Reynolds number.

The boundary layers on models in hypersonic tunnels are mostly, if not entirely, laminar. However, it is not clear that tripping the boundary layer is the answer to the problem of obtaining comparable flow fields over the model in the tunnel and the full-scale vehicle in the flight. In flight at hypersonic speeds, the full-scale vehicle is likely to have long runs of laminar flow if it has reasonably smooth surfaces. Reynolds numbers as high as 7×10^7 without transition have been reported on rockets. This highlights the difficulty of predicting the location where the transition will occur on the model in flight and consequently where or if a boundary layer trip should be used. The usual practice at present is to test models without transition strips in hypersonic tunnels. If it is found that the smooth model has extensive boundary layer separation at some point at which it is not expected on the vehicle in flight, then a transition strip may be tried as a means of eliminating separation.

3.10.6   Force Measurements

Force measurements in hypersonic tunnels are similar to those of subsonic and supersonic tunnels. However, there are a few problems in hypersonic tunnel force measurements which do not exist in the lower-speed tunnels. They are the following.

- Models will get heated up during the tests, since hypersonic tunnels use heated air.

- It is essential to ensure that the model heating and the heated air do not affect the electrical signals from the strain gauge balance.

- There are possibilities of significant temperature effects on balance readout at temperatures well below those for which the cement holding the gauges to the flexures fails.

- With the model at an angle of attack, surface heating rates on the model will be higher on the windward side than on the leeward side. Air circulating from the model base through the balance cavity will also heat the gauges on one side of the balance more than the gauges on the other side.

These cases of uneven heating of balance are not taken care of by temperature compensation of the bridges of the balance. Keeping the balance temperature essentially constant at a near ambient value during the test is the remedy for the variable balance temperature problem.

In addition to effects on balance readings, uneven heating on the windward and leeward model surfaces may cause model distortion of significant magnitude, especially when the length to diameter ratio of the model is large. This effect is usually avoided by cooling the model. In intermittent operating tunnels it may be eliminated by increasing model wall thickness or by using a material such as Invar, which has a low coefficient of thermal expansion.

Low model loads at high Mach numbers are another problem in hypersonic force tests. Aerodynamic loads in some cases may be considerably less than the model weight. This poses a problem in balance design. The balance must be strong enough to hold the model but it must also be weak enough to be sensitive to loads smaller than the model weight. Under these conditions, there is likelihood of continuous low-frequency oscillations of the model during a test. These oscillations may cause inertia loads to become a significant portion of the aerodynamic loads to be measured and satisfactory data cannot be obtained unless the data acquisition system is equipped with suitable electronic filtering.

3.11   Flow Visualization

Schlieren system for high-speed tunnels is often designed for passing the light through the test-section two times using a "double-pass" system. This is accomplished by using a circular arc mirror adjacent to one wall of the test-section and a light source and mirror focal point as close together as possible on the opposite side of the test-section, in order to increase the system sensitivity. However, it is found that obtaining good schlieren pictures of the flow around a model when the test-section pressures are less than about 1 mm mercury (absolute) is difficult, even with the double pass system. Pressures below 1 mm mercury are common in wind tunnels operating at Mach 8 and above.

To obtain better flow visualization at low test-section pressures, the air in the flow field of the model may be ionized using an electric current (used by the Jet Propulsion Laboratory). An electrode is placed a few centimeters upstream and a few centimeters above the model in the test-section. A potential of 5000 volts direct current is established between the electrode and model with a current flow of 0.4 amp. The flow of current ionizes the flow field, with the result that shock waves are clearly shown in regular photographs and are much more clearly visible in schlieren photographs than in schlieren photographs taken without ionization. Care must be taken to interlock (possibly with a low pressure switch) the power system to prevent injury to personnel.

3.12 Hypervelocity Facilities

These are experimental aerodynamic facilities that allow testing and research at velocities considerably above those achieved in the wind tunnels discussed in the previous sections of this chapter. The high velocities in these facilities are achieved at the expense of other parameters, such as Mach number, pressure, and/or run time.

From the discussions on supersonic and hypersonic tunnels, it is obvious that the aerodynamic problems of high-speed flight are not completely answered by tests in these facilities, where the tunnel operating temperature is only high enough to avoid liquefaction. Also, we know that, if the static temperatures and pressures in the test-section of a wind tunnel have to be equal to those at some altitude in the atmosphere and at the same time that the velocity in the wind tunnel equals the flight velocity of a vehicle at that altitude, then the total temperatures and pressures in the wind tunnel must be quite high. It is important to keep the static temperature, static pressure, and velocity in the test-section the same as those in the actual flight condition, since only then will the temperature and pressure in the vicinity of the model (behind shock waves and in boundary layers) correspond to conditions for the vehicle in flight.

Having the proper temperature and pressure in the vicinity of the model is considered important since at high temperatures, the characteristics of air are completely different from those at low temperatures. Experimental facilities which have been developed to simulate realistic flow conditions at high speeds and which are used extensively for high-speed testing are

- Hotshot tunnels.

- Plasma jets.

- Shock tubes.

- Shock tunnels.

- Light gas guns.

Though it is not our aim to discuss these facilities in this book, let us briefly look at them to get an idea about the facilities which are expected to dominate the experimental study in the high-speed regime in the future.

3.12.1 Hotshot Tunnels

Hotshot tunnels are devices meant for the generation of high-speed flows with high temperatures and pressures for a short duration. The high temperatures and pressures required at the test-section are obtained by rapidly discharging a large amount of electrical energy into an enclosed small volume of air, which then expands through a nozzle and a test-section. The main parts of a hotshot tunnels are shown schematically in Figure 3.50.

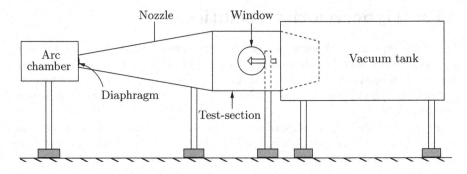

Figure 3.50: Main parts of hotshot tunnel

The arc chamber is filled with air at pressures up to 270 MPa. The rest of the circuit is evacuated and kept at low pressures at the order of a few microns. The high- and low-pressure portions are separated by a thin metallic or plastic diaphragm located slightly upstream of the nozzle throat. Electrical energy from a capacitance or inductance energy storage system is discharged into the arc chamber over a time interval of a few milliseconds. The energy added to the air causes an increase in its temperature and pressure, and this makes the diaphragm rupture. When the diaphragm ruptures, the air at high temperature and pressure in the arc chamber expands through the nozzle and establishes a high-velocity flow. The high-velocity flow typically lasts for 10 to 100 milliseconds, but varies continuously during the period. The flow variation is due to a decay of pressure and temperature in the arc chamber with time. The high-velocity flow is terminated when the shock that passed through the tunnel in starting the flow is reflected from a downstream end of the vacuum tank and arrives back upstream at the model.

Currently the common operating conditions of hotshot tunnels are about 20 MPa, 4000°C, and 20 Mach and above, although there is much variation between facilities. Data collection in hotshot tunnels is much more difficult than in conventional tunnels because of the short run times.

3.12.2   Plasma Arc Tunnels

Plasma arc tunnels are devices capable of generating high-speed flows with very high temperature. They use a high-current electric arc to heat the test gas. Unlike hotshot tunnels, plasma arc tunnels may be operated for periods of the order of many minutes, using direct or alternating current. Temperatures of the order of 13,000°C or more can be achieved in the test gas.

A typical plasma arc tunnel consists of an arc chamber, a nozzle usually for a Mach number less than 3, an evacuated test-chamber into which the nozzle discharges, and a vacuum system for maintaining the test-chamber at a low pressure, as shown in Figure 3.51.

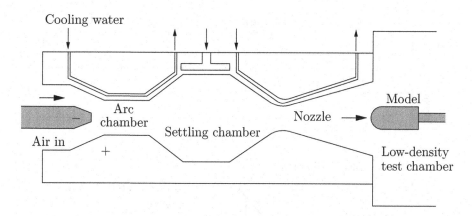

Figure 3.51: Schematic of plasma arc tunnel

In the plasma arc tunnel, a flow of cold test gas is established through the arc chamber and the nozzle. An electric arc is established through the test gas between an insulated electrode in the arc chamber and some surface of the arc chamber. The electric arc raises the temperature of the test gas to an ionization level, rendering the test gas as a mixture of free electrons, positively charged ions, and neutral atoms. This mixture is called "plasma" and it is from this that the plasma arc tunnel gets its name.

Plasma tunnels operate with low stagnation pressures of the order of 700 kPa or less, with gases other than air. The enthalpy level of the test gas, and consequently the temperature and velocity in a given nozzle, are higher for a given power input when the pressure is low. Argon is often used as the test gas since high temperature and high degree of ionization can be achieved with a given power input, also the electrode will not get oxidized in argon environment.

Mostly, plasma arc tunnels are used for studying materials for reentry vehicles. Surface material ablation tests, which are not possible in low-temperature tunnels or high-temperature short duration tunnels, can be made. These tunnels can also be used for "magneto-aerodynamics" and plasma chemistry fields to study the electrical and chemical properties of the highly ionized gas in a flow field around a model.

3.12.3 Shock Tubes

The shock tube is a device to produce high-speed flow with high temperatures, by traversing normal shock waves which are generated by the rupture of a diaphragm which separates a high-pressure gas from a low-pressure gas. Shock tube is a very useful research tool for investigating not only the shock phenomena, but also the behavior of the materials and objects when subjected to very high pressures and temperatures. A shock tube and its flow process are shown schematically in Figure 3.52.

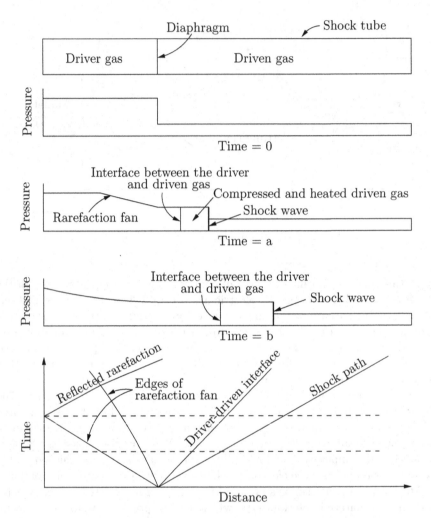

Figure 3.52: Pressure and wave diagram for a shock tube

The diaphragm between the high- and low-pressure sections is ruptured and the high-pressure driver gas rushes into the driven section, setting up a shock wave which compresses and heats the driven gas. The pressure variation through the shock tube at the instant of diaphragm rupture and at two short intervals later are shown in Figure 3.52. The wave diagram simply shows the position of the important waves as a function of time.

When the shock wave reaches the end of the driven (low-pressure) tube, all of the driven gas will have been compressed and will have a velocity in the direction of shock wave travel. Upon striking the end of the tube, the shock gets reflected and starts traveling back upstream. As it passes through the driven gas and brings it to rest, additional compression and heating is accomplished.

The heated and compressed gas sample at the end of the shock tube will retain its state except for heat losses until the shock wave reflected from the end of the tube passes through the driver gas-driven interface and sends a reflected wave back through the stagnant gas sample, or the rarefaction wave reflected from the end of the driver (high-pressure) section reaches the gas sample. The high temperature gas samples that are generated make the shock wave useful for studies of the chemical physics problems of high-speed flight, such as dissociation and ionization.

3.12.4 Shock Tunnels

Shock tunnels are wind tunnels that operate at Mach numbers of the order 25 or higher for time intervals up to a few milliseconds by using air heated and compressed in a shock tube. Schematic diagram of a shock tunnel, together with wave diagram, is shown in Figure 3.53.

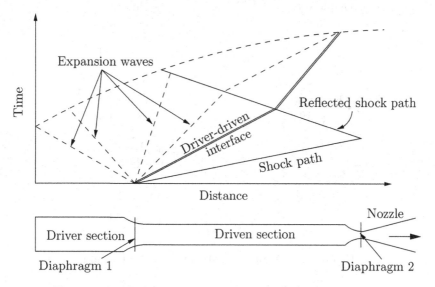

Figure 3.53: Schematic of shock tunnel and wave diagram

As shown in the figure, a shock tunnel includes a shock tube, a nozzle attached to the end of the driven section of the shock tube, and a diaphragm between the driven tube and the nozzle. When the shock tube is fired and the generated shock reaches the end of the driven tube, the diaphragm at the nozzle entrance is ruptured. The shock is reflected at the end of the driven tube and the heated and compressed air behind the reflected shock is available for operation of the shock tunnel. As the reflected shock travels back through the driven section, it travels only a relatively short distance before striking the contact surface, it will be reflected back towards the end of the driven section. When the reflected shock reaches the end of the driven section, it will result in a change in pressure and temperature of the gas adjacent to the end of the driven section. If the change in the conditions of the driven gas is significant, the flow in the nozzle will be unsatisfactory and the useful time will be terminated. The stagnation pressure and temperature in shock tunnels are about 200 MPa and 8000 K, respectively, to provide test times of about 6.5 milliseconds.

3.12.5   Gun Tunnels

The gun tunnel is quite similar to the shock tunnel, in operation. It has a high-pressure driver section and a low-pressure driven section with a diaphragm separating the two, as shown in Figure 3.54.

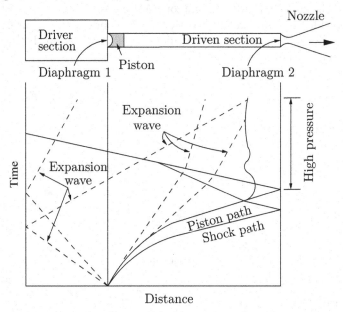

Figure 3.54: A gun tunnel and its wave diagram

A piston is placed in the driven section, adjacent to the diaphragm, so that when the diaphragm ruptures, the piston is propelled through the driven tube, compressing the gas ahead of it. The piston used is so light that it can be accelerated to velocities significantly above the speed of sound in the driven gas. This causes a shock wave to precede the piston through the driven tube and heat the gas. The shock wave will be reflected from the end of the driven tube to the piston, causing further heating of the gas. The piston comes to rest with equal pressure on its two sides, and the heated and compressed driven gas ruptures a diaphragm and flows through the nozzle.

As can be inferred, gun tunnels are limited in the maximum temperature that can be achieved by the piston design. The maximum temperatures normally achieved are about 2000 K. Run times of an order of magnitude higher than the shock tunnels are possible in gun tunnels. In general, the types of tests that can be carried out in gun tunnels are the same as those in the hotshot tunnels and the shock tunnels.

3.13 Ludwieg Tube

The concept of the *Ludwieg tube* was first proposed in 1955 by Hubert Ludwieg, a German scientist. The beauty of the Ludwieg tube is that, it provides clean supersonic/hypersonic flow with relative ease and low cost. Some of the well-established Ludwieg tube facilities are the following.

- Ludwieg Tube Group-GALCIT at Caltech: A Mach 2.3 facility is located at the California Institute of Technology, in Pasadena, CA, USA. The facility has a test time of 80 milliseconds.

- The Boeing/AFOSR Mach 6 quiet tunnel at Purdue University. This tunnel has a 9.5-inch exit diameter, runs for about 10 seconds, about once an hour for about US $10/short (built during 1995-2001).

- The Ludwieg tube tunnel at Marshall Space Flight Center. This 32-inch Ludwieg tube built at Marshall Space Flight Center is a long tube of constant diameter for storing air at 50 atm. The run times were short, but for a time duration of half a second or less the model is bathed in airflow that was constant in pressure and temperature and displayed very little turbulence.

- The Ludwieg tube tunnel at Hypersonic Technology Gottingen, and the large number of tunnels fabricated and supplied by HTG to different universities in Germany. These tunnels have test-section sizes varying from 250 mm diameter to 500 mm diameter and Mach number range from 5 to 12.

The significant characteristics of Marshall's Ludwieg tube was the high Reynolds number achieved–roughly three times that in conventional existing wind tunnels. This capability found immediate application in basic fluid dynamic research as well as for the determination of aerodynamic forces acting on launch vehicles. However, the Ludwieg tube had limited use in testing winged aircraft because of high stresses encountered and the consequent distortion of models.

3.13.1 Operating Principle of the Ludwieg Tube (Koppenwallner, 2000)

Basically Ludwieg tube tunnel is a blowdown type hypersonic tunnel. Due to its special fluid dynamic features no devices are necessary to control pressure or temperature during the run. It thus can be regarded as an "intelligent blowdown facility." The test gas storage occurs in a long charge tube which is, by a fast acting valve, separated from the nozzle, test-section, and the discharge vacuum tank. Upon opening the valve an unsteady expansion wave travels with the speed of sound a_T into the charge tube. This wave accelerates the gas along its way, to a tube Mach number M_T given by the area ratio of the tube/nozzle throat. During the forward and backward traveling time of the wave a constant steady flow to the expansion nozzle is established with pressure and temperature determined by a one-dimensional unsteady expansion process. When the wave reflecting from the end-wall reaches the nozzle throat the valve is closed and the test is finished. Thus, the length of the tube L_T and the speed of sound a_T of the charge tube gas determines the run time t_R of the tunnel. It is given by

$$t_R = 2\frac{L_T}{a_T}$$

From opening to closing of the valve only a short column of the charge tube gas has been discharged. To start a new test or shot, the charge tube to be refilled and the dump tank must be re-evacuated.

The optimum run time and interval between runs achieved with optimized valve operated Ludwieg tunnel are

Run time t_R: 0.1 to 0.5 second
Interval times t_I: 200 to 600 seconds

The operating principle of a typical Ludwieg tube is explained with a wave traveling diagram in the distance-time frame in Figure. 3.55.

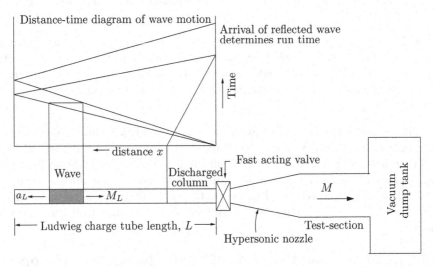

Figure 3.55: The principle and the elements of a hypersonic Ludwieg tunnel (Koppenwallner, 2000)

The area ratio between sonic throat and Ludwieg charge tube determines that the subsonic Ludwieg tube Mach number M_L is

$$\frac{A^*}{A_L} = \left(\frac{\gamma+1}{2}\right)^{\left[\frac{\gamma+1}{2(\gamma-1)}\right]} M_L \left(1 + \frac{\gamma-1}{2} M_L^2\right)^{-\left[\frac{\gamma+1}{2(\gamma-1)}\right]}$$

where A_L is the charge tube cross-sectional area, A^* is nozzle throat area, and γ is the specific heats ratio of the gas in the charge tube. Unsteady one-dimensional expansion relates the flow conditions behind the expansion wave to flow conditions of the charged Ludwieg tube. The following are the working relations.

Pressure ratio is given by

$$\frac{p_L}{p_{L0}} = \left(1 + \frac{\gamma-1}{2} M_L\right)^{\frac{-2\gamma}{\gamma-1}}$$

Temperature ratio is given by

$$\frac{T_L}{T_{L0}} = \left(1 + \frac{\gamma-1}{2} M_L\right)^{-2}$$

where subscript 'L' refers to the state of the gas behind the expansion wave and '$L0$' refers to the stagnation state of the gas in the charge tube.

Speed of sound is given by

$$\frac{a_L}{a_{L0}} = \left(1 + \frac{\gamma-1}{2} M_L\right)^{-1}, \quad \text{with} \quad a_{L0} = \sqrt{\gamma R T_{L0}}$$

The measurement time t_m available is given by

$$t_m = \frac{2L_L}{a_L} = 2\frac{2L_L}{a_{L0}} \frac{1}{1 + M_L} \left(1 + \frac{\gamma-1}{2} M_L\right)^{\frac{\gamma+1}{2(\gamma-1)}}$$

During the test time t_M a gas column length L_D is discharged with velocity V_L from the tube of length L_L. The subscript '0' refers to properties corresponding to stagnation condition. In a first approximation this discharged gas column length L_G is given by

$$L_G = V_L \times t_m = M_L \times a_L \times t_m = 2 \times L_L M_L a_L / a_{L0}$$

The discharged gas column length decreases with tube Mach number, which usually is kept between 0.05 and 0.15 (Koppenwallner, 2000). Thus, approximately 10–30 percent of the charge tube gas is discharged during one shot.

The Ludwieg tube principle can also be used for high subsonic or transonic wind tunnels. As the principle requires a discharge through a critical throat area A^* this throat is now placed behind the test-section. The throat can be combined with the fast acting valve and the ratio of valve opening area A^* to Ludwieg tube cross-sectional area A_L, determines the tube M_L and finally the test-section Mach number M.

3.13.2   Some Specific Advantages and Disadvantages of the Ludwieg Tube

We have seen that the Ludwieg tube tunnel is basically a short duration blow-down type tunnel. The idealized discharge process from a Ludwieg tube and the discharge from a pressurized sphere of equal volume are compared in Figure 3.56. Each pressure step in the discharge from Ludwieg tube results from a forward and backward running expansion wave.

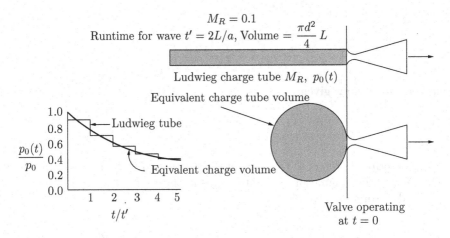

Figure 3.56: Discharge from a Ludwieg tube and from an equivalent spherical pressure vessel (Koppenwallner, 2000)

3.13.2.1   Advantages of Ludwieg Tube Tunnel Compared to Standard Blowdown Tunnels

1. The Ludwieg tube tunnel requires extremely short start and shut-off time.

2. No regulation of temperature and pressure during run is necessary. No throttle valve upstream of the nozzle is necessary. The gas dynamic principle regulates the temperature and pressure.

3. From points 1 and 2 above follows an *impressive economy*, since there exists no waste of mass and energy of flow during tunnel start and shut-off.

4. Due to elimination of the pressure regulation valve, the entrance to the nozzle can be kept clean. This ensures low turbulence level in the test-section.

5. Ludwieg tube is well suited for transient heat transfer tests.

6. There is no *unit Reynolds number effect* in the Ludwieg tube, as in the case of many facilities.

3.13.2.2 Disadvantages of the Ludwieg Tube Tunnel

Like any facility, the Ludwieg tube tunnel also has some disadvantages. They are

1. Short run time–of the order of 0.1 to 0.5 second.

2. Limitation of maximum stagnation temperature due to heater and charge tube material.

3.13.3 Hypersonic Simulation Requirements

Wind tunnel tests are usually meant for simulating the aerothermodynamic phenomena of free flight atmosphere in a scale test. A true simulation of the free flight conditions can in principle only be achieved if the dynamic similarity between the free flight and scale tests are established. This requires that the scale test must be based on the similarity laws for the problems to be investigated. In principle, the following flow phenomena need to be investigated at hypersonic conditions.

- Compressibility of the air.

- Viscous effects like friction and heat conduction.

- Chemistry due to dissociation of air molecules at high temperatures for flight velocities of the order of Mach 8.

It is now accepted to divide the simulation into the following regimes (Koppenwallner, 2000).

3.13.3.1 Mach–Reynolds–Simulation for Compressibility and Viscous Effects

Geometric similarity: *Model scale* $\quad S = L_{WT}/L_{FL}$

Compressibility: *Mach number* $\quad M_{WT} = M_{FL}; \quad \left(\frac{V}{a}\right)_{WT} = \left(\frac{V}{a}\right)_{FL}$

Viscous effects: *Reynolds number* $\quad Re_{WT} = Re_{FL}; \quad \left(\frac{\rho V L}{\mu}\right)_{WT} = \left(\frac{\rho V L}{\mu}\right)_{FL}$

Wall stagnation temperature ratio: $\quad \left(\frac{T_w}{T_0}\right)_{WT} = \left(\frac{T_w}{T_0}\right)_{FL}$

where the subscripts WT and FL refer to wind tunnel test model and actual body in flight, respectively. If we keep $V_{WT} = V_{FL}$ and $\mu_{WT} = \mu_{FL}$, then viscous similarity requires a scale test at increased density $\rho_{WT} = \rho_{FL}$. Thus, the model loads will increase with *1/scale* factor. If $V_{WT} = V_{FL}$ is not the case, the viscosity shall have the same and identical power law in both wind tunnel test and actual flight

$$\frac{\mu(T_0)}{\mu(T_\infty)} = \left(\frac{T_0}{T_\infty}\right)^\omega$$

where $\omega_{WT} = \omega_{FL}$.

3.13.3.2   Simulation of Real Gas Effects

This requires the simulation of the dissociation and recombination process of molecules behind the shock wave. In order to reach the high temperature behind shock waves in the scale test also, the first requirement is to duplicate the flight velocity.

- Duplication of velocities: $V_{WT} = V_{FL}$

- For chemical similarity the *Damkohler numbers* for forward (dissociation) and backward (recombination) have to be simulated.

Forward reaction: $(Da, f)_{WT} = (Da, f)_{FL}$ $\left(\frac{V}{\rho L} \frac{1}{k_f}\right)_{WT} = \left(\frac{V}{\rho L} \frac{1}{k_f}\right)_{FL}$

Backward reaction: $(Da, b)_{WT} = (Da, b)_{FL}$ $\left(\frac{V}{\rho^2 L} \frac{1}{k_b}\right)_{WT} = \left(\frac{V}{\rho^2 L} \frac{1}{k_b}\right)_{FL}$

As the backward reaction rate contains $\rho^2 L$ this is not compatible with viscous similarity which requires ρL to be simulated. Thus, a complete chemistry simulation cannot be combined with viscous flow simulation. This is the *high enthalpy simulation dilemma* (Koppenwallner, 2000).

3.13.3.3   Industrial Tunnels

Depending on the simulation task (i.e., small missiles or re-entry vehicles), the test-section diameter should range between 500 and 1000 mm.

3.13.3.4   Research Tunnels

These tunnels are meant for studying and exploring basic phenomena and validating theoretical methods and codes. Therefore, a full Reynolds number simulation on large models of flight vehicles will not be necessary. The tunnels shall however have the capability to reproduce all important local phenomena on a smaller scale. This can be achieved with small facilities having the capability to simulate local phenomena at various Reynolds numbers. Adequate test-sections for these tunnels range from 150 to 500 mm in diameter.

3.13.4 Pressure Measuring System

The pressures involved in the hypersonic flows generally are from orders of mega pascals to fractions of torr. Therefore, the pressure measuring device preferably should have a wide range.

One of the important temperatures to be measured is the charge tube temperature. We encounter a maximum of about 1400 K. For this any standard thermocouple made up of suitable "dissimilar" metals will be sufficient. In addition to this if the temperature on the model surface is of interest then one has to select thermocouple foils.

Example 3.12

A Ludwieg tube of length 60 m and diameter 200 mm is charged with air at 1100 K. Assuming the specific heats ratio for the air as 1.4 and the choked throat area as 0.0302 m^2, determine the run time of the Ludwieg tunnel run by this tube.

Solution

Given, $L = 60$ m, $T_{L0} = 1100$ K, $d_L = 200$ mm, $A^* = 0.0302$ m^2.

The Ludwieg tube cross-sectional area is

$$A_L = \frac{\pi d_L^2}{4}$$

$$= \frac{\pi \times 0.2^2}{4}$$

$$= 0.0314 \, \text{m}^2$$

The relation between the subsonic Ludwieg tube Mach number M_L and $\dfrac{A^*}{A_L}$ is

$$\frac{A^*}{A_L} = \left(\frac{\gamma+1}{2}\right)^{\frac{\gamma+1}{2(\gamma-1)}} M_L \left(1 + \frac{\gamma-1}{2}M_L^2\right)^{\frac{-(\gamma+1)}{2(\gamma-1)}}$$

For the given tube

$$\frac{A^*}{A_L} = \frac{0.0302}{0.0314} = 0.9618$$

For this area ratio, from the above relation, we get

$$M_L = 0.8$$

The temperature ratio in terms of M_L is

$$\frac{T_L}{T_{L0}} = \left(1 + \frac{\gamma - 1}{2} M_L\right)^{-2}$$

Therefore,

$$T_L = \frac{1100}{(1 + 0.2 \times 0.8)^2}$$

$$= 817.5\,\text{K}$$

$$a_L = \sqrt{1.4 \times 287 \times 817.5}$$

$$= 573.12\,\text{m/s}$$

Therefore, the tunnel run time t_R becomes

$$t_R = 2\frac{L_T}{a_L}$$

$$= 2 \times \frac{60}{573.12}$$

$$= \boxed{0.21\,\text{s}}$$

Example 3.13

If the density ratio across the shock at the nose of a vehicle flying at a very high hypersonic speed is 18, assuming the fluid as perfect gas, determine the value of specific heats ratio.

Solution

Given, $\rho_2/\rho_1 = 18$. For a shock in a perfect gas, the density ratio across it, when the Mach number is very large is

$$\frac{\rho_2}{\rho_1} = \frac{\gamma + 1}{\gamma - 1}$$

Thus,

$$18 = \frac{\gamma + 1}{\gamma - 1}$$

$$18(\gamma - 1) = \gamma + 1$$

$$17\gamma = 19$$

$$\gamma = \frac{19}{17}$$

$$= \boxed{1.12}$$

3.14   Summary

Wind tunnels are devices which provide air streams flowing under controlled conditions so that models of interest can be tested using them. From an operational point of view, wind tunnels are generally classified as *low-speed*, *high-speed*, and *special purpose tunnels*.

Low-speed tunnels are those with test-section speeds of less than 650 kmph. In low-speed tunnels, the predominant factors influencing the tunnel performance are inertia and viscosity. The effect of compressibility is negligible for these tunnels. Thus, if the Reynolds number of the experimental model and full-scale prototype are equal, any difference in viscosity becomes unimportant.

Tunnels with a test-section speed of more than 650 kmph are called *high-speed tunnels*. The predominant aspect in high-speed tunnel operation is that the influence of compressibility is significant. This means that, in high-speed flows it is essential to consider the *Mach number* as a more appropriate parameter than velocity. A lower limit of *high-speed* flow might be considered to be the flow with Mach number approximately 0.5 (about 650 kmph), at standard sea level conditions.

Based on the test-section Mach number, M, range, the high-speed tunnels are classified as follows.

- $0.8 < M < 1.2$ *Transonic tunnel*

- $1.2 < M < 5$ *Supersonic tunnel*

- $M > 5$ *Hypersonic tunnel*

All modern wind tunnels have four important components; the *effuser*, the *working* or *test-section*, the *diffuser*, and the *driving unit*.

Effuser is a converging passage in which the flow gets accelerated from rest at the upstream end of it to the required conditions at the test-section. *Test-section* is the section with uniform flow of desired quality, where the model to be studied is placed and tested. *Diffuser* is used to reconvert the kinetic energy of the air stream, leaving the test-section, into pressure energy, as efficiently as possible. *Driving unit* is the device to supply power to compensate for the losses encountered by the air stream in the tunnel circuit. In low-speed tunnels this usually takes the form of a fan or propeller.

Special purpose tunnels are these with layouts totally different from that of low-speed and high-speed tunnels. Some of the popular special purpose tunnels are: *spinning tunnels, free-flight tunnels, stability tunnels*, and *low-density tunnels*.

The power loss in a low-speed tunnel is due to the losses in cylindrical parts, guide vanes at the corner, diffuser, contraction cone, honey comb, screens, etc., losses in test-section (jet losses in case of open jet), and in the exit in case of open circuit tunnels.

The loss of energy is expressed in terms of static pressure drop Δp, in the dimensionless form, called pressure drop coefficient K, as follows

$$K = \frac{\Delta p}{q}$$

where q is the dynamic pressure of the flow. In terms of the test-section velocity V_0, the above equation can be rewritten as

$$K_0 = K \left(\frac{D_0}{D} \right)^4$$

where D_0 is the test-section diameter, and D is the local tunnel diameter.

The section energy loss ΔE in the tunnel may be expressed as,

$$\Delta E = K_0\, q_0\, A_0\, V_0 = K_0\, \frac{1}{2}\, \rho\, A_0 V_0^3$$

where A_0 is the test-section area, and A is the local cross-sectional area.

The ratio of the energy of air stream at the test-section to the input energy to the driving unit is a measure of the efficiency of a wind tunnel. It is nearly always greater than unity, indicating that the amount of stored energy in the wind stream is capable of doing work at a higher rate than what it is doing in a wind tunnel, before being brought to rest. The energy ratio ER is in the range 3 to 7 for most closed-throat wind tunnels.

$$
\begin{aligned}
ER &= \frac{\text{Kinetic energy of jet}}{\text{Energy loss}} \\[2mm]
&= \frac{\frac{1}{2}\rho A_0 V_0^3}{\sum K_0\, \frac{1}{2}\rho A_0 V_0^3} \\[2mm]
&= \frac{1}{\sum K_0}
\end{aligned}
$$

The power to drive a low-speed wind tunnel varies as the cube of the test-section velocity. Although this rule is not valid for the high-speed regime, the

implication of rapidly increasing power requirements with increasing test-speed is true for high-speed tunnels also.

High-speed tunnels are generally grouped into intermittent and continuous operation tunnels, based on the type of operation. The intermittent blowdown and induction tunnels are normally used for Mach numbers from 0.5 to about 5.0, and the intermittent pressure-vacuum tunnels are normally used for higher Mach numbers. The continuous tunnel is used throughout the speed range.

The advantages of blowdown-type wind tunnels are that they are the simplest among the high-speed tunnel types and most economical to build, large size test-section and high Mach numbers (up to M = 4) can be obtained, and constant blowing pressure can be maintained, and running time of considerable duration can be achieved. The major disadvantages of blowdown tunnels are that the charging time to running time ratio will be very high for large size tunnels, stagnation temperature in the reservoir drops during tunnel run, thus changing the Reynolds number of the flow, an adjustable throttling valve between reservoir and settling chamber is necessary for constant stagnation pressure operation, starting load is high, and Reynolds number of flow is low due to low static pressure in the test-section.

Commonly employed reservoir pressure range is from 100 to 300 psi for blowdown tunnel operations. As large as 2000 psi is also used where space limitations require. In induction-type tunnels a vacuum created at the downstream end of the tunnel is used to establish flow in the test-section. The *advantages* of induction tunnels are that the stagnation pressure and stagnation temperature are constants, no oil contamination in air due to the pump being at the downstream end, and starting and shutdown operations are simple. The *disadvantages* of induction type supersonic tunnels are that the size of the air drier required is very large, since it has to handle a large mass flow in a short duration, vacuum tank size required is also very large, high Mach numbers ($M > 2$) are not possible because of large suction requirements for such Mach numbers, and Reynolds number is very low since stagnation pressure is atmospheric.

The main *advantages* of continuous supersonic wind tunnels are the following. Better control over the Reynolds number possible, since the shell is pressurized, only a small capacity drier is required, test conditions can be held the same over a long period of time, the test-section can be designed for high Mach numbers ($M > 4$) and large size model, and starting load can be reduced by starting at low pressure in the tunnel shell. Disadvantages of continuous supersonic tunnels are that the power required is very high, temperature stabilization requires large size cooler, compressor drive must be designed to match the tunnel characteristics and design, and operations are more complicated.

The model size for a supersonic tunnel is dictated by the test-rhombus formed by the incident and reflected shocks.

The power loss in a continuous supersonic wind tunnel is due to friction and expansion losses in contraction cone, test-section, guide vanes and cooling system, due to shock wave (in diffuser supersonic part), model and support system drag.

Supersonic wind tunnel diffuser is a device to convert the kinetic energy of a flow to pressure energy.

For continuous and intermittent supersonic wind tunnels the energy ratio ER may be defined as follows.

For continuous tunnel

$$ER = \frac{KE \text{ at the test-section}}{\text{Work done in isentropic compression per unit time}}$$

For intermittent tunnel

$$ER = \frac{(KE \text{ in test-section})(\text{Time of tunnel run})}{\text{Energy required for charging the reservoir}}$$

In a supersonic wind tunnel, the second throat is used to provide isentropic deceleration and highly efficient pressure recovery after the test-section.

Usually three methods of operation are adopted for blowdown tunnel operation. They are, constant Reynolds number operation, constant pressure operation, and constant throttle operation.

For a given settling chamber pressure and temperature, the running time is the shortest for constant throttle operation, the longest for constant Reynolds number operation, and in between the above two for constant pressure operation.

For optimum performance of a tunnel in terms of running time t, the pressure in the settling chamber has to be the minimum pressure, $p_{b\text{min}}$, required for the run. For a given test-section Mach number M there is a p_b minimum in the settling chamber, given by the pressure ratio relation. The Reynolds number in the test-section depends on this p_b value and constant Reynolds number operation is possible only if p_{bi} value is so chosen that as t increases both p_0 and p_b reach $p_{b\text{min}}$ value simultaneously (to result in an optimum constant Reynolds number). If $p_{bi} > p_{b\text{opt}}$ then reservoir pressure will become equal to p_b at some instant, and then constant Reynolds number operation is not possible. If $p_{bi} < p_{b\text{opt}}$, then $p_b = p_{b\text{min}}$ will be reached at time t when $p_0 > p_b$, and supersonic operation will not be possible further.

Blowdown supersonic wind tunnels are usually operated with either constant dynamic pressure (q) or constant mass flow (m). For constant q operation, the only control necessary is a pressure regulating valve (PRV) that holds the stagnation pressure in the settling chamber at a constant value. For constant mass flow run, the stagnation temperature and pressure in the settling chamber must be held constant.

Hypersonic tunnels are tunnels with test-section Mach number above 5. Generally they operate with stagnation pressures in the range from 10 to 100 atmosphere and stagnation temperatures in the range from 50°C to 2000°C. Contoured nozzles which are more often axially symmetric are used in hypersonic tunnels. Models that can be tested in hypersonic tunnels are usually larger than those meant for test in supersonic tunnels. The model frontal area can go up to 10 percent of the test-section cross-sectional area. For hypersonic operations axisymmetric nozzles are better suited than two-dimensional nozzles. Calibration of wind tunnel test-section to ensure uniform flow characteristics everywhere in the test-section is an essential requirement in wind tunnel operation.

The instrumentation and calibration of low-speed tunnels involve the speed setting, flow direction determination, measurement of turbulence level, determination of flow quality, and determination of flow field in the wake of any model.

The actual Reynolds number of the test-section is equivalent to a much higher free field Reynolds number. This increase is called the *turbulence factor*, defined as

$$Re_e = TF \times Re_c$$

where the subscript e stands for effective Reynolds number, TF is the turbulence factor, and Re_c is the measured Reynolds number in the tunnel test-section.

The turbulence may be measured with turbulence sphere, pressure sphere, and hot-wire anemometer.

One of the most troublesome problems associated with wind tunnels is *tunnel surging*. It is *a low-frequency vibration in velocity that may run as high as 5 percent of the dynamic pressure q*. Surging is mainly due to the separation and reattachment of flow in the diffuser.

Wind tunnel balance is a device to measure the actual forces and moments acting on a model placed in the test-section stream. They are broadly classified into wire type, strut type, platform type, yoke type, and strain gauge type. A wind tunnel balance should be capable of measuring the various loads (1, 2, 3, 4, 5, or 6 components) acting on the model, with very a high degree of accuracy. The range of loads and accuracy for the balance are determined from the tunnel size, test-section velocity, model size, configuration, etc.

The variation of static pressure along the test-section produces a drag force known as *horizontal buoyancy*. It is usually small in closed test-section, and negligible in open jets.

The pressure of the lateral boundaries around the test-section (the walls of the test-section) produces *solid blocking, wake blocking*, and an alteration to the local angle of attack along the span.

Supersonic tunnels operate in the Mach number range of about 1.4 to 5.0. Maximum model cross-section areas up to 4 percent of the test-section area is quite common for supersonic tunnels. Typical pressure gradients in the test-section are small, and the buoyancy effects are usually negligible.

At subsonic speeds, the Mach number decreases and the pressure increases with condensation, whereas at supersonic speeds the reverse is taking place.

Calibration of a supersonic wind tunnel includes determining the flow Mach number throughout the range of operating pressure of each nozzle, determining flow angularity, and determining an indication of turbulence level effects.

Pitot pressures are measured by using a pitot tube. A pitot tube will normally have an inside to outside diameter ratio of 0.5 to 0.75, and a length aligned with the air stream of 15 to 20 tube diameters.

Supersonic flow static pressure measurements are much more difficult than the measurement of pitot and static pressures in a subsonic flow. The primary problem in the use of static pressure probes at supersonic speeds is that the probe will have a shock wave (either attached or detached shock) at its nose, which causes a rise in static pressure. If the probe consists of a conical tip followed by a cylinder, the air passing the shoulder would be expanded to a pressure below the stream static pressure, then as the distance from the shoulder is increased, the pressure on the probe will approach the true static pressure of the stream.

The flow angularity in a supersonic tunnel is usually determined by using either a cone or wedge yaw-meters.

Measurements with a hot-wire anemometer demonstrate that there are high-frequency fluctuations in the air stream of supersonic tunnels that do not occur in free air. These fluctuations, broadly grouped under the heading of "turbulence," consist of small oscillations in velocity, stream temperature (entropy), and static pressure (sound).

The test-section noise is defined as pressure fluctuations. Noise may result from unsteady settling chamber pressure fluctuations due to upstream flow conditions. It may also be due to weak unsteady shocks originating in a turbulent boundary layer on the tunnel wall. Test-section noise can be detected by either hot-wire anemometry measurements or by high-response pitot pressure measurements.

Supersonic tunnels are usually started by operating a quick-operating valve, which causes air to flow through the tunnel. In continuous operation tunnels, the compressors are normally brought up to the desired operating speed with air passing through a bypass line. When the operating speed is reached, a valve in the bypass line is closed, which forces the air through the tunnel. In blowdown tunnels a valve between the pressure storage tanks and the tunnel is opened to operate the tunnel.

The primary effects of Reynolds number in supersonic wind tunnel testing are on drag measurements. The pressure drag and drag due to lift are essentially independent of model scale or Reynolds number, and can be evaluated from wind tunnel tests of small models. But the skin friction and base drag are influenced by Reynolds number.

Hypersonic tunnels operate in the Mach number range of 5 to 10 or more. Models which are larger than those that can be tested in supersonic tunnels can be tested in hypersonic tunnels. Hypersonic tunnel models sometimes have frontal area as high as 10 percent of the test-section area. The air used in a hypersonic tunnel is heated to avoid liquefaction during expansion to the high Mach number and corresponding low temperatures and to facilitate heat transfer studies. The calibration procedure for hypersonic tunnel test-section is generally the same as that of a supersonic tunnel. The boundary layers at nozzle wall are much thicker and subject to larger changes in thickness than in supersonic tunnels, due to operating pressure and temperature. Also, the real gas effects make the test-section Mach number very sensitive to total temperature. Further, significant axial temperature gradients may exist in the settling chamber, with the temperature decaying as the nozzle throat is approached. In addition to axial gradients, serious lateral temperature gradients in the settling chamber are present. These must be eliminated before uniform flow can be achieved in the test-section.

Flow angularity in a hypersonic tunnel is usually determined by cones with included angle in the range of 20 to 90 degrees. The large contraction ratios of hypersonic tunnels have a tendency to reduce the turbulence percentage level in the test-section to insignificant values. Therefore, there is no need to determine the turbulence level of hypersonic tunnels.

At hypersonic speeds the boundary layers are relatively thicker and prone to separate in the presence of adverse pressure gradients, than at supersonic speeds. Also, there is likely to be intense interaction between shock waves and boundary layers. Under these conditions, loads on the model can no longer be considered simply as those due to an inviscid flow field which exerts pressure through the boundary layer and onto the model surface. Since the boundary layer primarily depends on the Reynolds number, we can say that the complete flow field around a vehicle at hypersonic speeds is dependent to a significant extent on the Reynolds number. Thus, force and moment coefficients in addition to drag are likely to be influenced significantly by the Reynolds number. In flight at hypersonic speeds, the full-scale vehicle is likely to have long runs of laminar flow if it has reasonably smooth surfaces. Reynolds numbers as high as 7×10^7 without transition have been reported on rockets.

Force measurements in hypersonic tunnels are similar to those of subsonic and supersonic tunnels. Low model loads at high Mach numbers are a problem in hypersonic force tests. Aerodynamic loads in some cases may be considerably less than the model weight. This poses a problem in balance design. The balance must be strong enough to hold the model, but it must also be weak enough to be sensitive to loads smaller than the model weight. Under these conditions, there is likelihood of continuous low-frequency oscillations of the model during a test. These oscillations may cause inertia loads to become a significant portion of the aerodynamic loads to be measured and satisfactory data cannot be obtained unless the data acquisition system is equipped with suitable electronic filtering.

For better visualization at low test-section pressures, the air in the flow field of the model may be ionized using an electric current (used by the Jet Propulsion laboratory).

Hypervelocity facilities are experimental facilities that allow testing and research at velocities considerably above those achieved in the supersonic and hypersonic tunnels. Experimental facilities which have been developed to simulate realistic flow conditions at high speeds and which are used extensively in high-speed testing are the hotshot tunnels, plasma jets, shock tubes, shock tunnels, and light gas guns.

Hotshot tunnels are devices meant for the generation of high-speed flows with high temperatures and pressures for a short duration.

Plasma arc tunnels are devices capable of generating high-speed flows with very high temperature. Temperatures of the order of $13,000°C$ or more can be achieved in the test gas.

A *shock tube* is a device to produce a high-speed flow with high temperatures, by traversing a normal shock wave which is generated by the rupture of a diaphragm separating a high-pressure gas from a low-pressure gas.

Shock tunnels are wind tunnels that operate at Mach numbers of the order 25 or higher for time intervals up to a few milliseconds by using air heated and compressed in a shock tube.

A *gun tunnel* is quite similar to the shock tunnel, in operation. A piston placed in the driven section is propelled through the driven tube, compressing the gas ahead of it.

Basically a Ludwieg tube tunnel is a blowdown-type hypersonic tunnel. Due to its special fluid dynamic features, no devices are necessary to control the pressure or temperature during the run. It thus can be regarded as an "intelligent blowdown facility." The test gas storage occurs in a long charge tube which is, by a fast acting valve, separated from the nozzle, test-section, and the discharge vacuum tank. Upon opening of the valve, an unsteady expansion wave travels with the speed of sound a_T into the charge tube. This wave accelerates the gas along its way, to a tube Mach number M_T given by the area ratio of the tube/nozzle throat. During the forward and backward traveling time of the wave, a constant steady flow to the expansion nozzle is established with pressure and temperature determined by one-dimensional unsteady expansion process. Upon return of the end wall reflected wave at the nozzle throat, the valve is closed and the test is finished.

Exercise Problems

3.1 If the velocity fluctuations in a subsonic wind tunnel test-section are isotropic, with a magnitude of 0.5 percent of the flow velocity, determine the turbulence number of the flow.

[Answer: 0.29]

3.2 Flow in the test-section of a subsonic wind tunnel is at an angle to the tunnel axis. Because of this angularity a pitot-static probe measures the test-section velocity as 19.9 m/s, instead of the correct value of 20 m/s. Determine the test-section dynamic pressure that would be indicated by the pitot-static probe, in mm of water, and the flow angularity. The stagnation pressure and temperature of the test-section are standard sea level values.

[Answer: 24.75 mm, 5.73°]

3.3 An open circuit subsonic wind tunnel runs by drawing standard sea level air. If a U-tube mercury manometer connected to a pressure tap in the test-section wall measures 260 mm suction, (a) calculate the test-section velocity, and (b) estimate the error in the velocity, calculated by assuming the test-section flow as incompressible.

[Answer: (a) 255.45 m/s, (b) 6.87 percent]

3.4 The contraction cone of an open-circuit subsonic wind tunnel draws air from standard sea level atmosphere and accelerates it to the test-section velocity of 75 m/s. If the flow velocity at the inlet is 4 m/s, find the contraction ratio of the contraction cone and the test-section Mach number.

[Answer: 18.31, $M = 0.22$]

3.5 The drag of a turbulence sphere in a subsonic wind tunnel test-section is measured to be 0.15 N. If the test-section speed is 45 m/s and the stagnation state is standard sea level condition, find the sphere diameter and the turbulence factor of the tunnel.

[Answer: 0.023 m, 5.46]

3.6 A Mach 1.8 wind tunnel of square test-section is run by air at a stagnation state of 6 atm and 330 K. If the area of the first throat is 0.1 m^2, (a) find the test-section Reynolds number, based on the height of the square test-section, and (b) the limiting minimum area of the second throat required to start the tunnel.

[Answer: (a) 2.64×10^7, (b) 0.123 m^2]

3.7 A supersonic tunnel is designed for Mach 1.6, with isentropic theory. In the actual operation, the friction causes 2 percent error to the test-section Mach number. When the test-section pressure is 143 kPa, (a) what will be the error in the test-section total pressure measured by a pitot probe? (b) If the mass flow rate per unit throat area is 1412.5 kg/(m^2 s), find the test-section velocity.

[Answer: (a) -3.45 percent, (b) 481.89 m/s]

3.8 An open-circuit subsonic wind tunnel of test-section area 1 m^2 draws atmospheric air at 1 atm and 30°C. (a) If a model of 4 percent blockage is tested at a test-section speed of 90 m/s, determine the change in mass flow rate per unit area, caused by the model blockage. (b) If the model blockage is 5 percent, how much would be the change in the mass flow rate per unit area?

[Answer: (a) 4.17 percent, (b) 5.26 percent]

3.9 A supersonic tunnel is run with a settling chamber pressure of 4.25 atm. If the test-section pressure is 1 atm, determine (a) the test-section dynamic pressure, and (b) the compressibility correction factor for this dynamic pressure.

[Answer: (a) 181574.4 Pa, (b) 1.814]

3.10 In an open-circuit subsonic wind tunnel, run by drawing sea level air, a pressure sphere is used to measure the turbulence factor. The test-section speed and temperature are 60 m/s and 22°C, respectively, and the resultant pressure difference between the front orifice and the four equally spaced orifices at 22° from the theoretical rear stagnation point is 98550 Pa. Determine (a) the turbulence factor of the tunnel and (b) the sphere diameter.

[Answer: (a) 1.2, (b) 64.87 mm]

3.11 The contraction cone of an open-circuit subsonic wind tunnel, with inlet and exit diameters of 3 m and 0.6 m, respectively, delivers flow at 70 m/s. The friction coefficient for the contraction cone is 0.005 and its length is 1.2 times the inlet diameter. When the tunnel draws air from standard sea-level atmosphere, (a) determine the pressure loss in the contraction cone and (b) the pressure drop coefficient for the contraction cone. (c) If the test-section is of diameter 0.3 m and length 1.5 m, determine the pressure drop coefficient for the test-section. Treat the flow to be incompressible. (*Hint:* For cylindrical sections for Reynolds number around 1.5×10^6, the friction factor can be taken as 0.0116).

[Answer: (a) 17.98 Pa, (b) 0.005, (c) 0.058]

3.12 A yaw sphere is used to measure the flow angularity in the test-section of an open-circuit subsonic wind tunnel, which draws air from standard sea-level atmosphere. Determine the yaw-head for the sphere when the test-section flow of velocity 50 m/s is at 3° yaw to the tunnel axis, when the tunnel draws air from standard sea-level atmosphere.

[Answer: 0.035/degree]

3.13 The diffuser of a subsonic tunnel has 5° semi-divergence angle. The inlet and exit diameters of the diffuser are 0.3 m and 1 m, respectively. Determine the total loss coefficient for the diffuser, if the friction coefficient is 0.03.

[Answer: 0.1863]

3.14 A Mach 2 blowdown tunnel of test-section diameter 200 mm is run by a storage tank of volume 10 m^3, with air at 15 atm and 30°C. The pressure ratio required for running the tunnel is 3.3 and the pressure loss in the pressure regulating valve is 20 percent. Assuming the temperature in the storage tank is invariant during the tunnel run and the polytropic index $n = 1.4$, determine the running time for the blowdown tunnel.

[Answer: 7.42 s]

3.15 At the entrance of an open-circuit subsonic wind tunnel, the flow velocity along the tunnel axis is 1.3 m/s and near the wall it is 1.1 m/s. If the honeycombs have to make the velocity uniform within 10 percent of the entry velocities difference, before entering the contraction cone, determine the number of honeycomb stages required, when the velocity drop caused by each honeycomb stage is 30 percent of the velocity entering a honeycomb.

[Answer: 7 stages]

3.16 A correctly expanded Mach 2 nozzle, run by a settling chamber at 8 atm, discharges air into an isentropic diffuser. If the total pressure at the outlet of the diffuser is 6 atm, determine the diffuser efficiency.

[Answer: 82.25 percent]

3.17 The pressure distribution over a two-dimensional wedge of vertex angle 10° and length 100 mm is to be measured at Mach 1.6. If the blockage due to the model is 6 percent of the test-section area of a Mach 1.6 wind tunnel, can this model be tested in the tunnel?

[Answer: yes.]

3.18 If the dynamic pressure in the test-section of a Mach 1.7 wind tunnel is 100 kPa, determine the total pressure of the test-section flow.

[Answer: 244 kPa]

3.19 If the model blockage should not cause more than 5 percent change of test-section Mach number, (a) determine the maximum permissible model blockage for a Mach 0.5 test-section. Also, find whether this blockage will change with change of Mach number in the subsonic range. (b) Determine the maximum permissible model blockage for a Mach 2 test-section. Also, find whether this blockage will change with change of Mach number in the supersonic range.

[Answer: (a) 3.36 percent, for a subsonic tunnel the blockage causing 5 percent of change (increase) in test-section Mach number decreases with increase of Mach number and increases with decrease of Mach number, (b) 7.88 percent, for a supersonic tunnel the blockage causing 5 percent of change (decrease) in test-section Mach number increases with increase of Mach number and decreases with decrease of Mach number.]

3.20 A Mach 2 blowdown tunnel of square test-section of area 0.0225 m² is run by a settling chamber at 8 atm and 300 K. (a) If an induction type tunnel of square test-section has to be designed for the same Mach number and mass flow rate, what should be the test-section height, when it runs at standard sea level atmosphere? (b) If the tunnel has to operate at correctly expanded state, what should be the backpressure?

[Answer: (a) 420 mm, (b) 12949.3 Pa]

3.21 Determine the compressor pressure ratio of a Mach 1.7 tunnel run with a diffuser with isentropic efficiency of 0.95. Also, find the energy ratio of the tunnel.

[Answer: 1.067, $ER = 19.57$]

3.22 The total pressure at the exit of a supersonic wind tunnel diffuser is 80 percent of the inlet total pressure. (a) If the efficiency of the tunnel is 0.7, determine the compressor pressure ratio, and (b) the test-section Mach number, using design consideration.

[Answer: (a) 1.786, (b) 4.55]

3.23 Determine the maximum possible contraction ratio for the diffuser of a Mach 2.8 tunnel.

[Answer: 1.363]

3.24 A supersonic tunnel is run by a compressor with 2 stages. The compressor pressure ratio required to run the tunnel is 1.3. If the power required to run the compressor is 40 hp, determine the mass flow rate in the test-section, running with stagnation temperature 360 K.

[Answer: 1.08 kg/s]

3.25 If the overexpansion level of a Laval nozzle, run by a settling chamber at 5 atm and discharging to sea level atmosphere, is 36.1 percent, determine the design Mach number of the nozzle, treating the flow through the nozzle as adiabatic and reversible.

[Answer: 2.0]

3.26 A Mach 3 supersonic tunnel of test-section area 0.0025 m^2 is run with stagnation state of 15 atm and 320 K. Determine (a) the power required to start the tunnel, and (b) the running power required.

[Answer: (a) 2749.06 hp (b) 549.8 hp]

3.27 (a) If the specific power required to run a closed circuit supersonic tunnel, maintaining the test-section total temperature at 430 K, is 149.2 kJ/kg, determine the test-section Mach number. (b) If the mass flow rate through the test-section of area 0.002 m^2 has to be 0.8 kg/s, what should be the total pressure in the settling chamber? (c) Determine the running power required. Assume the compression in the compressor is isentropic.

[Answer: (a) 2.82, (b) 732.1 kPa, 160 hp]

3.28 A wind tunnel draws atmospheric air at 20°C and 101.3 kPa by a large fan located near the exit of the tunnel. If the air velocity in the tunnel is 80 m/s, determine the pressure in the tunnel.

[Answer: 97.46 kPa]

Chapter 4

Flow Visualization

4.1 Introduction

Visualization of fluid flow motion proved to be an excellent tool for describing and calculating flow properties in many problems of practical interest, in both subsonic and supersonic flow regimes. It may seem surprising that a real grasp of the nature of aerodynamic motions was gained only in the last few decades, but the invisibility of air is a hindrance to understanding its motion completely. Fortunately, researchers in this field have developed many techniques to visualize the motion of air. Smoke streaks can be introduced into the flow to indicate not only its direction, but also whether it is smooth or disturbed. Small tufts of wool or even fine strands of hen-feathers or cat's whiskers can also be utilized to show direction and oscillations in a flow field. In the flow of water, aluminum powder sprinkled at the liquid surface will indicate local motions which can be photographed. Flow in water can also be studied by streams of hydrogen bubbles released by electrolysis from electrically charged thin wires arranged across it. If attention is focused on the layer of air near a surface, chemical films can be applied on which the air will create patterns by evaporation, showing the direction and steadiness of the flow. In supersonic flows, the air density changes are sufficiently large to allow the air to be photographed directly, using optical systems which are sensitive to density changes. In this chapter we will study some of the widely used popular techniques which are often employed for fluid flow analysis.

4.2 Visualization Techniques

The general principle for flow visualization is to render the "fluid elements" visible either by observing the motion of suitable selected foreign materials added to the flowing fluid or by using an optical pattern resulting from the variation of the optical properties of the fluid, such as refractive index, due to the variation of the properties of the flowing fluid itself. A third class of visualiza-

205

tion technique is based on a combination of the above two principles. Each of these groups of techniques is generally used for incompressible, compressible, and low-density gas flow, respectively (Madeleine Coutanceau and Jean-Rene Defaye, 1991). Since our intention is to consider only continuum flows, we will be discussing only the first two groups of flow visualization techniques. Some visualization techniques popularly used to study flow problems of practical interest are the following.

- _Smoke flow visualization_ is one of the popular techniques used in low-speed flow fields with velocities up to about 30 m/s. Smoke visualization is used to study problems like boundary layers, air pollution problems, design of exhaust systems of locomotives, cars, ships, topographical influence of disposal of stack gases, etc.

- _Tufts_ are used to visualize flow fields in the speed range from 40 to 150 m/s. This technique is usually employed to study boundary layer flow, wake flow, flow separation, stall spread, and so on.

- _Chemical coating_ is used to visualize flow with speeds in the range from 40 to 150 m/s. Boundary layer flow, transition of the flow from laminar to turbulent nature, and so on are usually described by this visualization technique.

- _Interferometer_ is an optical technique to visualize high-speed flows in the ranges of transonic and supersonic Mach numbers. This gives a qualitative estimate of flow _density_ in the field.

- _Schlieren_ technique is used to study high-speed flows in the transonic and supersonic Mach number ranges. This again gives only a qualitative estimate of the _density gradient_ of the field. This is used to visualize faint shock waves, expansion waves, etc.

- _Shadowgraph_ method is yet another flow visualization technique meant for high-speed flows with transonic and supersonic Mach numbers. This is employed for fields with strong shock waves.

4.2.1 Smoke Tunnel

Flow visualization with smoke is generally done in a smoke tunnel. It is a low-speed wind tunnel carefully designed to produce a uniform steady flow in the test-section with negligible turbulence. Smoke streaks are injected along the freestream or on the surface of the model for visualizing flow patterns. White dense smoke is used for this purpose. When a beam of light is properly focused on the smoke filaments, the light gets scattered and reflected by the smoke particles making them distinguishably visible from the surroundings. A smoke tunnel is generally used for demonstrating flow patterns such as flow around bodies of various shapes, flow separation, etc.

The quality of the flow pattern depends on the quality of the smoke used. For good results, properties of the smoke chosen should have the following qualities.

- The smoke should be white, dense, nonpoisonous, and noncorrosive.

- Smoke should have nearly the same density as that of the surrounding air, hence the smoke filaments are not appreciably influenced by gravity.

- Smoke particles should not disturb the flow in the wind tunnel by formation of deposits on the surface of the models or block the tubes used for smoke injection.

- Production of smoke should be easily and readily controllable.

4.2.1.1 Smoke Production Methods

There are many methods as well as materials used for producing smoke. Some of the popular methods in use are the following.

4.2.1.2 Titanium Tetrachloride

It is a liquid which when exposed to moist air produces copious fumes of titanium dioxide along with hydrochloric acid. This smoke though very dense is somewhat toxic and forms deposits and hence, it is not recommended for flow visualization, except for very short periods of use. Stannic chloride also has the same property as titanium tetrachloride.

4.2.1.3 Ammonia and Hydrochloric Acid

When ammonia vapor is passed over hydrochloric acid, white fumes of ammonium chloride are formed. The corrosive nature of the fume and the tendency to form white deposits make this method of smoke production undesirable, except on rare occasions.

4.2.1.4 Wood Smoke

Smoke can be produced by partial combustion of damp wood shavings. However, this smoke is rich in tar and water vapor. It has to be initially filtered thoroughly by passing through a filter and a water condenser, as shown in Figure 4.1. White and dense smoke can be produced by this method. But handling the apparatus is somewhat cumbersome since the burnt wood shavings have to be removed often. By choosing proper quality of wood, smoke having an agreeable aroma could be obtained. If one can afford it, sandalwood shavings are recommended for a good aroma. Addition of a small quantity of liquid paraffin will appreciably increase the optical density of the smoke.

4.2.1.5 Kerosene Smoke Generator

In this method, smoke is produced by evaporation and atomization of kerosene in an air stream. The system is compact and electrically operated. Smoke can be

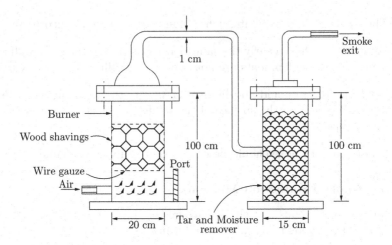

Figure 4.1: Wood smoke generator

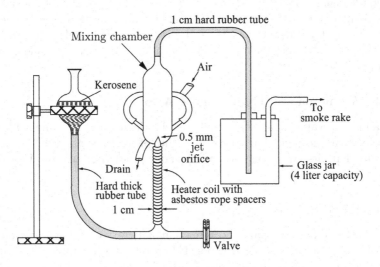

Figure 4.2: Kerosene smoke generator

generated within a few minutes. No solid deposits are formed. A typical kerosene smoke generator is shown in Figure 4.2.

The kerosene smoke generator consists of a reservoir, an electrically heated glass tube to the top of which is attached a narrow nozzle and a mixing chamber. Kerosene is heated to form vapor which emerges through the nozzle in the form of a jet inside the mixing chamber. Two air jets impinging on the nozzle opening atomizes the hot kerosene vapor forming white smoke. The smoke is taken out through the outlet provided at the top of the mixing chamber.

The electrical heater is made by winding heating wire on the glass cylinder. Wires are wound using asbestos rope in between as the insulator. A rheostat is connected in series to adjust the heating to the desired value.

In the operation of a kerosene smoke generator the following procedure has to be followed.

- Fill the reservoir with pure kerosene.

- Adjust the reservoir height until the kerosene level in the boiler glass tube reaches 8 to 10 cm below the nozzle.

- Switch on the heater. If the rheostat is adjusted properly the kerosene should boil within 5 minutes.

- Open the valve admitting air into the mixing chamber.

- Adjust the valve until dense white smoke appears. If the airflow is too much kerosene vapor will condense without forming smoke. Any condensed kerosene formed inside the mixing chamber should be removed through the drain provided by the side of the nozzle.

- After running the apparatus for some time, the reservoir can be raised to a height which just prevents spurting of the hot oil through the orifice. The interior of the mixing chamber should always appear dry.

Overheating and insufficient airflow can cause the formation of a black carbon deposit, especially at the nozzle, eventually blocking the orifice. In that case, the current supply to the heater has to be reduced. The blocked orifice can be opened by introducing a fine steel wire through the smoke outlet tube. Addition of a small quantity of liquid paraffin with kerosene will enhance the density of the smoke. The rubber tubes connecting the reservoir and the boiler glass tube should be thick and of high quality. This tube should be checked periodically for deterioration and replaced whenever necessary. Constant attention is required while operating the apparatus.

A large volume (approximately 4 liters) reservoir is provided, between the smoke generator and the tunnel smoke rake, to suppress fluctuations formed during the smoke generation process. This reservoir also acts as a condenser for the smoke. Some air pressure is always needed for the formation of the smoke and if the smoke formed is in excess of that required, it should be let out. With proper maintenance and careful handling, a smoke generator of this type can serve for many years without breakdown.

The production of steady thin smoke filaments in the test-section is possible only when the flow velocity is small, say 2 to 5 m/s. At high velocities, the wake lines formed at the smoke rakes become turbulent. It is also essential that the tunnel is kept away from natural air draughts.

Proper lighting arrangement is necessary to observe the smoke patterns especially while taking photographs of the smoke pattern. Lighting from above and below the tunnel is found to be the best arrangement. Deep shadows should be avoided if photographs of the flow patterns are to be taken and the lighting has to be adjusted carefully for this purpose.

The smoke filament lines coming out should be made to flow through a properly designed tunnel-section in order to make flow visualizations over models of interest. These tunnels are called smoke tunnels.

4.2.2   Design of Smoke Tunnel

Smoke tunnel is basically an open circuit, low-speed tunnel. The flow through the tunnel is induced by the suction of atmospheric air through the test-section, using a simple axial fan. Schematic diagram of a typical smoke tunnel is shown in Figure 4.3.

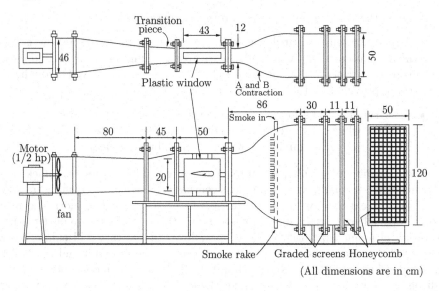

(All dimensions are in cm)

Figure 4.3: Smoke tunnel

The major components of the tunnel are: a double contraction, the test-section, the diffuser, the axial fan and motor assembly, the smoke rake, and the smoke generator. The size of the test-section shown in Figure 4.3 is 100 mm in width, 300 mm in height, and 450 mm in length. A double contraction made up of two parts, both two-dimensional, is used. The first section (A) reduces the lateral size from 400 mm to 100 mm, keeping the longitudinal dimension constant and the second section (B) reduces only the longitudinal dimension from 1200 mm to 300 mm. Thus at the end of the contraction the total area gets reduced by a factor of sixteen. The entry section of the smoke tunnel is fitted with a 12.5 mm × 12.5 mm honeycomb followed by two graded screens spaced 150 mm apart. The honeycomb is meant to break the large eddies entering the tunnel from outside into smaller size and the screens are meant for reducing the turbulence level. A contraction ratio of sixteen is sufficient to obtain a relatively smooth flow, with low turbulence level, in the test-section although tunnels with contraction ratios as high as 20 are also in use.

All four sides of the test-section should be provided with clear perspex windows. The top window is meant for lighting the smoke filaments from the top while the large side windows, which are detachable, are used for visualizing the flow and for fixing the model. A 2400-mm-long conical diffuser is fixed to the downstream end of the test-section with a rectangular to circular transition piece in between. The flow is sucked at the diffuser end by a four bladed axial fan coupled directly to a $\frac{1}{2}$ hp d.c. motor whose speed can be controlled by varying its voltage.

The smoke rake consists of a number of small tubes connected to the main tube in parallel, as shown in Figure 4.4.

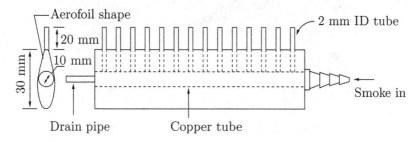

Figure 4.4: Smoke rake

An airfoil shape finish is given to the rake to obtain clean steady smoke filaments. The rake is fixed in the second contraction at a suitable position, where there is enough favorable pressure gradient to suppress vortex formation behind the airfoil section of the rake. A drain screw is provided at the bottom of the main hollow cylinder to remove the condensed kerosene, whenever it accumulates. The inlet of the smoke rake is connected to the smoke generator through a rubber tube.

4.2.2.1 Application of Smoke Visualization

Many interesting flow patterns can be demonstrated in a smoke tunnel and two specific patterns visualized in a smoke tunnel are listed below:

Flow over an Airfoil

Make an airfoil with its thickness equal to the width of the smoke tunnel test-section, with a spindle. The material used should preferably be soft and the surface should be highly polished and finished with black color by painting or blackodizing the surface. Fix the airfoil through one of the holes on the back plastic window such that one of the smoke lines touches the nose of the airfoil. When the airfoil is at zero angle of attack the smoke filament will smoothly flow over the airfoil and leave the trailing edge smoothly. Now, change the angle of attack by rotating the airfoil by a few degrees every time. At one stage the separation will become clearly visible and move upstream on the airfoil as the angle of attack is further increased.

Smoke flow pattern over a symmetrical airfoil at an incidence is shown in Figure 4.5.

Figure 4.5: Smoke pattern over an airfoil

From Figure 4.5 it is seen that, for an incidence more than stalling angle, the flow separation is right at the nose of the airfoil. Further, in the separated zone the smoke filaments have got mixed due to the vortex motion, resulting in a hasty flow pattern. Likewise, many interesting problems of practical importance can be studied with smoke visualization.

Flow over a Blunt Body

Smoke pattern over a blunt body is shown in Figure 4.6. It is seen that the flow reaches the stagnation point at the face of the body. Also, the streamlines begin to deflect well ahead of the body, negotiate the body contour and flow parallel to the wall, up to the rear end. This is because, the low-speed incompressible flow receives information about the presence of the body, through the acoustic waves which are traveling faster than the flow. Owing to this advance information, the streamlines could be able to adjust to move smoothly around the body. But at the rear end, there is an abrupt change in the body contour, with a sharp corner. Because of this, the flow gets separated. The smoke filaments cannot retain their shape any more and get mixed up with wake. This is a typical demonstration of the limitation of smoke visualization that the technique will not be able to give details about separated flows.

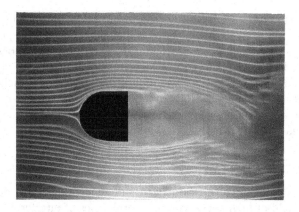

Figure 4.6: Smoke pattern over a blunt body

4.3 Compressible Flows

For visualizing compressible flows, optical flow visualization techniques are commonly used. *Interferometer*, *schlieren*, and *shadowgraph* are the three popularly employed optical flow visualization techniques for visualizing shocks and expansion waves in supersonic flows. They are based upon the variation of the refractive index, which is related to the fluid density by the Gladstone–Dale formula and consequently to the pressure and velocity of the flow. For making these variations visible, three different classes of methods mentioned above are generally used. With respect to a reference ray, that is, a ray which has passed through a homogeneous field with refractive index n, the

- *Interferometer* makes visible the *optical phase changes* resulting from the relative retardation of the disturbed rays.

- *Schlieren* system gives the *deflection angles of the incident rays*.

- *Shadowgraph* visualizes the *displacement experienced by an incident ray* which has crossed the high-speed flowing gas.

These optical visualization techniques have the advantage of being non-intrusive and thereby in the supersonic regime of flow, avoiding the formation of unwanted shock or expansion waves. They also avoid problems associated with the introduction of foreign particles which may not exactly follow the fluid motion at high-speeds, because of inertia effects. However, none of these techniques gives information directly on the velocity field. The optical patterns given by interferometer, schlieren, and shadowgraph, respectively, are sensitive to the *flow density*, *its first derivative*, and *its second derivative*. For quantitative evaluation, interferometry is generally chosen because this evaluation is based upon the precise measurement of fringe pattern distortion instead of the not so precise measurement of change in photographic contrast, as in schlieren and shadowgraph. However, schlieren and shadowgraph visualizations being useful and less expensive, are often used to visualize flow patterns, especially at supercritical Reynolds numbers. In particular, they clearly show shock waves and, when associated with ultrashort duration recordings, they also show the flow structure. Although these optical techniques are simple in principle, they are rather difficult to implement. High precision and high optical quality of the setup components, including the wind tunnel test-section windows, are required for proper visualization with these techniques.

4.3.1 Interferometry

Interferometer is an optical method most suited for qualitative determination of the density field of high-speed flows. Several types of interferometer are used for the measurement of the refractive index, but the instrument most widely used for density measurements in gas streams (wind tunnels) is that attributed to Mach and Zhender.

The fundamental principle of the interferometer is the following. From the wave theory of light we have

$$C = f\lambda \tag{4.1}$$

where C is the velocity of propagation of light, f is its frequency, and λ is its wavelength.

From corpuscular properties of light, we know that when light travels through a gas the velocity of propagation is affected by the physical properties of the gas. The velocity of light in a given medium is related to the velocity of light in vacuum through the index of refraction n, defined as

$$\frac{C_{\text{vac}}}{C_{\text{gas}}} = n \tag{4.2}$$

The value of refractive index n is 1.0003 for air and 1.5 for glass.

The Gladstone–Dale empirical equation relates the refractive index n with the density of the medium as

$$\frac{n-1}{\rho} = K \tag{4.3}$$

where K is the Gladstone–Dale constant, and is constant for a given gas and ρ is the gas density.

4.3.1.1 Formation of Interference Patterns

Figure 4.7 shows the essential features of the *Mach–Zhender interferometer*, schematically.

Light from the source is made to pass through lens L_1 which renders the light parallel. The parallel beam of light leaving the lens passes through a monochromatic filter. The light wave passes through two paths; 1-2-4 and 1-3-4, before falling on the screen, as shown in the figure. The light rays from the source

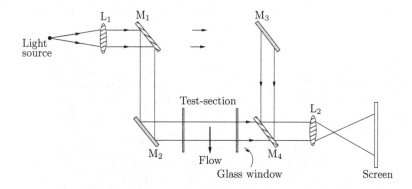

Figure 4.7: Mach–Zhender interferometer

are divided into two beams by the half-silvered mirror M_1. The two beams, after passing through two different paths (the lengths of paths being the same), recombine at lens L_2 and get projected on the screen. The difference between the two rays is that one (1-3-4) has traveled through room air while the other (1-2-4) has traveled through the test-section. When there is no flow through the test-section, the two rays having passed through identical paths are in phase with each other and recombine into a single ray. Thus, a uniform patch of light will be seen on the screen. Now, *if the density of the medium of one of the paths is changed (say increased) then the light beam passing through will be retarded and there will be a phase difference between the two beams.* When the magnitude of the phase difference is equal to $\lambda/2$, the two rays interfere with each other giving rise to a dark spot on the screen. Hence, if there is appreciable difference in the density the picture on the screen will consist of dark and white bands, the phase difference between the consecutive dark bands being equal to unity. A interferogram of a two-dimensional supersonic jet is shown in Figure 4.8.

It is seen that, far away from the jet axis, the fringes (dark and white bands) are parallel, indicating that the flow field is with uniform density (in this case, the zone is without flow). The mild kinks in the fringes are the location of density change. Those who are familiar with free jet structure can easily observe the barrel shock, the Mach disk, and the reflection of the barrel shock from the Mach disk.

4.3.1.2 Quantitative Evaluation

Before attempting any quantitative evaluation of an interferogram, it is imperative to understand the "bands" on the picture. If the optical path of each ray in one leg of the interferometer is equal to the optical path of the corresponding ray in the other leg, the split beams coming together at M_4 in Figure 4.7 will reinforce each other so as to render the screen uniformly bright. This also will occur if the optical path lengths are different by an integer number of wave-

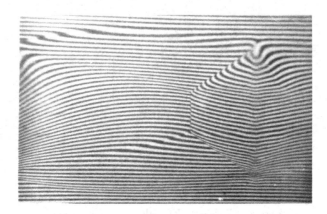

Figure 4.8: Interferogram of a two-dimensional supersonic jet at $M = 1.62$ (H. Shames, *Mechanics of Fluids*, McGraw-Hill, 1962)

length. Let us assume that this is the condition of the interferometer when there is no flow in the test-section. When there is flow in the test-section, the optical path lengths of each ray through the test-section may change depending on the density change encountered by the ray traversing the flow. If a ray going through the test-section has an optical path length which is an integer plus one-half of wavelength difference from the corresponding ray through the other leg of the interferometer, there will be complete destructive interference of the two rays as they emerge from the mirror M_4. This will give rise to a "black band" on the screen. Similarly, a complete constructive interference will result in a "white band" on the screen. Therefore, there is a possibility of a range of partially constructive and partially destructive interference, giving rise to a "gray zone" on the screen. The picture then will be a series of black and white fringes with a variation of hue between the fringes.

At this stage, we must note that a slight rotation of the mirrors M_1 and M_4 about their vertical axes will result in a series of equally spaced vertical fringes. In practice, this is the initial setting usually taken, since the use of such an initial setting makes it easier to ascertain the amount of retardation associated with each fringe when there is a flow present and in this way permits an easier means of determining the density field of the flow.

With this background we can now attempt to evaluate the interferogram quantitatively. We know that, on the dark bands of Figue 4.8, the light waves passing through the test-section are out of phase with those which pass through the room air in the compensating chamber by $1/2, 3/2, 5/2, \ldots$ of the wavelength λ_{room} of light in the room atmosphere. Therefore, the light beams passing through the adjacent dark bands of the test-section are out of phase by $1 \, \lambda_{\text{room}}$. Hence, if a represents the fluid lying in one dark band, and b the fluid in an adjacent dark band, the difference in time for a light beam to pass through a as compared to that passing through b is given by

$$t_b - t_a = \frac{\lambda_{\text{room}}}{C_{\text{room}}} \tag{4.4}$$

where t_a is the time for the light to pass through region of density a and t_b is the time required for the light to pass through region of density b. Let C_a and C_b be the velocities of propagation of light through regions a and b and L be the length of the test-section along the light direction.

We know that the frequency f of a given monochromatic light is constant. Therefore,

$$f = \frac{V}{\lambda} = \frac{C_{\text{room}}}{\lambda_{\text{room}}} = \frac{C_a}{\lambda_a} = \frac{C_b}{\lambda_b} = \frac{C_{\text{vac}}}{\lambda_{\text{vac}}} \tag{4.5}$$

The difference in travel time given by Equation (4.4) may also be expressed in terms of difference in speed of light in the test-section, using Equation (4.5), as

$$t_b - t_a = \frac{1}{f} = \frac{\lambda_{\text{vac}}}{C_{\text{vac}}}$$

Also,

$$t_b - t_a = \frac{L}{C_b} - \frac{L}{C_a} = \frac{\lambda_{\text{vac}}}{C_{\text{vac}}} \tag{4.6}$$

where L is the test-section width.

The velocity of light in a given medium is related to the velocity of light in vacuum through the index of refraction n, defined as

$$n \equiv \frac{C_{\text{vac}}}{C}, \quad n_a \equiv \frac{C_{\text{vac}}}{C_a}, \quad n_b \equiv \frac{C_{\text{vac}}}{C_b} \tag{4.7}$$

i.e.,

$$n_b - n_a = C_{\text{vac}} \left(\frac{1}{C_b} - \frac{1}{C_a} \right) = \frac{C_{\text{vac}}}{L} \frac{\lambda_{\text{vac}}}{C_{\text{vac}}}$$

That is,

$$n_b - n_a = \frac{\lambda_{\text{vac}}}{L} \tag{4.8}$$

Now, the index of refraction may be connected to the gas density through the empirical Gladstone–Dale equation (Equation (4.3)) to result in

$$\boxed{\rho_b - \rho_a = \frac{\lambda_{\text{vac}}}{LK}} \tag{4.9}$$

The right-hand side of this equation can easily be computed from the dimension of the test-section, the color of the monochromatic light used, and the value of K for air. The density in the low-speed flow upstream of the nozzle throat may be found by measuring the temperature and pressure in that region. With that region as a reference, the density on each dark band in the nozzle may be computed from Equation (4.9). This kind of interferogram is also termed "infinite-fringe," which signifies that the light field is uniform in the absence of flow through the test-section. Although, in principle it is possible to compute the density field quantitatively, as discussed above, using interferograms, the accuracy of this procedure using the infinite-fringe interferogram will not be high unless the optical components are extraordinarily accurate. In fact, this is one of the major hurdles in the use of this technique for the quantitative evaluation of compressible flow fields.

4.3.2 Fringe-Displacement Method

This method is used when a more accurate quantitative estimate is required. This is just a modified version of the infinite-fringe technique described above. Let us consider the interferometer arrangement shown in Figure 4.7. Let the mirror M_3 be rotated through a small angle with respect to mirror M_1. The two rays of light which were in phase at M_1 will now be out of phase at the screen. Thus, the image on the screen (with no flow in the test-section) will consist of alternate white and dark bands, uniformly spaced, with each fringe lying parallel to the axis of rotation. The spacing of successive dark fringes may be shown to be equal to $\lambda/2\delta$, where δ is the difference in the angles of rotation between the two splitters (i.e., mirrors M_1 and M_3).

Now, assume that the air density in the test-section is increased uniformly. This will result in a uniform displacement of all the wavefronts passing through the test-section. This displacement in turn will cause the interference bands on the screen to shift in a direction normal to the bands, even though the bands will remain parallel and uniformly spaced. The fringe shift is a measure of density change in the test-section. It can be shown that,

$$\rho_2 - \rho_1 = \frac{\lambda_{\text{vac}}}{LK}\frac{l}{d} \tag{4.10}$$

where ρ_1 and ρ_2 are the density at the initial reference condition and density in the test-section, respectively, d is the distance between the dark fringes in the reference condition, and l is the distance shifted by a dark fringe in passing from condition 1 to condition 2.

When there is flow in the test-section, non-uniform fringe shift will occur corresponding to the density field, the resultant fringes will be curved. Equation (4.10) may then be applied at each point in the flow. If both flow and no-flow photographs are taken, Equation (4.10) may be used to determine the density change at each point, with respect to the no-flow density.

4.3.3 Schlieren System

The schlieren method is a technique for visualizing the density gradients in a transparent medium. Figure 4.9 shows a typical schlieren arrangement, usually employed for supersonic flow visualization.

Light from a source is collimated by the first lens and then passed through the test-section. It is then brought to a focus by the second lens and projected

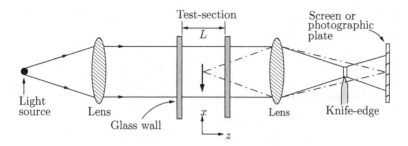

Figure 4.9: Schlieren system

on the screen. At the focal point of the second lens, where the image of the source is formed, a knife–edge (which is an opaque object) is introduced to cut off part of the light. The screen is made to be uniformly illuminated by the portion of the light escaping the knife–edge, by suitably adjusting it to intercept about half the light, when there is no flow in the test-section. For the sake of simplicity, for instance, let us assume the test-section to be two-dimensional, with each light ray passing through a path of constant air density. When flow is taking place through the test-section, the light rays will get deflected, since any light ray passing through a region in which there is a density gradient normal to the light direction will be deflected as though it had passed through a prism. In other words, if the medium in the test-section is homogeneous (constant density) the rays from the source will continue in their straight line path. If there is density gradient in the medium, the rays will follow a curved path, *bending towards the region of higher density and away from the region of lower density.* Therefore, depending on the orientation of the knife–edge with respect to the density gradient, and on the sign of the density gradient, more or less of the light passing through each part of the test-section will escape the knife–edge and illuminate the screen. Thus, *the schlieren system makes density gradients visible in terms of intensity of illumination.* A photographic plate at the viewing screen records density gradients in the test-section as different shades of gray.

Let us assume that the flow through the test-section is parallel and in the xy-plane. Let the light be passing through the test-section in the z-direction. From theory of light it is known that *the speed of a wavefront of light varies inversely with the index of refraction of the medium through which the light travels.* Therefore, a given wavefront will rotate as it passes through a gradient in the refractive index n. Hence, the normal to the wavefront will follow a curved path. This effect is stated earlier in other words as "the ray will follow a curved path bending towards the region of higher density and away from the region of lower density." In such a case, the radius of curvature R of the light ray is proportional to $1/n$. It can be shown that

$$\frac{1}{R} = \text{gradient } n$$

The total angular deflection ϵ of the ray in passing through the test-section of width L is therefore given by

$$\epsilon = \frac{L}{R} = L \text{ grad } n$$

Resolving this into Cartesian components, we have

$$\epsilon_x = L\frac{\partial n}{\partial x} \qquad \epsilon_y = L\frac{\partial n}{\partial y}$$

Using Equation (4.3), these equations can be expressed as

$$\boxed{\epsilon_x = LK\frac{\partial \rho}{\partial x}} \tag{4.11}$$

$$\boxed{\epsilon_y = LK\frac{\partial \rho}{\partial y}} \tag{4.11a}$$

From Equations (4.11) and (4.11a) it is seen that the schlieren is sensitive to the first derivative of the density.

Referring to Figure 4.9 it can be visualized that, if the knife–edge is aligned normal to the flow, i.e., in the y-direction, only deflection ϵ_x will influence the light passing the knife–edge. Therefore, only density gradients in the x-direction will be made visible, and the gradients in the y-direction will not be visible. Similarly, if the knife–edge is aligned parallel to x-direction, only the gradients in the y-direction will be visible. A typical schlieren picture of a free jet is shown in Figure 4.10.

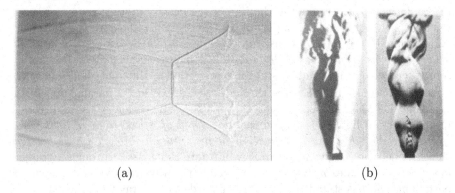

(a)   (b)

Figure 4.10: Schlieren picture: (a) Picture of a supersonic free jet; (b) picture of Bunsen flame with knife–edge vertical (left) and knife–edge horizontal (right)

At this stage, we should note that the schlieren lenses must not only be of high optical quality but also must have large diameters and long focal lengths.

The large diameter is necessary to cover the required portion of the flow field, which is often large in size (say 200 mm in diameter). The long focal length is necessary in order to get the "required" precision and image size. Further, the schlieren lens should be free of chromatic and spherical aberrations. Also, the astigmatism must be as small as possible.

In experiments where the region under study has a large cross-section as in the case of many modern wind tunnels, it is difficult to obtain lenses of sufficient diameter and focal lengths, and at the same time with the required optical properties. Even if such lenses are made specially for such use they will prove to be extremely expensive. As a result concave mirrors have been widely used. They are comparatively free from chromatic aberration and mirrors of large diameters and long focal lengths are much easier to grind and correct than lenses. A twin-mirror schlieren system that gives good resolving power is shown in Figure 4.11. The mirrors C and E are a carefully matched pair. Usually they are made of glass and their front surfaces are parabolized to better than one tenth of a wavelength of light. The excellence of their optical quality bears a direct relation with the image quality produced. Also, due to their size (often more than 300 mm in diameter) and weight they must be carefully mounted to avoid distortions.

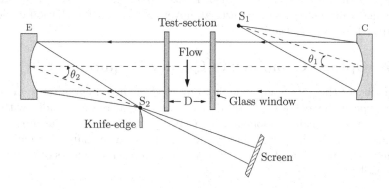

Figure 4.11: Twin-mirror schlieren system

In the schlieren setup arrangement, it is essential that the angle θ_1 must be approximately equal to angle θ_2 and their value should be as small as possible although angles up to about 7° are used successfully to obtain flow visualization of acceptable quality. The distance between the mirrors is not critical but it is a good practice to make it greater than twice the focal length of the mirrors. Also, the optical system beyond S_2 is simplified if the distance from the disturbance to be observed at test-section D to the mirror E is greater than the focal length of E. The parallel rays entering the region D are bent by the refractive index gradient and are no longer parallel to the beam from C and hence, cannot be focused by the second mirror unless the distance from D to the second mirror E is greater than the focal length of E.

The image of the test-section flow field (with the model) focused at the focal point at S_2 will diverge and proceed further. This image can be made to fall on a flat screen. The clarity of the image can be modified by adjusting the knife–edge. Proper adjustment of the knife–edge can result in sharp images of the shock (or compression) and expansion waves prevailing in the flow to fall on the screen. A still or video camera can record the image on the screen. When video camera is used, the image can be made to fall on the camera lens. This will avoid the parallax error associated with capturing the image from the screen with a still camera kept at an angle from the screen, without cutting the light rays from S_2.

4.3.4   Range and Sensitivity of the Schlieren System

Let us assume that the contrast on the screen is increased by reducing the size of the image. That is, the knife–edge is made to cut off most of the light, any ray deflecting beyond a certain limit will be completely cut off by the knife–edge, and further deflection will have no effect on the contrast. This means that the range gets limited. Increase in sensitivity affects the range of density gradient for which the system could be used. The contrast or sensitivity requirement depends on the problem to be studied. Hence, to adjust the contrast the knife–edge is generally mounted on a vertical movement so that its position could be altered with respect to the image.

4.3.5   Optical Components Quality Requirements

The quality of the optical equipment to be used in the schlieren setup depends on the type of investigation carried out. The cost increases rapidly with the quality of the optical components. The vital components are the mirror and the light source. Now, optical quality mirrors are easily available. The following specifications are sufficient to meet the visualization requirements of a 200-mm-diameter flow field.

4.3.5.1   Schlieren Mirrors

- Two parabolic mirrors of 200 mm diameter.

- Focal length of the mirrors about 1.75 m.

- Thickness of the mirror glass about 25 mm.

The reflecting surface of the mirrors is ground to an accuracy of 1/4 wavelength of sodium light and aluminized. Parabolic mirrors are the most suitable even though they are more expensive than spherical mirrors which also will serve the purpose. It is important to note here that, though optical finish of $\lambda/4$ is good enough for visualization of shock waves, if it is the aim to study the structure of the flow field (e.g., shear layers in a free jet, etc.) with ultra short schlieren photography, mirror surface finish of the order of $\lambda/20$ is essential.

4.3.5.2 Light Source

- Small intense halogen lamp of 30 watts is commonly used.

- Mercury vapor lamp of suitable intensity (say 200 watts) may also be employed.

- Provision to vary the intensity of light will prove to be useful.

4.3.5.3 Condenser Lens

Any condenser lens pair generally used for projection systems will be sufficient. This need not be of very high quality.

4.3.5.4 Focusing Lens

This lens is positioned in the schlieren system in such a way that a flow field is focused on the screen. An ordinary double convex lens can be used.

4.3.5.5 Knife-Edge

Any straight, sharp-edged, opaque object mounted on an adjustable stand will be sufficient to serve as knife–edge. Schlieren technique is generally used only for qualitative work, even though in principle it can be used for quantitative work. If quantitative measurements are to be done the density of the image has to be measured and this can be done with a photo densitometer. This instrument contains a photo cell and it is scanned over the photographic film of the schlieren image. By properly adjusting the exposure time the brightness of the pattern on the photographic print can be made proportional to the brightness of the schlieren system. The effect of knife–edge on the image obtained with schlieren is evident from Figure 4.10(b).

4.3.5.6 Color Schlieren

If the knife–edge which is kept at the focal point of the second mirror is replaced by a colored filter containing different colors, the image formed on the screen will have different colors depending on which way the beam bends. The contrast in the ordinary black and white schlieren will now be represented by colors. Usually the colors red, yellow, and green are used. These filters are of 1 or 2 mm in width and placed side by side. When there is no flow the image of the source is allowed to fall on the yellow portion of the filter. Now the image on the viewing screen will be completely yellow. When the density gradient is introduced the image gets displaced and falls partly on the neighboring filter thus altering the color on the screen. In the three filter color schlieren screen the color indicates the size of the density gradient also. The color effect described can also be achieved with a dispersion prism placed at the knife–edge location.

4.3.5.7 Short Duration Light Source

To study unsteady phenomena such as flow over a moving object or turbulent fluctuations in the wake of body or the mixing shear layer in a jet flow field, it is often necessary to take short duration exposures of the schlieren image to arrest (record) the unsteadiness in the photograph. For this the duration of exposure required is of the order of one or two microseconds or even less. An ordinary shutter in conjunction with a continuous light source is limited to an exposure time of not more than 1/1000 seconds. Therefore, to obtain shorter exposures the light sources capable of emitting light of very short duration should be employed. A condenser discharge type electric spark unit is commonly used for this purpose. Sparks of durations of the order of microseconds could be obtained by condenser discharge.

A spark light source of short duration can be made in the laboratory. A low inductance conductor and a high voltage D.C. source are the vital components of this unit. The simplest form of spark unit consists of two electrodes separated by an air gap and the electrodes are connected to the terminals of the capacitor. The discharge time depends on the value of the time constant of the system, that is, on the CR value, where C is the capacitance and R is the resistance which includes the effect of the inductance of the circuit. The total energy of the discharge is $\frac{1}{2}CV^2$, where V is the voltage. The energy discharge during the sparking gets partly dissipated into heat and the rest goes into electromagnetic radiation including the visible range. To obtain a discharge of very short duration it is necessary to use high voltage with a small capacitor having negligible inductance. In addition, the resistance of the overall circuit should be kept at a minimum by mounting the spark gap assembly directly on the condenser electrodes. A D.C. voltage of 10 kV and a capacitor of 0.1 microfarad is sufficient to make a good spark source. If carefully designed a discharge time of 1 microsecond could easily be achieved.

A spark source should be capable of being triggered whenever needed. This is generally done by using a hydrogen thyratron capable of conducting high current as well as high voltage. Sometimes a simple third ionizing electrode is employed since the hydrogen thyratrons are expensive and have a short life time. The electric circuit for the above system is shown in Figure 4.12. The electrode is used to reduce the resistance in the path between the two main electrodes by ionization. For the schlieren system a line source is needed and this could be achieved by placing two electrodes between two glass plates, as shown in Figure 4.12. The glass plates confine the spark instead of allowing it to wander around. The electrodes can be steel, nickel, aluminum, or even tungsten. Aluminum electrodes produce high-intensity light and increase the duration due to after glow. The after glow is in the low frequency of the visible spectrum and can be filtered out to some extent by using special optical filters.

In general, the schlieren method is used either for the detection of small refractive index gradients or for the quantitative measurement of these gradients. For the detection of small gradients the apparatus described in Figure 4.11 in which the deviation of the light ray ϵ gives rise to the relative light intensity

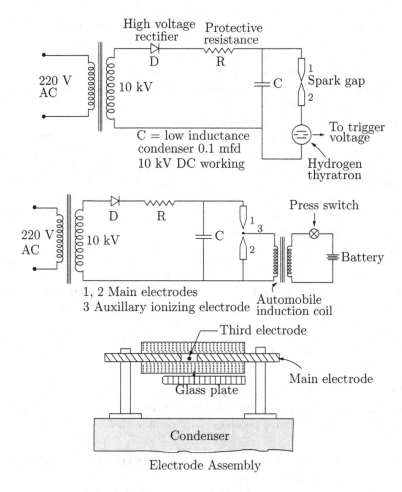

Figure 4.12: Schlieren spark source circuit

change $\Delta I/I$ on the photographic plate, is almost universally used for studying phenomena in gas dynamics. The method may also be made qualitative, but careful attention must be paid to the several variables in the experimental arrangement. If the disturbance is to be photographed, the arrangement must give a maximum contrast between the images of the undisturbed and disturbed regions and at the same time the photograph must be dense enough to be measurable by photometric means. In most cases high contrast photographic plates are preferable even though they are somewhat slower. In order to calibrate the system, a known refractive index gradient such as a small glass prism may be inserted in some corner of the median plane of the test-field zone which allows a check on the formulae used for the optical system. The sensitivity, i.e., $\Delta I/I$ depends directly upon the brightness of the image of S_1 and S_2 and upon its uniformity of illumination. This of course requires as bright and uniform a source as possible to start with and an optical system which sacrifices no more light than necessary. Rectangular sources are usually superior. For proper adjustment the knife–edges or slits should be of high quality and should be mounted in such a manner that they can be raised and lowered, rotated or moved forward or backward by micrometer adjustments. Also, the mirrors should be accurately adjustable. The mounting of all components should be rigid. The sensitivity also increases directly with the focal length of the schlieren mirror or lens.

When the schlieren method is applied to the study of disturbances such as density gradients in a supersonic wind tunnel in which the flow is two-dimensional, with the flow in the x-direction and the light beam in the z-direction, the component of the gradient of the refractive index in the z-direction vanishes. The index of refraction in air for sodium light can be expressed in terms of density, by the relation (Ladenburg, 1954)

$$n = 1 + 0.000293 \, \frac{\rho}{\rho_{\mathrm{NTP}}} \qquad (4.12)$$

where ρ_{NTP} is the density at 1 atm and $0\,^\circ\mathrm{C}$. The components of the angular deflection ϵ_x and ϵ_y in the x and y directions, respectively, are given by

$$\epsilon_x = \int c \, \frac{\partial \rho}{\partial x} \, dz, \qquad \epsilon_y = \int c \, \frac{\partial \rho}{\partial y} \, dz$$

where c is a constant.

When the component of the density gradient in the direction of the light beam does not vanish (as in the case of a three-dimensional flow field), the interpretation of the pictures becomes more complicated.

4.3.6 Sensitivity of the Schlieren Method for Shock and Expansion Studies

So far we have discussed the various aspects of the technical and application details of the schlieren method. The emphasis was laid mainly on the qualitative aspects of the flows with density gradients, such as the supersonic flow over an object. Now we can ask a question, whether the schlieren method is capable of detecting every density gradient irrespective of the intensity of the gradient or is there any threshold below which it is not possible to detect the disturbance with the schlieren method? Let us try to get an answer for this question. Let us assume that our interest now is to know that *under what condition will an oblique shock become visible?* Assume the knife–edge to be parallel to the front of the oblique shock and, as a typical value, the angle between the shock and the optic axis $\theta = 1^\circ = 0.0175$ radians. For the present arrangement, it can be shown that all the light incident upon the shock is refracted and the amount of light reflected is negligible. Let subscripts 0, 1, and 2 refer to stagnation state at $20\,^\circ\mathrm{C}$, and states upstream and downstream of the shock, respectively. For this flow field, Snell's law of refraction leads to the relation (Ladenburg, 1954)

$$1 + 0.000293 \left(\frac{273}{293}\right) p_0 \left(\frac{\rho_1}{\rho_0} - \frac{\rho_2}{\rho_0}\right) = 1 - \epsilon \tan\theta \qquad (4.13)$$

where ϵ is the angular deflection of the light ray due to the presence of the shock. Let $\epsilon = 10^{-5}$ radians (which is a typical value). Equation (4.13) becomes

$$\rho_2 - \rho_1 = p_0 \frac{\tan\theta}{27.3\, p_0} \qquad (4.14)$$

The relation between the shock angle β and Mach angle μ can be written as

$$\sin\beta = \frac{M_n}{M_1} = M_n \sin\mu \qquad (4.15)$$

where M_n is the component of upstream Mach number M_1, normal to the oblique shock. The ratio of the static to stagnation densities upstream of the shock is

$$\frac{\rho_1}{\rho_0} = \left(1 + \frac{\gamma-1}{2} M_1^2\right)^{-\frac{1}{\gamma-1}}$$

With $\gamma = 1.4$, the density ratio becomes

$$\frac{\rho_1}{\rho_0} = \left(1 + \frac{M_1^2}{5}\right)^{-\frac{5}{2}}$$

The ratio of downstream to upstream densities of the shock wave is

$$\frac{\rho_2}{\rho_1} = \frac{6M_n^2}{M_n^2 + 5}$$

$$\frac{\rho_2 - \rho_1}{\rho_0} = \frac{\rho_2 - \rho_1}{\rho_1} \frac{\rho_1}{\rho_0} = \frac{5(M_n^2 - 1)}{M_n^2 + 5} \frac{\rho_1}{\rho_0} = \frac{\tan\theta}{27.3\, p_0} \qquad (4.16)$$

provided the knife–edge is parallel to the shock front. Therefore, for the given values of θ, p_0, and M_1, the minimum difference $(\beta - \theta)$, which will give a visible schlieren effect can be determined.

If the density change across the oblique shock is small, then $\beta - \theta$ becomes small. Further, $M_n^2 = 1 + \delta$, with $\delta \ll 1$. For this case, we can show that,

$$\beta - \mu = \frac{3}{5} \frac{\rho_0}{\rho_1} \frac{\tan\theta}{27.3\, p_0 \sqrt{M_1^2 - 1}} \qquad (4.17)$$

If $\epsilon = 10^{-5}$, the knife–edge is parallel to the front of the shock wave, and $\theta = \beta - \mu = 1°$, the minimum stagnation pressure (for some given Mach numbers) at which shock can be visualized with schlieren (at the above assumed rather favorable condition) is given in Table 4.1.

Table 4.1 Minimum stagnation pressures

M	1.5	2.0	2.5	3.0	3.5	4.0	4.5	5.0
$p_{0\ min,atm}$	0.050	0.055	0.073	0.102	0.145	0.205	0.287	0.396

6.0	7.0	8.0	9.0	10.0
0.715	1.216	1.959	3.015	4.465

4.3.7   Shadowgraph

In our discussions on the schlieren system we have seen that the positions of the image points on the viewing screen are not affected by deflections of light rays in the test-section. This is because *the deflected rays are also brought to focus in the focal plane, and the screen is uniformly illuminated when the knife–edge is not inserted into the light beam.* On the other hand, if the screen is placed at a position close to the test-section, the effect of ray deflection will be visible. This effect, termed *shadow effect*, is illustrated in Figure 4.13.

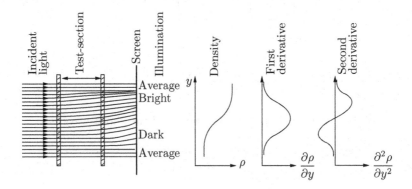

Figure 4.13: The shadow effect

On the screen there are bright zones where the rays crowd closer and dark zones where the rays diverge away. At places where the spacing between the rays is unchanged, the illumination is normal even though there has been refraction. Thus, the shadow effect depends not on the absolute deflection but on the relative deflection of the light rays, that is, on the rate at which they converge or diverge on coming out of the test-section.

A shadowgraph consists of a light source, a collimating lens, and viewing screen, as shown in Figure 4.14.

Let us assume that the test-section has stagnant air in it and that the illumination on the screen is of uniform intensity. When flow takes place through the test-section the light beam will be refracted wherever there is a density gradient. However, *if the density gradient everywhere in the test-section is constant,*

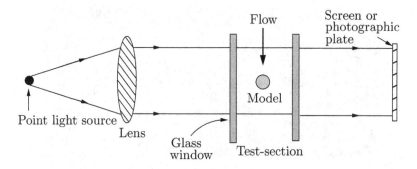

Figure 4.14: Shadowgraph system

all light rays would deflect by the same amount, and *there would be no change in the illumination of the picture on the screen. Only when there is a gradient in density gradient will there be tendency for light rays to converge or diverge.* In other words, *the variations in illumination of the picture on the screen are proportional to the second derivative of the density.* For a two-dimensional flow the increase of light intensity can be expressed as

$$\Delta I = k \left(\frac{\partial^2 \rho}{\partial x^2} + \frac{\partial^2 \rho}{\partial y^2} \right) \tag{4.18}$$

where k is a constant and x and y are the coordinates in a plane normal to the light path.

Therefore, the shadowgraph is best suited only for flow fields with rapidly varying density gradients. A typical shadowgraph of a highly underexpanded circular sonic jet is shown in Figure 4.15. Since the jet is underexpanded, the waves present in the field would be strong enough to result in a large density gradient across them. One such wave termed *Mach disk*, normal to the jet axis, is seen in the field. The Mach disk is essentially a normal shock and hence, the shock has positive and negative rate of change of density gradient across it. Therefore, the shock is made up of a dark line followed by a bright line in the shadow picture, in accordance with the *shadow effect*.

4.3.8 Comparison of Schlieren and Shadowgraph Methods

As we saw in Section 4.3.3, the theory shows that the schlieren technique depends upon the first derivative of the refractive index (flow density) while the shadowgraph method depends upon its second derivative. Consequently, in phenomena where the refractive index varies relatively slowly, the schlieren method is to be preferred to the shadowgraph method, other things being equal. On the other hand, the shadow method beautifully brings out the rapid changes in the index of refraction. The shadow method also has the advantage of greater simplicity and somewhat wider possible application. The two methods therefore

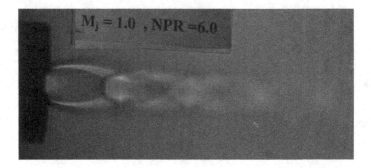

Figure 4.15: Shadowgraph of an underexpanded sonic jet operating at nozzle pressure ratio 6

supplement each other and both should be used wherever possible. Fortunately, in many cases the same apparatus or optical parts can be used for both the methods by simple rearrangement and without too much effort on the part of the experimenter. In addition to the first and second derivatives, the refractive index can also be obtained by integration. However, whenever possible, it is preferable to measure the density directly rather than obtaining it from its derivative. For this reason it is clear that the schlieren and shadow methods should be supplemented by the interference method, which gives the refractive index directly.

4.3.9 Analysis of Schlieren and Shadowgraph Pictures

The power of schlieren and shadowgraph are illustrated in this section with the demonstration of Mach disk formation in a sonic jet, using the schlieren and shadowgraph pictures.

As we saw in this chapter, the schlieren system uses light from a single collimated source shining on, or from behind, a target object. Variations in refractive index caused by density gradients in the fluid distort the collimated light beam. This distortion creates a spatial variation in the intensity of the light, which can be visualized directly with a shadowgraph system.

In the schlieren system, the collimated light is focused with a lens, and a knife–edge, placed at the focal point, is positioned to block about half the light. In a flow of uniform density this will simply make the photograph half as bright. However in a flow with density variations the distorted beam focuses imperfectly, and parts which have been focused in an area covered by the knife–edge are blocked. The result is a set of lighter and darker patches corresponding to positive and negative fluid density gradients in the direction normal to the knife–edge. A schlieren system measures the first derivative of density in the direction of the knife–edge. In other words, the schlieren method visualizes the distribution of fluid density within a fluid, as fluid density controls the index of refraction. Regions of density gradient deflect light beams, shifting their position on the image plane. The relative change in light intensity can be used to infer the original density and flow field.

A shadowgraph system measures the second derivative of density. As we know, the density of a fluid varies with temperature and pressure, and the index of refraction changes with fluid density. Variations in the refractive index deflect or phase shift the light passing through the fluid. If a screen is placed opposite to the light source, these effects create shadows on the screen creating an image called a shadowgraph. A traditional shadowgraph collapses a three-dimensional field into a planar image.

If the fluid flow is uniform the image will be steady, but any turbulence will cause scintillation, the shimmering effect, such as that seen on hot surfaces on a sunny day. To visualize instantaneous density profiles, a short duration flash (rather than continuous illumination) may be used.

Now let us examine the waves prevailing in the flow field of a Mach 1 circular jet from a convergent nozzle, run with nozzle pressure ratios (NPRs), 2.0, 2.5, 3.0, 3.5, and 4.0, captured with schlieren and shadowgraph systems.

4.3.9.1 Schlieren and Shadowgraph Pictures

The schlieren pictures of the sonic jet at different NPRs considered here are shown in Figures 4.16(a) to 4.16(k).

The schlieren picture of the sonic flow that comes out through a circular convergent nozzle run by a high-pressure tank at a pressure, p_0, of 2 atm, and discharges into an environment at atmospheric pressure, p_a, is shown in Figure 4.16(a).

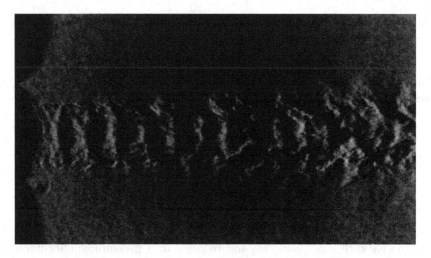

Figure 4.16a: Schlieren picture of Mach 1 jet at NPR 2

Figure 4.16(a) is the schlieren picture of a Mach 1 circular jet of air, from a convergent nozzle, at nozzle pressure ratio, NPR, 2. The NPR (p_0/p_a) is the ratio of the stagnation pressure at the nozzle inlet to the pressure of the environment to which the jet is discharged. The pressure of the environment to which the jet is discharged is referred to as *backpressure*. A sonic jet of air is correctly expanded with the nozzle exit pressure p_e equal to the backpressure p_b, when NPR = 1.89, as per insentropic theory. Therefore, p_e for the convergent nozzle run at NPR 2 will be greater than p_b. The state with $p_e > p_b$ is termed *underexpanded*. The underexpansion level for NPR 2 can be calculated as follows.

The isentropic pressure-Mach number relation is

$$\frac{p_0}{p_e} = \left(1 + \frac{\gamma - 1}{2} M_e^2\right)^{\gamma/(\gamma-1)}$$

For $M_e = 1$,

$$\frac{p_0}{p_e} = \left(1 + \frac{\gamma - 1}{2}\right)^{\gamma/(\gamma-1)}$$

For air, the specific heats ratio $\gamma = 1.4$, therefore,

$$\frac{p_0}{p_e} = \left(1 + \frac{1.4 - 1}{2}\right)^{1.4/(1.4-1)}$$

$$= 1.2^{3.5}$$

$$= 1.89$$

Thus, for NPR 1.89,

$$p_{ec} = \frac{p_0}{1.89}$$

where p_{ec} is the pressure at the nozzle exit, for correctly expanded sonic jet. Similarly, for NPR 2

$$p_{eu} = \frac{p_0}{2}$$

where p_{eu} is the pressure at the nozzle exit, for underexpanded sonic jet. This gives the level of underexpansion as

$$\frac{p_{ec}}{p_{eu}} = \frac{p_0/1.89}{p_0/2}$$

$$= 2/1.89$$

$$= 1.0566$$

Thus, the underexpansion level is 5.66%. The jet issuing out with pressure higher than the backpressure should expand to come to a pressure equilibrium with the surroundings. This physical requirement establishes an expansion fan at the nozzle exit, as seen in Figure 4.16(a).

The waves seen in Figure 4.16(a) and the flow process through these waves can be represented schematically, as shown in Figure 4.16(b).

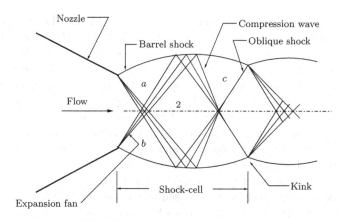

Figure 4.16b: Waves in an underexpanded sonic jet.

It is important to note that the pressure ratio across the nozzle, $p_a/p_{01} = 0.5$, is less than the isentropic choking pressure ratio of 0.528. Therefore, the flow leaving the nozzle would be choked and the exit velocity will be sonic. Also, the choked flow leaving the nozzle is underexpanded and finds a large space to expand further. Therefore, the flow on free expansion can attain a Mach number corresponding to the pressure ratio of 0.5. For this pressure ratio, isentropic expansion would result in flow Mach number of about 1.05. This simply implies that the underexpanded flow exiting the nozzle at Mach 1 will pass through an expansion fan and attain Mach 1.05. At zones a and b the flow experiences only half the expansion compared to zone 2, and hence the Mach number in zones a and b would be less than the Mach number that the flow would attain at zone 2, after passing through double expansion. Soon after leaving the nozzle the sonic flow passes through the expansion fan and becomes supersonic. But at every point in the expansion fan the flow has a different supersonic Mach number. Also, since the flow at the vertex of the expansion fan is turned almost suddenly to move away from the nozzle axis, it has to be turned towards the axis to become axial at a downstream distance. This turning is caused by the shock which is essentially an oblique shock. But on either side of the axis, for a two-dimensional flow and around the axis for axisymmetric flow, the shape of the shock assumes the form of a "barrel" and thus is referred to as a *barrel shock*. Outside of the barrel shock, the flow is a mixture of supersonic and subsonic Mach numbers. The expansion rays get reflected from the free boundary as compression waves, as illustrated in Figure 4.16(b), since the reflection of a wave from a free boundary is *unlike*. The supersonic flow in zone a gets decelerated on passing through these reflected compression waves and hence the flow Mach number at c becomes less than that at a. These compression waves coalesce to form shock waves that cross each other at the axis, proceed to the barrel shock and reflect back as expansion waves, as shown in Figure 4.16(b). The distance from nozzle exit to the first kink location on the barrel shock, where the shocks get reflected as expansion waves, is called a *shock-cell*, in jet literature (Rathakrishnan, 2010).

The wave pattern described above (Figure 4.16(b)) can be clearly seen in the schlieren picture of the sonic underexpanded jet, at NPR 2, shown in Figure 4.16(a). The flow traversed by the expansion rays experiences reduced density and thus the first derivative of the density, $\partial \rho / \partial n$, where n is the unit vector normal to the expansion rays, becomes smaller than that ahead of the ray. This causes darkness in the flow image. The reflected expansion rays join and form a compression wave, thus increases the flow density, leading to higher value of $\partial \rho / \partial n$, resulting in a brighter zone in the schlieren image. Thus, the bright and dark zones in the schlieren image correspond to the zones of flow traversed by the expansion and compression waves, respectively.

Shadowgraph picture of the sonic jet, run at NPR 2, is shown in Figure 4.16(c). As we know, the shadowgraph senses the second derivative, $\partial^2 \rho / \partial n^2$ of the density. Thus only the waves across which the density change is large (i.e., the second derivative of density is of considerable magnitude) will be visible in a shadowgraph picture. In accordance with this, the barrel shock forming the first shock cell is seen in Figure 4.16(c). In this picture, the bright spots are the zones of flow traversed by compression waves, since right rays bend towards the zone of higher density.

Schlieren picture of the sonic jet run with NPR 2.5 is shown in Figure 4.16(d). For this sonic jet, NPR 2.5 corresponds to an underexpansion level of about 32% (2.5/1.89 = 1.323). Therefore, the expansion fan at the nozzle exit is stronger than that for NPR 2. Thus, the larger density decrease caused by the expansion at the nozzle exit results in a distinct dark zone just behind the nozzle exit. The expansion rays getting reflected, from the inner boundary of the jet, as compression waves coalesce forming a stronger compression front (i.e., a stronger oblique shock) than NPR 2, as seen in Figure 4.16(d). Also, the first shock-cell has become longer and the subsequent shock-cells have become more prominent than NPR 2. Also, the oblique shocks formed by the reflected compression waves, from the opposite edges, cross-over at the jet center line, as seen in Figure 4.16(d). The zones traversed by the compression and expansion waves are seen as distinct bright and dark zones, respectively, in the picture.

Shadowgraph of the sonic jet, run at NPR 2.5, is shown in Figure 4.16(e). This picture clearly shows the barrel shocks and the width and length of the shock-cells present in the jet core. It is seen that, compared to the schlieren picture (Figure 4.16(d)), the shadowgraph picture (Figure 4.16(e)) is more convenient to quantify the length and the number of shock cells. However, the schlieren picture scores over the shadowgraph picture in giving the details of the expanse of the flow zones traversed by the compression waves and expansion fans. The shadowgraph picture for NPR 3.5, shown in Figure 4.16(i), clearly shows the shock cells, the waves inside the cells, as well as the barrel shocks bounding the cells and the shock cross-over points in the jet core.

Schlieren picture of the jet at NPR 3.0 is shown in Figure 4.16(f). The underexpansion level for NPR 3 is 63%. It is interesting to see from Figure 4.16(f) that, with increase of underexpansion level, the expansion at the nozzle exit becomes stronger, as indicated by the longer conical dark zone at the nozzle exit. Also, the oblique shocks formed by the reflection of the expansion rays from the jet boundary as compression waves have become weaker than those for NPR 2.5. The nonlinearity of the shocks after the cross-over point become higher. All the shock-cells as well as the low-density zones due to the passing of the flow through the expansion fans in the jet core have become longer.

Figure 4.16c: Shadowgraph picture of Mach 1 jet at NPR 2.

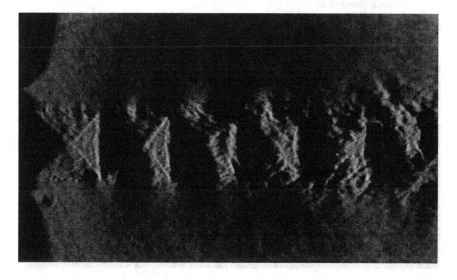

Figure 4.16d: Schlieren picture of Mach 1 jet at NPR 2.5.

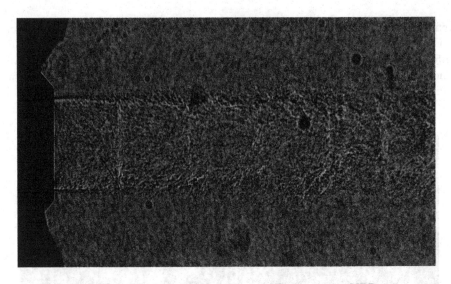

Figure 4.16e: Shadowgraph picture of Mach 1 jet at NPR 2.5

Shadowgraph picture of Mach 1 jet at NPR 3.0 is shown in Figure 4.16(g). All the features of the compression waves, showing stronger barrel shock, weaker oblique shocks, the higher nonlinearity of the shocks after the shock cross-over point, and the distinct borders of the shock-cells are clearly seen at NPR 3. Indeed, even approximate quantification of the angles of the shocks in the first 2 shock cells are possible with this picture in Figure 4.16(g). The conical area (seen as dark portion) seen in this picture are the flow zones (of low density) traversed by the expansion fans formed by the reflection of compression waves from the shock boundary.

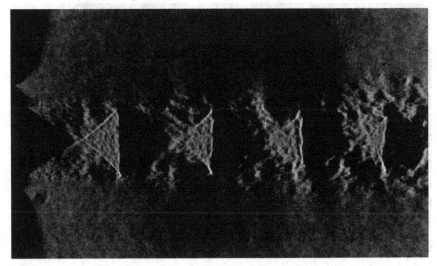

Figure 4.16f: Schlieren picture of Mach 1 jet at NPR 3.0

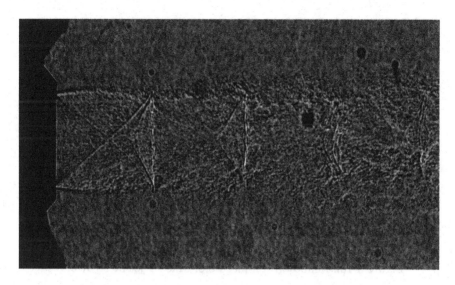

Figure 4.16g: Shadowgraph picture of Mach 1 jet at NPR 3.0

With further increase in the underexpansion level to 85%, corresponding to NPR 3.5, the expansion at the nozzle exit becomes stronger and the nonlinearity of the shocks after the cross-over point becomes higher, as seen from the schlieren picture of the jet at NPR 3.5, shown in Figure 4.16(h).

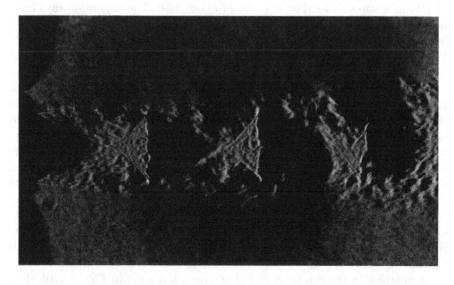

Figure 4.16h: Schlieren picture of Mach 1 jet at NPR 3.5

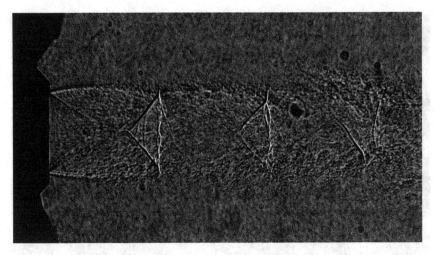

Figure 4.16i: Shadowgraph picture of Mach 1 jet at NPR 3.5

The shadowgraph picture for NPR 3.5, shown in Figure 4.16(i), clearly shows the shock cells, the waves inside the cells, as well as the barrel shocks bounding the cells and the shock cross-over points in the jet core.

Schlieren picture of the jet at NPR 4, establishing an underexpansion level of 116.4%, is shown in Figure 4.16(j). For the jet, this is a highly underexpanded state. Therefore, the jet on issuing out of the nozzle should expand through a strong expansion fan to reduce the pressure to the level of p_b. Thus there is a powerful expansion fan angle about 19.9°. That is, the flow is turned away from the jet axis, by 19.9° on either side. The sonic jet, on expansion through this fan attains a supersonic Mach number of about 1.56. The expansion rays from one side pass through the expansion rays from the other end. Thus the flow passing through the middle portion of the jet, around the jet axis is traversed by two expansion fans of equal strength and thus becomes supersonic flow of Mach number much higher than 1.56.

A shadowgraph picture of the jet at NPR 4 is shown in Figure 4.16(k). An interesting feature seen in the schlieren and shadowgraph pictures of the jet at NPR 4 is the shock cross-over point, present in the jet at the lower NPRs, in the first shock-cell has become a disk-like structure, normal to the jet axis. The shocks on reaching the disk reflect back, instead of passing through each other, as seen in Figures 4.16(j) and 4.16(k). The disk normal to the jet axis is called *Mach disk*. The NPR corresponding to an underexpansion level of 100%, at which the shock cross-over point becomes a Mach disk may be regarded as the border between the moderately and highly underexpanded states. For sonic jets this limiting NPR seems to be around 4. Thus, for jets of higher Mach number, the NPR at which the Mach disk will form will increase with increase of Mach number.

Comparison of the schlieren and shadowgraph images in Figure 4.16(a) to 4.16(k), clearly reveals the effect of the first and second derivatives of the density, sensed by the schlieren and shadowgraph techniques. In the schlieren images, both the waves and the influence of the waves on the flow field, in the form of low-density zones (dark patches) and the increased perturbation caused by the entropy change (in the form of disturbed nature of the flow) are clearly visible. In other words, the dark patches in the schlieren picture are the zones with low pressure (low density) and the rough zones are the zones traversed by the compression waves, across which entropy increases. But in the shadowgraph images, only the lines (or waves) across which there is large change in the density gradient are seen. Therefore, if the aim is to study a field with feeble changes in the density (i.e., a field with weak compression and expansion waves), schlieren is more suitable than the shadowgraph.

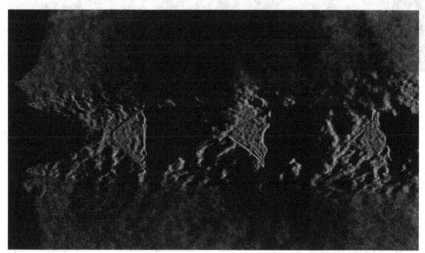

Figure 4.16j: Schlieren picture of Mach 1 jet at NPR 4.0

It is interesting to note that the values of the Mach angle, μ, and Prandtl–Meyer function, ν, measured from the optical images of the sonic flow from the nozzle, at the nozzle pressure ratios of 2.5, 3.0, 3.5, and 4.0 compare closely with the isentropic theoretical values.

The values of fully expanded isentropic Mach number, M, for NPRs 2.5, 3.0, 3.5, and 4.0 are 1.23, 1.36, 1.465, and 1.56, respectively. The values of Mach angle, μ, and Prandtl–Meyer function, ν, for these Mach numbers, based on isentropic theory, are listed in the table below.

NPR	M	μ°	ν°
2.5	1.23	54.39	4.3
3.0	1.36	47.33	7.84
3.5	1.465	43.23	10.9
4.0	1.56	39.87	13.68

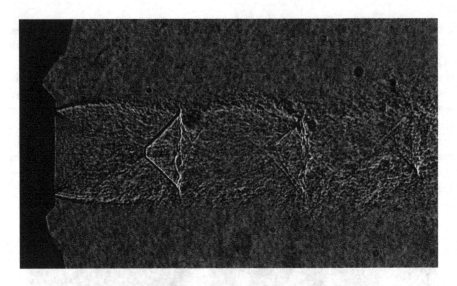

Figure 4.16k: Shadowgraph picture of Mach 1 jet at NPR 4.0

From the values of μ listed above, the fan angle for the expansion of the flow from Mach 1.23 to Mach 1.36 is

$$
\begin{aligned}
\theta_{\text{fan}} &= \mu_{1.23} - \mu_{1.36} \\
&= 54.39 - 47.33 \\
&= 7.06°
\end{aligned}
$$

Similarly, the fan angle for the expansion of the flow from Mach 1.36 to Mach 1.465 is

$$
\begin{aligned}
\theta_{\text{fan}} &= \mu_{1.36} - \mu_{1.465} \\
&= 47.33 - 43.23 \\
&= 4.1°
\end{aligned}
$$

and the fan angle for the expansion of the flow from Mach 1.465 to Mach 1.56 is

$$
\begin{aligned}
\theta_{\text{fan}} &= \mu_{1.465} - \mu_{1.56} \\
&= 43.23 - 39.87 \\
&= 3.36°
\end{aligned}
$$

Now, let us compare these theoretical values to the values of Mach angle measured from the schlieren and shadowgraph pictures and the fan angles calculated from these measured values of μ. The measured values are listed below.

NPR	μ°
2.5	≈ 55
3.0	≈ 47
3.5	≈ 43
4.0	≈ 40

The theoretical and experimental fan angles are compared below.

NPR	$\theta_{\text{exptl.}}$ (deg.)	θ_{theory} (deg.)
2.5		
	7.06	8
3.0		
	4.1	4
3.5		
	3.36	3
4.0		

It is seen that, the theoretical and experimental values of the fan angles agree fairly well.

4.4 Particle Image Velocimetry

Particle image velocimetry (PIV) is an optical method of flow visualization. It can be used to obtain the instantaneous velocities and the related properties in fluid flows. The fluid is seeded with tracer particles which are sufficiently small so that they can flow with the fluid at the same speed. The fluid with seeding particles is illuminated so that particles are visible. The motion of the seeding particles is used to calculate the speed and direction of the flow being studied.

The main difference between PIV and laser Doppler velocimetry is that PIV produces two-dimensional or even three-dimensional vector fields, while the laser Doppler velocimetry measures only the velocity at a point. In PIV, the particle concentration is such that it is possible to identify individual particles in an image. When the particle concentration is so low that it is possible to follow an individual particle it is called *particle tracking velocimetry*. While *laser speckle velocimetry* has to be used for cases where the particle concentration is so high that it is difficult to observe individual particles in an image.

A PIV apparatus consists of a camera (usually a digital camera with a CCD chip), a strobe or laser with an optical arrangement to limit the physical region illuminated (usually a cylindrical lens to convert a light beam to a line), a synchronizer to act as an external trigger for the control of the camera and laser, the seeding particles, and the fluid under investigation. A fiber optic cable or liquid light guide may connect the laser to the lens setup. Usually PIV software needs to be used to post-process the optical images.

4.4.1   Historical Background

Even though adding particles to a fluid to observe its flow has been in practice for a long time, there was no systematic technique to quantify the observed pattern of the flow until the early 20th century. Ludwig Prandtl was the first to use particles to study fluids in a more systematic manner, in the early 20th century.

Laser Doppler Velocimetry predates PIV as a laser-digital analysis system, because of its capability to measure the velocity at a specific point in a flow field. It can be considered as the two-dimensional PIV's immediate predecessor. PIV finds its roots in laser speckle velocimetry, developed in the late 1970s. In the early 1980s it was found that it was advantageous to decrease the particle concentration down to levels where individual particles could be observed. At these particle densities it was further noticed that it was easier to study the flows if they were split into many very small "interrogation" areas that could be analyzed individually to generate one velocity for each area. The images were usually recorded using analogue cameras. Very large computing power is required for the analysis of these images. With the increasing power of computers and widespread use of CCD cameras, digital PIV has become a powerful technique for fluid flow studies.

4.4.2   PIV System

The primary components of PIV equipment are: *(i) seeding particles, (ii) camera, (iii) laser and optics*, and *(iv) synchronizer*.

4.4.2.1   Seeding Particles

The seeding particle is an important component of the PIV system. It is essential to ensure that the particles be able to match the fluid properties reasonably well. This is essential for the PIV analysis to be considered accurate. Ideal particles should have the same density as the fluid system being used, and be spherical. These particles are called *microspheres*. While the actual particle choice is dependent on the nature of the fluid, generally for macro PIV investigations they are glass beads, polystyrene, polyethylene, aluminum flakes, or oil droplets. Refractive index for the seeding particles should be different from the fluid which they are seeding, so that the laser sheet incident on the fluid flow will reflect off of the particles and be scattered towards the camera.

Seeding particles are typically of a diameter in the order of 10 to 100 micrometers. The particles should be small enough so that *response time* of the particles to the motion of the fluid is reasonably short to accurately follow the flow, yet large enough to scatter a significant quantity of the incident laser light. For some experiments involving combustion, seeding particle size may be smaller, in the order of 1 micrometer, to avoid the quenching effect that the inert particles may have on flames. Due to the small size of the particles, the particle's motion is dominated by stokes drag and settling or rising effects. In a model where particles are modeled as spherical (microspheres) at a very low Reynolds number, the ability of the particles to follow the fluid flow is inversely proportional to the difference in density between the particles and the fluid, and also inversely proportional to the square of their diameter. The scattered light from the particles is dominated by *Mie* scattering and so is also proportional to the square of the particle's diameters. Thus the particle size needs to be balanced to scatter enough light to accurately visualize all particles within the laser sheet plane, but small enough to accurately follow the flow. Also, the seeding mechanism needs to be designed so as to seed the flow to a sufficient degree without overly disturbing the flow.

4.4.2.2 Camera

For the PIV analysis on the flow, two exposures of laser light are required upon the camera from the flow. Originally, with the inability of cameras to capture multiple frames at high speeds, both exposures were captured on the same frame and this single frame was used to determine the flow. A process called autocorrelation was used for this analysis. However, as a result of autocorrelation the direction of the flow becomes unclear, as it is not clear which particle spots are from the first pulse and which are from the second pulse. Faster digital cameras using CCD[1] or CMOS[7] chips has been developed that can capture two frames at high speed with a few hundred ns (one billionth of a second (10^{-9} or $1/1,000,000,000$ second)) difference between the frames. This has allowed each exposure to be isolated on its own frame for more accurate cross-correlation analysis. The limitation of typical cameras is that this fast speed is limited to a pair of shots. This is because each pair of shots must be transferred to the computer before another pair of shots can be taken. Typical cameras can only take a pair of shots at a much slower speed.

4.4.2.3 Laser and Optics

For macro PIV setups, lasers are vital due to their ability to produce high-power light beams with short pulse durations. This yields short exposure times for each frame. The Nd:YAG lasers, which are commonly used in PIV setups,

---

[1]The first digital cameras used CCD (charged coupling devices) to turn images from analogue light signals into digital pixels. They're made through a special manufacturing process that allows the conversion to take place in the chip without distortion. This creates high-quality sensors that produce excellent images. But, because they require special manufacturing, they are more expensive than their newer CMOS counterparts.

[7]CMOS (complementary metal oxide semiconductor) chips use transistors at each pixel to move the charge through traditional wires. This offers flexibility because each pixel is treated individually. Traditional manufacturing processes are used to make CMOS. It's the same as creating microchips. Because they're easier to produce, CMOS sensors are cheaper than CCD sensors. Because CMOS technology came after CCD sensors and is cheaper to manufacture, CMOS sensors are the reason that digital cameras have dropped in price.

emit primarily at 1064 nm wavelength and its harmonics (532, 266, etc.). For safety reasons, the laser emission is typically bandpass filtered to isolate the 532 nm (nanometer) harmonics (this is green light, the only harmonic that can be seen by the naked eye). A fiber optic cable or liquid light guide might be used to direct the laser light to the experimental setup.

The optics of a typical PIV system consist of a spherical lens and cylindrical lens combination. The cylindrical lens expands the laser into a plane while the spherical lens compresses the plane into a thin sheet. This is critical as the PIV technique cannot generally measure motion normal to the laser sheet and so ideally this is eliminated by maintaining an entirely two-dimensional laser sheet. The minimum thickness depends on the order of the wavelength of the laser light and occurs at a finite distance from the optics setup (the focal point of the spherical lens). This is the ideal location to place the study area of the experiment. Selection of correct lens for the camera is vital for proper focus and visualization of the particles within the study area.

4.4.2.4　Synchronizer

The synchronizer acts as an external trigger for both the camera(s) and the laser. While analogue systems in the form of a photosensor, rotating aperture, and a light source have been used in the past, most systems in use today are digital. The synchronizer, controlled by a computer, can dictate the timing of each frame of the CCD camera's sequence in conjunction with the firing of the laser to within 1 ns (nanosecond) precision. Thus the time between each pulse of the laser and the placement of the laser shot in reference to the camera's timing can be accurately controlled. Knowledge of this timing is critical as it is needed to determine the velocity of the fluid in the PIV analysis. Standalone electronic synchronizers, called digital delay generators, offer variable resolution timing from as low as 250 ps (picosecond, i.e., 10^{-12} second) to as high as several ms (millisecond). With up to eight channels of synchronized timing, they offer the means to control several flash lamps and Q-switches[1] as well as provide for multiple camera exposures.

4.4.3　Analysis

The frames are split into a large number of interrogation areas, or windows. It is then possible to calculate a *displacement vector* for each window with the help of *signal processing* and *autocorrelation* or *cross-correlation* techniques. This is converted to a velocity using the time between laser shots and the physical size of each pixel on the camera. The size of the interrogation window should be chosen to have at least 6 particles per window on average.

The synchronizer controls the timing between image exposures and also permits image pairs to be acquired at various times along the flow. For accurate

---

[1]Q-switching is a technique for obtaining energetic short pulses from a laser by modulating the intracavity losses. This technique is mainly applied for the generation of nanosecond pulses of high energy and peak power with solid-state bulk lasers.

PIV analysis, it is ideal that the region of the flow that is of interest should display an average particle displacement of about 8 pixels. This is a compromise between a longer time spacing which would allow the particles to travel further between frames, making it harder to identify which interrogation window traveled to which point, and a shorter time spacing, which could make it overly difficult to identify any displacement within the flow.

The scattered light from each particle should be in the region of 2 to 4 pixels across on the image. If too large an area is recorded, particle image size drops and peak locking might occur with loss of subpixel precision. There are methods to overcome the peak locking effect, but they require some additional work.

4.4.4 Advantages

• PIV is a nonintrusive technique. The added tracers (if they are properly chosen) generally cause negligible distortion of the fluid flow.
• Optical measurement avoids the need for pitot tubes, hot-wire anemometers, or other intrusive flow measurement probes. The method is capable of measuring an entire two-dimensional cross section (geometry) of the flow field simultaneously.
• High speed data processing allows the generation of a large number of image pairs which may be analyzed in real time or at a later time, and a high quantity of almost continuous information may be gained.
• Subpixel displacement values allow a high degree of accuracy, since each vector is the statistical average for many particles within a particular tile. Displacement can typically be accurate down to 10% of one pixel on the image plane.

4.4.5 Disadvantages

• In some cases the particles, due to their higher density, will not perfectly follow the motion of the fluid (gas/liquid). For instance, if experiments are done in water it is easily possible to find very cheap particles (for example, plastic powder with a diameter of about 60 μm) with the same density as water. If the density still does not fit, the density of the fluid can be tuned by increasing/decreasing its temperature. This leads to slight changes in the Reynolds number, so the fluid velocity or the size of the experimental object has to be changed to account for this.
• PIV methods in general will not be able to measure components along the z-axis (towards/away from the camera). These components might not only be missed, they might also introduce an interference in the data for the x/y-components caused by parallax. These problems do not exist in stereoscopic PIV, which uses two cameras to measure all three velocity components.
• Since the resulting velocity vectors are based on cross-correlating the intensity distributions over small areas of the flow, the resulting velocity field is a spatially averaged representation of the actual velocity field. This will affect the accuracy of spatial derivatives of the velocity field, vorticity, and spatial correlation functions that are often derived from PIV velocity fields.

4.4.6 Complex PIV Setups

Some of the complex PIV setups are the following.

(i) Stereoscopic PIV
(ii) Dual plane stereoscopic PIV
(iii) Micro PIV
(iv) Holographic PIV
(v) Scanning PIV
(vi) Tomographic PIV, and
(vii) Thermographic PIV.

4.4.6.1 Stereoscopic PIV

This utilizes two cameras with separate *viewing angles* to extract the z-axis displacement. Both cameras must be focused on the same spot in the flow and must be properly calibrated to have the same point in focus.

In fluid mechanics, displacement within a unit time in the x, y, and z directions are commonly defined by the variables U, V, and W. As was previously described, basic PIV extracts the U and V displacements as functions of the in-plane x and y directions. This enables calculations of the U_x, V_y, U_y, and V_x velocity gradients. However, the other 5 terms of the velocity gradient tensor are unable to be found from this information. The stereoscopic PIV analysis also grants the z-axis displacement component, W, within that plane. Not only does this grant the z-axis velocity of the fluid at the plane of interest, but two more velocity gradient terms can be determined: W_x and W_y. The velocity gradient components U_z, V_z, and W_z cannot be determined. The velocity gradient components form the tensor:

$$\begin{bmatrix} U_x & U_y & U_z \\ V_x & V_y & V_z \\ W_x & W_y & W_z \end{bmatrix}$$

4.4.6.2 Dual Plane Stereoscopic PIV

This is an expansion of stereoscopic PIV by adding a second plane of investigation directly offset from the first one. Four cameras are required for this analysis. The two planes of laser light are created by splitting the laser emission with a beam splitter into two beams. Each beam is then polarized orthogonally with respect to one another. They are transmitted through a set of optics and used to illuminate one of the two planes simultaneously.

The four cameras are paired into groups of two. Each pair focuses on one of the laser sheets in the same manner as single-plane stereoscopic PIV. Each of the four cameras has a polarizing filter designed to only let pass the polarized scattered light from the respective planes of interest. This creates a system by which two separate stereoscopic PIV analysis setups are run simultaneously with only a minimal separation distance between the planes of interest.

This technique allows the determination of the three velocity gradient components which a single-plane stereoscopic PIV could not calculate: U_z, V_z, and W_z. With this technique, the entire velocity gradient tensor of the fluid at the two-dimensional plane of interest can be quantified. One difficulty here is that the laser sheets should be maintained close enough together so as to approximate a two-dimensional plane, yet offset enough that meaningful velocity gradients can be found in the z-direction.

4.4.6.3 Micro PIV

With the use of an epifluorescent microscope[1], microscopic flows can be analyzed. Micro PIV makes use of fluorescent particles that excite at a specific wavelength and emit at another wavelength. Laser light is reflected through a dichroic mirror, travels through an objective lens that focuses on the point of interest, and illuminates a local volume. The emission from the particles, along with reflected laser light, shines back through the objective lens, the dichroic mirror and through an emission filter that blocks the laser light. This PIV draws its two-dimensional analysis properties from the planar nature of the laser sheet, micro PIV utilizes the ability of the objective lens to focus on only one plane at a time, thus creating a two-dimensional plane of viewable particles.

Micro PIV particles have the size of the order of several hundred nm in diameter, thus they are extremely susceptible to Brownian motion. Thus, a special ensemble averaging analysis technique must be utilized for this technique. The cross-correlation of a series of basic PIV analyses are averaged together to determine the actual velocity field. Thus, only steady flows can be investigated. Special preprocessing techniques must also be utilized since the images tend to have a zero-displacement bias from background noise and low signal:noise ratios.

4.4.6.4 Holographic PIV

Holographic PIV (HPIV) encompasses a variety of experimental techniques which use the interference of coherent light scattered by a particle and a reference beam to encode information of the amplitude and phase of the scattered light incident on a sensor plane. This encoded information, known as a *hologram*, can then be used to reconstruct the original intensity field by illuminating the hologram with the original reference beam via optical methods or digital approximations. The intensity field is interrogated using 3-D cross-correlation techniques to yield a velocity field.

Off-axis HPIV uses separate beams to provide the object and reference waves. This setup is used to avoid speckle noise generated from the interference of the two waves within the scattering medium, which would occur if they

---

[1]This is a microscope equipped with a high-intensity light source (usually a mercury arc lamp) that emits light in a broad spectrum from visible through ultraviolet.

were both propagated through the medium. An off-axis experiment is a complex optical system comprising numerous optical elements.

In-line holography is another approach that provides some unique advantages for particle imaging. Perhaps the largest of these is the use of forward scattered light, which is orders of magnitude brighter than scattering oriented normal to the beam direction. Additionally, the optical setup of such systems is much simpler because the residual light does not need to be separated and recombined at a different location. The in-line configuration also provides a relatively easy extension to apply CCD sensors, creating a separate class of experiments known as digital in-line holography. The complexity of such setups shifts from the optical setup to image post-processing, which involves the use of simulated reference beams.

A variety of issues degrade the quality of HPIV results. The first set of issues involves the reconstruction itself. In holography, the object wave of a particle is typically assumed to be spherical; however, due to Mie[1] scattering theory, this wave is a complex shape which can distort the reconstructed particle. Another issue is the presence of substantial speckle noise which lowers the overall signal-to-noise ratio of particle images. This effect is of greater concern for in-line holographic systems because the reference beam is propagated through the volume along with the scattered object beam. Noise can also be introduced by the impurities in the scattering medium, such as temperature variations and window blemishes. Because holography requires coherent imaging, these effects are much more severe than traditional imaging conditions. The combination of these factors increases the complexity of the correlation process. In particular, the speckle noise in a HPIV recording often prevents traditional image-based correlation methods from being used. Instead, single particle identification and correlation are implemented, which set limits on particle number density.

In the light of these issues, it is obvious that HPIV is too complicated and error-prone to be used for flow measurements. However, many impressive results have been obtained with all holographic approaches.

4.4.6.5   Scanning PIV

By using a rotating mirror and a high-speed camera and correcting for geometric changes, PIV can be performed nearly instantly on a set of planes throughout the flow field. Fluid properties between the planes can then be interpolated. Thus, a quasi-volumetric analysis can be performed on a target volume. Scanning PIV can be performed in conjunction with the other two-dimensional PIV methods described to approximate a three-dimensional volumetric analysis.

---

[1]The term *Mie theory* is sometimes used for this collection of solutions and methods; it does not refer to an independent physical theory or law. More broadly, "Mie scattering" suggests situations where the size of the scattering particles is comparable to the wavelength of the light, rather than much smaller or much larger.

4.4.6.6 Tomographic PIV

Tomographic PIV is based on the illumination, recording, and reconstruction of tracer particles within a 3-D measurement volume. The technique uses several cameras to record simultaneous views of the illuminated volume, which is then reconstructed to yield a discretized 3-D intensity field. A pair of intensity fields are analyzed using 3-D cross-correlation algorithms to calculate the 3-D, 3-C (three-component) velocity field within the volume. The technique was originally developed by Elsinga et al. in 2006.

The reconstruction procedure is a complex under-determined inverse problem. The primary complication is that a single set of views can result from a large number of 3-D volumes. Procedures to properly determine the unique volume from a set of views are the foundation for the field of tomography. In most tomographic PIV experiments, the multiplicative algebraic reconstruction technique (MART) is used. The advantage of this pixel-by-pixel reconstruction technique is that it avoids the need to identify individual particles.

At least four cameras are needed for acceptable reconstruction accuracy, and best results are obtained when the cameras should be placed at approximately 30° normal to the measurement volume (Elsinga et al., 2006).

Tomographic PIV has been applied to a broad range of flows, such as the structure of a turbulent boundary layer/shock wave interaction (Elsinga et al., 2006), the vorticity of a cylinder wake (Humble et al., 2009), rod-airfoil aeroacoustic experiments (Scarano and Poelma, 2009), and even in aquatic predator-prey interactions (Violato, et al., 2010).

4.4.6.7 Thermographic PIV

Thermographic PIV is based on the use of thermographic phosphors as seeding particles. The use of these thermographic phosphors permits simultaneous measurement of velocity and temperature in a flow.

Thermographic phosphors consist of ceramic host materials doped with rare-earth or transition metal ions, which exhibit phosphorescence when they are illuminated with UV-light. The decay time and the spectra of this phosphorescence are temperature sensitive and offer two different methods to measure temperature. The decay time method consists of the fitting of the phosphorescence decay to an exponential function and is normally used in point measurements, although it has been demonstrated in surface measurements. The intensity ratio between two different spectral lines of the phosphorescence emission, tracked using spectral filters, is also temperature-dependent and can be employed for surface measurements.

The micrometer-sized phosphor particles used in thermographic PIV are seeded into the flow as a tracer and, after illumination with a thin laser light sheet, the temperature of the particles can be measured from the phosphorescence, normally using an intensity ratio technique. It is important that the particles are of small size so that they not only follow the flow satisfactorily but they also rapidly assume its temperature. For a diameter of 2 μm, the thermal slip between particle and gas is as small as the velocity slip.

Illumination of the phosphor is achieved using UV light. Most thermographic phosphors absorb light in a broad band in the UV and therefore can be excited using a YAG:Nd laser. Theoretically, the same light can be used both for PIV and temperature measurements, but this would mean that UV-sensitive cameras are needed. In practice, two different beams originated in separate lasers are overlapped. While one of the beams is used for velocity measurements, the other is used to measure the temperature.

The use of thermographic phosphors offers some advantageous features including ability to survive in reactive and high temperature environments, chemical stability, and insensitivity of their phosphorescence emission to pressure and gas composition. In addition, thermographic phosphors emit light at different wavelengths, allowing spectral discrimination against excitation light and background.

Thermographic PIV has been demonstrated for time averaged (Adhikari and Longmire, 2013) and single shot (Kim et al., 2011) measurements. Recently, time-resolved high speed (3 kHz) measurements (Omrane et al., 2008) have been successfully performed.

4.4.7 Applications

PIV has been applied to a wide range of flow problems, varying from the flow over an aircraft wing in a wind tunnel to vortex formation in prosthetic heart valves. Three-dimensional techniques have been sought to analyze turbulent flow and jets.

Rudimentary PIV algorithms based on cross-correlation can be implemented in a matter of hours, while more sophisticated algorithms may require a significant investment of time.

PIV can also be used to measure the velocity field of the free surface and basal boundary in granular flows such as those in shaken containers (Fond et al., 2012), tumblers (Abram et al., 2013), and avalanches. This analysis is particularly well-suited for nontransparent media such as sand, gravel, quartz, or other granular materials that are common in geophysics. This PIV approach is called "granular PIV." The setup for granular PIV differs from the usual PIV setup in that the optical surface structure which is produced by illumination of the surface of the granular flow is already sufficient to detect the motion. This means one does not need to add tracer particles in the bulk material.

4.4.8 Visualization with PIV

For quantitative flow visualization techniques, a flow is visualized by seeding the fluid with small particles that allow to follow the instantaneous changes of the flow. The region in the flow to be researched is illuminated, mostly by a sheet of light. This light-sheet is generated by an expanding laser beam by means of a cylindrical lens or due to the projection of the beam on a rotating hexagonal mirror. The instantaneous flow is recorded at least twice by very short light flashes with a separation time (Δt) in between that is known. The flashes have to be sufficiently short in order to image the particle shape. Too long illumination times will result in streaks of the particle images that will not allow to determine the exact particle location in the fluid. Several illuminations result in trace patterns of each particle on the image plane. Analyzing these particle traces results in local particle displacements. As the separation time between the recordings is known, the analysis directly results in velocities. Mainly two different techniques are employed: particle tracking velocimetry (PTV) and particle image velocimetry (PIV). The difference is in the density of the seeding. The particle density is often represented by two dimensionless parameters: the image particle density NI and source density NS. The source density represents whether the particle images are overlapping ($NS > 1$) or can be recognized individually ($NS < 1$). NI represents the ratio of the length of the particle tracks between successive illuminations and the distances between individual particles.

4.4.9 Vortex Formation

As an application of PIV for vizualizing some important aspects of the flow process, such as the formation of vortices and their sensitivity to flow Reynolds number, the vortices formed at the edge of a free jet were visualized with PIV.

Free jet issuing from a rectangular nozzle was visualized using a water jet facility in the Fluid Engineering Laboratory of Professor Shouichiro Iio, Department of Environmental Science and Technology, Shinshu University, Nagano, Japan. The schematic layout of the water supply system and water jet facility are shown in Figure 4.17(a).

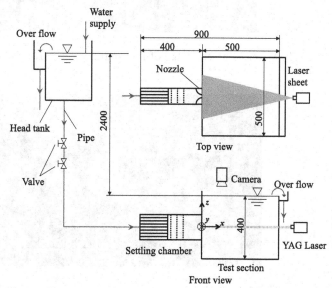

Figure 4.17a: Experimental apparatus (all dimensions in mm)

The experimental nozzle used in the present investigation is a rectangular nozzle. The dimensions of the nozzle exit are 10 mm in width, w, 150 mm in length, h, and the contraction ratio is 10.0. The Reynolds numbers of the jet issuing out of the nozzle, based on nozzle exit width, are 500, 1000, 1500, respectively. Visualizations were made in the test-section, which is 500 mm wide and length 400 mm deep. The photographs shown here were captured from movies recorded via high-speed digital video camera at a frame rate of 500 fps for a total time of 6 seconds. Using a high-speed video camera and a laser sheet light for laser induced fluorescence (LIF) method reduced background noises. The coordinate system is fixed at the nozzle center, x-axis aligned with the flow direction, the y-axis in the span-wise direction, and the z-axis in the vertical direction, which is also perpendicular to the flow.

Some of the vital details about vortex formation due to the differential shear at the interface of the jet edge and the surrounding environment which is stagnant, can be directly inferred from the PIV pictures given in Figures 4.17(b) to 4.17(d), at different Reynolds numbers.

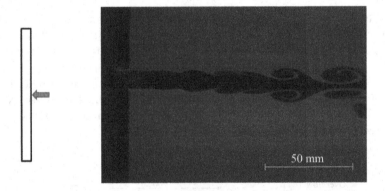

Figure 4.17b: Vortices at the edge of rectangular jet at $Re = 500$

Figure 4.17c: Vortices at the edge of rectangular jet at $Re = 1000$

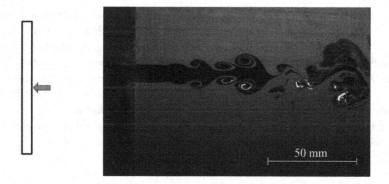

Figure 4.17d: Vortices at the edge of rectangular jet at $Re = 1500$

At Reynolds number 500, as seen from Figure 4.17(b), the jet just shows a tendency for vortex formation, in the near field. Only at about $2.5W$, where W is the jet width at the nozzle exit, is there a bubble formation. A proper vortex formation is taking place only after $5.5W$. The vortices on either side of the jet grow in size and a necking is formed around the jet axis at that location. Downstream of the first necking, the vortices stagger and the neck becomes narrower. Beyond this location, the vortices break into a number of smaller vortices. The formation of bulb, vortices, necking and breaking of large vortices into smaller structures advance and move closer to the nozzle exit, as the Reynolds number increases, as seen from Figures 4.17(c) and 4.17(d).

It is interesting to note that over the specified axial distance of 50 mm, marked on Figures 4.17(b) to 4.17(d), the flow structure is strongly influenced by the Reynolds number. The vortices over this length show a tendency to stagger at $Re = 500$, become staggered and get fragmented at $Re = 1000$, and exhibit a wake-like structure at $Re = 1500$.

4.5 Water Flow Channel

Water flow channel is a simple duct through which water is made to flow with uniform velocity over a length of the channel. The uniform flow portion can be used to visualize flow past objects. Schematic diagram and a pictorial view of a water flow channel, of rectangular cross-section are shown in Figures 4.18(a) and 4.18(b), respectively. Water from the chamber spills over the inclined plate and is conditioned using an array of wire-meshes, before reaching the test-section. It is essential to ensure that the flow quality is fairly uniform in the test-section. For this a color dye may be injected at specified locations across the test-section width. If the dye streaks in the test-section are parallel and smooth, the quality of flow can be taken as good for any visualization study. The velocity of the flow may be measured using the floating-particle method technique, over the length of the test-section.

The uniformity of the flow may also be checked by observing a floating body (say, a tiny bit of paper) moving from the beginning to the end of test-section of the channel. If the test-section flow is uniform, a tiny floating object will travel straight and parallel to the channel floor and sidewalls. After ensuring flow uniformity, flow velocity can be calculated by measuring the time required for the floating object to cover a certain specified distance along the test-section length. A set of measurements has to be made for each speed and the average has to be taken as the test-section flow speed. With this flow speed, the Reynolds number for a characteristic length can be calculated as follows.

For test conducted for a flat plate of chord 25 mm in the water channel at flow speed 33 mm/s and temperature 20°C (Takama et al., 2008).

The Reynolds number is given by

$$\mathrm{Re} = \frac{\rho V c}{\mu}$$

where ρ is the density, V is the velocity of the flow, c is a characteristic length, and μ is the viscosity coefficient.

At 20°C, the density and viscosity coefficient for water are (Rathakrishnan, 2012)

$$\rho = 998\,\mathrm{kg/m^3}$$

$$\mu = 1.002 \times 10^{-3}\,\mathrm{kg/(m\ s)}$$

Thus

$$
\begin{aligned}
\mathrm{Re} &= \frac{\rho V c}{\mu} \\[2mm]
&= \frac{998 \times (33 \times 10^{-3}) \times (25 \times 10^{-3})}{1.002 \times 10^{-3}} \\[2mm]
&= 821.7
\end{aligned}
$$

The desired Reynolds number for a given test model can be arrived at by adjusting the test-section velocity, of course within the limitation of the water table.

Many interesting aspects of flow physics which cannot be visualized with smoke, tuft, and surface coating can be visualized comfortably with a water flow table. Indeed, using hydraulic analogy even some aspects of supersonic and even hypersonic flow can be studied by matching the Froude number of the water channel flow to the desired Mach number. Hydraulic analogy details and application are available in Chapter 6.

To show the potential of water flow channel as a specific problem of positioning twin-vortex behind objects and manipulation of these vortices has been demonstrated here.

A thorough understanding of the mechanism of generating vortices of desired size is of high-value in many applications requiring mixing, such as control of jet mixing, mixing of fuel and oxidizer in combustion chambers, and so on.

Interestingly, it is possible to demonstrate the generation and control of vortex size, with a simple low-cost water flow channel. The feasibility of generating vortices of different size from flat and arc plates and the sensitivity of the vortex size to the plate geometry is demonstrated here.

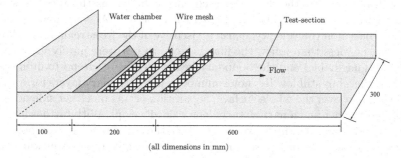

(a) Schematic diagram with the dimensions of the water flow channel designed by the author

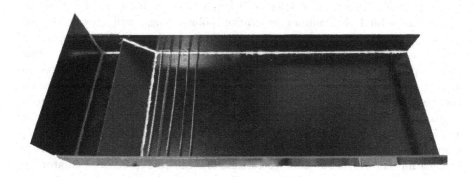

(b) A photographic view of the fabricated water flow channel[1]

Figure 4.18: Water flow channel

In the case of flow issuing out of a passage, the size of a vortex formed is dictated by the radius of curvature of the exit geometry from which it is shed (Takama et al., 2008). For example, the size of the vortices at the exit of a circular nozzle is proportional to the nozzle exit radius. Thus, vortices formed at the periphery of a free jet exiting a circular nozzle will be of uniform size, because the azimuthal radius of curvature of the exit is uniform. A free jet exiting an elliptical nozzle will encounter vortices of a large number of sizes with those at the extremities of the minor axis being the largest and those at the extremities of the major axis being the smallest, because of the continuous variation of the azimuthal radius of curvature from an end of major axis to an

---

[1]This channel was fabricated with plastic sheets by Dr. Watanabe, Suzuki Lab, University of Tokyo, Kashiwa Campus and Ethirajan Rathakrishnan, in 2011.

end of minor axis. Thus, the radius of curvature of the exit dictates the vortex size, in the case of flow from a closed passage.

One of the extensively studied vortex fields is that behind a circular cylinder. The way in which the wake area varies relative to fluid speed may be considered with reference to the flow past a circular cylinder, as the speed is slowly increased from zero (Houghton and Carruthers, 1982). At very low Reynolds numbers (based on the velocity, density, viscosity of the freestream and cylinder diameter), i.e., less than unity, the flow behaves as if it were purely viscous (inertia effects negligible) and the boundary layers effectively extend to infinity. At slightly higher, but still low Reynolds numbers, true boundary layers form which remain laminar over the whole surface. Separation occurs on either side near the rear of the cylinder and a narrow turbulent wake develops. With further increase of the Reynolds number, in the range 10 to 60, the laminar separation points on either side of the rear of the streamwise diameter move rapidly outwards and forward to points near the opposite ends of a transverse diameter. This results in a corresponding increase in wake width and consequent form drag.

At some stage, for a value of Reynolds number somewhere between 60 and 140, a pair of symmetrical vortices will begin to develop on either side of the center line behind the laminar separation points. These will grow with time (at the particular Reynolds number), continuously stretching downstream until a stage is reached when they become unsymmetrical and the system breaks down, one vortex becomes detached and moves away downstream. In practice perfect symmetry and absolute smoothness is not possible and this causes the vortices to be of different size and shed differentially. This makes the vortex formation and shedding become alternative with a frequency for a given shape and initial conditions. The alternative vortex motion behind the cylinder is the popular Karman vortex street. The subsequent wake motion, which is typically seen within the Reynolds number range from 140 to 50,000, is oscillatory in character. This motion was investigated by Theodor von Karman in the first decade of the 20th century and he showed that a stable system of vortices will be shed alternately from the laminar separation point on either side of the cylinder. Thus, a standing vortex will generate in the region behind the separation point on one side, while a corresponding vortex on the other side will break away from the cylinder and move downstream in the wake. When the attached vortex reaches a particular strength, it will in turn break away and a new vortex will begin to develop again on the second side and so on.

Following this work by Karman, a lot of research is being done on twin vortex and wake with vortices (Houghton and Carruthers, 1982, Hemant Sharma et. al., 2008). But the mechanism of formation of the vortices behind shapes other than circular cylinder is still a gray area for research in spite of their effective applications in flow control devices. One such study addressing the formation of twin vortex and wake behind a flat and arc plates has been investigated by Takama et al. (2008) in the water flow channel shown in Figure 4.18. The velocity of the uniform flow was measured by the floating technique. The measurement error is $\pm 5\%$.

Velocity profile exhibiting uniformity is shown in Figure 4.19. As shown in Figure 4.19, the uniformity of flow is fairly good, except near the side walls. In this investigation, the turbulence level was not measured and quantified. But the flow condition upstream of flat and arc plates was kept the same. The state of identical flow condition is good enough for the present study since the results were analyzed only for the relative size of the vortices behind the plates. Test models in the form of flat and arc plates were used in the study.

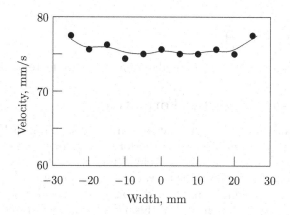

Figure 4.19: Velocity distribution across the width of the test-section

The plate placed in the test-section was aligned normal to the flow. Once the alignment was done, some time was given for the flow around the plate to get established. After that the dye was introduced at an upstream location. The dye mixes with the flow, tracing the fluid elements. This enables the visualization of the vortices behind the plate. The flow motion was continuously videoed (SONY, DCR-TRV50) right from dye injection time. The recorded videos were played and the required information about the formation and size of the vortices were discerned from the video.

To identify the position, the dye was introduced from a far downstream location where there is no reverse flow. The injection location was shifted upstream progressively. This procedure enabled the finding of the location at which the reverse flow begins. The entire sequence was recorded by video. After the experiments, the video pictures were analyzed to measure the semi-major axis. The plate length (L) was taken as the characteristic length scale for calculating the Reynolds number and non-dimensionalizing the vortex size. Three flat plates of length 10 mm, 15 mm, and 20 mm and four arc-plates of the same length 14 mm were studied. For the arc plates, the width (W) was varied keeping the length L constant, as shown in Figure 4.20. Four arc-plates of L/W = 11.2, 5.6, 3.1, and 1.7 were studied. Even though the blockage effect appears significant, the streamlines through the gap between the plate edge and the channel side-wall were not disturbed and continued to flow parallel to the channel wall. This aspect was studied for every model tested. The flow Reynolds numbers studied were in the range from 350 to 1850.

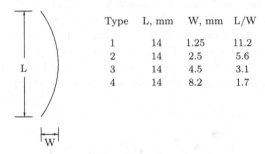

Type	L, mm	W, mm	L/W
1	14	1.25	11.2
2	14	2.5	5.6
3	14	4.5	3.1
4	14	8.2	1.7

Figure 4.20: Schematic diagram of the arc-plate used

4.5.1 Physics of Vortex Formation

The vortex formation behind the flat and arc plates, shown in Figure 4.20, is dictated by the pressure-hill around the forward stagnation location and the shape of the gradual (arc) geometry behind the edges. It is important to note that, in addition to the shape of the forward face and pressure-hill, the possibility of vena-contracta just downstream of the plate edge could also have an effect on the vortex formation, even though this aspect is not addressed in this work.

For the flat plate, the edges are sharp without any gradual change in curvature. Therefore, the flow negotiating the pressure-hill at the front face of the plate finds a sudden increase of area just after the plate edge. The sudden increase in area causes the flow to encounter diffuser effect. The diffuser effect causes the flow downstream of the edge to get decelerated. This deceleration causes increase in pressure, and the pressure increase assists the movement of fluid elements towards the suction zone at the plate base. Thus, when the flow just begins to pass over the plate, the combination of the diffuser effect and the low-pressure at the plate base result in the flow rolling around and turning toward the back face of the plate. This abrupt turn of the flow causes formation of vortex. From the two edges of the plate two vortices are formed. Figure 4.21 shows four stages of vortex formation behind a flat plate. In the fourth stage the twin vortex formed is seen. This formation is at Reynolds number 969. This is far higher than the twin vortex formation Reynolds number of 60-140 for a circular cylinder. The reason for this increased Reynolds number at which twin vortices form behind a flat plate is that the flow behind the plate encounters a higher suction caused by the large vortices behind the plate compared to its cylindrical counterpart. This suction induces a reverse flow, which inducts the mass from a location downstream of the vortices to flow in the reverse direction and fill the region at the plate back. At the same time the flow from the edges also turns towards the region at the plate back. Thus the flow turning towards the plate back rolls and forms vortices behind the plate. These large size vortices act as efficient entrainers and scoop the mass from the zone at the plate back surface (Rathakrishnan, 2012). Due to this a low pressure region is formed at the rear. Once this low pressure level reaches a certain minimum at which the pressure at the base is lower than that downstream of the vortices, the flow just behind the vortex begins to flow towards the plate, establishing a reverse flow. During this process, depending on the plate length and flow Reynolds number, the vortex formation behind the plate is due to the upstream flow rolling at the edges or due to the reverse flow. Thus there appears to be a limiting situation at which the vortex formation is dominated by the reverse flow. At this stage of reverse flow domination, the upstream flow rolling-in is pushed away. However, in both the cases the top (left when viewed in the flow direction) vortex is clockwise, and the bottom (right when viewed in the flow direction) vortex is counter clockwise. Few steps of twin vortex formation behind the plate visualized with water table are shown in Figures 4.21(a) to 4.21(d). The twin vortex at the rear of the flat plate shown in Figures 4.21(c) and 4.21(d) indicates the dominant role played by the reverse flow on the vortex formation.

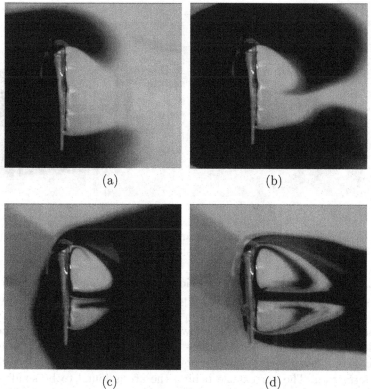

(a) (b)

(c) (d)

Figure 4.21: Sequence of vortex formation behind a flat plate

To get an insight into the effect of geometrical guidance to the flow from the edges of the plate to the base, a plate in the form of an arc termed *arc-plate* was studied by keeping the concave face facing the flow. Also in this case, the twin vortices at the rear and the vortex size were studied. A typical shape of the twin vortices behind an arc-plate and a flat plate of equal chord, at the same Reynolds number is shown in Figures 4.22(a) and 4.22(b). It is interesting to see that the vortices get elongated (aspect ratio of the ellipse increased) when there is a faring form at the edge to the base. In this case the pressure-hill is bound to be stronger than that for a flat plate. Therefore, the concavity of the geometry would be greatly reduced by this pressure-hill zone. Because of this the incoming flow would encounter a better turning than over a flat plate. This enables the flow to negotiate the arc without any abrupt change in the direction, up to the edges. Furthermore, downstream of the edge, the faired geometry enables the flow to proceed smoothly to some distance toward the base. Thus the rolling of the flow towards the arc base center is delayed compared to the abrupt turning behind a flat plate. This causes the vortex to be weaker than that behind the plate. This process makes the mass from the upstream resist the reverse flow better, resulting in an elongated shape of the twin vortices, as seen in Figure 4.22(a).

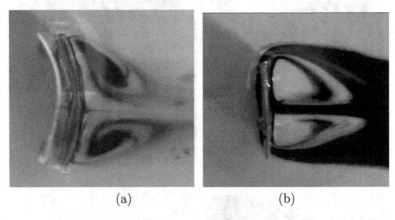

(a)   (b)

Figure 4.22: Vortices behind (a) an arc plate and (b) a flat plate of equal chord, at Reynolds number 930

From the above discussions it is evident that the sharp and smooth edges strongly influence the formation of twin vortices. For flat and arc plates, the twin vortex formation continues to Reynolds numbers which are much larger than that for a circular cylinder. The low pressure caused by the vortices due to the roll-in of the upstream flow induces a reverse flow towards the plate base. In the case of a flat plate, the reverse flow seems to be a dominant driver of the vortex formation. The vortex size behind the arc is found to be smaller than that behind a comparable flat plate at identical flow conditions. It implies that the smooth base reduces the strength of the reverse flow. Thus in general, it appears that there is a limiting situation at which the twin vortex formation is dominated by the reverse flow, not by the upstream flow rolling at the edge. Furthermore, it is seen that the size of the vortex can be controlled with plate geometry.

4.6 Summary

Visualization of fluid flow motion is proved to be an excellent tool for describing and calculating flow properties in many problems of practical interest, in both subsonic and supersonic flow regimes.

The general principle for flow visualization is to render the "fluid elements" visible either by observing the motion of suitable selected foreign materials added to the flowing fluid or by using an optical pattern resulting from the variation of the fluid optical properties (such as refractive index) due to the variation of the properties of the flowing fluid itself. A third class of visualization technique is based on a combination of the above two principles.

Smoke flow visualization is one of the popular techniques used in low-speed flow fields with velocities up to about 30 m/s.

Tufts are used to visualize flow fields in the speed range from 40 to 150 m/s.

Chemical coating is used to visualize flow with speeds in the range from 40 to 150 m/s.

Interferometer is a technique to visualize high-speed flows in the ranges of transonic and supersonic Mach numbers.

Schlieren is used to study high-speed flows in the transonic and supersonic Mach number ranges.

Shadowgraph is yet another flow visualization technique meant for high-speed flows with transonic and supersonic Mach numbers.

Flow visualization with smoke is generally done in a smoke tunnel. It is a low-speed wind tunnel, carefully designed to produce an uniform steady flow in the test-section with negligible turbulence. For good results, the smoke should be white, dense, nonpoisonous, and non-corrosive. Smoke should have nearly the same density as that of the surrounding air, hence the smoke filaments are not appreciably influenced by gravity. Smoke particles should not disturb the flow in the wind tunnel by formation of deposits on the surface of the models or block the tubes used for smoke injection.

For visualizing compressible flows, *interferometer, schlieren,* and *shadowgraph* are the three popularly employed optical flow visualization techniques. Interferometer makes visible the optical phase changes resulting from the relative retardation of the disturbed rays. Schlieren system gives the deflection angles of the incident rays. Shadowgraph visualizes the displacement experienced by an incident ray which has crossed the high-speed gas flow.

The quality of the optical equipment to be used in the schlieren setup depends on the type of the investigation carried out. The cost increases rapidly with the quality of the optical components. The vital components are the mirrors, and the light source.

Interferometer is an optical method most suited for qualitative determination of the density field of high-speed flows.

In general, the schlieren method is used either for the detection of small refractive index gradients or for the quantitative measurement of these gradients.

The shadowgraph is best suited only for flow fields with rapidly varying density gradients.

The theory shows that the schlieren technique depends upon the first derivative of the refractive index (flow density) while the shadowgraph method depends upon its second derivative. Consequently, in phenomena where the refractive index varies relatively slowly, the schlieren method is to be preferred to the shadowgraph method, other things being equal. On the other hand, the shadow method beautifully brings out the rapid changes in the index of refraction. The shadow method also has the advantage of greater simplicity and somewhat wider possible application. The two methods therefore supplement each other and both should be used wherever possible.

Particle image velocimetry (PIV) is an optical method of flow visualization. It can be used to obtain the instantaneous velocities and the related properties in fluid flows.

The main difference between PIV and laser Doppler velocimetry is that PIV produces two-dimensional or even three-dimensional vector fields, while the laser Doppler velocimetry measures only the velocity at a point. In PIV, the particle concentration is such that, it is possible to identify individual particles in an image.

Water flow channel is a simple duct through which water is made to flow with uniform velocity over a length of the channel. The uniform flow portion can be used to visualize flow past objects.

Exercise Problems

4.1 Find the limiting minimum diameter of the tubes of the smoke rake for the smoke tunnel with test-section speed 4 m/s, if the stagnation state corresponds to the standard sea level condition.

[Answer: 0.73 mm]

4.2 What will be the percentage change in the value of Gladstone–Dale constant in air if the density is increased to 10 times the sea level value? Assume the refractive index of air as 1.000293 at both the pressures.

[Answer: −90 percent]

Chapter 5

Hot-Wire Anemometry

5.1 Introduction

The word anemometer simply means an instrument meant for the measurement of wind or air velocity. Maybe hot-wire was initially thought of as an instrument to measure air speed and therefore, was named *hot-wire anemometer*. But in the present context the word anemometer may be considered as inaccurate, since the hot-wires and hot-films are used, besides measurement in air, for measurements in other fluids such as fresh water, salt water, oil, mercury, blood, and so on. However, it should also be realized that, what is important for an instrument is its capability for accurate, quick, and reliably repeatable measurements in a simple and less expensive manner rather than the name by which it is designated. This chapter is concerned with the fundamental principles of hot-wire anemometry as applied to the study of turbulence and unsteady laminar flows in wind streams.

5.2 Operating Principle

Basically, hot-wire anemometer is a *thermal transducer*. Simply stated, the principle of operation of a hot-wire anemometer is as follows: *"The heat transfer from a fine filament which is exposed to a cross flow varies with variation in the flow rate."* That is, when an electric current is passed through a fine filament which is exposed to a cross flow, the heat transfer from the filament varies as the flow rate varies. This in turn causes variation in the heat balance of the filament. The filament is made from a material which possesses a *temperature coefficient of resistance*, that is *if the temperature of the filament varies, so also does its resistance*. The variation of resistance is monitored by various electronic methods which give signals related to the variation in flow velocity or flow temperature. The hot-wire can therefore be used for measuring instantaneous velocities and temperatures at a point in a flow field.

Generally the following two modes of operation are followed in hot-wire anemometry.

- The first is the *constant current mode*. Here the current flow through the hot-wire is kept constant and variations in the wire resistance caused by the fluid flow are measured by monitoring the voltage drop variations across the filament.

- The second is the *constant temperature mode*. Here the hot-wire filament is placed in a feedback circuit which tends to maintain the hot-wire at a constant resistance and hence at a constant temperature. Fluctuations in the cooling of the filament are seen as variations in the current flow through the hot-wire.

Based on the type of electronic instruments used, the hot-wire anemometers may be classified into three categories. They are

- The *constant temperature anemometer*, which supplies the sensor a heating current that varies with the fluid velocity, in order to maintain constant sensor resistance and thus, constant temperature.

- The *constant current anemometer*, which supplies a constant heating current to the sensor. The variation in the sensor resistance with fluid flow velocity change causes voltage drop variation across the sensor.

- The *pulsed wire anemometer*, which measures velocity by momentarily heating a wire to heat the fluid around it. This spot of heated fluid is convected downstream to a second wire that acts as a temperature sensor, the time of flight of the hot spot is inversely proportional to the fluid velocity.

5.3   Hot-Wire Filaments

The vital component in a hot-wire anemometer system is the hot-wire filament. The hot-wire filament may be regarded as an infinitely long, straight cylinder in cross flow. Large number of empirical heat transfer relations have been proposed for this kind of problems. The following is one such empirical relation evolved by Kramer (Hinze, *Turbulence*, 1987), which gives satisfactory results for many gases and liquids:

$$Nu = 0.42 \, Pr^{0.20} + 0.57 \, Pr^{0.33} \, Re^{0.50} \tag{5.1}$$

where

$$Nu \; = \; \frac{hd}{k} \qquad \text{is } \textit{the Nusselt number}$$

$$Pr \; = \; \frac{\mu C_p}{k} \qquad \text{is } \textit{the Prandtl number}$$

$$Re \; = \; \frac{\rho U d}{\mu} \qquad \text{is } \textit{the Reynolds number}$$

where h is the film coefficient which is the heat flux leaving the wire surface per unit temperature difference ($\Delta\theta$) between the wire and the freestream, d is hot-wire diameter, k is thermal conductivity of the gas, μ is dynamic viscosity of the gas, ρ is density of the gas, β is coefficient of expansion of the gas, C_p is the specific heat of the gas at constant pressure, and U is velocity of flow past the wire.

Equation (5.1) is valid for air and diatomic gases in the Reynolds number range (Hinze, *Turbulence*, 1987)

$$0.01 < Re < 10000$$

Gas properties occurring in the dimensionless groups in Equation (5.1) refer to the film temperature θ_f given by

$$\theta_f = \frac{\theta_w + \theta_g}{2} = \theta_g + \frac{\theta_w - \theta_g}{2} = \theta_g + \frac{\Delta\theta}{2} \qquad (5.2)$$

where θ_w is temperature of the wire, and θ_g is temperature of the gas. For air as the gas and a temperature difference of $\Delta\theta = 100°C$, $(\mu/\rho)_f \approx 0.20\,\text{cm}^2/\text{s}$, therefore, Ud should be

$$0.002 < Ud < 2000\,\text{cm}^2/\text{s}$$

when the wire diameter is 5 micron, this would give

$$4 < U < 4 \times 10^6\,\text{cm/s}$$

This velocity range is wide enough for many practical applications of hot-wire anemometry.

At this stage, we may ask the question whether it is reasonable to ignore free-convection and radiation effects on the heat transfer from the hot-wire, as is done in the assumption for Equation (5.1). The answer to this question is the following.

It is well established that the free-convection effect depends mainly on the value of the group $Gr \times Pr$ and this effect may be neglected for $Re > 0.5$ if

$$Gr \times Pr = \frac{g\,C_p\,\rho^2\,\beta\,d^3\,\Delta\theta}{\mu\,k} < 10^{-4}$$

where Gr is called *Grashof number* and is defined as

$$Gr = \frac{g\,\rho^2\,\beta\,d^3\,\Delta\theta}{\mu^2}$$

For air and a wire of 5 μ diameter, $Gr \times Pr$ is of the order of 10^{-6}, i.e., far below the limiting value.

Radiation effects in heat transfer to the ambient air are negligibly small under operating conditions where the wire temperature does not exceed 300°C. Thus, the thermal radiation from wires of 5-micron diameter can be safely neglected for wire temperatures below 300°C.

The heat per unit time transferred to the ambient gas from a wire of length l at a uniform temperature of θ_w is

$$h\,\pi\,d\,l\,(\theta_w - \theta_g)$$

This may be expressed as

$$\frac{h\,d}{k}\,k\pi\,l\,(\theta_w - \theta_g)$$

i.e.,

$$Nu\,\pi\,k\,l\,(\theta_w - \theta_g)$$

Substituting for N_u from Equation (5.1), we get

$$\pi\,k\,l\,(\theta_w - \theta_g)\left[0.42\,Pr^{0.20} + 0.57\,Pr^{0.33}\,Re^{0.50}\right]$$

For thermal equilibrium, this heat loss per unit time must be equal to the heat generated per unit time by the electric current through the hot-wire; i.e., it must be equal to $I^2\,R_w$, where I is electric heating current and R_w is total electric resistance of the wire. Thus, for thermal equilibrium of the wire we have

$$I^2\,R_w = C_c\,\pi\,k\,l\,(\theta_w - \theta_g)\left[0.42\,Pr^{0.20} + 0.57\,Pr^{0.33}\,Re^{0.50}\right]$$

where C_c is a conversion constant. Usually the left-hand side is obtained in joule per second, and the right-hand side is obtained in calories per second. In this case, $C_c = 4.2$ (Hinze, 1987). The gas properties occurring on the right-hand side of this equation are still functions of θ_f, i.e., θ_g and $(\theta_w - \theta_g)$. They usually give rise to second-order effects.

The temperature dependence of the electric resistance of the wire gives an effect of first order, it is on this effect that the use of a hot-wire as an anemometer is based. The important property of a wire, namely the *temperature coefficient of resistivity* allows the interpretation of voltage fluctuations in terms of velocity fluctuations. This temperature dependence may be expressed as

$$R_w = R_0\left[1 + C\,(\theta_w - \theta_0) + C_1\,(\theta_w - \theta_0)^2 + \cdots\right] \tag{5.3}$$

where R_w is the wire resistance at temperature θ_w, R_0 is wire resistance at a reference temperature θ_0, and C and C_1, etc., are temperature coefficients of electrical resistivity. Typical values for these constants are

For platinum: $C = 3.5 \times 10^{-3} \, \text{K}^{-1}$, $\quad C_1 = -5.5 \times 10^{-7} \, \text{K}^{-2}$

For tungsten: $C = 5.2 \times 10^{-3} \, \text{K}^{-1}$, $\quad C_1 = 7.0 \times 10^{-7} \, \text{K}^{-2}$

A heat balance of hot-wire filament may be stated as

$$H_g = H_T + H_A \tag{5.4}$$

where H_g is the heat generated per unit time by joule heating, H_T is heat transferred per unit time to the fluid, and H_A is heat accumulation per unit time. The heat loss at the ends of the hot-wire filament has been neglected here. From Equation (5.3), we get,

$$\Delta\theta = \frac{R_w - R_0}{R_0 C} \tag{5.5}$$

Note that, the reference state is taken as the gas state for defining $\Delta\theta$ in Equation (5.5).

The film coefficient h in Equation (5.1), by definition, is

$$h = \frac{H_T}{A_s \Delta\theta} \tag{5.6}$$

where A_s is the hot-wire surface area. Combining Equations (5.1), (5.5) and (5.6), we obtain

$$H_T = (R_w - R_g)(A_1 + B_1 \sqrt{U}) \tag{5.7}$$

where

$$A_1 = \frac{0.42 \, k \, A_s}{R_g \, C \, d} \left(\frac{\mu C_p}{k} \right)^{0.2} \tag{5.8}$$

$$B_1 = \frac{0.57 \, k \, A_s}{R_g \, C \, d} \left(\frac{\mu C_p}{k} \right)^{0.33} \left(\frac{\rho d}{\mu} \right)^{0.5} \tag{5.9}$$

For a given hot-wire and fluid at a given operating temperature, A_1 and B_1 can be regarded as constants.

The accumulation of heat H_A in the hot-wire is

$$H_A = C_w \, v \, \frac{d\theta_w}{dt} \tag{5.10}$$

where C_w is the specific heat of the wire material and v is its volume. Assuming the gas temperature to be a constant, $d\theta_w/dt$ can be replaced by $d\Delta\theta/dt$, and using Equation (5.5) the H_A can be expressed as

$$H_A = c \, \frac{dR_w}{dt} \tag{5.11}$$

where $c = C_w v / R_g C$. This is a constant associated with the thermal capacity of the wire. Using Equations (5.7) and (5.11), Equation (5.4) can be expressed as

$$I^2 R_w = (R_w - R_g)(A_1 + B_1 \sqrt{U}) + c \frac{dR_w}{dt} \tag{5.12}$$

where I is the electric current passing through the hot-wire. For steady conditions, Equation (5.12) reduces to

$$\boxed{I^2 R_w = (R_w - R_g) \left(A_1 + B_1 \sqrt{U} \right)} \tag{5.13}$$

This equation, popularly known as *King's relation* was first derived by King by assuming potential flow about the wire and making many drastic assumptions about the circumferential distribution of heat flux. It is adequately accurate for analyzing the behavior of hot-wire systems, even though it is not precise and its application as a calibration law is questioned in literature.

Now, by defining a parameter called *resistance ratio*, $R = R_w / R_g$, Equation (5.13) can be rewritten as

$$I^2 = \frac{R-1}{R} \left(A_1 + B_1 \sqrt{U} \right) \tag{5.14}$$

For a fixed value of R, Equation (5.14) becomes

$$I^2 = \left(A + B \sqrt{U} \right) \tag{5.15}$$

where A and B are constants to be determined from an experimental plot.

The variation of I^2 with $\sqrt{U}$, using Equation (5.15) is as shown in Figure 5.1, is called the King's law plot.

The constants A and B represent the intercept and slope, respectively, of a straight line of best fit. At very low velocities Equation (5.15) will break down because of natural convection effects.

From Figure 5.1 it is seen that all lines are passing through the same point Q, and the slope of the curve is proportional to $(R-1)/R$. For $R \to \infty$, the asymptotic curve $I^2 = A_1 + B_1 \sqrt{U}$ is obtained. This corresponds to *burn-out* of the wire since the temperature becomes infinite. This is the physical situation at which the rate of heat removal cannot any longer balance the heat generated.

The two basic modes of operating a hot-wire, (a) the constant current mode, and (b) the constant temperature mode, are shown as their trajectories for perturbations in U about the operating point C defined by the mean value of current and velocity in Figure 5.1.

Further, it is interesting to note from Figure 5.1 that, for the constant current operation, reduction in velocity U below a certain limit will result in the burn-out of the wire. This is indicated by point A'. This is one of the disadvantages of the constant current operation, since even the switching off of the wind tunnel could destroy the wire.

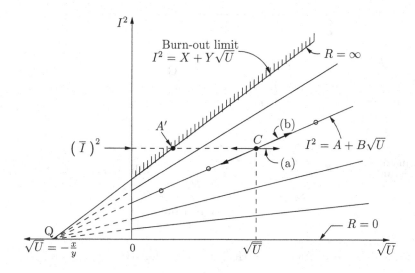

Figure 5.1: (a) Constant current trajectory; (b) constant temperature trajectory

5.4 Constant Current Hot-Wire Anemometer CCA

Even though the constant current hot-wire anemometer is rarely used for measuring velocity fluctuations except in heat transfer experiments these days, it is illustrative to study the CCA. The CCA was in the forefront until the mid 1960s. From then onwards, the CCA system has been replaced to a large extent by the superior constant temperature anemometer, CTA. However, much of the theory associated with CCA forms the basis for the more complex CTA theory. Also, the CCA is still considered to be the best instrument for the measurement of temperature fluctuations in a gas.

5.4.1 Mean Velocity Measurements

Figure 5.2 shows a classical constant current hot-wire anemometer circuit. The current is kept constant in a simple manner by connecting a large resistance R_s in series in the heating circuit, as shown in Figure 5.2, such that the change in current due to change in wire resistance is small.

The resistance R_s should be at least 1000 times larger than the wire resistance R_w. The wire resistance ratio is first set by setting R_b. By adjusting R_s and observing the galvanometer G, the bridge is brought to balance. This is done for different velocities and a static constant-temperature calibration curve of $\overline{I}^2$ versus $\sqrt{\overline{U}}$ is plotted. The slope of the straight line of best fit gives the constant B_1 in Equation (5.13). The intercept of this curve gives the constant A_1. The initial current in the system is adjusted such that the hot-wire resistance

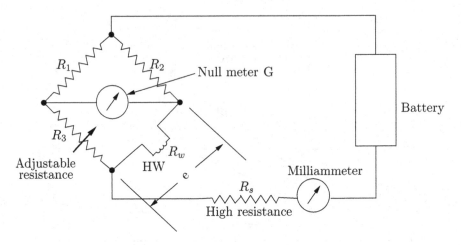

Figure 5.2: Heating circuit for CCA

with no flow is nearly 1.5 times the cold resistance R_g. That is, the balancing resistance is kept at $1.5R_g$ and the current is adjusted until the galvanometer indicates null reading. As the flow velocity increases R_w will decrease. For each velocity the bridge has to be balanced and the value of R_w noted down. Now, using Equation (5.13), *the flow velocity U, which is the only unknown in the equation can be computed.*

Basically hot-wire is placed in the flow and the wire resistance is measured using the balancing arm resistance R_3 in the Wheatstone bridge in Figure 5.2. For the measurement of current, the accuracy of an ordinary panel-type ammeter is inadequate. It is usually computed by measuring the voltage drop across a fixed one ohm resistance connected in series in the circuit. A precision null type potentiometer is employed for this purpose. Modern digital voltmeters which are comparable in quality as well as cost to the potentiometer may also be used for this measurement.

5.4.2 Fluctuating Velocity Measurements

The current across the hot-wire is kept constant but the voltage varies. It can be shown that, this voltage can be used for measuring the fluctuating velocities using an amplifier and a root mean square (RMS) voltmeter.

From the King's relation, Equation 5.13, we have

$$\frac{I^2 R_w}{R_w - R_g} = A_1 + B_1\sqrt{U} \qquad (5.16)$$

where I and R_g are constants, and A_1 and B_1 also can be regarded as constants. Strictly speaking, A_1 and B_1 are constants only for CTA and for CCA they vary with flow velocities. Therefore, the assumption of treating A_1 and B_1

as constants for CCA is not acceptable. However, this unacceptable assumption is made and still the relation is claimed to yield reasonable results. This state of the art of 1960s is continued even now since after the introduction of CTA, CCA is seldom used for turbulence measurements. This may be the reason for the continuation of 1960s state of the art of the problem. As we mentioned in the introductory remarks, it is instructive to examine the constant current anemometer, even though now it is rarely used for measuring velocity fluctuations except in heat transfer experiments. With this aspect in mind let us also assume that the constants A_1 and B_1 are invariants even in CCA.

For a turbulent flow, the velocity is made up of mean and fluctuational components. For instance, the velocity U in Equation (5.16) may be expressed as

$$U = \overline{U} + u' \tag{5.17}$$

where $\overline{U}$ is the time averaged velocity, u' is the fluctuating velocity, and U is the instantaneous velocity. In a similar manner, it can be argued that the hot-wire resistance R_w is given by

$$R_w = \overline{R_w} + r'_w \tag{5.18}$$

where $\overline{R_w}$ and r'_w are the mean and the fluctuating components of the wire resistance.

Substitution of Equations (5.17) and (5.18) into Equation (5.16) yields

$$\frac{I^2 \left(\overline{R_w} + r'_w\right)}{\left(\overline{R_w} + r'_w - R_g\right)} = A_1 + B_1 \sqrt{\overline{U} + u'} \tag{5.19}$$

The left-hand side of Equation (5.19) may be rearranged as

$$
\begin{aligned}
\frac{I^2 \left(\overline{R_w} + r'_w\right)}{\left(\overline{R_w} + r'_w - R_g\right)} &= I^2 \left[\frac{\overline{R_w} + r'_w - R_g + R_g}{\overline{R_w} + r'_w - R_g}\right] \\[2mm]
&= I^2 \left[1 + \frac{R_g}{\left(\overline{R_w} - R_g\right) + r'_w}\right] \\[2mm]
&= I^2 \left[1 + \frac{R_g}{\left(\overline{R_w} - R_g\right)} \left\{1 + \frac{r'_w}{\left(\overline{R_w} - R_g\right)}\right\}^{-1}\right]
\end{aligned}
$$

But, $r'_w / \left(\overline{R_w} - R_g\right) \ll 1$ and therefore, expanding the term $\left\{1 + \frac{r'_w}{\left(\overline{R_w} - R_g\right)}\right\}^{-1}$ and neglecting high-order terms, we get

$$\frac{I^2\left(\overline{R_w}+r'_w\right)}{\left(\overline{R_w}+r'_w-R_g\right)} = I^2\left[1+\frac{R_g}{\left(\overline{R_w}-R_g\right)}\left\{1-\frac{r'_w}{\left(\overline{R_w}-R_g\right)}\right\}\right]$$

$$= I^2\left[1+\frac{R_g}{\left(\overline{R_w}-R_g\right)}-\frac{R_g r'_w}{\left(\overline{R_w}-R_g\right)^2}\right]$$

$$= I^2\left[\frac{\overline{R_w}}{\left(\overline{R_w}-R_g\right)}-\frac{R_g r'_w}{\left(\overline{R_w}-R_g\right)^2}\right]$$

$$= \frac{I^2\overline{R_w}}{\left(\overline{R_w}-R_g\right)}\left[1-\frac{R_g r'_w}{\overline{R_w}\left(\overline{R_w}-R_g\right)}\right]$$

The right-hand side of Equation (5.19) becomes

$$A_1 + B_1\sqrt{\overline{U}+u'} = A_1 + B_1\sqrt{\overline{U}\left(1+u'/\overline{U}\right)}$$

where $u'/\overline{U} \ll 1$, since the fluctuating velocity u' is very small compared to the mean velocity $\overline{U}$, thus, neglecting the terms smaller than $u'/\overline{U}$, we have

$$A_1 + B_1\sqrt{\overline{U}+u'} = A_1 + B_1\sqrt{\overline{U}}\left(1+\frac{1}{2}\frac{u'}{\overline{U}}\right)$$

Hence, Equation (5.19) becomes

$$\frac{I^2\overline{R_w}}{\left(\overline{R_w}-R_g\right)}\left[1-\frac{R_g r'_w}{\overline{R_w}\left(\overline{R_w}-R_g\right)}\right] = A_1 + B_1\sqrt{\overline{U}}\left(1+\frac{1}{2}\frac{u'}{\overline{U}}\right) \quad (5.20)$$

Equating the fluctuating quantities on the left- and right-hand sides of Equation (5.20), we get

$$-\frac{I^2 R_g r'_w}{\left(\overline{R_w}-R_g\right)^2} = B_1\frac{u'}{2\sqrt{\overline{U}}}$$

The fluctuating voltage $e' = I\,r'_w$. Using the above equation e' can be expressed as

$$e' = -\frac{\left(\overline{R_w}-R_g\right)^2}{2IR_g}B_1\frac{u'}{\sqrt{\overline{U}}}$$

On the right-hand side all the quantities except u' are constants. Therefore, grouping the constants together as $1/S$, the relation for fluctuating velocity becomes

$$\boxed{u' = Se'}$$ (5.21)

where

$$S = -\frac{2IR_g\sqrt{\overline{U}}}{\left(\overline{R_w} - R_g\right)^2 B_1}$$

In the derivation of Equation (5.21), $\left(u'/\overline{U}\right)$ is assumed to be small. The error introduced by this assumption is negligible: less than 1 to 2 percent, for $\left(u'/\overline{U}\right)$ less than 0.10, which is true for most of the turbulent flows. From Equation (5.21) it is evident that *the fluctuating voltage is a direct measure of velocity fluctuation.*

5.4.3 Thermal Inertia of Hot-Wire

Even though the diameter of the hot-wire is only a few microns and its length is only one or two millimeters, the thermal inertia, namely *the inability of the wire to get heated up or cooled down fully in time with the velocity fluctuations*, of the wire sets a limit to the frequency response which is inadequate for measurement in turbulent flows. The remedy to this problem is to increase the amplifier gain in such a manner that the gain compensates the loss. This amplifier is known as a *compensating amplifier*. Schematic diagram of the amplification circuit required for a CCA system is shown in Figure 5.3.

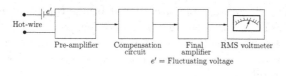

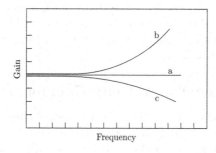

Figure 5.3: (a) Gain of uncompensated amplifier; (b) gain of the amplifier after compensation; (c) frequency response of the hot-wire signal due to thermal inertia

It can be shown mathematically that the increase in gain needed is

$$\left[1 + (2fM)^2\right]^{\frac{1}{2}}$$

where f is the frequency of fluctuation and M is the time constant (a measure of thermal inertia) of the wire. A capacitance resistance network, shown in Figure 5.3, has the same attenuation characteristics and when this circuit is inserted in the amplifier in a suitable manner, the output voltage can be made to fluctuate as though the wire is not subjected to thermal inertia. A hot-wire in conjunction with a compensated amplifier can be operated in the frequency range of 0 to 20 kHz or even more.

5.4.4   RMS Measurements of the Fluctuating Velocities

The level of turbulence is described by the root mean square (RMS) value of the fluctuations

$$\left(\sqrt{\overline{u'^2}}, \ \sqrt{\overline{v'^2}}, \ \sqrt{\overline{w'^2}}\right)$$

in turbulence measurements. The average value of the fluctuating velocity is always zero in a stationary random process. The fluctuating voltage obtained at the output end of the compensating amplifier is squared electronically and averaged over a period, usually of the *order of ten seconds*. RMS voltmeters for random signals are specially manufactured for this application.

In the discussion on the calibration of CCA it was seen that the constants A_1 and B_1 are obtained by calibrating the hot-wire at constant temperature. It is justified since the voltage perturbation at the top of the bridge due to velocity fluctuations is described by a single pole (Perry, 1982). If current is injected across the bridge, the voltage perturbation at bridge top is again described by a single pole. This enables setting up the computer network with a current injection. This also enables the determination of output voltage sensitivity to velocity for constant current operation *even though the wire has been calibrated at constant temperature.*

At this stage, it is important to note that the CCA cannot be used for large velocity perturbations since the output becomes distorted because of the weighting given to the higher harmonics by the compensator. The higher harmonics arise from nonlinearities.

5.4.5   Measurement of Velocity Components

We know that the turbulent fluctuations are always three-dimensional. These fluctuational components are usually measured by suitable configuration of the hot-wire probe. A single wire probe with the hot-wire element placed perpendicular to the main flow is used for measuring the velocity U, in the direction of main flow. For the measurement of the intensity of the turbulent velocity components u', v', and w' the directional sensitivity of the hot-wire is made use of. A wire kept inclined at an angle to the flow will respond only to the component of the velocity normal to the wire. Two identical hot-wires fixed at right angles and placed in a flow with one of the wires inclined at 45° to the main stream will sense $(u' \cos 45° + v' \sin 45°)$ and $(u' \cos 45° - v' \sin 45°)$, one of the wires responding to the first group and the other to the second. Since $\sin 45° = \cos 45° = \frac{1}{\sqrt{2}} = k$ (say), we can write

$$(\text{wire})_1 = k\,(u' + v') \quad \text{and} \quad (\text{wire})_2 = k\,(u' - v')$$

Subtracting the output of these two wires, we get

$$(\text{wire})_1 - (\text{wire})_2 = 2\,v'\,k \tag{5.22}$$

In a similar manner, u' and w' can be obtained. When the outputs of the wires are squared, averaged and subtracted, we obtain

$$\overline{(\text{wire})}_1^2 - \overline{(\text{wire})}_2^2 = \overline{k^2}\left[\left(u'^2 + v'^2 + 2u'v'\right) - \left(u'^2 + v'^2 - 2u'v'\right)\right]$$

$$= 4\,k^2\,u'\,v' \tag{5.23}$$

The Reynolds stress

$$(-\,\rho\overline{u'^2},\ -\rho\overline{u'v'},\ -\rho\overline{u'w'}),\ (-\,\rho\overline{v'^2},\ -\rho\overline{v'u'},\ -\rho\overline{v'w'}),\ (-\,\rho\overline{w'^2},\ -\rho\overline{w'u'},\ -\rho\overline{w'v'})$$

is obtained this way. The hot-wire probe in cross-wire construction employed for the above measurement is shown schematically in Figure 5.4.

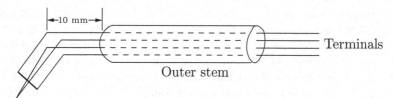

Figure 5.4: Hot-wire probe with cross-wire construction

The wires used in cross-wire probes should have the same diameter, length, and resistance. In practice, making such probes is difficult and requires a lot of caution and experience.

5.4.6 Measurement of Temperature by Constant Current Method

The constant current hot-wire anemometer may also be employed for measuring fluid temperature. The hot-wire may be used for both steady-state temperature measurements and temperature fluctuations measurements. In temperature measurement with hot-wire, the dimensionless parameter namely the *overheat ratio a*, defined as

$$a = \frac{\text{The resistance of the heated wire at its operating temperature}}{\text{The resistance of the sensor at the temperature of the ambient fluid}}$$

plays a dominant role.

5.4.7   Measurement of Steady-State Temperature

A calibration curve is made between the hot-wire resistance and fluid temperature by simply measuring the wire resistance at different temperatures of the fluid. Now, the temperature of a stream may simply be measured by recording the hot-wire resistance when it is kept at the stream. The corresponding temperature can be read from the calibration graph.

For continuous measurement of steady-state temperature, both the constant current and constant temperature hot-wire anemometers can be used, in principle. However, at low overheat ratios, drift in the measurement due to the stability requirement of the constant current source in the constant current anemometer poses a serious problem.

5.4.8   Measurement of Temperature Fluctuations

It can be shown that a reduction in the sensor temperature would cause a decrease in velocity sensitivity for both constant current and constant temperature hot-wire anemometers. Because of this effect, fluctuating temperature measurements are made using probes having a *low value of overheat ratio*. However, for stability of the hot-wire system, the overheat ratio must be as high as possible. Therefore, it is important to identify an optimum value of overheat ratio, for accurate measurements of fluctuating temperatures. Even though both CCA and CTA can be used for fluctuating temperature measurements, the CCA is preferred for this measurement since it has comparatively better stability.

The principles of temperature measurements given here in this section are only the outline of the technique used for such measurements. For complete details about temperature measurement with hot-wire one may refer to books fully devoted to hot-wire anemometers like A.E. Perry (1982) and Charles G. Lomas (1986).

5.5   Constant Temperature Hot-Wire Anemometers

In constant temperature hot-wire anemometer systems, the wire resistance R_w is always kept constant and not allowed to fluctuate. The current is made to decrease or increase as soon as the hot-wire changes its resistance due to flow velocity fluctuations. The current adjustment is automatically done by employing a servo amplifier which has a feedback capacity for a frequency range from

0 to many kHz. A schematic diagram of typical constant temperature hot-wire anemometer is shown in Figure 5.5. Basically it consists of a Wheatstone bridge, a feedback electronic servo amplifier, a high gain amplifier, and an RMS voltmeter.

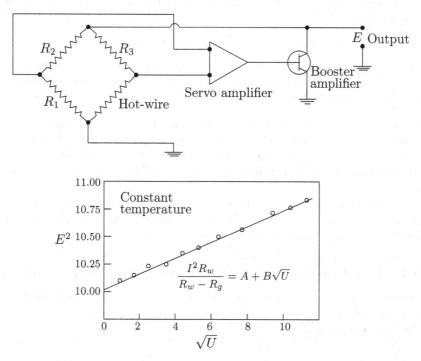

Figure 5.5: Constant temperature hot-wire anemometer circuit and calibration plot

The following are the advantages of CTA.

- Since the feedback takes place almost instantaneously keeping the wire temperature and hence its resistance at a predetermined value, no compensator is needed.

- The compensation for the thermal inertia of the filament is continuously adjusted automatically as its operating point varies. Thus, when taking a traverse of a jet, wake, or boundary layer with a CTA, there is no need for a special calibration and compensator setting for each $\overline{U}$.

- In spite of the nonlinear calibration, it is possible in certain cases to apply an inversion formula to the output signal for reconstructing the input wave form of velocity even for large signals.

The CTA has become more popular than CCA for the following reasons.

- The CCA is calibrated at constant temperature, and then used in a pseudo-constant current mode.

- The CCA calibration has to be static since to obtain a constant temperature calibration, each operating point of the bridge has to be adjusted manually.

- The CTA is used in the same way as it is calibrated.

- The temperature of the wire is maintained approximately constant automatically by a feedback circuit and this makes it possible to calibrate the system dynamically.

Using modern solid state devices, the physical size of the servo amplifiers can be made very small. However, when very high frequencies beyond 20 kHz are involved a constant current system has to be resorted to due to limitations posed by the feedback system. Such high frequencies are involved only in supersonic flows.

In reality, the constant temperature anemometer does not hold the wire exactly at constant temperature. The wire resistance has to vary by a small amount so as to give the amplifier a *feedback* signal.

5.5.1   Relation between Flow Velocity and Output Voltage

In a constant temperature hot-wire anemometer system, the flow velocity and the output voltage of the servo amplifier are related as follows:

From Equation 5.13, we have

$$\frac{I^2 R_w}{(R_w - R_g)} = A_1 + B_1 \sqrt{U} \tag{5.24}$$

where R_w and R_g are constants. Therefore, $I = E/R_w$, where E is the voltage. The above equation may be expressed as

$$\frac{E^2}{R_w^2} \frac{R_w}{(R_w - R_g)} = A_1 + B_1 \sqrt{U} \tag{5.25}$$

or

$$\boxed{E^2 = A + B \sqrt{U}} \tag{5.26}$$

where

$$A = A_1 R_w (R_w - R_g)$$

and

$$B = B_1 R_w (R_w - R_g)$$

In Equation (5.26) A and B are constants. When E^2 is plotted against $\sqrt{U}$, a straight line is obtained with A as the intercept and B as the slope.

The fluctuating voltage e' is related to the fluctuating velocity u' in the following manner.

Let $U = \overline{U} + u'$ and $E = \overline{E} + e'$, where $\overline{U}$ and $\overline{E}$ are the mean velocity and voltage, respectively.

In the above relations, instantaneous or local velocity U and instantaneous voltage E are expressed as the sum of the respective mean and fluctuating quantities. Further, $e' = R_w dI$ and $dU = u'$, for $u' \ll U$.

Differentiation of Equation (5.24) results in

$$\frac{2\,I\,R_w\,dI}{R_w - R_g} = \frac{B_1 dU}{2\sqrt{\overline{U}}}$$

i.e.,

$$\frac{4\,I\,e'}{R_w - R_g} = \frac{B_1}{\sqrt{\overline{U}}}u'$$

Solving for u', and writing I as E/R_w, we get

$$u' = \frac{4E}{R_w\,(R_w - R_g)}\frac{\sqrt{\overline{U}}}{B_1}\,e'$$

or,

$$\boxed{u' = c_1\,e'} \tag{5.27}$$

where $c_1 = \dfrac{4\,E\sqrt{\overline{U}}}{B_1\,R_w\,(R_w - R_g)}$

Squaring and taking the mean of either side of Equation (5.27) yields

$$\boxed{\overline{u'^2} = c\,\overline{e'^2}} \tag{5.28}$$

The fluctuating voltage is measured as $\sqrt{\overline{e'^2}}$ in the RMS voltmeter and therefore, the fluctuational voltage can be determined from Equation (5.28), since c is a constant for a given flow condition.

The other components of mean and fluctuational velocity components can be measured with CTA, following the same procedure described for CCA in Section 5.4.

5.6   Hot-Wire Probes

The most important component in a hot-wire system is the hot-wire probe or hot-wire filament. The hot-wire element which is usually made of *platinum* or *tungsten* wire of *about* 1 mm *long and about* 5 micrometers *in diameter*. The wire is soldered to the tips of two needles, as shown in Figure 5.6, which forms the probe.

Platinum-rhodium alloy wire is also used instead of pure platinum to give extra tensile strength to the hot-wire element. For laboratory turbulence measurements *wires with diameters* in the range 4 to 8 microns and *length* 1 or 2

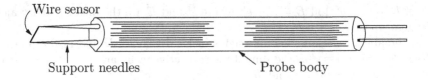

Figure 5.6: Hot-wire probe

mm are used. Platinum wire can be *soldered* with the support needles, whereas the tungsten wire has to be *welded*.

The platinum wire is sold in the form of *woolaston wire*, which is nothing but a fine platinum wire (or platinum-rhodium alloy wire) coated with a thick film of silver. The increase in wire diameter due to the coating facilitates easier handling and soldering of the wire to the tips of the support needles. The required portion of the wire is etched with dilute nitric acid. Silver gets dissolved in nitric acid, leaving the platinum bare. Etching can be done by allowing a small jet of acid to impinge on the woolaston wire, through the tip of a burette. Etching can be hastened and made uniform by the passage of a small amount of electric current between the jet and the woolaston wire. A typical etching unit will be as shown in Figure 5.7.

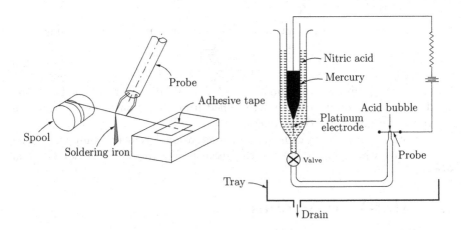

Figure 5.7: Etching unit

Tungsten wire is directly welded to the probe terminals using a special welding setup. Tungsten has less resistance than platinum for wires having the same dimensions and hence its sensitivity to flow velocity is less. However, the tensile strength of tungsten is much superior compared to platinum.

5.7 Hot-Wire Bridge for Classroom Demonstration

In the earlier sections of this chapter we saw the procedure for the measurement of mean and fluctuational velocities and temperatures of fluid flow with a hot-wire anemometer. Even though the principles involved in these measurements are fairly simple, the detailed measurement of velocity and temperature fluctuations requires knowledge of electronics and the reader is recommended to refer to books specializing in such measurements. Further, quantitative research with hot-wire anemometers becomes expensive because of the electronic circuitry requirements. However, for the demonstration of a hot-wire principle in the classrooms, a simple hot-wire bridge can easily be built within the laboratory.

Consider the Wheatstone bridge circuit shown in Figure 5.8.

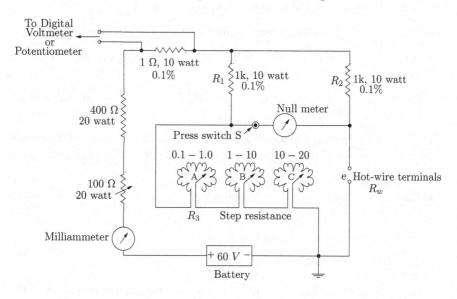

Figure 5.8: Wheatstone bridge circuit

In the circuit R_1, R_2, the hot-wire resistance R_w and the adjustable resistance R_3 form the four arms of the Wheatstone bridge. A 60-volt battery with a 100-ohm rheostat is used to supply the heating current to the hot-wire. The variable resistance R_3 which is used for nulling the current is made into three parts A, B, and C, which is further subdivided into 10 parts at each stage so that any resistance in the range 0 to 20 ohms could be obtained with an accuracy of 0.1 ohm. The current flow through the hot-wire is measured in terms of the voltage drop across a standard one-ohm resistance, using an electronic voltmeter.

5.7.1   Hot–Wire Bridge Operating Procedure

- Connect the battery and keep the rheostat resistance at its maximum value.

- Connect the hot-wire to the bridge circuit.

- Adjust the rheostat until the millivoltmeter shows a reading of 2 to 3 milliamperes.

- Now press the switch S and adjust resistance R_3 until the null meter reads zero.

- This value of R_3 is the resistance of the hot-wire at room temperature, that is R_g.

- Place the hot-wire in the flow stream whose turbulence level is not high (a low turbulence wind tunnel is best suited for this).

- Measure the flow velocity with a pitot-static tube.

- Increase the current through the hot-wire until the current meter reads 50 milliamperes (for a 5-micron wire).

- Balance the bridge by pressing S and adjusting R_3. Note the value of R_3 for that flow velocity; $R_3 = R_w$

- Connect the digital voltmeter across the standard one ohm resistance and read the value. This voltage is the same as the current in amperes, that is I.

- Repeat the experiment at different velocities of the stream and note down the flow velocity as well as R_w.

- Switch off S and disconnect the battery.

- Plot $\sqrt{U}$ versus $I^2 R_w/(R_w - R_g)$

This plot is the calibration curve for the hot-wire used. To visualize the turbulence of the flow stream, the output from the hot-wire may be connected to an oscilloscope. Currently, equipment with built-in fast Fourier transform (FFT) analyzers is readily available. Using these gadgets, the turbulence patterns may be recorded and the recorded signals can be analyzed for quantitative results like frequency, intermittency, and the associated properties of turbulence.

5.7.2   A Note of Caution

At this stage, we must realize that the hot-wire anemometer will prove to be an extremely useful instrument to get an insight into the turbulence characteristics, if properly made standard hot-wire probes are used. In particular, the fastening (welding and soldering) of the hot-wire filament to the support needles has to be done with utmost care. Otherwise, the equipment will give spurious results which will be nowhere near reality. For instance, if 5 micron wire is soldered to a needle point by hand soldering, the soldering bubble formed at the needle tip is many orders of magnitude larger than the wire dimension. Further, two such bubbles are formed on either end of the wire and therefore, whatever the probe measures as turbulence will be the disturbance due to the solder and this by no means can be taken as the turbulence of the stream.

5.8 Effect of Compressibility

From the fundamentals of fluid mechanics we know that, compressible flows are those in which both density and temperature changes associated with flow velocity are significant. This effect of velocity change accompanied by density and temperature changes is called the *compressibility effect*. Here our intention is to make some remarks about the effect of compressibility of the fluid on the measurements of turbulence velocity fluctuations with hot-wire anemometer. We are making these remarks only briefly because of the complexity of the processes affecting the response of hot-wire anemometer. Also, due to this complexity, the problem has not yet been solved experimentally in a satisfactory manner.

The fundamental principle on which the hot-wire anemometer functions, namely, that the hot-wire is sensitive to variation in heat transfer is the basic cause for compressibility problems. In a high-speed turbulent flow, the velocity fluctuations may be accompanied by density variations and temperature variations of sufficient intensity so as to affect the response of the hot-wire anemometer. Therefore, the instantaneous voltage fluctuation e across the wire may be visualized to compose of three components due to variations of velocity e_u, variations of density e_ρ, and variations of temperature e_θ, i.e.,

$$e = e_u + e_\rho + e_\theta$$

Also, we know that the velocity, density, and temperature fluctuations are not independent. Further, they are related through the pressure fluctuations, by the mass conservation (continuity) equation, the energy conservation equation or the first law of thermodynamics, and the constitutive equation of the fluid. Also, any velocity field may be decomposed locally into an irrotational mode, a solenoidal vorticity mode, and a harmonic mode. The harmonic mode is irrotational and solenoidal, but for the present case, Morkovin (Hinze, 1987) suggested including it in the effect of the vorticity mode, since it will be indistinguishable from rotational flow by means of a single hot-wire anemometer.

The irrotational mode appears to be identical to the isentropic sound field. Therefore, as early as 1953 Kovasznay (Hinze, 1987) proposed for a correct interpretation of the response of a hot-wire anemometer to make a distinction between a sound mode, vorticity mode, and an entropy mode. Let p be the pressure fluctuation associated with sound mode, ω be the vorticity fluctuation, and ζ be the entropy fluctuation. With these the instantaneous value of voltage

e may be written as

$$e = s_p\, p + s_\omega\, \omega + s_\zeta\, \zeta$$

where, s_p, s_ω, and s_ζ are the corresponding sensitivities of the hot-wire anemometer. Pressure sensitivity s_p includes the effects of velocity, temperature, and density, s_ω includes the effects of velocity and temperature, and s_ζ includes the effects of temperature and density. They depend in general on the parameters occurring in the general relation Equation (5.1), and on the variable, mainly temperature sensitive, fluid properties. Since mean-square values of random variables are measured with electronic equipment, we conclude that the mean-square value of voltage fluctuations $\overline{e^2}$ contains six unknowns namely, the three mean square values $\overline{p^2}$, $\overline{\omega^2}$, and $\overline{\zeta^2}$, and the three mean cross products $\overline{p\omega}$, $\overline{p\zeta}$, and $\overline{\omega\zeta}$ or their corresponding correlation coefficients. In principle, it is possible to obtain these six unknowns experimentally, e.g., by making measurements at six different overheat temperatures. But this is almost impractical and in fact so far no such complete set of calibrations has been made. However, under certain conditions and depending on the type of flow it will be possible to neglect some of the terms. This is because the fluctuation amplitudes of one mode are much smaller than those of the other modes, or because correlations between two modes are sufficiently small to warrant their neglect. This, for instance, is often the case between the sound mode and the other two modes.

Detailed analysis of the effect of compressibility is not the intention here. The above material is presented to highlight such effects. For a complete discussion of the application of hot-wire anemometry to measurements in high-speed turbulent flow, based upon the above concepts, one may refer to the AGARDO-graph No. 24, 1956, by Morkovin.

5.9   Limitations of Hot-Wire Anemometer

Even though the hot-wire anemometer is considered to be one of the most useful instruments for measuring turbulence, it has many limitations. Some of these limitations have become apparent in the course of our discussions in the preceding sections. These involve

- A nonlinear character of heat transfer with respect to velocity and temperature.

- An onset of practical limitations due to the complex nature of the heat transfer between the wire and fluid, in compressible flows.

- A limitation set by the resolution power in space, that is, in the direction of the wire, due to its finite length.

- A limitation set by resolution power in time, in the flow direction, due to the finite time constant of the hot-wire.

Many other limitations which are both fundamental and applied in nature have been investigated by numerous researchers in this field. A fine collection of such limitations has been presented by Hinze (1987).

Although the hot-wire, as of today, cannot be treated as a completely dependable instrument, it is definitely a useful and powerful instrument since it gives an estimate of some of the quantities associated with one of the complex and little understood topics in science, namely turbulence. However, we should realize at this stage that for an understanding of turbulence one has to be very very careful in using the hot-wire anemometer. In fact, we must have a fairly good command over the fundamentals of fluid dynamics before even touching the hot-wire anemometer, otherwise, whatever we do will prove to be futile both to the field as well as to ourselves.

5.10 Summary

Hot-wire anemometer is a *thermal transducer*. The principle of operation of hot-wire anemometer is that *the heat transfer from a fine filament which is exposed to a cross flow varies with variation in the flow rate*. The hot-wire filament is made from a material which possesses a *temperature coefficient of resistance*, that is, *if the temperature of the filament varies, so also does its resistance*.

Generally two modes of operation are followed in hot-wire anemometry. They are the (i) *constant current mode*. Here the current flow through the hot-wire is kept constant and variations in the wire resistance caused by the fluid flow are measured by monitoring the voltage drop variations across the filament, and (ii) the *constant temperature mode*, in which the hot-wire filament is placed in a feedback circuit which tends to maintain the hot-wire at a constant resistance and hence at a constant temperature, and the fluctuations in the cooling of the hot-wire filament are seen as variations in the current flow through the hot-wire.

The vital component in a hot-wire system is the hot-wire filament. The hot-wire filament may be regarded as an infinitely long, straight cylinder in a cross flow. The empirical heat transfer relation which governs the heat transfer from a hot-wire is

$$Nu = 0.42\,Pr^{0.20} + 0.57\,Pr^{0.33}\,Re^{0.50}$$

For thermal equilibrium of the wire we have

$$I^2\,R_w = C_c\,\pi\,k\,l\,(\theta_w - \theta_g)\left[0.42\,Pr^{0.20} + 0.57\,Pr^{0.33}\,Re^{0.50}\right]$$

The temperature dependence of the electric resistance of the wire gives an effect of first order, it is on this effect that the use of a hot-wire as an anemometer is based. The important property of a wire, namely the *temperature coefficient of resistivity* allows the interpretation of voltage fluctuations in terms of velocity fluctuations. This temperature dependence may be expressed as

$$R_w = R_0\left[1 + C\,(\theta_w - \theta_0) + C_1\,(\theta_w - \theta_0)^2 + \cdots\right]$$

For steady conditions, the heat balance relation for the hot-wire becomes

$$I^2 R_w = (R_w - R_g)\left(A_1 + B_1\sqrt{U}\right)$$

This equation, popularly known as *King's relation*, is adequately accurate for analyzing the behavior of hot-wire systems.

Constant current hot-wire anemometer is rarely used for measuring velocity fluctuations except in heat transfer experiments these days. However, much of the theory associated with CCA forms the basis for the more complex CTA theory. Also, the CCA is still considered to be the best instrument for the measurement of temperature fluctuations in a gas. Basically hot-wire is placed in the flow and the wire resistance is measured using the balancing arm resistance R_s in the Wheatstone bridge. For the measurement of current, the accuracy of an ordinary panel-type ammeter is inadequate.

The current across the hot-wire is kept constant but the voltage varies. It can be shown that this voltage can be used for measuring the fluctuating velocities using an amplifier and a root mean square (RMS) voltmeter.

The turbulent fluctuations are usually measured by suitable configuration of the hot-wire probe. A single wire probe with the hot-wire element placed perpendicular to the main flow is used for measuring the velocity U, in the direction of main flow. For the measurement of the intensity of the turbulent velocity components u', v', and w', the directional sensitivity of the hot-wire is made use of. The constant current hot-wire anemometer may also be employed for measuring fluid temperature. The hot-wire may be used for the measurement of both steady-state temperature and temperature fluctuations.

In constant temperature hot-wire anemometer systems the wire resistance R_w is always kept constant and not allowed to fluctuate. The current is made to decrease or increase as soon as the hot-wire changes its resistance due to flow velocity fluctuations. The current adjustment is automatically done by employing a servo amplifier which has a feedback capacity for a frequency range from 0 to many kHz.

The CTA has become more popular than CCA because of the following reasons.

- The CCA is calibrated at constant temperature, and then used in a pseudo-constant current mode.

- The CCA calibration has to be a static one since to obtain a constant temperature calibration, each operating point of the bridge has to be adjusted manually.

- The CTA is used in the same way as it is calibrated.

- The temperature of the wire is maintained approximately constant automatically by a feedback circuit and this makes it possible to calibrate the system dynamically.

Using modern solid state devices, the physical size of the servo amplifiers can be made very small. However, when very high frequencies beyond 20 kHz are involved a constant current system has to be resorted to due to limitations posed by the feedback system. Such high frequencies are involved only in supersonic flows.

The hot-wire element is usually made of *platinum* or *tungsten* wire of about 1 mm long and about 5 micron in diameter. The wire is soldered to the tips of two needles, which forms the probe.

Platinum-rhodium alloy wire is also used instead of pure platinum to give extra tensile strength to the hot-wire element. For *laboratory turbulence measurements* wires with diameters in the range 4 to 8 micron and length 1 or 2 mm are used. Platinum wire can be *soldered* with the support needles whereas tungsten wire has to be *welded*.

The platinum wire is sold in the form of *woolaston wire* which is nothing but a fine platinum wire (or platinum-rhodium alloy wire) coated with a thick film of silver. The increase in wire diameter due to the coating facilitates the easier handling and soldering of it to the tips of the support needles.

Tungsten wire is directly welded to the probe terminals using a special welding setup. Tungsten has less resistance than platinum for wires having the same dimensions and hence its sensitivity to flow velocity is less. However, the tensile strength of tungsten is much superior compared to platinum.

The hot-wire anemometer will prove to be an extremely useful instrument to get an insight into the turbulence characteristics, if properly made standard probes are used. Especially the fastening (welding and soldering) of the hot-wire filament to the support needles has to be done with utmost care. Otherwise, the equipment will give spurious results which will be nowhere near reality.

Compressible flows are those in which both density and temperature changes associated with flow velocity are significant. This effect of velocity change accompanied by density and temperature changes is called the *compressibility effect*. The fundamental principle on which the hot-wire anemometer functions, namely, that the hot-wire is sensitive to variation in heat transfer is the basic cause for compressibility problems. In a high-speed turbulent flow, the velocity fluctuations may be accompanied by density variations and temperature variations of sufficient intensity so as to affect the response of the hot-wire anemometer. Therefore, the instantaneous voltage fluctuation e across the wire may be visualized to compose of three components due to variations of velocity e_u, density e_ρ, and temperature e_θ, i.e.,

$$e = e_u + e_\rho + e_\theta$$

The density and temperature fluctuations are not independent, and are related through the pressure fluctuations, by the mass conservation (continuity) equation, the energy conservation equation, or the first law of thermodynamics, and the constitutive equation of the fluid. Now any velocity field may be decomposed locally into an irrotational mode, a solenoidal vorticity mode, and a harmonic mode. The harmonic mode is irrotational and solenoidal, but for the present case, Morkovin (Hinze, 1987) suggested including it in the effect of the vorticity mode, since it will be indistinguishable from rotational flow by means of a single hot-wire anemometer.

The irrotational mode appears to be identical to the isentropic sound field. So, as early as 1953 Kovasznay (Hinze, 1987) proposed for a correct interpretation of the response of a hot-wire anemometer to make a distinction between a sound mode, vorticity mode, and an entropy mode.

Even though hot-wire anemometer is considered to be a most useful instrument for measuring turbulence, it has many limitations.

- A nonlinear character of heat transfer with respect to velocity and temperature.

- An onset of practical limitations due to the complex nature of the heat transfer between wire and fluid, in compressible flows.

- A limitation set by the resolution power in space, that is, in the direction of the wire, due to its finite length.

- A limitation set by resolution power in time, in the flow direction, due to the finite time constant of the hot-wire.

Exercise Problems

5.1 A hot-wire filament of diameter 3 micron and length 1 mm is used in constant temperature mode in an air stream at 40 m/s, 1 atm, and 30°C. If the wire temperature is maintained at 45°C, determine the heat transfer rate from the wire.

[Answer: 2.186 mW]

5.2 A constant temperature hot-wire of diameter 5 micron, maintained at 40°C, is used for measurement in an air stream at 32°C, assuming that the free-convection heat transfer effects are negligible. Is this assumption justified?

[Answer: Justified]

Chapter 6

Analogue Methods

6.1 Introduction

From our studies on basic fluid mechanics, we know that what we have understood is only very little and the rest of the fluid flow phenomena are highly complex and even giving a qualitative description of them is not possible with the presently acquired knowledge. The scientific community always looks for more advanced techniques to understand and solve such problems. In Chapter 4, we saw that flow visualization plays a dominant role by way of paving the path for the development of techniques for solving important but complex problems, such as separated flows, jets, and so on. In fact in many flow situations, the information given by flow visualization cannot be obtained by any other technique or instrument. However, these visualizations lack authentic mathematical procedure. They simply serve as a tool to understand certain complex flow fields in a simple manner. To overcome this shortcoming of visualization, certain methods were developed to study the fluid flow problems without actually going into the complexities involved. They are termed *analogue methods*. In analogue methods, fluid flow problems are solved by setting up another physical system, such as an electric field, for which the basic governing equations are of the same form with corresponding boundary conditions as those of the fluid flow. The solution of the original problem may then be obtained experimentally from measurement on the analogous system. Some of the well-known analogy methods for fluid flow problems are the *Hele-Shaw analogy, electrolytic tank,* and surface waves in a *ripple tank.*

6.2 Hele-Shaw Apparatus

The Hele-Shaw apparatus produces a flow pattern which is similar to that of potential flow. It is an analogy experiment known as Hele-Shaw analogy. The flow in the apparatus is actually a highly viscous flow between two parallel plates with a very small gap between them. In this flow the inertia force is negligible compared to the viscous force. Under this condition the flow equation has the same form as that of Euler's potential flow, however it does not satisfy the no-slip wall boundary condition. There is slip at the wall. Many interesting phenomena pertaining to potential flow can be observed using this apparatus. The flow through Hele-Shaw apparatus is two-dimensional and incompressible. In Hele-Shaw apparatus, the viscous flow of a liquid between two closely spaced plates may be shown to simulate the streamlines in the flow of a frictionless inviscid fluid. The Hele-Shaw flow is a low Reynolds number flow which has wide application in flow visualization apparatus because of its surprising property of reproducing the streamlines of potential flows (i.e., infinite Reynolds number flows).

The Hele-Shaw apparatus is shown schematically in Figure 6.1.

This equipment basically consists of two parallel plates made of thick, transparent (glass or plastic) plates clamped together along the edges with a narrow(about 1 mm) space in between them. The assembly of these two transparent sheets kept parallel with an uniform narrow space between them is provided with two small tanks of rectangular cross-section at the top end, as shown in Figure 6.1. The tanks are connected to the rectangular slit formed by the transparent sheets by a set of small holes (about 1-mm diameter) arranged in a row, as shown in Figure 6.1. The holes from the two tanks are arranged to occupy alternate locations for communication. The other end of the rectangular slit is made to terminate in a circular hole, by gradually narrowing it after a specified distance from the tanks at the top. One of the tanks is filled with water and the other with a dye (say potassium permanganate). Initially the circular passage at the bottom end of the apparatus is closed and the apparatus is kept vertical. Once the passage is opened, flow of water and dye takes place through the rectangular passage. Within a short duration, a flow field of water with uniform streamlines of dye in it is established in the narrow passage between the transparent plates.

The water as well as the dye kept in the tanks are essentially fed to the apparatus by gravity. Clean tap water is good enough for the main flow. A dilute solution of potassium permanganate can be used as the dye for visualization. The uniform flow field established in the rectangular slit of Hele-Shaw apparatus can be used as the test-section for visualizing flow over many objects of practical interest.

For flow visualizing the with Hele-Shaw apparatus the procedure described below may be employed.

- Mount the Hele-Shaw apparatus in a vertical position, as shown in Figure 6.1.

- Place the model of interest (say a rectangular body), which has the same thickness as the slit, at the middle of the test-section.

- Fill up the tanks with water and dye with the pinch cock closed.

- Connect the drain tube of the apparatus to a measuring jar.

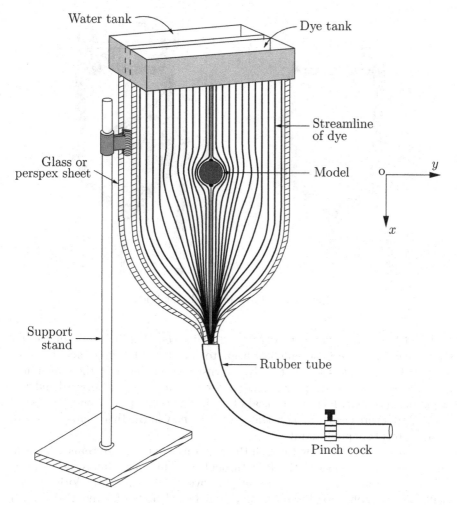

Figure 6.1: Hele-Shaw apparatus

- Open the pinch cock slowly. Water and dye will start flowing through the slit, establishing a flow field around the rectangular model.

- The flow field with the dye streaks as the streamlines is flowing around the rectangular model may be photographed or sketched on a plain paper by projecting a light to pass through the transparent sheets and making the image fall on the paper placed on the other side of the apparatus.

The above procedure has to be repeated for visualizing flow pattern around any other model of interest.

Photographs of flow patterns around a rectangular body kept at two orientations in Hele-Shaw apparatus are shown in Figure 6.2.

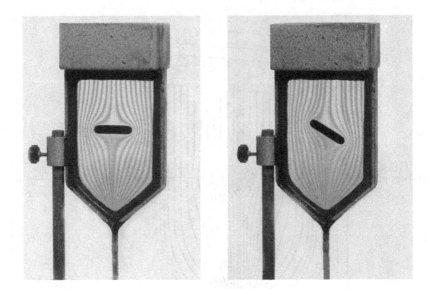

Figure 6.2: Flow past a blunt body

From the flow patterns in Figure 6.2, it is seen that the uniform parallel streamlines in the far upstream field begin to deviate as they approach the body, in order to negotiate the blunt body. Downstream of the body the streamlines gradually turn to become parallel and uniform again. Also, the forward and rear stagnation points, which are typical of an inviscid flow past a body, are seen. Thus, the picture clearly demonstrates that the flow in the Hele-Shaw apparatus in analogous to a potential flow.

The velocity of the flow through the apparatus may be determined by measuring the volume rate of the flow through it. This may simply be done by collecting the flow through the drain plug over a time interval. With this velocity, the Reynolds number of the flow may be calculated for the model under study. Many interesting steady, two-dimensional potential flow patterns can be demonstrated using the Hele-Shaw apparatus.

From the above discussion it is evident that the Hele-Shaw apparatus is simple equipment capable of yielding many interesting flows of practical importance, provided it is used efficiently. It is a matter of common sense to realize that the models kept in the slit must not allow any flow to take place between the model side surface and the apparatus wall. This is to make sure that the flow is strictly two-dimensional.

6.2.1 Basic Equations of Hele-Shaw Analogy

We have seen that the Hele-Shaw apparatus produces a flow pattern which is similar to that of potential flow. Further, it can be mathematically shown that, under Hele-Shaw flow conditions, the governing equations of the flow are of the same form as that of Euler's potential flow, however it does not satisfy no-slip condition at the wall. There is slip at the wall. The flow through the apparatus, as can be seen, is two-dimensional and incompressible.

The Navier–Stokes equations for steady incompressible flow are

$$u\frac{\partial u}{\partial x} + v\frac{\partial u}{\partial y} + w\frac{\partial u}{\partial z} = -\frac{1}{\rho}\frac{\partial p}{\partial x} + \nu\left[\frac{\partial^2 u}{\partial x^2} + \frac{\partial^2 u}{\partial y^2} + \frac{\partial^2 u}{\partial z^2}\right]$$

$$u\frac{\partial v}{\partial x} + v\frac{\partial v}{\partial y} + w\frac{\partial v}{\partial z} = -\frac{1}{\rho}\frac{\partial p}{\partial y} + \nu\left[\frac{\partial^2 v}{\partial x^2} + \frac{\partial^2 v}{\partial y^2} + \frac{\partial^2 v}{\partial z^2}\right] \qquad (6.1)$$

$$u\frac{\partial w}{\partial x} + v\frac{\partial w}{\partial y} + w\frac{\partial w}{\partial z} = -\frac{1}{\rho}\frac{\partial p}{\partial z} + \nu\left[\frac{\partial^2 w}{\partial x^2} + \frac{\partial^2 w}{\partial y^2} + \frac{\partial^2 w}{\partial z^2}\right]$$

where u, v, and w are the velocity components along the x, y, and z directions, respectively. The x-coordinate is along the flow, z in the direction of gap between the two plates, and y is the transverse direction, as shown in Figure 6.1. The velocity component $w = 0$, since there is no flow normal to the side wall plates, and hence z-component in the Navier–Stokes equation vanishes. In the Hele-Shaw flow the inertia force is negligible compared to the viscous force and hence the left-hand side of x and y-momentum equations can be neglected. Similarly, in the viscous terms only $\partial^2 u/\partial z^2$ and $\partial^2 v/\partial z^2$ are significant since the other gradients are comparatively small. Therefore, x and y-momentum equations reduce to

$$\frac{\partial p}{\partial x} = \mu\frac{\partial^2 u}{\partial z^2}$$

$$\frac{\partial p}{\partial y} = \mu\frac{\partial^2 v}{\partial z^2} \qquad (6.2)$$

Also, we know that, for a two-dimensional laminar flow in a channel, the velocity profile is given by the relation

$$u = u_{\text{max}}\left(1 - \frac{z^2}{c^2}\right)$$

where $2c$ is the channel width, that is the distance between the two plates in the Hele-Shaw apparatus, and u_{max} is the maximum of the x-component of velocity u. The above velocity profile may also be written as

$$u = \frac{3}{2}u_m\left(1 - \frac{z^2}{c^2}\right) \qquad (6.3)$$

where u_m is the mean value of the u velocity component. Similarly, the velocity along the transverse direction can be expressed as

$$v = \frac{3}{2}v_m\left(1 - \frac{z^2}{c^2}\right) \qquad (6.4)$$

Using Equations (6.3) and (6.4) into Equations (6.2), we get

$$\frac{\partial p}{\partial x} = -\frac{3\mu}{c^2}u_m$$

$$\frac{\partial p}{\partial y} = -\frac{3\mu}{c^2}v_m$$

$$(6.5)$$

Differentiating the x and y components of Equations (6.5) with respect to y and x, respectively, and comparing the equal terms, we obtain

$$\frac{\partial u_m}{\partial y} = \frac{\partial v_m}{\partial x} \tag{6.6}$$

In terms of stream function, the velocity components are

$$u = \frac{\partial \psi}{\partial y}$$

and

$$v = -\frac{\partial \psi}{\partial x}$$

Assume (for the time being) that, $u = u_m$ and $v = v_m$. From Equation (6.6), we get

$$\frac{\partial^2 \psi}{\partial y^2} = -\frac{\partial^2 \psi}{\partial x^2}$$

i.e.,

$$\frac{\partial^2 \psi}{\partial x^2} + \frac{\partial^2 \psi}{\partial y^2} = 0 \tag{6.7}$$

Equation (6.7) represents a steady, two-dimensional, irrotational, and incompressible flow. We arrived at Equation (6.7), starting from full Navier–Stokes equations and simplifying them for Hele-Shaw flow. Hence, it may be concluded that the viscous incompressible flow through Hele-Shaw apparatus is equivalent to potential flow.

Example 6.1

(a) Design and fabricate a Hele-Shaw apparatus. Discuss the theory. Calibrate the apparatus and ensure that the flow in the empty test-section of the apparatus is uniform. (b) After ensuring the flow uniformity, visualize the potential flow pattern (i) around a circular cylinder, without any sharp corner, (ii) around a square plate possessing 4 sharp corners, and (iii) around a triangular plate with 3 sharp corners; with the vertex up and vertex down orientations.

Solution

(a) Design and Fabrication

A Hele-Shaw device is a unique facility, capable of generating flow analogous to potential flow. Hele-Shaw demonstrated that, when a liquid is made to flow in a narrow gap (say, around 1 mm) between two large flat plates, the flow becomes analogous to inviscid flow. Thus, Hele-Shaw technique is an analogy method, which renders the flow of a viscous fluid, such as water, to behave as a potential flow. This example presents the design, fabrication and testing of Hele-Shaw device, carried out in Suzuki Laboratory, in the Graduate School of Frontier Sciences, University of Tokyo, Kashiwa Campus, Japan[1]. The frames were made of hylam material. The side windows were made of transparent plastic sheet.

Design Procedure

In the present design, the gap for the flow is kept at 1.5 mm. The side plates are given a shape, as shown in Figures E6.1 and E6.2. All the dimensions are in millimeters and the angles marked in Figures E6.1, E6.3 and E6.5 are in degrees.

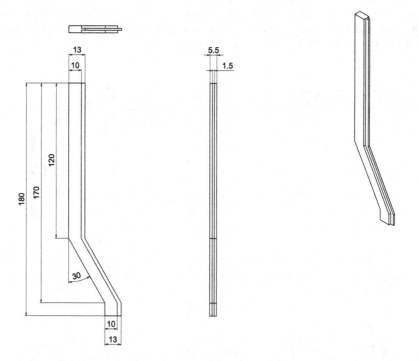

Figure E6.1: Schematic diagram of side frame

---

[1]The fabrication and tests were done by Dr. Yasumasa Watanabe, a doctoral student of Professor Kojiro Suzuki, Department of Aeronautical and Astronautical Engineering, University of Tokyo.

A photographic view of the side plate is shown in Figure E6.2.

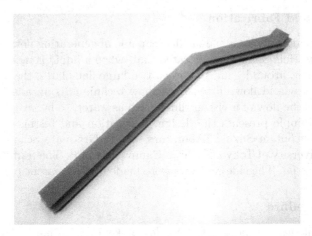

Figure E6.2: A pictorial view of side frame

The assembly drawing of the setup, which includes 2 side plates, the transparent windows, two tanks at the top, a bottom part which accommodates the rectangular shape of the device bottom, and has a circular opening at the opposite end is shown in Figures E6.3 and E6.4.

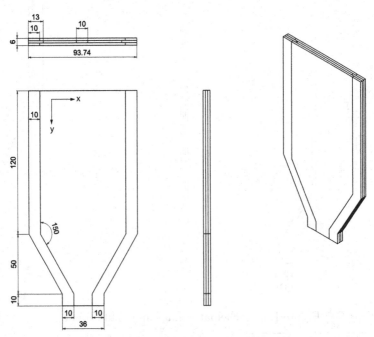

Figure E6.3: Hele-Shaw device assembly

An artistic view of the device is shown in Figure E6.4.

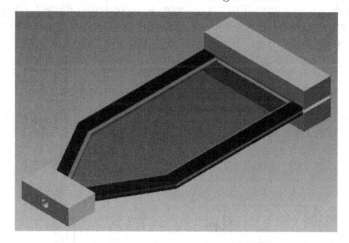

Figure E6.4: Hele-Shaw device assembly

The dimensions of the side (transparent) window are shown in Figure E6.5.

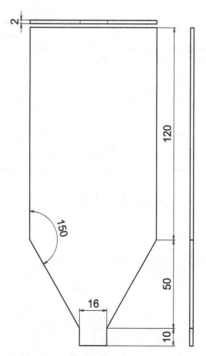

Figure E6.5: Side window

The details of the bottom piece, which has a rectangular slot at the top and terminating to a circular hole at the bottom, is shown in Figure E6.6.

Drawings of the tanks and a photographic view of the fabricated tanks, with holes, are shown in Figure E6.7. The holes were of 1-mm diameter with 1-mm gap in between them. The holes of one tank are staggered compared to the holes of the other tank.

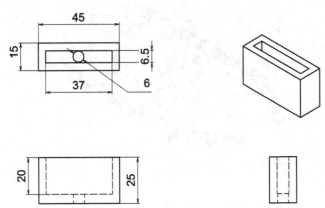

Figure E6.6: Schematic diagram of the bottom piece

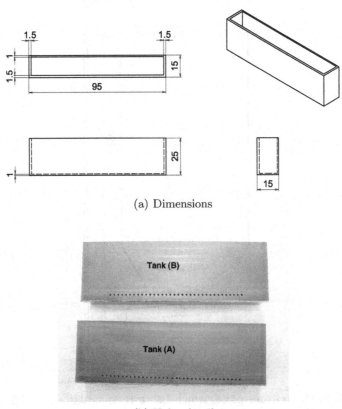

(a) Dimensions

(b) Holes detail

(c) A photographic view of the tanks

Figure E6.7: Tanks details (a) schematic diagram, (b) pictorial side view, (c) pictorial view from the top

An artistic view of the side plates with transparent windows is shown in Figure E6.8.

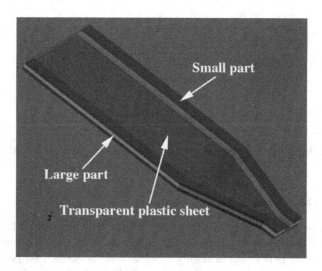

Figure E6.8: An artistic view of the side plates with transparent windows

A photographic view of the Hele-Shaw apparatus, designed and fabricated, is shown in Figure E6.9. The tube at the bottom of the device is connected to a flow control valve. The control valve opening can be adjusted to vary the flow speed in the test-section of the Hele-Shaw device.

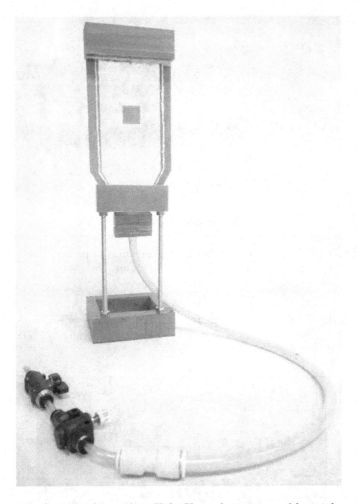

Figure E6.9: A view of complete Hele-Shaw device assembly, with a square
plate model in the test-section

Theory of Hele-Shaw Flow

The Hele-Shaw apparatus produces a flow pattern which is similar to that of potential flow. It is an analogy experiment known as Hele-Shaw analogy. The flow
in the apparatus is actually a highly viscous flow between two parallel plates
with a very small gap between them. In this flow the inertia force is negligible compared to the viscous force. Under this condition the flow equation has
the same form as that of Euler's potential flow, however it does not satisfy the
no-slip wall boundary condition. Indeed, there is slip at the wall. Many interesting phenomena pertaining to potential flow can be visualized (observed) using
this apparatus. The flow through the Hele-Shaw apparatus is two-dimensional
and incompressible. In the Hele-Shaw apparatus, the viscous flow of a liquid

between two closely spaced plates may be shown to simulate the streamlines in the flow of a frictionless inviscid fluid. The Hele-Shaw flow is a low Reynolds number flow which has wide application in flow visualization studies because of its surprising characteristics of resembling the streamlines of potential flows (i.e., infinite Reynolds number flows).

A view of the Hele-Shaw apparatus is shown in Figure E6.9. This equipment basically consists of two parallel plates made of thick, transparent (glass or plastic) plates clamped together along the edges with a narrow (about 1 or 1.5 mm) space between them. The assembly of these two transparent sheets, kept parallel with an uniform narrow space between them, is provided with two small tanks of rectangular cross-section at the top end, as seen in Figure E6.9. The tanks are connected to the rectangular slit formed by the transparent sheets by a set of small holes (about 1-mm diameter) arranged in a row, as shown in Figure E6.7. The holes from the two tanks are arranged to occupy alternate locations for communication. The other end of the rectangular slit is made to terminate in a circular hole, by gradually narrowing it, after a specified distance from the tanks at the top. One of the tanks is filled with water and the other with a dye (say water color). Initially the circular passage at the bottom end of the apparatus is closed and the apparatus is kept vertical. Once the passage is opened, flow of water and dye takes place through the rectangular passage. Within a short duration, a flow field of water with uniform streamlines of dye in it is established in the narrow passage between the transparent plates.

The water as well as the dye kept in the tanks are essentially fed to the apparatus by gravity. Clean tap water is good enough for the main flow. A dilute solution of fountain pen ink or printing ink or water color paint can be used as the dye for visualization. The uniform flow field established in the rectangular slit of the Hele-Shaw apparatus can be used as the test-section for visualizing potential flow past many objects of practical interest.

For flow visualization with Hele-Shaw apparatus the procedure described below may be employed.

• Mount the Hele-Shaw apparatus in a vertical position, as shown in Figure E6.9.

• Place the model of interest (say a rectangular body), which has the same thickness as the slit, at the middle of the test-section.

• Fill one of the tanks with water and the other with the dye, with the flow regulating valve in closed position.

• Connect the drain tube of the apparatus to a measuring jar.

• Open the flow regulating valve slowly. Water and dye will start flowing through the slit, establishing a flow field around the rectangular model.

• The flow field with the dye streaks as the streamlines flowing around the rectangular model may be photographed or sketched on a plain paper by projecting a light to pass through the transparent sheets and making the image fall on the screen or paper placed on the other side of the apparatus.

The above procedure has to be repeated for visualizing flow pattern around any other model of interest.

(b) Flow Uniformity Check

First step in the visualization with a Hele-Shaw apparatus is to ensure that the flow in the device is uniform. This can easily be tested by operating the device without any model in the test-section. The flow pattern in the empty test-section of the present device is shown in Figure E6.10.

Figure E6.10: Streamlines in the test-section

The dye used is printing ink. The black and white streaks are the ink and water streak lines, respectively. It is heartening to see that the streak lines are absolutely smooth and parallel, ensuring that the flow in the test-section is perfectly uniform. This certifies that the device fabricated is of good standard.

Once the flow uniformity in the test-section is studied and certified, any model to be studied can be placed in the test-section. It is essential to ensure that the model thickness is exactly the same as the gap between the window plates. The model has to be placed properly, at the middle of the test-section. Once the model is positioned in the test-section, the flow regulating valve can be opened gently to establish the flow. The flow coming out of the device can be collected for a known period of time. The volume collected over a known time period will give the flow velocity through the device.

The cross-section of the device is A, say. Then the volume flow rate becomes

$$\dot{V} = A \times V$$

Thus, if t is the time over which the volume collected is $\mathbb{V}$ we have the flow velocity as

$$V = \frac{\mathbb{V}}{t \times A}$$

Now, the Reynolds number of the flow is given by

$$\text{Re} = \frac{\rho V L}{\mu}$$

where ρ and μ are the density and viscosity coefficient of water, respectively, and L is the characteristic length.

Thus, the velocity of the flow through the apparatus may be determined by measuring the volume rate of the flow through it. This may simply be done by collecting the flow through the drain plug over a time interval. With this velocity, the Reynolds number of the flow may be calculated for the model under study. Many interesting steady, two-dimensional potential flow patterns can be demonstrated using the Hele-Shaw apparatus.

Reynolds Number Calculation

The Reynolds number calculated for the flow past a circular cylinder of diameter 15 mm, placed in the test-secton of the Hele-Shaw device fabricated, at a particular opening position of the flow regulating valve, is given below.

The cross-sectional area of the test-section, A, is

$$A \;=\; \text{width} \times \text{thickness}$$

$$=\; 67.74 \times 1.5$$

$$=\; 101.61 \, \text{mm}^2$$

Volume of water and dye mixture collected in 90 seconds is 50 milliliter.

Thus the flow velocity in the test-section is

$$V \;=\; \frac{\text{volume flow rate}}{\text{area}}$$

$$=\; \frac{(50/90) \times 10^{-6}}{101.61 \times 10^{-6}}$$

$$=\; 5.47 \times 10^{-3} \, \text{m/s}$$

For water, taking the density, ρ and viscosity coefficient, μ, as 10^3 kg/m^3 and 1×10^{-3} kg/(m s), respectively, we have the Reynolds number of the test-section flow, based on the cylindrical disk of diameter 15 mm as

$$Re = \frac{\rho V d}{\mu}$$

$$= \frac{10^3 \times (5.47 \times 10^{-3}) \times (15 \times 10^{-3})}{1 \times 10^{-3}}$$

$$= 82$$

From the above discussion it is evident that the Hele-Shaw apparatus is a simple device capable of yielding many interesting flows of practical importance, provided it is used efficiently. It is a matter of common sense to realize that the models kept in the slit must not allow any flow to take place between the model side surface and the apparatus wall. This is to make sure that the flow is strictly two-dimensional.

Basic Equations of Hele-Shaw Analogy

We have seen that the Hele-Shaw apparatus produces a flow pattern which is similar to that of potential flow. Further, it can be mathematically shown that, under Hele-Shaw flow conditions, the governing equations of the flow are of the same form as that of Euler's potential flow, however it does not satisfy no-slip conditions at the wall. There is slip at the wall. The flow through the apparatus, as it can be seen, is two-dimensional and incompressible. It is shown in Section 6.2, starting from full Navier-Stokes equations and simplifying them for Hele-Shaw flow, that the flow in the Hele-Shaw apparatus represents a steady, two-dimensional, irrotational, and incompressible flow. Hence, it may be concluded that the viscous incompressible flow through the Hele-Shaw apparatus is equivalent to potential flow.

Flow Past Some Shapes

Ensuring that the flow through the test-section of the Hele-Shaw device is uniform and parallel, as seen in Figure E6.10, some specific models with and without sharp corners were placed in the test-section, and the streamline patterns around the models were visualized, using black printing ink as the dye. Flow past a circular disk, a square plate, and a triangular plate, in two orientations was visualized. The streamlines around these objects studied are given in Figures E6.11 to E6.14.

(i) Flow Past a Circular Cylinder

The streamlines around the circular disk, shown in Figure E6.11, exhibit a pattern identical to potential flow past a two-dimensional circular cylinder. It is seen that the flow is symmetrical about both horizontal and vertical axes of the cylinder. The forward and rear stagnation points are clearly seen. This flow pattern clearly demonstrates that the flow in the Hele-Shaw device is analogous to potential flow.

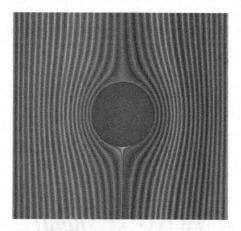

Figure E6.11: Flow past a circular disk

(ii) Flow Past a Square Plate

Flow pattern around a square plate, visualized in the Hele-Shaw device, is shown in Figure E6.12. It is interesting to see that, as in the case of flow past the circular disk, the flow past the square plate is also symmetrical about both horizontal and vertical axes. In spite of the presence of sharp corners, the flow negotiates the object and flows around it, without separation. The forward and rear stagnation points are clearly seen. There is no wake behind the object, as if the flow is inviscid.

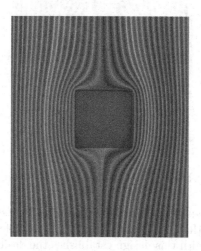

Figure E6.12: Flow past a square plate

(iii) Flow Past an Equilateral Triangular Plate

Flow past an equilateral triangular plate, with its vertex up and base up, studied in the Hele-Shaw device is shown in Figures E6.13 and E6.14, respectively. It is heartening to see that the flow pattern, for both the orientations of the triangular plate, exactly duplicates that of potential flow past the object.

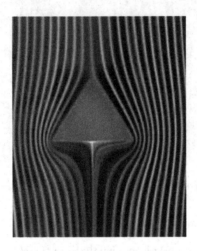

Figure E6.13: Flow past a triangular plate with vertex up

Figure E6.14: Flow past a triangular plate with vertex down

From the above visualization of flow past objects, with and without sharp corners, it is evident that the present design of the Hele-Shaw device is good and the device made with this design establishes the desired flow field, which is analogous to potential flow.

6.3 Electrolytic Tank

The electrolytic tank is another popular analogue technique used for solving potential flow problems. In other words, it is used to solve the Laplace equation. Basically it is an analogy method. It makes use of the fact that the equations governing incompressible potential flow and distribution of the electrical potential lines are the same, in their form, to establish an analogy between the two fields. We know that the potential flows are irrotational and the velocity potential ϕ and stream function ψ are related as follows.

$$u = \frac{\partial \phi}{\partial x} = \frac{\partial \psi}{\partial y}$$

$$v = \frac{\partial \phi}{\partial y} = -\frac{\partial \psi}{\partial x}$$

where u and v are the x and y components of flow velocity, respectively.

For two-dimensional incompressible flows, the potential and stream functions can be expressed in differential form as

$$d\phi = u\,dx + v\,dy$$

$$d\psi = u\,dy - v\,dx$$

The potential function satisfies the Laplace equation, i.e.,

$$\frac{\partial^2 \phi}{\partial x^2} + \frac{\partial^2 \phi}{\partial y^2} = 0$$

or

$$\nabla^2 \phi = 0 \tag{6.8}$$

For potential flows the stream function ψ satisfies the irrotationality condition,

$$\frac{\partial v}{\partial x} - \frac{\partial u}{\partial y} = 0$$

From basic fluid mechanics we know that nothing like potential flow exists in reality. It is only an idealization to simplify the flow fields to make them amenable for solutions. Therefore, it is not possible to produce an inviscid (potential) flow. If we want to solve the potential flow problems all that is necessary is to solve the Laplace equation with appropriate boundary conditions.

It can be shown that, when an electric current flows in a sheet of conductor of thickness t and constant specific resistance σ, the components of current intensity i at a point (x, y) are given by the equations

$$i_x = -\frac{1}{\sigma}\frac{\partial A}{\partial x}$$

$$i_y = -\frac{1}{\sigma}\frac{\partial A}{\partial y}$$

$$i_y = \frac{1}{t}\frac{\partial B}{\partial x}$$

$$i_x = -\frac{1}{t}\frac{\partial B}{\partial y}$$

where A denotes the electrical potential and B the electric current function. With the above equations, A and B may be expressed in the differential form as

$$dA = -\sigma(i_x\,dx + i_y\,dy) \tag{6.9}$$

$$dB = -t(i_x\,dy - i_y\,dx) \tag{6.10}$$

From Equations (6.9) and (6.10) it is seen that the electrical potential A and the current function B also satisfy the Laplace equation and irrotationality condition, respectively. That is, in this case both the flow of current and the inviscid fluid flow follow identical equations, as shown below.

For *fluid flow*,

$$d\phi = u\,dx + v\,dy$$

$$d\psi = u\,dy - v\,dx$$

For *electric current flow*,

$$dA = -\sigma(i_x\,dx + i_y\,dy)$$

$$dB = -t(i_x\,dy - i_y\,dx)$$

Therefore, A and B can be identified with ϕ and ψ, respectively. Hence, i_x corresponds to u and i_y corresponds to v.

From the above discussion, it is clear that the flow pattern around a body immersed in a perfect fluid can be determined by finding the passage of the current through a uniform sheet of conductor, except within the boundary corresponding to the body immersed in the fluid. If the potential function ϕ is identified with the electrical potential A, then the equipotential (voltage) lines will become potential lines in the flow field and the equal current lines will become the streamlines. In practice it is easy to measure or plot equipotential lines rather than current lines. The streamlines have to be drawn orthogonal to the potential lines.

A typical electrolytic tank with an airfoil model and the measured streamlines and potential lines around the model are shown in Figure 6.3.

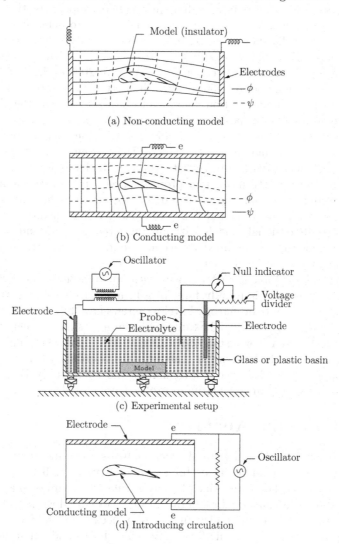

(a) Non-conducting model

(b) Conducting model

(c) Experimental setup

(d) Introducing circulation

Figure 6.3: Electrolytic tank

The other way of solving the flow field is as follows: If the current function B is identified with potential function ϕ, then A corresponds to the stream function ψ. For this case the constant potential lines can be obtained straightaway. This requires a slightly different arrangement. Since the contour of the body is a streamline, the model has to be made of a good conductor. The streamlines and potential lines obtained with both ways around an airfoil is shown in Figure 6.3.

Usually the sheet of conductor used in this experiment is an *electrolyte*. The electrolyte is used instead of a metallic conductor because of its reasonably high resistance, so that large potential difference can be applied to the two ends without passing much current. In addition, it is easy to get a uniform sheet of electrolyte on a flat-bottomed basin since liquid finds its own level. Direct current is generally avoided since it gives rise to polarization and electrolysis. An alternating current of low frequency, say 1000 to 2000 cycles per second, is used. The voltage may be from 50 to 100 volts between the two terminals. It is preferable to keep the dissipation below 20 to 30 watts so that the electrolyte does not get heated appreciably.

The electromotive force (e.m.f) at any point in the tank is measured by dipping the probe and measuring the voltage. To get accurate readings a nulling device in conjunction with an accurate voltage divider is generally used, as shown in Figure 6.3. By using two probes spaced apart at a small interval (say 1 mm) and measuring the voltage difference between them, it is possible to obtain the derivative of ϕ and hence the velocity. The facility described so far is for two-dimensional models without any circulation around them. This electrolytic tank can be modified to accommodate axisymmetric models and circulation can also be introduced.

The electrolytic tank is just a basin with squared sides and a large flat bottom. The tank base is usually built of a firm surface such as a thick glass plate or a slab of granite or marble. Leveling screws have to be provided to keep the bottom surface horizontal. The tank can be large or small, depending on the model size to be used. For instance, it can be 2 m× 1 m in size and about 10 to 15 cm in height. The electrolyte generally used is ordinary tap water with a pinch of copper sulfate dissolved in it to increase the conductivity. For measuring the voltage, a 1-mm-diameter platinum probe may be used.

6.4   Hydraulic Analogy

The analogy between shallow water flow with a free-surface and two-dimensional gas flow has been found to be useful for qualitative as well as quantitative study of high-speed flows. In principle, any regulated stream of shallow water can be used for this analogy. In laboratories it is usually done in a water flow table (channel). This technique is valued highly because of the fact that many practical problems in supersonic flows, involving shocks and expansion waves, which require a sophisticated and expensive wind tunnel and instrumentation for their analysis may be studied in an inexpensive manner with a simple water flow channel facility.

The analogy may be used to understand many challenging problems of practical interest, such as transonic unsteady flows past wings, transient phenomena in high-speed flow and shock wave interaction, and so on. Simulation of such flows in a wind tunnel is highly complicated and expensive and hence preliminary information explaining the basic nature of such flows will be of immense use in solving such problems. In such situations, the analogy channel (or tank) proves itself handy and helps us by providing preliminary information and a qualitative picture of the flow field.

The essential feature of this analogy is that *the Froude number of shallow water flow with a free-surface is equivalent to a gas stream with Mach number equal to that Froude number.*

6.4.1 Theory of the Analogy

The basic governing equations of flow of an incompressible fluid with a free-surface, in which the depth of the flow is small compared to its surface wave length, forms the shallow water theory. The similarity between the governing equations of motion of a two-dimensional isentropic flow of a perfect gas and the two-dimensional shallow water flow forms the analogy.

6.4.1.1 Shallow Water Flow

Consider a continuous flow (without any hydraulic jump) of a perfect liquid in the absence of any external forces like electromagnetic forces, viscous forces, surface tension, etc., except gravitational force. For this flow, neglecting the variation of pressure (if any) on the free-surface and treating the flow as two-dimensional with constant energy, we can write the basic governing equations as follows.

The continuity equation is

$$\frac{\partial h}{\partial t} + \frac{\partial (hu)}{\partial x} + \frac{\partial (hv)}{\partial y} = 0 \qquad (6.11)$$

The momentum equations along the flow (x) and transverse (y) directions, respectively, are

$$\frac{\partial u}{\partial t} + u\frac{\partial u}{\partial x} + v\frac{\partial u}{\partial y} = -\frac{g}{2}\frac{1}{h}\frac{\partial (h^2)}{\partial x}$$

$$(6.12)$$

$$\frac{\partial v}{\partial t} + u\frac{\partial v}{\partial x} + v\frac{\partial v}{\partial y} = -\frac{g}{2}\frac{1}{h}\frac{\partial (h^2)}{\partial y}$$

where u and v are the velocity components along x and y directions, respectively, g is gravitational acceleration, and h is the depth of the fluid stream.

Now, let us define a fictitious pressure p_f and a fictitious density ρ_f as

$$p_f = \int_0^h p\,dz$$

$$= \int_0^h \rho_w gz\,dz = \rho_w \frac{g}{2} h^2$$

$$\rho_f = \rho_w h$$

where ρ_w is the density of water and z is the coordinate in the depth direction. Introduction of these expressions for p_f and ρ_f into Equations (6.11) yields

$$\frac{\partial}{\partial t}\left(\frac{\rho_f}{\rho_w}\right) + \frac{\partial}{\partial x}\left(\frac{\rho_f}{\rho_w}u\right) + \frac{\partial}{\partial y}\left(\frac{\rho_f}{\rho_w}v\right) = 0$$

This simplifies to

$$\frac{\partial}{\partial t}(\rho_f) + \frac{\partial}{\partial x}(\rho_f u) + \frac{\partial}{\partial y}(\rho_f v) = 0 \tag{6.13}$$

since, ρ_w is a constant.

Similarly, introducing p_f and ρ_f, Equations (6.12) can be reduced to

$$\frac{\partial u}{\partial t} + u\frac{\partial u}{\partial x} + v\frac{\partial u}{\partial y} = -\frac{1}{\rho_f}\frac{\partial}{\partial x}(p_f)$$

$$\frac{\partial v}{\partial t} + u\frac{\partial v}{\partial x} + v\frac{\partial v}{\partial y} = -\frac{1}{\rho_f}\frac{\partial}{\partial y}(p_f)$$

$$\tag{6.14}$$

6.4.1.2 Gas Flow

Consider a two-dimensional isentropic flow of a perfect gas in the absence of any external forces such as gravitational force, electromagnetic force, etc. The basic equations for this flow are the following.

The continuity equation is

$$\frac{\partial}{\partial t}(\rho_g) + \frac{\partial}{\partial x}(\rho_g u) + \frac{\partial}{\partial y}(\rho_g v) = 0 \tag{6.15}$$

and the momentum equations are

$$\frac{\partial u}{\partial t} + u\frac{\partial u}{\partial x} + v\frac{\partial u}{\partial y} = -\frac{1}{\rho_g}\frac{\partial p}{\partial x}$$

$$\frac{\partial v}{\partial t} + u\frac{\partial v}{\partial x} + v\frac{\partial v}{\partial y} = -\frac{1}{\rho_g}\frac{\partial p}{\partial y}$$

$$\tag{6.16}$$

where ρ_g is the density of the gas.

Comparing Equations (6.15) and (6.16) with Equations (6.13) and (6.14), it can be seen that the shallow water flow equations are similar to the gas flow equations. In terms of ρ_w and ρ_f, the fictitious pressure p_f can be expressed as

$$p_f = \rho_w \frac{g}{2} h^2$$

$$= \frac{g}{2} \frac{\rho_w}{\rho_w^2} \rho_f^2$$

$$= \frac{g}{2} \frac{\rho_f^2}{\rho_w^2} \rho_w$$

$$\frac{p_f}{\rho_f^2} = \frac{g}{2\rho_w} = \text{constant}$$

This expression corresponds to the polytropic process equation for gases with the polytropic index $n = 2$. This imposes a further condition to be satisfied for complete analogy between these two streams of water and gas. Here we should note that the value of the index n can never be equal to 2 for isentropic flow of gases. The index becomes γ, namely the isentropic index, which is the ratio of specific heats C_p and C_v. Also, we know that γ varies from 1 to 1.67 only. Therefore, $n = 2$ required for hydraulic analogy is bound to introduce some error in the quantitative results obtained with hydraulic analogy. In spite of this shortcoming, hydraulic analogy is found to yield results with reasonable accuracy. This aspect as well as the ease with which the results for gas flows at high speeds can be obtained with this technique make it attractive, despite the above-mentioned drawback.

It can be seen from the definition of fictitious pressure and density that the gas pressure is proportional to square of water stream depth h and the gas density is directly proportional to the water stream depth. That is, we can express that

$$\frac{p}{p_0} = \frac{h^2}{h_0^2} \qquad (6.17)$$

$$\frac{\rho}{\rho_0} = \frac{h}{h_0} \qquad (6.18)$$

By perfect gas state equation, we have

$$p = \rho R T$$

Using this in Equation (6.17), we obtain

$$\frac{T}{T_0} = \frac{h}{h_0} \qquad (6.19)$$

In Equations (6.17) to (6.19), p_0, ρ_0, T_0, and h_0 are the stagnation pressure, stagnation density, stagnation temperature of gas, and the stagnation depth of water stream, respectively.

In gas flow, the velocity of propagation of a small pressure disturbance is the speed of sound "a." The corresponding parameter in hydraulic flow is "a_w" the velocity of propagation of a surface disturbance with a large wave length compared to the water depth. The wave speeds a and a_w may be expressed as

$$a = \left(\frac{\gamma p}{\rho}\right)^{\frac{1}{2}}$$

$$a_w = \left(2\frac{g}{2}\frac{\rho_w h^2}{\rho_w h}\right)^{\frac{1}{2}}$$

$$= (gh)^{\frac{1}{2}}$$

We know from our basic studies on fluid flows that it is possible to identify a group of dimensionless parameters between any two dynamically similar flows. Therefore, from the above discussion on gas and water flows, it is possible to identify the Mach number and Froude number as the nondimensional similarity parameters, respectively, for gas and water streams. That is, for the gas and water streams to be dynamically similar, the Mach number of the gas stream and the Froude number of the water stream must be equal. The Mach number M and Froude number Fr may be expressed as

$$M = \frac{V_g}{a} \tag{6.20}$$

$$F_r = \frac{V_w}{\sqrt{gh}} \tag{6.21}$$

where V_g and V_w are the velocities of gas and water streams, respectively.

- When $M < 1$ the flow is termed *subsonic.*

- When $M > 1$ the flow is termed *supersonic.*

- When $Fr < 1$ the flow is termed *subcritical.*

- When $Fr > 1$ the flow is termed *supercritical.*

Further, there exists a degree of similarity between the flow patterns in the water and gas streams. In supersonic flows, $\mu_g = \sin^{-1}(1/M)$ is the smallest angle with which disturbance prevails in the flow with a given Mach number M, where μ_g is the Mach angle. The analogous quantity in shallow water flow is $\psi_w = \sin^{-1}(1/F_r)$, where ψ_w is called wave angle. At this stage, it must be noted that in the above analysis of similarity between shallow water and gas streams, nowhere was the assumption of *zero vorticity* made. The expression for vorticity in a two-dimensional gas flow as well as in a water flow can be written as

$$\text{curl} \, V = k \left(\frac{\partial v}{\partial x} - \frac{\partial u}{\partial y}\right) \tag{6.22}$$

where V is the resultant velocity, u and v are the velocity components along x and y directions, respectively, and k is the directional vector in the z-axis direction. Hence, in principle it is possible to compare rotational flow in the two media. However, in practice the real fluid effects complicate the comparison. From the above discussion it is evident that, with $\gamma = 2$, the governing equations for the gas flow are identical to those for shallow water stream. The analogy holds for flows with Mach numbers smaller and greater than 1. The analogous parameters and flow fields are summarized in Table 6.1.

Table 6.1 Flow analogy

Item	Two–dimensional gas flow	Liquid flow with free-surface in gravity field
Nature of the flow medium	Hypothetical gas with $\gamma = 2$	Incompressible fluid (e.g., water)
Side boundaries geometrically similar	No specific boundary	Side boundary vertical bottom horizontal
Analogous magnitudes	Velocity, V/V_∞	Velocity, V/V_∞
	Pressure, p/p_0	Square of velocity depth ratio, $(h/h_0)^2$
	Temperature, T/T_0	Water depth ratio, h/h_0
	Density ratio, ρ/ρ_0	Water depth ratio, h/h_0
	Speed of sound, a	Wave velocity, $\sqrt{gh}$
	Mach number, V/a	Froude number, $V/\sqrt{gh}$
	Subsonic flow	Streaming water
	Supersonic flow	Shooting water
	Compressive shock	Hydraulic jump

From Table 6.1 it is seen that all the general relationships of isentropic gas flow can be applied in the hydraulic analogy merely by putting $\gamma = 2$ and using this table of correspondence.

6.5 Hydraulic Jumps (Shocks)

It is known that in "shooting" water under certain conditions, the velocity may decrease over short distances and the water depth suddenly increases. An unsteady motion of this type is known as *hydraulic jump*. Hydraulic jumps occur only in shooting water, that is, in water streams with flow velocity greater than the wave propagation velocity. To have a better understanding of hydraulic jump, let us examine flow through a large sluice gate. Let us imagine the forward water ahead of the gate to be at rest, and that from behind there arrives the front of a water wave which arose from the opening of the sluice gate. If the wave were very small, it would move forward with the basic velocity $(gh_1)^{\frac{1}{2}}$. Since however, it has finite height $(h_2 - h_1)$, it moves, to a first approximation, with the velocity

$$u_1 = [g\,(h_1 + h_2)\,/2]^{\frac{1}{2}}\,(h_2/h_1)^{\frac{1}{2}} \qquad (6.23)$$

where subscripts 1 and 2 refer to states ahead of and behind the water wave. This velocity is much larger than $(gh_1)^{\frac{1}{2}}$ or $(gh_2)^{\frac{1}{2}}$. In the present flow system, water may be considered as moving with the velocity u_1, with respect to the wave. That is, apparently the wave is made to remain at rest in space. Now the water ahead of the wave flows with velocity u_1, and it is greater than $(gh_1)^{\frac{1}{2}}$. From the above argument it can be seen that the hydraulic jump will remain stationary only in shooting water, such as flow through a sluice gate. If the wave existed in streaming water, it would, because of its propagation velocity, which in this case is larger than the flow velocity, travel upstream. There would be the usual outflow from upper to lower level without *shock*. Here the term *shock* will be used interchangeably with *hydraulic jump*, and naturally has nothing to do with the compressibility of the water. A *shock* (or hydraulic jump) *in which the wave front is normal to the flow direction is called a right hydraulic jump*. It naturally has the property that the propagation velocity of the shock wave relative to the water is equal and opposite to the water velocity ahead of the jump.

The second kind of *hydraulic jump which is along a line oblique to the flow direction is called slant hydraulic jump*. Let the water flow from left to right out of an open sluice gate. The water depth decreases and the velocity increases. The water flows from a constant upper water level into a basin with constant lower water level. Since the difference in head is greater than one third of the upper water depth, the water after escaping from the sluice receives a larger velocity than the basic wave propagation velocity, so that it shoots. It is thus possible for it to accelerate so rapidly that the water surface of the flow becomes lower than the lower water level. There is a portion of the flow for which there is considerable pressure rise over a short distance. In this flow, however, the jump does not take place normal to the velocity, but instead along a line oblique to the flow direction, and we have a slant jump.

The slant jump, like the right jump, occurs only in shooting water. In order to be able to give a simple numerical treatment of the slant hydraulic jump, we make the assumption that the motion is entirely unsteady; i.e., the water jumps suddenly along a line-the jump line-from the lower water level to the level after the jump. The simplest case of such a jump is obtained if a parallel flow is deflected by an angle. Similar deflection in a gas stream with supersonic velocity will result in an oblique shock. Flow past such shocks are treated exhaustively in standard books on gas dynamics. Here, however, for the shock of the shooting water, the analogy with a compressible gas flow for $\gamma = 2$ no longer strictly holds.

At this stage, it is essential to realize that most of the assumptions made in the establishment of analogy cease to be valid at discontinuous flows like hydraulic jump. For instance, the assumption that the vertical velocity or acceleration is negligibly small is not valid when a jump occurs, since the vertical component of velocity is considerable at a jump. But equations of continuity and momentum are still valid on either side of the discontinuity.

In supercritical shallow water flows, discontinuities in the form of "hydraulic jump" occur. This discontinuity results in sudden rise in water depth and change in velocity. At a jump, surface tension and viscosity act as equilibrating forces and they cannot be neglected. Energy losses occur at hydraulic jump, which are dissipated through heat losses in turbulence at strong jumps and undulations in weak jumps.

In supersonic gas flows shocks occur. In such a case viscosity and heat conduction effects are no longer negligible. A part of it is used to increase the internal energy while the rest affects the dynamic properties of the flow. The flow process across a shock is no longer isentropic but adiabatic.

Since the flows with shocks or jumps are not isentropic it might be expected that the quantitative correspondence breaks down. The entropy change, of course, is third order in shock strength so that the weak shocks are nearly isentropic. In weak jumps also energy losses are negligibly small. These changes are of the third order with reference to the difference in depth across the jump. Hence, the quantitative similarity should hold to a good approximation.

6.5.1   General Equations for Attached Oblique Shocks

For an oblique shock with shock angle β and flow turning angle θ in a gas flow the Mach number after the shock may be expressed as (Rathakrishnan, 1995)

$$M_2^2 \sin^2(\beta - \theta) = \frac{1 + \dfrac{\gamma - 1}{2} M_1^2 \sin^2 \beta}{\gamma M_1^2 \sin^2 \beta - \dfrac{\gamma - 1}{2}} \tag{6.24}$$

where the subscripts 1 and 2, respectively, refer to conditions upstream and downstream of the shock wave.

The density ratio across the shock can be expressed as

$$\frac{\rho_2}{\rho_1} = \frac{(\gamma + 1) M_1^2 \sin^2 \beta}{(\gamma - 1) M_1^2 \sin^2 \beta + 2} \tag{6.25}$$

or

$$\frac{\rho_2}{\rho_1} = \frac{\tan \beta}{\tan(\beta - \theta)}$$

The pressure ratio across the shock is given by

$$\frac{p_2}{p_1} = 1 + \frac{2\gamma}{\gamma + 1}\left(M_1^2 \sin^2 \beta - 1\right) \tag{6.26}$$

In terms of density ratio, the pressure ratio may also be expressed as

$$\frac{p_2}{p_1} = \frac{\dfrac{\gamma + 1}{\gamma - 1}\dfrac{\rho_2}{\rho_1} - 1}{\dfrac{\gamma + 1}{\gamma - 1} - \dfrac{\rho_2}{\rho_1}}$$

6.5.2   General Equations for Slant (oblique) Attached Hydraulic Jumps

Let subscripts 1 and 2 refer to conditions immediately before and after the jump, respectively. The Froude numbers $Fr_1 (= V_{1w}/\sqrt{gh_1})$ and $Fr_2 (= V_{2w}/\sqrt{gh_2})$ and the depth ratio h_2/h_1 can be expressed as

$$Fr_1 = \frac{1}{8\sin^2\beta} \left[\frac{\tan\theta\left(1 - 2\tan^2\beta\right) - 3\tan\beta}{\tan\theta - \tan\beta} - 1 \right]^{\frac{1}{2}} \tag{6.27}$$

$$Fr_2 = \frac{h_1}{h_2} \frac{1}{\sin\,(\beta-\theta)} \left[\frac{1}{2}\left(1 + \frac{h_2}{h_1}\right) \right] \tag{6.28}$$

$$\frac{h_2}{h_1} = \frac{1}{2}\left[\left(1 + 8\,Fr_1^2\sin^2\beta_w\right)^{\frac{1}{2}} - 1 \right] \tag{6.29}$$

From the general equations for the oblique shocks in gas and jump in liquid, it is obvious that these equations are similar.

6.5.3   Limitation of the Analogy

It is essential to understand the limitations and their influence on the accuracy of the analogy in order to assign the significance of the data obtained. Let us see the limitations of the hydraulic analogy in some detail.

6.5.3.1   Two–Dimensionality

The analogy is valid for one-dimensional and two-dimensional flows only. But the two-dimensionality limitation is not a serious drawback since most of the problems in gas dynamics are of two-dimensional nature and even flows which are not strictly two-dimensional can be approximated as two-dimensional flows without introducing significant error.

6.5.3.2   Specific Heats Ratio

The hydraulic analogy is well established for a gas with $\gamma = 2$. The influence of the valve of γ, with reference to the analogy, varies considerably according to the type of the flow. For moderate subsonic speeds or for uniform streams with small perturbations, the pressure distribution as given by various linearized theories, like Prandtl–Glauert rule, Karman–Tsien theory, etc., is independent of the specific heats ratio γ. Hence, the influence of γ does not come into the picture at all. Even for high-speed flows or for large changes in velocity, the effect of γ is not so large as to change the order of magnitude of the numerical results. For supersonic flows, the effect of γ varies with the shock strength. In transonic, flows the effect of γ can be taken care of by making use of the transonic similarity laws.

However, for one-dimensional flows it is well established that by altering the water flow channel cross-section, flow conditions corresponding to gases with different values of γ can be obtained. Flow in rectangular section corresponds to flow of a gas with $\gamma = 2$. Flow in a triangular section corresponds to $\gamma = 1.5$, and flow in a parabolic section corresponds to $\gamma = 1.4$.

6.5.3.3 Velocity of Wave Propagation

The effect of ignoring the surface tension and the vertical acceleration imposes a limitation on the choice of water depth. If the ratio of water stream depth to the wave length, h/λ, is not significantly small then the analogy fails, since for such cases there is no constant corresponding to the speed of sound in the isentropic gas flow. In fact, if the wave velocity C varies with wave length λ, not only does the analogy break down but even the concept of wave velocity itself losses all physical significance. In that case, C should be termed *phase velocity* rather than a *wave velocity*, since it is neither the rate of propagation of energy nor the wave form. Then, no given wave form can be propagated without a continuous change in shape even though an overall motion proceeds at a constant velocity known as the group velocity. From a physical point of view this so-called group velocity is always more significant than the wave velocity, since group velocity is directly related to the rate of propagation of energy itself. The wave velocity is of importance only when it coincides with the group velocity. The individual waves of a given wave length forming a grouped disturbance move more rapidly than the group velocity of the disturbance itself, but these individual waves die out as they approach the front of the group, their energy being continuously transferred to the disturbance itself. It is only when the group velocity coincides with the wave velocity that a given velocity can be propagated without change in form as in two-dimensional, supersonic flow, at constant velocity.

It can be shown that, the group velocity is given by

$$\left(C - \lambda \frac{dC}{d\lambda} \right)$$

where the group velocity for a finite combination of waves is given by the intercept of the C-axis, by tangent to the curve C versus λ. At a given λ, it is then obvious that the capillary ripples having a small λ travel with group velocity greater than C. On the other hand, gravity waves with higher λ must have that the group velocity is exactly less than one half of the wave velocity in deep water ($h >> \lambda$), no matter how large the model is. Only as the water becomes more shallow, the group velocity approaches C, which in turn approaches the constant value $a_w = (gh)^{1/2}$ which is independent of wave length for all λ, except for all those corresponding to the extremely small capillary ripples, which may be considered as an additional source of energy dissipation.

6.5.3.4 Vertical Accelerations

To establish the analogy, it has been assumed that the water flow is two-dimensional. However, this would appear to rule out the vertical component of velocity and changes in water depth and lead to the contradiction that accelerations are not permitted. The answer to this seeming paradox is that the analogy refers to a limiting case, where the vertical components of velocity and acceleration are negligibly small. In practice it means that the water depth must be very small and suggests the use of large models compared to water depth.

Wherever there are abrupt/rapid changes in water depth, as in shocks or Prandtl–Meyer expansions, the vertical motions are of first-order importance. Most of the discrepancies between the water channel and water-tunnel experiments seem to be associated with the fact that vertical motions are ignored in the analogy.

The direct effect of vertical accelerations, that is the surface undulations following the initial overshooting are very important for deciding on the model size. The errors arising as a consequence of neglecting vertical accelerations are proportional to 'h' and $(Fr)^2$. The error also depends on the shape and the size of the model and varies inversely to the size of the model.

6.5.3.5 Viscosity and Heat Conductivity

In the above discussions, the viscosity in both gas and water streams was ignored for the purpose of establishing analogy. To account for this it may be proposed that tests on both flows should be carried out at the same Reynolds number. But Reynolds numbers in the hydraulic analogy tank are usually one order of magnitude lower than that in wind tunnels. Further, boundary layers in liquids at low-speeds do not behave the same way as those in compressible gaseous flow at high speeds. However, tests conducted in water and air at quite different Reynolds numbers result in the same flow pattern, provided the Mach number–Froude number matching is done for analogy.

Therefore, if the flow pattern is sensitive to viscous effects, as in the case of shock boundary layer interaction, the water channel can provide at best only qualitative information. Experience has shown that no improvement in the analogy can be achieved by accounting for the displacement and momentum thickness of the boundary layer either on the model or on the channel floor.

6.5.3.6 Surface Tension

The water surface acts as a stretched membrane and the disturbances at the surface are propagated as capillary waves of small wave lengths and large propagation speeds. These waves of course are not a part of the analogy.

In water channel experiments the capillary waves introduce error in the height measurements and in the interpretation of the wave patterns from the photographs. The effect of capillary waves may be made insignificant by increasing the depth of the model size. Another undesirable feature of the surface tension is the meniscus formation where the free-surface of the liquid comes in contact with the boundary of the model. This affects the depth measurements around the model. To reduce surface tension it is an usual practice to add 0.1 by volume of lauryl isoquinolinium bromide to the water.

6.5.3.7 Appropriate Technique for Different Types of Problems

Choice of the best suited experimental setup for the simulation of a given flow depends on the following factors.

1.*Nature of motion*

- Steady motion.

- Unsteady motion.

2.*Speed regime*

- Subsonic.

- Transonic.

- Supersonic.

3.*Type of result required*

- Flow pattern.

- Aerodynamic coefficients.

- Pressure distribution over the model.

These are the three vital factors that should be considered for the choice of suitable experimental technique. The stationary model and flowing water technique is well suited for steady subsonic and supersonic flow problems. Any type of steady flow can easily be set up using this technique and depth measurement can easily be carried out.

For unsteady problems in subsonic and supersonic flows, towed model technique is more appropriate, since it is easier to vary the model velocity than that of the stream. Steady and unsteady transonic flows are best investigated by towed model technique because of large flow distributions that occur with the flowing water technique, at a Froude number in the range around unity.

It is easier to obtain the depth distribution in flowing water technique, whereas the problem is complicated in a towed model technique. Forces and moments can be directly obtained in the towed model method, whereas the same has to be obtained only through the measurement of depth distribution in a flowing water method. Flow visualization can be done equally well using either method.

6.5.4 Depth Measurement

The type of instrument used for depth measurement depends on whether the depth measurement is taken in the flow field or on the wall. Again the instrumentation is different for flowing water technique and towed model technique. Different types of methods used for depth measurement depending on the type of technique used are given in Figure 6.4.

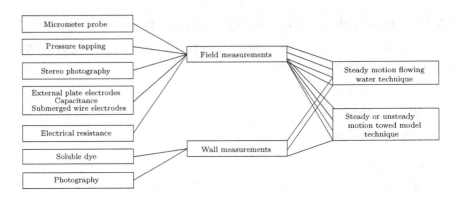

Figure 6.4: Depth measurement techniques

6.6   Velocity Measurement

The Froude number of any water stream is determined from the depth and velocity measurements of the freestream. In a towed model technique this becomes simple since the velocity of the flow is the velocity of the carriage. In a flowing model technique various methods are used to determine the freestream velocity of the flowing water. They are

1. Small Prandtl-type pitot tube and plain impact tubes are used to get the velocity profile in fast shallow water flows. From the profile the average velocity can be determined accurately.

2. Photograph technique using a timed multiple exposure. When small, light particles like aluminum foil are placed on the water surface, they adhere to the water surface and float downstream with the surface velocity. Their motion can be photographed using the shadowgraph technique. Illumination may be provided by a stroboscope calibrated to appropriate frequency so that two or three images of the same foil particle are recorded by the camera. The velocity on the surface can be obtained by direct measurement on the photographic plate and stroboscopic frequency. To obtain the average velocity of flow, the value of the surface velocity must be multiplied by a coefficient determined from corresponding velocity profile.

3. Hot-wire technique. For transient measurement this method is well suited.

6.7   Experimental Study

With our discussions about the hydraulic analogy in the previous sections, let us see the application aspects of hydraulic analogy. Applications are illustrated here by

- Studying the flow fields around hypercritical airfoils in steady and unsteady transonic regime of flow.

- Examining the velocity of the analogy to steady supersonic flows.

For investigating the flow field around airfoils qualitatively, two profiles, namely a shockless lifting airfoil and a quasi-elliptical nonlifting airfoil section have been chosen. The experiments have been carried out in a towing tank facility.

For demonstrating the quantitative aspects of the analogy, experiments have been done with a semi-wedge airfoil. From depth measurements the aerodynamic coefficients are computed and the flow pattern around the model wave recorded photographically. For this a water flow channel facility was used.

6.7.1 Towing Tank

A towing tank is essentially a channel in which water is made to flow in a controlled manner with any desired depth. Schematic of the towing tank used for the present study is shown in Figure 6.5.

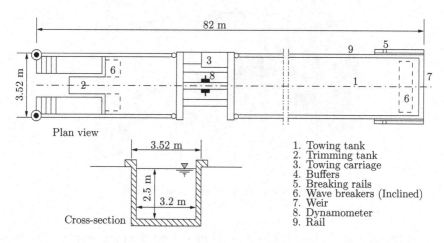

1. Towing tank
2. Trimming tank
3. Towing carriage
4. Buffers
5. Breaking rails
6. Wave breakers (Inclined)
7. Weir
8. Dynamometer
9. Rail

Figure 6.5: Towing tank

Two rails are provided for the carriage along and parallel to the length of the tank. The carriage is driven by four synchronous electric motors attached to each of the four wheels. The carriage speed can be varied over a wide range from a few mm/s to a few m/s, with these meters.

A chronograph is provided to record the actual speed of the carriage. From the time versus displacement plot the velocity and acceleration at any instant can be determined.

6.7.1.1   Flow Past Shockless Lifting and Nonlifting Airfoils

Flow fields around shockless lifting airfoil and quasi-elliptic airfoil models at zero angle of attack, when they are subject to transonic steady and unsteady flows are recorded photographically. A view of the experimental facility with model and camera mounted on the carriage is shown in Figure 6.6. The desired flow conditions are achieved by towing the model which is fixed to the platform of the tank by a fixture. Photographs of the flow patterns can be taken continuously by making the model move at required speed.

Figure 6.6: A view of the towing tank

The changes in the flow field when the model is subject to sudden acceleration can also be photographed.

Two airfoil models with chord lengths 100 mm and 120 mm were tested. In order to achieve shallow water flow the depth of the tank was maintained at a constant value of 202 mm. The angle of attack of the model was set to zero and the model was placed at the center of the tank. The chronograph in the tank carriage provided the record of time versus displacement, from which the actual velocity and acceleration were calculated.

The Froude number $Fr = V/\sqrt{gh}$ was determined from the speed of the carriage, obtained from the chronograph record and the average depth of water. In Section 6.4 it was shown that the hydraulic analogy results are true only for a fictitious gas with $\gamma = 2$ and it is important to correlate the results to air. Since the flow regime under study is transonic, the correlation is achieved in two different ways. The first method makes use of transonic similarity parameter K $= (M^2 - 1)/[t(\gamma + 1)]^{2/3}$ (Rathakrishnan, 1995) where K is a constant for a family of geometrically similar profiles. The second method makes use of the following streamline similarity property.

6.7.2 Streamline Similarity and Transonic Similarity Rule

Correlation of the results obtained for a fictitious gas with $\gamma = 2$ using hydraulic analogy to a perfect gas with $\gamma = 1.4$ may be carried out in two different ways, when the flow is transonic. The transonic similarity rule is quite often resorted to for this purpose, but still an effective method which may be applied to a wide range of Mach numbers is the streamline similarity rule. In order that the same range of the relative spacing is available in both cases, it is necessary that the relative streamline spacing is expressed in terms of the variable A_c/A_1, where A_c refers to area of the streamline corresponding to $M = 1$ and A is the local area.

An indication of the possibility of existence of the transonic similarity is afforded by a study of transonic similarity. According to the transonic similarity laws, developed for flows in which the local Mach numbers throughout the flow field are nearly equal to 1.0, the following substitution is made use of:

$$\xi = \frac{x}{a} \eta = \frac{y}{a} \left[(\gamma + 1)t \right]^{1/3} f(\xi, \eta) = \frac{\phi}{ca^*}$$

where x and y are rectangular coordinates, a^* is velocity of sound when $M = 1$, c is half-chord of the body, t is thickness to chord ratio, ϕ is velocity potential in physical plane ξ, η the natural coordinates, and $f(\xi, \eta)$ is nondimensional velocity potential.

The differential equation of motion becomes

$$\frac{\partial^2 f}{\partial^2 \eta^2} = 2K \frac{\partial f}{\partial \xi} \frac{\partial^2 f}{\partial \xi^2} \tag{6.30}$$

where $K = (1 - M_\infty) \left[(\gamma + 1)t \right]^{-2/3}$ and M_∞ is the freestream Mach number. For similar flows to prevail, K must remain constant when M_∞, γ and t are varied. In order to show that the corresponding relationship holds throughout the flow field, let,

$$K_1 = (1 - M_1) \left[(\gamma + 1)t \right]^{-2/3}$$

be a quantity analogous to the transonic similarity parameter K, but formed with local Mach number M_1.

It can be shown that

$$K_1 = -K^{-2/3} \frac{\partial f}{\partial \xi}$$

since for similar solutions, $\frac{\partial f}{\partial \xi}$ at a given point (ξ, η) in the flow field is independent of variations in M_∞, γ, and t, and the value of K_1 at the point (ξ, η) must also remain invariant.

Since K_1 is a function of three variables, M_1, γ, and t at point (ξ, η), it is permissible to assume an arbitrary definite relationship between two of the variables and adjust the third so that the similarity is maintained. With this idea in mind, let

$$t = \frac{C_1}{(\gamma + 1)^{1/4}}$$

where C_1 is an arbitrary constant. That is, $(1 - M_1)$ must vary directly as $(\gamma + 1)^{1/2}$.

The area Mach number relationship gives

$$\frac{A}{A_{\text{th}}} = \left(\frac{2}{\gamma + 1}\right)^{\frac{\gamma+1}{2(\gamma-1)}} \frac{\left[1 + \frac{\gamma - 1}{2}(1 - M^*)^2\right]^{\frac{\gamma+1}{2(\gamma-1)}}}{(1 - M^*)}$$

where $M^* = 1 - M_1$. If this expression is expanded in terms of M^* and terms contain powers of M^* greater than 2 are neglected, then the above relation reduces to

$$\frac{A}{A_{\text{th}}} = \left(\frac{2}{\gamma + 1} M^{*2} + 1\right)$$

As mentioned earlier, if streamline similarity is to be obtained the value of A/A_{th} at corresponding points in the field must remain unchanged when M^* is varied in accordance with the condition

$$M^{*2} = C_2(\gamma + 1)$$

where the value of C_2 depends on the value of A/A_{th} at the particular point under consideration or $(1 - M_1)$ varies directly as $(\gamma + 1)$. Then the streamline similarity is consistent with a special case of general transonic similarity rule. Hence, the correlation of Mach number can easily be done using the streamline similarity.

The experimental Froude numbers together with the correlated Mach numbers obtained with the above discussed methods are given in Table 6.2.

Table 6.2 Correlated Mach numbers in the transonic flow

Carriage speed m/s	Froude number (for $\gamma = 2$)	Correlated Mach number in air	
		Using transonic similarity rule	Using streamline similarity
0.85	0.603	0.674	0.632
0.99	0.702	0.750	0.748
1.13	0.801	0.834	0.830
1.50	1.064	1.060	1.060

6.8 Application of the Hydraulic Analogy to Supersonic Airfoils

6.8.1 Aerodynamic Forces on Airfoils

Predictions of aerodynamic force coefficients, namely C_L and C_D of wings in steady two-dimensional flow is of great importance in the field of aerodynamics. Extensive exact and approximate theories have been developed for this study. The purpose of the present experiment is to show the validity of the analogy to steady supersonic flow past airfoils.

The airfoil chosen is a half-wedge airfoil. The aerodynamic coefficients can be determined accurately, using "shock expansion" theory. The necessary condition for applying this theory is that the sharp edge model kept at an angle of attack in a supersonic stream must have the shock wave attached to the airfoil at the leading edge. The theory employs a step-by-step application of shock relations for shock waves and expansion waves.

6.8.2 Hydrodynamic Forces on Airfoils

When an airfoil is placed in a high-velocity shallow water stream of a fixed Froude number with free-surface, the resulting water depth distribution around the model will be analogous to pressure distribution around the same model in supersonic gas stream Mach number equal to the Froude number. The conditions under which the analogy will be valid are

- The hydraulic model should be geometrically similar to the airfoil.

- The Froude number of the water should be equal to the Mach number of the gas stream.

- The chord plane of the model must be perpendicular to the water channel bottom floor.

From hypercritical water flow, through the measurement of depth distribution, the hydrostatic pressure around the model can be determined. Hence, the *hydrodynamic lift*, *drag*, and *moment* acting on the model can be computed.

6.8.3 Measurements with a Semi-Wedge Airfoil

A semi-wedge airfoil of wedge angle 7°, shown in Figure 6.7, has been used for the study. The experiments were conducted in a water flow channel with a test-section cross-section of 300 mm × 130 mm.

The semi-wedge airfoil was tested at a Froude number of 2.13 at various

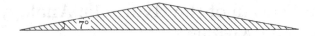

Figure 6.7: Semi-wedge airfoil

angles of attack. Using the properties of analogy, it can be shown that

$$C_L = 2\left(\frac{h_4}{h_1}\right)^2 \cos\alpha - \left(\frac{h_2}{h_1}\right)^2 \frac{\cos(\phi-\alpha)}{\cos\phi} - \left(\frac{h_3}{h_1}\right)^2 \frac{\cos(\phi+\alpha)}{\cos\phi} \quad (6.31)$$

$$C_D = 2\left(\frac{h_4}{h_1}\right)^2 \sin\alpha - \left(\frac{h_2}{h_1}\right)^2 \frac{\sin(\phi-\alpha)}{\cos\phi} - \left(\frac{h_3}{h_1}\right)^2 \frac{\sin(\phi+\alpha)}{\cos\phi} \quad (6.32)$$

$$C_M = \left(\frac{h_4}{h_1}\right)^2 - \left(\frac{h_2}{h_1}\right)^2 \frac{1}{4\cos^2\phi} - \left(\frac{h_3}{h_1}\right)^2 \frac{3}{4\cos\phi} \quad (6.33)$$

where C_L, C_D, and C_M are the lift drag and pitching moments, respectively, and α and ϕ, respectively, are the angle of attack and the semi-wedge angle. The subscripts 1, 2, 3, and 4 to the water stream depth h refer to the zones of the flow field around the airfoil shown in Figure 6.8.

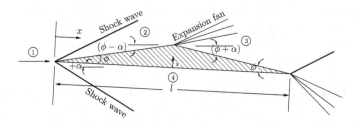

Figure 6.8: Wave pattern around a semi-wedge airfoil

By measuring the depths at various locations around the wedge the hydro-dynamic coefficients can be obtained from Equations (6.31) to (6.33).

6.9   Experimental Study

The experimental runs were made with a freestream Froude number of 2.13 keeping the semi-wedge airfoil model at $0°$, $\pm 3°$, $\pm 6°$, and $\pm 9°$ angles of attack. The Froude number required was obtained by a trial and error procedure, as follows. The volume flow in the water channel was measured for position of valve opening. Knowing the freestream depth and channel width, the Froude number was calculated. The position of the valve opening is adjusted to different flows. The adjustment was continued until a opening of valve resulted in a freestream with Froude number 2.13. The corresponding depth of the stream was 5 mm. This satisfies the condition for the flow to be shallow water stream and thus is an appropriate depth for the analogy. Depth measurements to the accuracy of ± 0.1 mm were made using a point gauge. A view of the experimental setup with depth gauge is shown in Figure 6.9.

Figure 6.9: A view of the experimental setup

The flow field around the semi-wedge model had been photographed with a flash camera. Using the measured values of the water stream depths, h_1, h_2, h_3, and h_4, in the different zones of the flow field around the airfoil shown in Figure 6.8, at angle of attack $\alpha = 0°$, $\pm3°$, $\pm6°$, and $\pm9°$, and freestream Froude number 2.13, the values of values of lift, drag, and pitching moment coefficients were calculated. The experimental values of C_L, C_D and C_M at different angles of attack are listed in Table 6.3.

Table 6.3 The experimental result of aerodynamic coefficients for semi-wedge airfoil with hydraulic analogy

Angle of attack	C_L	C_D	C_M
+ 9deg	0.354	0.065	0.097
+ 6deg	0.221	0.036	0.050
+ 3deg	0.087	0.017	0.082
0deg	− 0.024	0.017	0.021
− 3deg	− 0.032	0.021	− 0.032
− 6deg	− 0.254	0.044	− 0.087
− 9deg	− 0.425	0.099	− 0.151

Figure 6.10: Flow pattern around semi-wedge at $0°$ angle of attack and $M = 2.13$

The validity of this analogy can be checked by calculating the C_L, C_D, and C_M independently, with shock expansion theory and comparing the results of the two methods. For example, the shock expansion theory applied for the semi-wedge airfoil at $0°$ angle of attack yields $C_L = -0.022$, $C_D = 0.016$, and $C_M = 0.033$. It is seen that the results agree fairly closely with the analogy results. It can be noticed by performing theoretical calculations at higher values of α that the analogy results are in good agreement with theoretical results at low values of α, compared to its high values. The C_L, C_D, and C_M values for freestream Mach number of 2.13 and at different angles of attack of the experimental study, calculated with shock-expansion theory are given in Table 6.4.

Table 6.4 The aerodynamic coefficients for semi-wedge airfoil with
shock-expansion theory

Angle of attack	C_L	C_D	C_M
+ 9deg	0.312	0.061	0.021
+ 6deg	0.206	0.034	0.025
+ 3deg	0.091	0.019	0.029
0deg	− 0.022	0.016	0.033
− 3deg	− 0.135	0.026	0.037
− 6deg	− 0.249	0.047	0.043
− 9deg	− 0.364	0.082	0.049

The flow visualization pictures of the flow pattern around the semi-wedge, at $0°$ and $3°$ angles of attack are given in Figures 6.10 and 6.11, respectively.

It is seen from these flow patterns that all the features of supersonic gas flow field around a semi-wedge, such as the shock wave and the expansion fan,

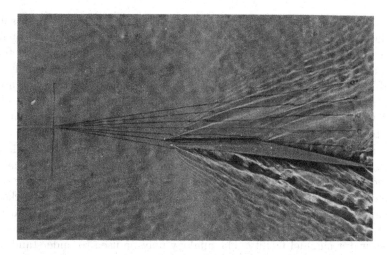

Figure 6.11: Flow pattern around semi-wedge at 3° angle of attack and $M = 2.13$

which can be viewed employing a schlieren technique in a gas flow, are exhibited by the analogous water flow field. It is important to realize here that, in addition to the compression and expansion waves, there are many waves caused by the disturbance. Therefore, the experimenter has to exercise extreme caution in interpreting the analogous gas dynamic waves around the model. Thus, a thorough knowledge of gas dynamics is essential to effectively utilize this inexpensive facility of high value.

6.10 Summary

In analogue methods, fluid flow problems are solved by setting up another physical system, such as an electric field, for which the basic governing equations are of the same form with corresponding boundary conditions as those of the fluid flow. The solution of the original problem may then be obtained experimentally from measurement on the analogous system. The well-known analogy methods for fluid flow problems are the *Hele-Shaw analogy, electric tank*, and surface waves in a *ripple tank*.

The Hele-Shaw apparatus produces a flow pattern which is similar to that of potential flow. It is an analogy experiment known as Hele-Shaw analogy. In Hele-Shaw apparatus, the viscous flow of a liquid between two closely spaced plates may be shown to reproduce the streamlines in the flow of a frictionless inviscid fluid. The Hele-Shaw apparatus produces a flow pattern which is similar to that of potential flow. Further, it can be mathematically shown that under Hele-Shaw flow conditions, the flow equation has the same form as that of potential flow, however it does not satisfy no-slip condition at the wall. There is slip at the wall.

The *electrolytic tank* is another popular analogue technique used for solving potential flow problems. In other words, it is used to solve the Laplace equation. It makes use of the fact that the equations governing incompressible potential flow and distribution of the electrical potential lines are the same, in their form, to establish an analogy between the two fields. The flow pattern around a body immersed in a perfect fluid can be determined by finding the passage of the current through a uniform sheet of conductor, except within the boundary corresponding to the body immersed in the fluid.

The analogy between shallow water flow with a free-surface and two-dimensional gas flow has been found to be useful for qualitative as well as quantitative study of high-speed flows. This technique is valued highly because of the fact that many practical problems in supersonic flows, involving shocks and expansion waves, which require a sophisticated and expensive wind tunnel and instrumentation for their analysis may be studied in an inexpensive manner with a simple water channel facility. The analogy may be used to understand many challenging problems of practical interest, such as transonic unsteady flows past wings, transient phenomena in high-speed flow and shock wave interaction, and so on.

The essential feature of this analogy is that *the Froude number of shallow water flow with a free-surface is equivalent to a gas stream with Mach number equal to that Froude number.* The analogy is valid for one-dimensional and two-dimensional flows only. But the two-dimensionality limitation is not a serious drawback since most of the problems in gas dynamics are of a two-dimensional nature and even flows which are not strictly two-dimensional can be approximated to two-dimensional flows without introducing significant error.

The hydraulic analogy is well established for a gas with $\gamma = 2$. The influence of the value of γ, with reference to the analogy, varies considerably according to the type of the flow. For moderate subsonic speeds or for uniform streams with small perturbations, the pressure distribution as given by various linearized theories, like Prandtl–Glauert rule and Karman–Tsien theory, etc., is independent of the specific heats ratio γ. Hence, the influence of γ does not come into the picture at all. Even for high-speed flows or for large changes in velocity, the effect of γ is not so large as to change the order of magnitude of the numerical results. For supersonic flows the effect of γ varies with the shock strength. In transonic flows the effect of γ can be taken care of by making use of the transonic similarity laws.

Wherever there are abrupt/rapid changes in water depth, as in shocks or Prandtl-Meyer expansions, the vertical motions are of first-order importance. Most of the discrepancies between the water channel and wind tunnel experiments seem to be associated with the fact that vertical motions are ignored in the analogy.

The direct effect of vertical accelerations, that is the surface undulations following the initial overshooting, are very important for deciding on the model size. The errors arising as a consequence of neglecting vertical accelerations are proportional to 'h' and $(Fr)^2$. The error also depends on the shape and the size of the model and varies inversely to the size of the model.

The viscosity in both gas and water streams was ignored for the purpose of establishing analogy. To account for this it may be proposed that tests on both flows should be carried out at the same Reynolds number. But Reynolds numbers in the hydraulic analogy tank are usually one order of magnitude lower than that in wind tunnels. Further, boundary layers in liquids at low-speeds do not behave the same way as those in compressible gaseous flow at high speeds. However, tests conducted in water and air at quite different Reynolds numbers result in the same flow pattern, provided the Mach number–Froude number matching is done for analogy.

The water surface acts as a stretched membrane and the disturbances at the surface are propagated as capillary waves of small wave lengths and large propagation speeds. These waves of course are not a part of the analogy.

Choice of the best suited experimental setup for the simulation of a given flow depends on the following factors.

1. *Nature of motion*

   • Steady motion.

   • Unsteady motion.

2. *Speed regime*

   • Subsonic.

   • Transonic.

   • Supersonic.

3. *Type of result required*

   • Flow pattern.

   • Aerodynamic coefficients.

   • Pressure distribution over the model.

These are the vital factors that should be considered for the choice of suitable experimental technique.

For unsteady problems in subsonic and supersonic flows the towed model technique is more appropriate, since it is easier to vary the model's velocity than that of the stream. Steady and unsteady transonic flows are best investigated by the towed model technique because of large flow distributions that occur with the flowing water technique, at a Froude number in the range around unity.

Chapter 7

Pressure Measurement Techniques

7.1 Introduction

To be a successful engineer, one must have command over both theory and experiment. Even among successful engineers, the innovative person is one who knows how to use the theory for experimentation and experiment for theoretical analysis. The purpose of this chapter is to discuss the science of instrument-building for experimentation in fluid flows. As we know, the fundamental quantities to be measured for a complete understanding of any fluid flow are the pressure, temperature, density, and the flow velocity and its direction. In fact, we will realize shortly that the direct measurement of density is impossible in most fluid flow problems, and the velocity can be determined from the measured static and total pressures. Therefore, it amounts to measuring only the pressure and temperature in most of the fluid flow experiments. Even though we talk of velocity measurement and their fluctuations, as in the case of turbulent flows, the basic principle underlying these measurements is sensing of pressure and temperature only. In the course of our discussion in this chapter, we will realize that the experimentation is more an art than science. In fact, it is correct to state that in experimentation, *anybody can make sophistication but it really calls for a genius to make simplification.*

7.1.1 Pressure Measuring Devices

The pressure measuring devices meant for measurements in fluid flow may broadly be grouped into manometers and pressure transducers. Various types of liquid manometers are employed depending upon the range of pressures to be measured and the degree of precision required. The U-tube manometers, multitube manometers, and Betz-type manometers are some of the popular liquid manometers used for pressure measurements in fluid flow. The pressure transducers used for fluid flow experimentation may be classified as electrical-type transducers, mechanical-type transducers, and optical-type transducers, based on the functioning principle of the sensor in the transducer.

7.1.2 Principle of the Manometer

The basic measuring principle of the liquid manometer is that *the pressure applied is balanced by the weight of a liquid column*. The sensitivity of the instrument depends on the density of manometric fluid used. Some of the commonly used manometric liquids are water, alcohol, and mercury. For pressure measurements in compressible flows, with high subsonic and supersonic Mach numbers, mercury is the suitable liquid, since fluids like water and alcohol show unmanageable variations in manometer column heights for the pressures associated with such speeds. In addition to these manometers, an accurate barometer is essential for pressure measurements, since in manometers the pressures are invariably measured in terms of a difference in pressure from some known reference. The most common reference is the local atmospheric pressure. Therefore, for pressures measured with reference to atmospheric pressure, conversion to absolute pressure requires that the local atmospheric pressure be known. The common mercury barometer is quite satisfactory for this purpose. When equipped with a suitable device for viewing the meniscus of the mercury column and reading the column height scale, a good barometer will allow measurement of atmospheric pressure with an accuracy of a fraction of a millimeter of mercury. This is usually quite adequate for the purpose of fluid flow pressure measurements.

7.2 Barometers

The basic principle of pressure measurement with barometer can easily be seen with the system shown in Figure 7.1(a). There is a reservoir containing a liquid. At the surface of this liquid atmospheric pressure p_a is acting. An open end glass tube is inserted into the liquid, as shown. The atmospheric pressure p_a will act down through the tube onto the surface of the liquid contained in the tube. Since only p_a acts at the liquid surface both outside and inside the glass tube, the liquid surface level throughout will be the same. It must be ensured that the tube bore is large enough that no capillary action due to surface tension occurs. Examine the system shown in Figure 7.1(b). The top of the glass tube is sealed and the tube has been evacuated to vacuum. Therefore, there is no pressure acting inside the tube. Since the atmospheric pressure p_a acts in all directions within the liquid, there is an unbalanced pressure in the upward direction inside the tube. The liquid is thus forced up in the tube until the weight of the liquid column above the surface level of the reservoir balances the atmospheric pressure p_a. If h is the height of the liquid column and ρ is the density of the liquid, then the pressure is given by

$$p_a = \rho g h$$

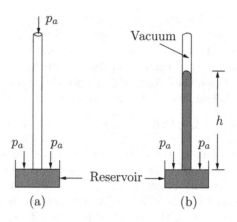

Figure 7.1: Simple barometer

This simply implies that $p \propto h$, since ρ and g are constants. The height of the liquid column established in the above manner can therefore be used as a measure of the atmospheric pressure. This is the measuring principle of the barometer, and was developed by Evangelista Torricelli (1608–1647), an Italian physicist. To create necessary vacuum above the surface of the liquid, the tube is firstly inverted and completely filled with liquid. The open end is now held closed, the tube is turned over so that the open end is at the bottom, and, with the end still held closed, it is immersed in the liquid in the reservoir. The end is then opened. The liquid in the tube now adjusts itself to balance the atmospheric pressure by falling down the tube until the required height is attained. A vacuum will thus be formed in the tube above the surface of the liquid. This vacuum is referred to as a Torricelli vacuum. The height to which the liquid column will rise in the tube depends upon the density of the liquid used. Let us assume the liquid to be water and the column height in the tube to be h m. The atmospheric pressure becomes

$$p_a = \rho g h$$

$$= 9.81 \times 10^3 \times h \text{ Pa}$$

since the density of water is 10^3 kg/m^3. In international standard atmosphere, the atmospheric pressure is 101325 Pa (1.01325 bar). Therefore,

$$101325 = 9.81 \times 10^3 \times h$$

Thus, the water column height of 1 atmosphere (at standard conditions) becomes

$$h = \frac{101325}{9.81 \times 10^3}$$

$$= 10.329 \text{ m}$$

A column height of 10.329 m of water is too large for measurement purposes. In actual measurements this order of column height is unmanageable. In order to reduce this column height, mercury is used in the barometer. As we know, mercury is 13.6 times heavier than water. Hence, the height of mercury column will be only 1/13.6 times the height of corresponding water column. Thus, the barometric column height, under standard sea level conditions, is

$$h = 10329/13.6 = \boxed{760 \text{ mm (approximately)}}$$

This height is convenient to measure. Even though barometric height is useful for measurement purposes, in practice we express pressure as a force per unit area. Usually pressure is expressed as N/m^2 or pascal (Pa) in SI units. An important aspect of practical interest which must be noted here is that the barometric height increases with increase of pressure. The mercury necessary to accommodate this increase in height has to come from the reservoir at the bottom of the barometer. Hence, the mercury level in the reservoir falls. Conversely, the mercury level in the reservoir increases when the barometric height falls, as a result of decrease in atmospheric pressure. In either case, with increase in or fall of level in the reservoir, the barometric height must be measured from the mercury surface level in the reservoir. To accommodate for the rise or fall of mercury level in reservoir, two different methods described below are generally used.

7.2.1 Syphon Barometer

The syphon barometer is essentially a U-tube of glass with one limb very much shorter than the other, as shown in Figure 7.2. The short limb is open to atmosphere and the long limb end is closed and a Torricelli vacuum is created in this limb. A common scale is used for both the limbs. If h_1 is the height of mercury column in the closed limb and h_2 is that at the open limb, then the barometric height $h = h_1 - h_2$. Thus, the atmospheric pressure is measured easily with the syphon barometer.

7.2.2 Fortin Barometer

The typical Fortin barometer is illustrated schematically in Figure 7.3. This is a reservoir or cistern-type barometer, as shown in the figure. The reservoir of mercury is contained in a metal tube A. The bottom of the reservoir, inside tube A, is made of a leather bag and this leather bag can be pushed up or down by rotating the adjusting screw B. The surface level of mercury in the reservoir can be observed through the observation tube C. The barometer tube passes up through a metal tube D and the surface level of the mercury at barometric height can be observed through a window in the scale tube E. A sliding vernier F is fitted across this window and the height of the vernier can be set by means of the vernier screw G. The zero of the scale is located at the bottom of the barometer. A pointer H is fitted to the top of the observation tube C and the tip

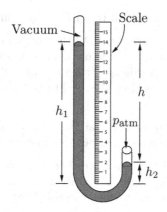

Figure 7.2: Syphon barometer

of the pointer is zero on the scale. The measuring technique of the barometer is as follows:

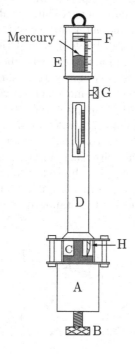

Figure 7.3: Fortin barometer

- The surface level of the mercury in the reservoir A is adjusted by means of the adjusting screw B until its surface just touches pointer H.

- Vernier screw G is now rotated until the bottom of the vernier F is in line with the surface meniscus of the mercury at the top of the barometer. In order to avoid error in this adjustment, scale tube E is usually provided with another window cut in the rear. The vernier is also made in the form of a tube.

- If the vernier bottom surface and the mercury meniscus are brought in line, then the vernier adjustment can be taken as accurate.

It is now possible to read the barometer height in millimeters. Note that the rise and fall in barometric height is readily accommodated in this barometer using the adjustment provided to the reservoir level by means of screw B. So long as the level is brought to the tip of pointer H the barometer reading is very accurate.

7.2.3   Aneroid Barometer

The basic element of aneroid barometer is the corrugated circular cell A, shown in Figure 7.4. The measuring principle of this barometer is illustrated in this figure. The corrugated cell is made of very thin metallic sheet and is hollow. The cell is partially evacuated. At the outer surface of the cell atmospheric pressure p_a is acting. The cell surface will be depressed if the atmospheric pressure increases and the cell surface will move outwards when the atmospheric pressure decreases. This movement is used as a measure of change of atmospheric pressure from one level to another. Since this movement is very small it is magnified by means of a lever mechanism B. A simple lever mechanism, appropriate for this application is illustrated in Figure 7.4, with a pointer moving over a scale C. In practice lever mechanisms of larger movement magnification would be required and the scale is usually circular with the pointer rotating about the center of the scale. Further magnification can be achieved by increasing the number of cells. The scale is usually calibrated in terms of millimeters of mercury and only that part of the scale on either side of the standard atmospheric pressure, 760 mm of Hg, is included.

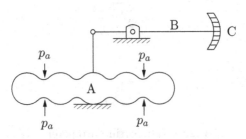

Figure 7.4: Aneroid barometer

7.3 Manometers

Manometers measure the difference between a known and an unknown pressure by observing the difference in heights of two fluid columns. Two popular types of manometers are illustrated in Figure 7.5.

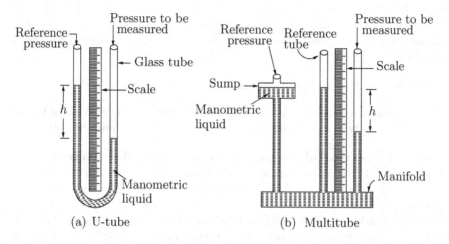

(a) U-tube   (b) Multitube

Figure 7.5: Manometers

As seen from Figure 7.5(a), a simple manometer consists of two vertical glass tubes joined together with a U-tube connection at the bottom. Each tube has a linear scale attached to it, which is usually marked off in millimeters. The tubes are filled with a fluid until the fluid level in the tubes is at the middle of the adjacent scales. A reference pressure is applied at the top of one of the tubes and the pressure to be measured is applied at the top of the second tube. The heights of the two columns of fluid will change until the difference between the heights, h, is equal to the pressure to be measured in terms of fluid column height. The commonly applied reference pressure for this type of manometer is the atmospheric pressure. However, in many cases the difference between the atmospheric and measured pressures will represent a very long column of manometer fluid than that which can be accommodated by the tubes. In such cases, the only way to use the manometer (other than changing the fluid) is to adjust the reference pressure so that a smaller fluid column height will be reached. However, this method has the disadvantage of adding an intermediate pressure measure. Instead of two separate scales for measuring the liquid levels at the two columns of U-tube manometer, a common scale fixed at the middle of U-tube, as shown in Figure 7.5, may also be used for measurement. In fact this has a specific advantage, since reading the liquid level in one of the tubes of the U-tube is adequate for pressure measurement if the scale is marked with zero at the middle and the graduations in increasing order from zero in both up and down directions. The liquid column in both the tubes must be set to zero before measurement, in this case. With this arrangement, measured column level in one tube will give the pressure to be measured, when it is doubled.

The multitube manometer, sketched in Figure 7.5(b), operates on the same principle as U-tube manometer. However, in this manometer a large cross-sectional area sump takes the place of the tube in the U-tube manometer, to which the reference pressure is applied. The sump level is used as the reference, and often a number of tubes are employed to form a multitube manometer. The sump and tube manometer has the following advantages over the U-tube manometer.

- It can be used for measuring more than one pressure at a time.

- The reference level can be adjusted so that only one scale needs to be read instead of two, to determine the manometric fluid column height.

Either of the two types of manometers can be constructed with tubes and scale that can be tilted. In this way an improvement in accuracy can be obtained, since a given distance along the scale will represent a smaller vertical height and consequently a smaller pressure.

These manometers can easily be fabricated in any laboratory. A typical U-tube and a multitube manometer constructed in the author's laboratory are shown in Figures 7.6 and 7.7, respectively.

Figure 7.6: U-tube manometer

Even though the liquid manometers are easy to build and accurate in measuring steady-state pressures, they have the following disadvantages.

- They are not suitable for very high or very low pressure measurements.

Figure 7.7: Multitube manometer

- They have very poor frequency response.

- A little dirt in a tube or a bubble in a line or presence of condensate changing the fluid specific gravity can produce anomalous readings.

However, if one can take care of the above-mentioned problems, a liquid manometer will prove to be an excellent device for pressure measurement in most practical situations, barring a few flow situations like flow in intermittent tunnels where accurate measurement of pressure with liquid manometers is very difficult because of the availability of a very short duration for measurements. Even these problems can be sorted out if one could arrange to connect an effective pinching mechanism (as the one at the top of the manometers in Figure 7.7), which could close all columns of the manometer at an appropriate time, thereby making the liquid level in different tubes stay where they are. After noting down the readings, the tubes can be opened again for the next measurement.

7.3.1 Inclined Manometer

When small pressure differences are involved in an experiment it becomes difficult to measure accurately the height of the liquid column in the vertical manometer. In such a case, the length of the liquid column could be increased by inclining the tube. If L is the length of the liquid column and h is the vertical height and α is the inclination with the horizontal then $h = L \sin \alpha$. By decreasing α to very low values, L can be increased considerably. However, in practice α is limited to around $30°$. This is because when the inclination is beyond the limiting value, the meniscus of manometric fluid will spread out and the surface tension effects will become predominant, resulting in large errors in the measurements.

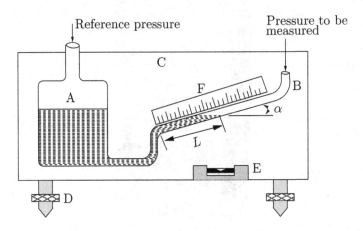

Figure 7.8: Inclined manometer

Inclined manometers are generally used for measuring very small pressure differences. The essential features of an inclined manometer are shown in Figure 7.8.

A large reservoir A is connected to an inclined manometer tube B. The cross-sectional area of the reservoir is very large compared to the inclined tube cross-sectional area. The assembly is suitably positioned on a vertical board C with the manometer tube at an angle α to the horizontal. The manometer can be set level by adjusting the leveling screws D at its base, in conjunction with the spirit level E. The scale F along the manometer tube is calibrated directly in terms of vertical height. The length of the scale, and hence the degree of accuracy of the scale, are controlled by angle α. For instance, let the manometer be required to make changes in vertical height up to 30 mm and let the slope be set such that $\alpha = 30°$. For this, the scale length L along the tube is given by

$$L = \frac{30}{\sin 30} = 60 \text{ mm}$$

Thus, a length of 60 mm on the scale can be calibrated to read 30 mm vertical height. The manometer liquid used is usually water.

The cross-sectional area of the reservoir is considerably larger than the cross-sectional area of the manometer tube. Hence, liquid level change in the reservoir will be very small compared to the change in level occurring in the manometer tube. Thus, the change of liquid level in the reservoir can be sensibly neglected and changes in pressure are therefore measured directly as changes along the manometer tube scale. The manometer is usually connected to the pressure tap by means of a rubber or plastic tube. The bore of the manometer tube using mercury or water should not be less than about 6 mm diameter to avoid the surface tension effects resulting in capillary action.

In low-pressure manometers, such as the inclined manometer, paraffin oil is sometimes used. Paraffin oil has a specific gravity of about 0.8 and thus, a larger change in level occurs when paraffin oil is used as manometric liquid instead of water. It also has less surface tension and can be used successfully in a tube of about 6 mm bore diameter.

7.3.2 Micro Manometer

Micro manometer is basically a precision U-tube manometer. It is a rugged instrument capable of measuring liquid column height with an accuracy of ± 0.1 mm. A typical micro manometer is shown in Figure 7.9.

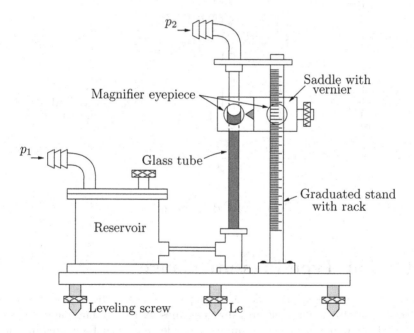

Figure 7.9: Micro manometer

From the figure it is seen that the micro manometer is a simple U-tube manometer with a vernier scale for precise measurement of liquid level. The construction details can be modified according to the user's desire. The pressure p_2 to be measured is applied to the manometer tube and a reference pressure (usually the ambient atmospheric pressure) p_1 is applied to the reservoir.

7.3.3   Betz Manometer

The Betz manometer is yet another U-tube manometer in a general sense. The fluid column carries a float and a graduated scale is attached to the float, as shown in Figure 7.10. The scale is transparent and the graduation markings could be projected on a screen. Thus, the manometer readings can be read directly. Even though it is a very convenient instrument to use, it is somewhat sluggish due to the large volume of the reservoir. The measurement range of this instrument is usually about 0 to 300 mm and its sensitivity is ±0.1 mm.

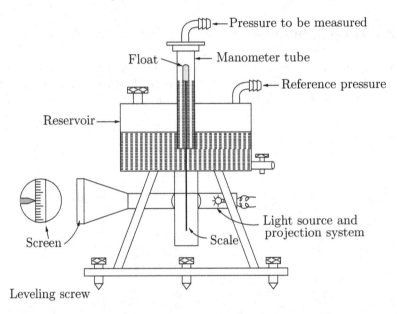

Figure 7.10: Betz manometer

7.4   Dial–Type Pressure Gauge

When large pressures of the order of mega pascals have to be measured, the liquid column required for such pressures becomes extremely large and measurement with manometers becomes unmanageable. For such high pressure measurements, dial-type pressure gauges are appropriate. Dial-type pressure gauges generally operate on the principle of expansion of a bellows or Bourdon tube which is usually an elliptical cross-sectional tube having a C-shape configuration. When pressure is applied to the tube, an elastic deformation, proportional to the applied pressure, occurs. The degree of linearity of the deformation depends on the quality of the tube material. The deflection drives the needle on a dial through mechanical linkages. Although pressure gauges of this type may be obtained with accuracies suitable for quantitative pressure measurements, they are not usually used for this purpose. They are primarily used only for visual monitoring of pressure in many pressure circuits. Schematic diagram of a typical Bourdon pressure gauge is shown in Figure 7.11.

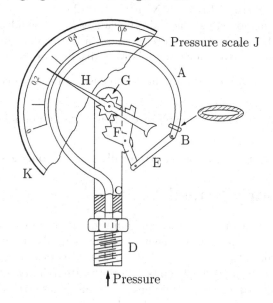

Figure 7.11: Bourdon pressure gauge

The vital element of this type of pressure measuring device is the tube A which is of elliptical cross-section and is bent into an arc of a circle, as shown. One end B of the tube is sealed while the other end C is open and is bent down. The open end C is fixed into a connecting union D. Any pressure applied is transmitted through a hole in this union into the inside of this tube A. When the pressure inside this tube increases the tube will tend to straighten. Conversely when the pressure decreases, the tube will tend to curl in. Change of pressure will therefore appear as a movement of the end B. The end B is connected to a quadrant gear F, by means of a link E. The quadrant gear engages with a small pinion gear G, onto which a pointer H is attached. Thus, any movement of the elliptical tube end B, due to change of pressure, will be transmitted to the quadrant gear, which will rotate the pinion and hence the pointer. The pointer travels over a calibrated pressure scale J. The entire system of mechanism is enclosed in a cylindrical case K. Such gauges can be obtained to measure pressures below, as well as above, atmospheric pressure. Very high pressures which cannot be measured by manometers, can be measured by this type of gauge. However, like manometers, the Bourdon pressure gauge also measures gauge pressure, with reference to the local atmospheric pressure. Therefore, the atmospheric pressure must be added to the gauge pressure in order to determine the absolute pressure.

The Bourdon pressure gauges are calibrated usually in units such as kilo newton/meter^2 (kN/m^2) or mega newton/meter^2 (MN/m^2). However, some gauges are calibrated in terms of bar (1 bar = 10^5 N/m^2). The advantage claimed here is that 1 bar is a pressure very nearly equal to the pressure of 1 standard atmosphere (0.101325 MN/m^2 = 1.01325 bar).

Compared to manometers, the dial-type gauges have the advantage of being easier to read. Also, they can measure pressure ranges well beyond those of the manometers. However, they have the following disadvantages.

- They must be calibrated periodically to ensure that they continue to read correctly.

- The manometers are less expensive when large numbers of pressures are to be read.

- They cannot be easily read electronically, like manometers.

7.5   Pressure Transducers

Pressure transducers can be designed and built for almost any pressure generally encountered in fluid flow measurements and they can also be used for remote indication. Pressure transducers are *electromechanical* devices that convert pressure to an electrical signal which can be recorded with a data acquisition system such as that used for recording strain gauge signals. These transducers are generally classified as *mechanical, electrical,* and *optical type.* Commonly used transducers employ elastic diaphragms (of various shapes) which are subjected to a displacement whenever pressure is applied. This movement is generally small and kept within the linear range, and this movement is amplified using mechanical, electrical, electronic, or optical systems. Schematic diagram of a typical diaphragm pressure capsule, with strain gauges mounted on the diaphragm, is shown in Figure 7.12.

The total strain produced on the diaphragm is proportional to the pressure applied. For a circular diaphragm, the deflection δ at the center is given by

$$\delta = \frac{3pa^4}{16Et^2}(1 - \mu_r^2)$$

where a and t are the radius and thickness of the diaphragm, respectively, μ_r is Poisson's ratio, E is Young's modulus, and p is the pressure applied.

It should be noted that the above relation holds good for small deflections only. For large deflections, corrugated diaphragms have to be used. Usually the diaphragms are made of *beryllium copper* or *phosphor bronze* sheets. Heat-treated *stainless steel* diaphragms are also employed for pressure measurements. A photographic view of some strain gauge transducers is shown in Figure 7.13.

The following are the advantages of pressure transducers over manometers and other pressure gauges.

- They provide signal proportional to the applied pressure which can be automatically recorded by any data acquisition system.

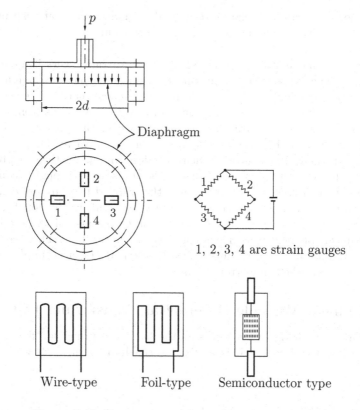

Figure 7.12: Pressure transducer (strain gauge type)

Figure 7.13: Pressure transducer

- They are relatively low volume devices and consequently respond more rapidly to pressure changes.

- They are small enough to be mounted inside wind tunnel models.

Their major disadvantage relative to a good manometer is that they must be calibrated, whereas a manometer with a known fluid can be considered as a pressure standard.

Because of the relatively high cost of pressure transducers in quantity, a scheme has been devised for using one transducer to measure a number of pressures-up to 48. This scheme of pressures communication uses a device known as "pressure scanni valve." In using the scanni valve, pressures from different pressure taps on a model are allowed to stabilize in the pressure tube lines leading from the model through the stator of the scanni valve. The rotor of the scanni valve is then turned through one revolution, communicating the pressures trapped in the connecting tubes in turn to the transducer through a communication slot. Likewise, all the tubes holding the trapped pressure are connected to the transducer and the respective pressures are read. Even though it is expensive, the scanni valve has a specific advantage because of the use of only one transducer for measuring a number of pressure readings. Thus, correction of measured data for any error introduced by the transducer is simple, since all the data will have the same error (if any).

7.5.1   Linear Variable Differential Transformer LVDT

Linear differential transformer works on the induction principle, like any other transformer, but with a movable core fixed to the diaphragm, as shown in Figure 7.14.

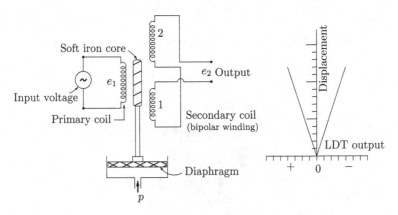

Figure 7.14: Linear variable differential transformer

As seen in Figure 7.14, the LVDT consists of a primary coil, two secondary coils, and a movable iron core. The secondary coils are connected in opposition. When the core is in null position between the two secondary coils the voltage induced gets canceled. Any movement of the core on either side gives an output proportional to the displacement from the null point. This instrument is capable of detecting displacements of the *order of a few microns*. Applying known pressures the associated displacements of the core and the corresponding outputs of the LVDT can be noted to make the calibration chart, as shown in the figure.

7.5.2 Capacitance Pickup

In this transducer the capacitance between two metallic plates with a dielectric in between is varied. One of the plates is fixed and the other one is connected to a pressure capsule. The change in capacitance is a measure of the pressure. A typical capacitance-type transducer is shown in Figue 7.15.

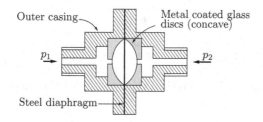

Figure 7.15: Capacitance pressure transducer

7.5.3 Optical–Type Pressure Transducer

Schematic sketch of an optical-type pressure transducer is shown in Figure 7.16. The pressure sensing element is a thin diaphragm which is highly polished on both sides. When the pressures p_1 and p_2 are equal the diaphragm remains flat and the width of the reflected light beams 'a' and 'b' is the same. Under this condition the differential output from the photocells is zero. But when the diaphragm gets deflected due to pressure difference, the width of the reflected beams is different. The amount of light falling on the photocells through the slits is different. This causes a differential output proportional to the difference between the pressures p_1 and p_2.

Example 7.1

Find the pressure that would be read by a mercury manometer connected to a static pressure tap located at the wall of a convergent nozzle where the flow Mach number is 0.8 and the nozzle is connected to a tank at a pressure of 3 atmospheres absolute (assume $\gamma = 1.4$, for the gas).

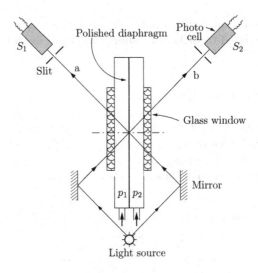

Figure 7.16: Optical type pressure transducer

Solution

The stagnation pressure p_0 of the stream is 3 atm. The Mach number of the stream is 0.8. By isentropic relation, we have

$$\frac{p_0}{p} = \left(1 + \frac{\gamma - 1}{2} M^2\right)^{\frac{\gamma}{\gamma - 1}}$$

where p is the static pressure. From isentropic table, for M = 0.8, we have

$$\frac{p_0}{p} = 1.524$$

Therefore,

$$p = \frac{3}{1.524} = 1.9685 \text{ atm}$$

i.e.,

$$p_{\text{gauge}} + p_{\text{atm}} = 1.9685 \text{ atm}$$

Therefore, $p_{\text{gauge}} = 0.9685$ atm. Let 1 atm = 760 mm of Hg. Thus,

$$p_{\text{gauge}} = 0.9685 \times 760 = 736.06 \text{ mm}$$

The manometer will measure the pressure as $\boxed{736.06 \text{ mm}}$ of Hg.

7.6 Pitot, Static, and Pitot-Static Tubes

It is well known that the measurement of both total and static pressure is essential to analyze fluid flows. The flow velocity can be determined using Bernoulli's equation, if these two pressures are known. Also, we know that both incompressible and compressible forms of the Bernoulli equation are simply relations between the total pressure, static pressure, and flow velocity. Now, let us see some of the popular probes which are used for measuring these pressures.

From our basic studies on fluid mechanics we know that pitot or total pressure is that pressure which results when a fluid stream is brought to rest isentropically. Also, measurement of total pressure is relatively easier than the static pressure. When a body is placed in a fluid stream and if it is possible to locate the exact point on it where the fluid is brought to rest then the pressure acting at that point will be the total pressure. Hence, this pressure can be measured by providing an orifice at that point and connecting it to a manometer. This forms the basis for pitot or total tube, which is the most popular instrument used for total pressure measurement. It is an experimentally established fact that the pressure indicated by a properly aligned open-ended tube facing a flow is exactly equal to the local total pressure, except for viscous effects at very low Reynolds numbers and for supersonic flows where there will be a detached shock positioned ahead of the tube.

Measurement of static pressure requires more attention than the pitot pressure measurement. In principle, for measuring static pressure with a probe placed in a flow, the static holes should be drilled at the location where the local pressure is the same as that of freestream static pressure upstream of the probe nose. In reality this condition cannot be achieved exactly, due to the pressure gradients caused by the presence of the probe.

A pitot-static tube is used for measuring the total and static pressures simultaneously. It is usually a blunt-nosed tube with an opening at its nose to sense the total pressure and a set of holes on the surface of the tube to sense the static pressure. Pitot, static, and pitot-static probes, made based on the above basic principles, are used extensively in experimental measurements. However, it is essential that a researcher use his discretion in applying appropriate correction to the measured pressure, to account for the errors due to compressibility, viscosity, misalignment, and so on.

7.6.1 Dynamic Head Measurement

In any flow field, basically there are three pressures which are of interest to gas dynamic studies. However, in fluid mechanics, we use yet a fourth one in parallel flow studies, which is termed the *geometric pressure*. This pressure is due only to the gravitational action on a static fluid having the same geometry as the actual flow.

The three pressures of primary interest in fluid dynamics are the total or pitot, static, and dynamic pressures. We saw that *the total pressure is that which results when a flow is decelerated to rest isentropically*. From this we can infer that, at a position in a flow where the flow velocity is zero, the total pressure and the undisturbed static pressure are identical.

The static pressure is that pressure which acts equally in all directions. The third pressure, namely the dynamic pressure, can be associated with the flow conditions at a point (say the Mach number) by taking the difference between the stagnation pressure and undisturbed static pressure.

In modern terminology yet another term namely, *velocity pressure*, is used. It is simply half of the product of fluid density and square of the speed; also termed *kinetic pressure*. In incompressible flow it is simply the difference between total and static pressures.

7.6.1.1 Incompressible Flow

For incompressible flows, the total, static, and dynamic pressures are linked together by Bernoulli's equation as

$$p_0 - p = \frac{1}{2} \rho V^2$$
$$= q \text{ (incompressible)}$$

where p_0, p, and $\frac{1}{2} \rho V^2$ are the total, static, and dynamic pressures, respectively. In any flow the total pressure will be the same everywhere when there are no losses. It is seen from the above equation that an increase of flow velocity results in a decrease of static pressure and vice versa. The dynamic pressure q is linked to the kinetic energy of the flow and *it has the same direction as that of the flow*. Also, we know that *the static pressure of a flow is that pressure which is acting normal to the flow direction*. Therefore, it follows from Bernoulli's equation that the total pressure also has a definite direction. Total pressure, as we have seen, is measured by a simple device called a pitot (total) probe. A pitot probe is simply a tube with a blunt-nosed open end facing the gas stream. The tube normally has an inside-to-outside diameter ratio of $1/2$ or $3/4$, and a length aligned with the gas stream of 15 to 20 tube outer diameters. The pressure hole is formed by the inside diameter of the tube at the blunt end, facing the flow. A typical pitot probe is shown in Figure 7.17.

The open-ended tube facing into the flow always measures the stagnation pressure it sees, in subsonic flows. But at supersonic speeds there will be a detached shock formed and positioned in front of the pitot probe nose. It implies that the tube does not measure the actual stagnation pressure but it only measures the stagnation pressure behind a normal shock. This new value is called *pitot pressure* and in modern terminology it refers to a supersonic stream.

This indicated pitot pressure p_{02}, the total pressure behind the normal shock, may be used for calculating the stream Mach number, as follows. For a supersonic stream with Mach number M_1, and p_1 and p_{01} as the static and stagnation pressures, respectively, by isentropic and normal shock relations, we have

$$\frac{p_{01}}{p_1} = \left(1 + \frac{\gamma - 1}{2} M_1^2\right)^{\frac{\gamma}{\gamma - 1}}$$

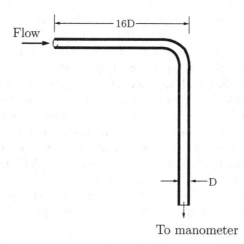

Figure 7.17: Pitot probe

$$\frac{p_{02}}{p_{01}} = \left(1 + \frac{2\gamma}{\gamma + 1}(M_1^2 - 1)\right)^{\frac{-1}{\gamma - 1}} \left[\frac{(\gamma + 1)M_1^2}{(\gamma - 1)M_1^2 + 2}\right]^{\frac{\gamma}{\gamma - 1}}$$

Multiplying these two equations, we obtain,

$$\frac{p_1}{p_{02}} = \frac{\left[\frac{2\gamma}{\gamma + 1}M_1^2 - \frac{\gamma - 1}{\gamma + 1}\right]^{\frac{1}{\gamma - 1}}}{\left(\frac{\gamma + 1}{2}M_1^2\right)^{\frac{\gamma}{\gamma - 1}}}$$

This relation is known as the *Rayleigh supersonic pitot formula*. Once the static pressure p_1 and pitot pressure p_{02} are known, M_1 can be calculated using this formula.

For a compressible flow, the pressures and the velocity are related by the equation

$$\frac{\gamma}{\gamma - 1}\frac{p}{\rho} + \frac{V^2}{2} = \frac{\gamma}{\gamma - 1}\frac{p_0}{\rho_0}$$

This is called the *compressible Bernoulli equation*.

Pressures measured by pitot probes are significantly influenced by very low Reynolds numbers based on probe diameter. However, this effect is seldom a problem in supersonic streams because a probe of reasonable size will usually have a Reynolds number well above 500, which is above the range of troublesome Reynolds number.

Measurement of static pressure in a supersonic flow is more difficult than in a subsonic flow. Though static pressure probes are not used extensively in supersonic flows, a good amount of study has been made for the development of accurate pressure probes for other applications. The major problem associated with the use of static pressure probes at supersonic speeds is that the probe will have a shock wave at its nose which will cause a "jump" in static pressure.

Now, let us see the measurement of static pressure in subsonic flows. If the probe has a conical nose tip followed by a cylindrical stem shoulder, the flow passing the shoulder will be accelerated to a pressure which is below the freestream static pressure. However, as the distance from the nose increases, the pressure on the probe will approach the true static pressure of the stream. The sharp-nosed tip of the probe facing the flow is closed and static pressure holes are drilled perpendicular to the tube axis at a specified distance from the nose. The included angle of the nose cone should be small for good results. Also, it is well established that static pressure hole located beyond eight times the outer diameter (8D) of the probe, from the nose, is good enough for reasonably accurate measurements. A typical "L" shaped static probe is shown in Figure 7.18.

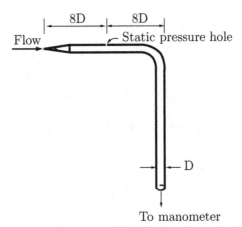

Figure 7.18: Static probe

Static pressure measurements are very sensitive to the inclination of the tube to the flow direction. Inclinations beyond 5° result in large errors in the measured pressures. In order to minimize this error, it is a common practice to have static probes with 4 holes in mutually perpendicular directions. In subsonic streams, when a static pressure probe is placed in the flow field, flow gets accelerated because of the horizontal portion of the probe and decelerated because of the vertical stem. Therefore, the static pressure holes have to be located at such a location that these effects cancel each other and the probe measures the correct pressure. Such an appropriate location found by experimental research is about $8D$, from the probe nose, as shown in Figure 7.18.

For measurement of static pressures in supersonic flow, the static pressure hole should be located beyond 18D from the nose and the velocity change associated with the presence of the probe should be kept to a minimum. For this the nose should be very sharp with an included angle less than 10°. Also, it is a usual practice to employ a straight probe (without vertical stem) for static pressure measurements in supersonic wind tunnels. The tip of the probe should be located at subsonic zone of the flow field (for instance upstream of the throat in a Laval nozzle) and hence there is no formation of shock because of the probe presence.

The dynamic pressure of a flow can be measured directly by a special probe called pitot-static probe, which is the combination of pitot and static probes. A typical pitot-static probe is shown in Figure 7.19.

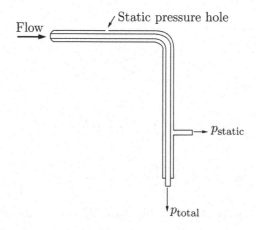

Figure 7.19: Pitot-static probe

For incompressible flows, the flow velocity can be calculated from dynamic pressure q, using the relation

$$V = \sqrt{\frac{2q}{\rho}}$$

It is seen that, the flow velocity V can be obtained from the incompressible Bernoulli equation. But for compressible flows, the above formula cannot be used for calculating the flow velocity, since in compressible flows, the relation $p_0 - p = q$ is not valid. We have to use the compressible form of the Bernoulli equation to connect the total, static, and dynamic pressures. Also, it is useful to note that, in the regime of compressible flow, the Mach number is more important than the velocity itself. For compressible flows, the dynamic pressure can be expressed in terms of Mach number M, as follows

$$q = \frac{1}{2}\rho V^2$$

$$= \frac{1}{2}\rho M^2 a^2$$

$$= \frac{1}{2}\rho M^2 \gamma RT$$

$$= \frac{\gamma p}{2}M^2$$

That is,

$$p = \frac{2}{\gamma M^2}q \tag{7.1}$$

By isentropic relation, we have

$$p_0 = p\left(1 + \frac{\gamma - 1}{2}M^2\right)^{\frac{\gamma}{\gamma - 1}}$$

$$p_0 - p = p\left[\left(1 + \frac{\gamma - 1}{2}M^2\right)^{\frac{\gamma}{\gamma - 1}} - 1\right]$$

Now, replacing p on the right-hand side of the above equation by Equation (7.1), we get

$$p_0 - p = \frac{2q}{\gamma M^2}\left[\left(1 + \frac{\gamma - 1}{2}M^2\right)^{\frac{\gamma}{\gamma - 1}} - 1\right] \tag{7.2}$$

For air at standard conditions $\gamma = 7/5$, therefore, the above equation can be simplified as

$$p_0 - p = q\frac{2}{\gamma M^2}\left[\left(1 + \frac{M^2}{5}\right)^{\frac{7}{2}} - 1\right]$$

Binomial expansion of this equation, retaining terms only up to fourth power in M, results in

$$p_0 - p = q\left(1 + \frac{M^2}{4} + \frac{M^4}{40}\right)$$

For $M = 0$, this equation reduces to the Bernoulli equation for incompressible flow. From this equation, we get the dynamic pressure q as

$$q = \frac{p_0 - p}{K} \tag{7.3}$$

where $K = (1 + \frac{M^2}{4} + \frac{M^4}{40})$ is called the correction coefficient for dynamic pressure in compressible flow. Equation (7.3), for dynamic pressure, is accurate enough up to $M = 2$. For Mach numbers up to 1 even without fourth power term in M this equation proves to be reasonably accurate. For Mach numbers more than 2, Equation (7.2) itself should be used for proper estimation of dynamic pressure with a pitot-static tube. At $M = 1$, with $K = 1$ (i.e., without compressibility correction) the error in q with Equation (7.3) is 28 percent, but with $K = (1 + M^2/4)$, the error in q is only 2.5 percent. At $M = 2$, with $K = 1 + \frac{M^2}{4} + \frac{M^4}{40}$ the error in q is 4 percent. Beyond $M = 2$, use of Equation (7.3) results in large error in q.

Example 7.2

Calculate the dynamic pressure of a flow with $V_\infty = 200$ m/s, $p_\infty = 1$ atm, and $T_\infty = 300$ K. What will be the percentage error if the flow is treated as incompressible?

Solution

For a compressible flow, the dynamic pressure is given by

$$q_\infty = \frac{p_0 - p_\infty}{K}$$

where $K = (1 + \frac{M^2}{4} + \frac{M^4}{40})$. For the given flow, the freestream Mach number is

$$M_\infty = \frac{V_\infty}{a_\infty} = \frac{200}{\sqrt{1.4 \times 287 \times 300}}$$

$$= 0.576$$

Therefore,

$$q_\infty = \frac{p_0 - p_\infty}{1 + 0.083 + 0.00275}$$

$$= \frac{p_0 - p_\infty}{1.08575}$$

By isentropic relation, $p_0 = p_\infty (1 + \frac{\gamma-1}{2} M^2)^{\frac{\gamma}{\gamma-1}} = 1.252$ atm. Therefore,

$$q_{comp} = \frac{(1.252 - 1)}{1.08575} = \boxed{0.232\,\text{atm}}$$

$$q_{incomp} = p_0 - p_\infty = 0.252\,\text{atm}$$

Thus, the error involved in assuming the flow to be incompressible is,

$$\text{error} = \frac{q_{comp} - q_{incomp}}{q_{comp}} \times 100$$

$$= \boxed{-8.62\,\text{percent}}$$

The negative sign indicates that the error introduced is negative, i.e., the assumption results in overestimation of dynamic pressure.

7.7 Pitot-Static Tube Characteristics

The pressure distribution along the surface of the horizontal stem of a pitot-static tube, in the absence of vertical stem is something similar to the pressure distribution on a streamlined body of revolution as shown in Figure 7.20.

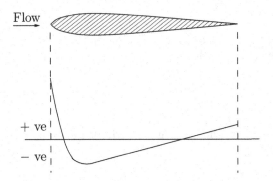

Figure 7.20: Pressure distribution on a streamlined body

For correct measurement of static pressure, the pressure tap must be located in the zone where the pressure acting at the probe surface is equal to the fluid stream static pressure. Hence, the appropriate position of the static hole will depend on the shape of the probe nose. The above argument is true only for those probes without a vertical stem. However, in most cases the static probes used are "L"-shaped, with a vertical stem. Therefore, it is essential to locate the measuring orifice, considering the effect of vertical stem also. The effects of nose shape and vertical stem on the position of static orifice was first studied by Ower and Johansen in 1925 (Ower and Pankhurst, 1977). They arrived at a balance of pressures due to nose and stem, as shown in Figure 7.21.

From the plot it is seen that, at about 6 diameters from the probe nose, the acceleration effect due to the horizontal portion and deceleration effect due to the vertical stem of the probe almost cancel each other. Therefore, 6 diameters from the nose may be taken as the appropriate location for the static pressure orifice. It can easily be seen that the position of pressure balance can be controlled by altering the distance between the stem and base of the nose, where the horizontal portion of the probe begins. Further, investigation in this line revealed that, if the vertical stem is located at 8 diameters behind the static pressure holes, the pressure at these holes will be the stream static pressure. This resulted in the National Physical Laboratory (NPL) standard hemispherical-nose pitot-static tube, shown in Figure 7.22. This consists of two quarter ellipses separated by a distance equal to the pitot-hole diameter d, the semi-major axis a, and the

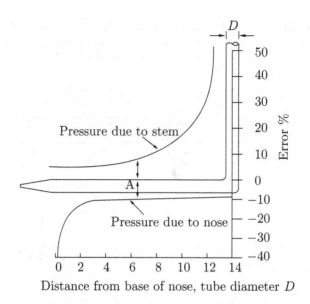

Figure 7.21: Pressures due to nose and stem of a static probe

semi-minor axis b being $2D$ and $(D - d)/2$, respectively, and the length of the nose equal to $2D$.

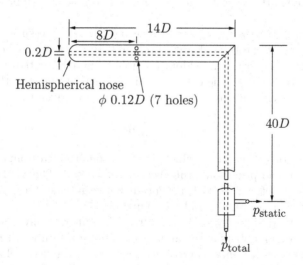

Figure 7.22: NPL standard pitot-static tube with hemispherical nose (for the normal size D = 8 mm)

The NPL pitot-static probe was treated as the standard for many years. The main cause for this may be that because of its hemispherical nose it is less prone to mechanical damage than the sharp nosed probe. Another popular pitot-static probe was that designed by Prandtl, shown in Figure 7.23.

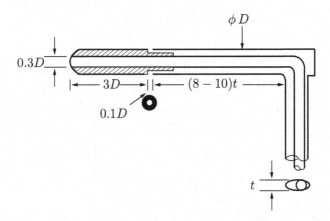

Figure 7.23: Pitot-static tube by Prandtl

Prandtl's probe is a round-nosed instrument with an annular slit instead of circular orifices for measuring static pressures. The slit has the advantage of symmetry and it is less likely to get blocked compared to small holes, in dust-laden gas streams. However, its sharp edges pose problems of reproducibility in manufacture.

A pitot-static tube is meant for measuring dynamic pressure, which is the difference between the total and static pressures. But in practice, because of the nature of static pressure, it is seldom possible to find a location for the static orifice, which will measure the correct static pressure. In most of the cases, the fabricated probe will indicate a pressure difference Δp given by

$$\Delta p = F\frac{1}{2}\rho V^2$$

where F is the correction factor which has to be determined from experimental calibration. A proper pitot-static tube is that with $F \approx 1$. The NPL and Prandtl probes have this factor close to 1, for speeds between 6 and 60 m/s, when they are made as per design (Ower and Pankhurst, 1977).

Following the designs by NPL and Prandtl, numerous investigations have been done by large numbers of researchers on pitot-static tube, with an emphasis to get F as close to unity as possible, over a wide range of velocities. Some such studies which resulted in the modified pitot-static probe design and which are widely accepted for practical applications are the following. It was pointed out by Kettle (1954) that tapered and hemispherical heads have pressure distributions with pronounced suction peaks near the nose, and that these could produce effects on the boundary layer along the tube which would vary with

Reynolds number and air stream turbulence. This high suction peak could cause separation of the boundary layer from the walls of the tube because of the steep adverse pressure gradient downstream of the suction peak. To eliminate this defect of the hemispherical and the tapered nose, Kettle designed an ellipsoidal head which had a much less pronounced suction peak near the nose over a wide range of flow speeds. A similar modified ellipsoidal nose was developed by Salter, Warsap, and Goodman (Ower and Pankhurst, 1977), with a diameteral section of the modified ellipsoidal nose consisting of two quarter ellipses separated by a distance equal to the pitot hole diameter, as shown in Figure 7.24.

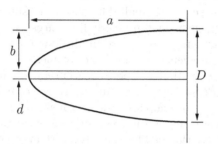

Figure 7.24: Modified ellipsoidal nose

This modified ellipsoidal nose has been adopted in the new NPL standard pitot-static tube, shown in Figure 7.25.

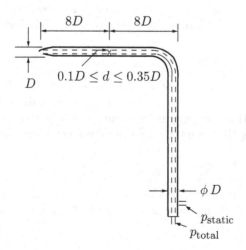

Figure 7.25: NPL pitot-static tube with modified ellipsoidal nose

This modified probe is less sensitive to slight imperfections in shape. Also, it has no scale effect on F over the speed range covered by the tests (13 to

60 m/s), for a tube of diameter about 8 mm. This corresponds to a Reynolds number range of about 6500 to 30,000.

7.7.1 Pitot-Static Tube Limitations

The practically constant value of the factor F of a pitot-static tube over a wide range of air speeds, along with the convenient reproducibility of the factor from one instrument to another of similar geometry made the instrument one of the most popular instruments for differential pressure or velocity measurements. However, like any instrument the pitot-static tube also has its limitations, especially in the measurement of small differential pressures experienced at low speeds. For measuring such small pressure differences, the requirement of ultra high sensitive manometers poses severe limitations on the use of pitot-static probes for such measurements. For instance, to measure the velocity head for an air speed of 0.6 m/s, within ± 1 percent accuracy, the manometer has to be sensitive to about 0.02 mm of water column. An instrument of this sensitivity has to be designed with special care. Such an instrument can only be used in laboratories. If we take 1 mm of water as a limiting sensitivity for a manometer, then the lowest air speed that can be measured with an accuracy of ± 1 percent, with pitot-static tube is about 4 m/s.

7.8 Factors Influencing Pitot-Static Tube Performance

The major factors influencing pitot-static tube measurements are turbulence, velocity gradient, viscosity, wall proximity, misalignment, and blockage effect.

7.8.1 Turbulence Effect

Even though the turbulent fluctuations of a flow have a definite effect on pitot-static probe readings, error introduced by this effect is generally small. Let u', v', and w' be the fluctuational velocity components and U be the mean velocity. Then we can write, using the incompressible Bernoulli equation, that

$$p_0 = p_s + \frac{1}{2}\rho U^2 + \frac{1}{2}\rho \left[\overline{u'^2} + \overline{v'^2} + \overline{w'^2} \right]$$

where p_0, p_s, and ρ, respectively, are the pitot pressure, static pressure, and density of the flow. The static pressure hole will read (Goldstein analysis; Ower and Pankhurst, 1977)

$$p_{st} = p_s + \frac{1}{2}\rho \left\{ \frac{1}{2} \left[\overline{v'^2} + \overline{w'^2} \right] \right\}$$

where p_{st} is the static pressure with turbulence and p_s is that without turbulence. Note that the second term on the right-hand side of the above equation lists the average of $\overline{v'^2}$ and $\overline{w'^2}$ only. This is because, $\overline{u'^2}$ is already absorbed in the p_{st} term. Assuming isotropic turbulence, we have, $\overline{u'^2} = \overline{v'^2} = \overline{w'^2}$. Therefore, we can write the Bernoulli equation as

$$p_0 = p_s + \frac{1}{2}\rho\left(U^2 + 3\overline{u'^2}\right)$$

and

$$p_{st} = p_s + \frac{1}{2}\rho\overline{u'^2}$$

$$p_0 - p_{st} = \frac{1}{2}\rho\left(U^2 + 2\overline{u'^2}\right)$$

It is seen from the above equation that the pitot-static probe will record $\frac{1}{2}\rho U^2\left(1 + 2\overline{u'^2}/U^2\right)$ instead of $\frac{1}{2}\rho U^2$. These results have been used extensively. The analysis presented here estimates only the first-order effects. To have a quantitative idea about the error introduced by turbulence, let us assume that, $\sqrt{\overline{u'^2}}/U = 0.10$, which is a reasonably high level of turbulence. Using this in the expression $\frac{1}{2}\rho U^2\left(1 + 2\overline{u'^2}/U^2\right)$, it can be seen that the error introduced by this turbulence is only 2 percent, which is not high for an experimental study. For a detailed analysis of turbulence on pitot-static probe measurements, including higher-order effects, the reader is encouraged to refer to literature specializing in this topic, like the studies of Barat (1958), and Bradshaw and Goodman (1968).

7.8.2 Velocity Gradient Effect

A small correction is essential for accuracy when there is a velocity gradient across the mouth of the pitot-static tube. For instance, measurements made in pipe flows experience this effect. In such situations, the effective center of the pitot tube is displaced from the geometric axis of the pipe towards the region of higher velocity, and hence the deduced local velocity at the true center of the pipe will be too high. However, for pitot tubes with sharp-edged conical noses the effective displacement is found to be negligible (Ower and Pankhurst, 1977).

For boundary layer studies, because the region in which measurements have to be made is thin, usually flattened pitot-tubes, with a mouth as shown in Figure 7.26 are employed.

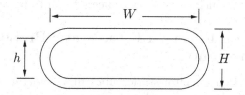

Figure 7.26: Flattened pitot-tube mouth

The displacement effect for such tubes, with h/H equal to 0.6 and W/H about 1.6, was found to be (Quarmby and Das, 1969)

$$Z/H = 0.19 \pm 0.01$$

where Z is the direction normal to the solid boundary. Because of the accompanying effects of turbulence, of viscosity when Reynolds number is low, and of wall proximity near a solid boundary, precise determination of displacement effect due to velocity gradient is difficult. However, the displacement effect is generally small and in practice this effect can be made negligible by choosing a probe of proper size.

7.8.3 Viscosity Effects

At low Reynolds numbers pitot-static tube measurements are severely influenced by viscosity. Figure 7.27 shows the effect of viscosity on pitot-static tube measurements. The errors due to low Reynolds number are illustrated in Figure 7.27, which shows the variation of pressure coefficient as a function of Reynolds number and probe W/H.

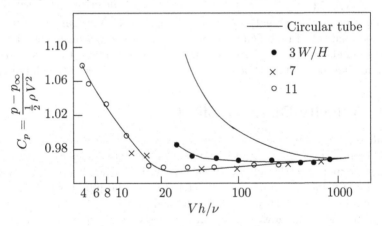

Figure 7.27: Viscous effect error for small pitot tubes

As seen from Figure 7.27, this error can be avoided by choosing bigger size tubes for pitot-static probes, thereby keeping the tube Reynolds number above the limiting value at which the viscosity effects begin to become significant.

7.8.4 Vibration Effect on Pitot-Static Probe Measurements

The effect of vibration on pitot-static tubes may sometimes be considerable to introduce significant errors in the measurements. These vibrations can be due to aerodynamic effect, such as vortex shedding, or mechanical reasons. The movement of the probe head and hence magnitude of the error will be the largest when there is resonance between the exciting frequency and the natural frequency of the probe. These vibrations can be both along the wind direction and in a transverse plane. *Aerodynamically excited vibrations are mainly of the transverse type, and are due to periodic shedding of vortices from alternate sides of the cylindrical stem of the probe.* The relation between the period of vortex shedding, probe stem diameter, and flow speed can be expressed as (Ower and Pankhurst, 1977)

$$V = 4.72\, nd + \frac{c\nu}{d}$$

where V is flow speed, n is vortex shedding frequency, d is tube diameter, c is a constant and has values of 21.2 and 12.7 for the Reynolds number ranges of 50–150 and 300–2000, respectively, and ν is kinematic viscosity. From this it follows that, for a probe of 8-mm diameter, the vortex shedding frequency, which is nearly proportional to the flow speed, is about 40 Hz at 1.5 m/s and about 160 Hz at 6 m/s.

The vibrations transmitted to the probe from *mechanical sources is likely to cause more problems than those due to vortex shedding*, especially when they make the probe head vibrate along the wind direction. If the vibration is assumed to be simple harmonic, the displacement x of the probe head at any instant may be written as $a \sin qt$, where a is the maximum amplitude of vibration and q is equal to $2\pi f$, where f is frequency of vibration. The velocity due to vibration is $aq \cos qt$, and the axial speed of pitot head relative to air speed is $V + aq \cos qt$. The corresponding dynamic pressure is

$$\frac{1}{2}\rho(V + aq\, \cos\, qt)^2$$

The average dynamic pressure p_{dyn} over a cycle is

$$p_{\mathrm{dyn}} = \frac{q}{2\pi} \int_0^{2\pi/q} \frac{1}{2}\rho(V + aq\, \cos\, qt)^2 dt$$

That is,

$$p_{\mathrm{dyn}} = \frac{1}{2}\rho\left(V^2 + \frac{a^2 q^2}{2}\right)$$

If f is the frequency of vibration, $q = 2\pi f$. It follows from this that for high frequencies and very large amplitudes of vibration the term aq may not be negligible compared to V, and the errors introduced by this vibration on pitot-static probe measurement may be significant. Therefore, probe vibration should be eliminated, for error free measurements, by either stiffening it or with suitable stay wires or other bracing which does not disturb the flow significantly.

7.8.5 Misalignment Effect

The pressure measured by a pitot or static probe will be in error if it is not aligned with the flow direction. But for small angles of misalignment the error is generally small, especially for pitot probes, which are far less sensitive to this effect than the static probes. For probes with vertical stem, angular deviations in the plane containing both the head and stem, namely *the pitch*, give slightly different results from the corresponding deviations about the axis of the stem termed *yaw*. Let us see this effect on the probes in incompressible and compressible flows separately in the following sections.

7.8.6   Blockage Effect

If the ratio of the flow field area, A_{flow}, to the probe projected area, A_{probe}, normal to the flow direction is more than 64, the blockage of the probe will not cause any appreciable change to the flow properties. Therefore, the limiting minimum size of the probe below which the area blockage due to the probe will not introduce any significant error in the measurement is $A_{\text{flow}}/A_{\text{probe}} \geq 64$ (Chiranjeevi Phanindra and Rathakrishnan, 2010).

7.9   Pitot Probes

A typical variation of the pitot pressure with yaw angle is shown in Figure 7.28. The ratio of pitot pressures measured with yaw (p_ψ) and without yaw (p_0) is presented as a function of yaw angle ψ.

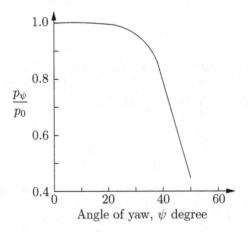

Figure 7.28: Effect of yaw on pitot pressure measurements

It is seen from this figure that pitot tube is *insensitive to yaw up to about 20 degrees*. For instance, at 20 degrees yaw the pressure measured is only about 1 percent less than that at zero yaw. This feature seems to be common for different types of pitot heads. In fact, pitot probes which are insensitive to about $\pm$ 63 degrees yaw also have been developed (Ower and Pankhurst, 1977).

7.10   Static Probes

As stated earlier, static probes are more sensitive to angular deviation than the pitot probes. The results of yaw effect on two NPL probes are shown in Figure 7.29.

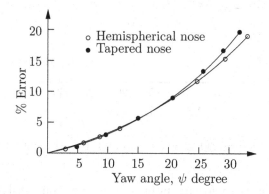

Figure 7.29: Yaw effect on static probe

It is seen from this figure that the sensitivity of static probe to yaw is well pronounced even at very small yaw angles.

7.11   Pitot-Static Probes

Having seen the effect of yaw on pitot and static probes independently, it may be useful to examine the effect of yaw on their combination, in the form of pitot-static tube. Such results are shown in Figure 7.30.

It is seen that the hemispherical nosed probe is preferable if the direction of flow is uncertain up to 30 degrees. The maximum error introduced by yaw in this range would be less than 5 percent of the dynamic pressure $\frac{1}{2}\rho V^2$ for hemispherical nose. If the direction of flow is known to be within 15 degrees, the ellipsoidal nose is more appropriate than the hemispherical or tapered nose.

7.12   Yaw Effect in Compressible Flow

The appreciable insensitivity of the pitot probe reading to misalignment in incompressible flow continues in compressible flow also, right up to supersonic regimes. That is, the probe can be set with sufficient accuracy for most measurements by eyesight. For supersonic flows, the yaw insensitivity is usually defined as *the angular misalignment at which the pitot reading has changed by 1 percent of the difference between the zero yaw pitot and static pressures of the undisturbed flow.* In incompressible flow this is 1 percent of the kinetic pressure of undisturbed flow. With the above definition (Ower and Pankhurst, 1977) the

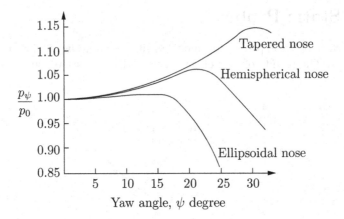

Figure 7.30: Yaw effect on pitot-static probe

value of yaw insensitivity for any pitot probe can be comfortably taken as ±11 degrees for compressible flow.

7.13 Static Pressure Measurement in Compressible Flows

Starting from incompressible regime of flow, up to Mach 0.8 a static pressure probe aligned with the flow direction continues to record the freestream static pressure correctly. Once the freestream Mach number reaches a level such that the flow near the probe becomes transonic, the pressure reading is severely affected by the waves generated. For further increase of freestream Mach number, there are shock wave formations on the probe and the pressure measured by the probe becomes something totally different from the actual value. Therefore, locating the static pressure orifice at proper location for measuring static pressure with reasonable accuracy poses serious problems in the probe design. In general, static probes with measuring orifice located sufficiently far downstream is suitable for measurement of static pressure in supersonic flows. A typical probe, shown in Figure 7.31, measures static pressure within ± 0.5 percent at Mach 1.6 (Ower and Pankhurst, 1977), if the pressure tapings are located more than 10 tube diameter downstream of the shoulder. This distance increases with increase of Mach number, in the supersonic regime.

The yaw sensitivity of this probe with string support is about 1 percent reduction in the measured pressure for 3 to 5 degrees yaw. If it is used with a stem support, the support should be at least 13 tube diameters away from the static orifice.

Even though investigations have been done on supersonic static probes, we can see from the fundamental nature of the supersonic stream that it is a flow

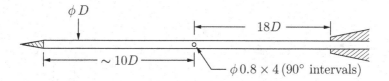

Figure 7.31: Supersonic static probe

stream dominated by waves in the form of Mach waves, expansion waves, and shock waves. It may be possible to develop a static probe for a particular Mach number, but development of a static probe for use at all supersonic Mach numbers with reasonable accuracy will prove to be a futile task. Therefore, in practice one can make use of the boundary walls of nozzles, like, Laval nozzle, which develops supersonic flow - this is mostly the case in the laboratories - to locate an orifice (on the nozzle wall) for measuring stream static pressure. For instance, a pressure tap located, as shown in Figure 7.32, at the wall is capable of measuring the stream static pressure almost exactly, provided the nozzle is properly made with smooth inner surfaces to generate a shock-free supersonic stream.

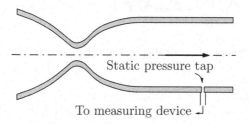

Figure 7.32: Laval nozzle

However, if the static pressure in a free jet issuing from a supersonic nozzle has to be measured, then one has to design and fabricate special probes for that purpose. One such special probe is the disc-probe, illustrated in Figure 7.33.

The disc probe shown in Figure 7.33 has a thin circular disc supported by a stem. The top surface of the disc is smooth and has a static pressure orifice located at its center. The probe has to be aligned to have its top flat surface perfectly in line with the flow. Such an arrangement will ensure a shock-free supersonic flow over the top surface and hence the static pressure measured may be taken as the correct pressure.

Static pressure in supersonic flows may also be measured by employing a knife-edge flat plate, shown in Figure 7.34. Here again the principle of measurement is the same as that of the disc probe. Like the disc-probe, the leading edge of the plate must be thin and sharp to ensure only an attached shock wave and

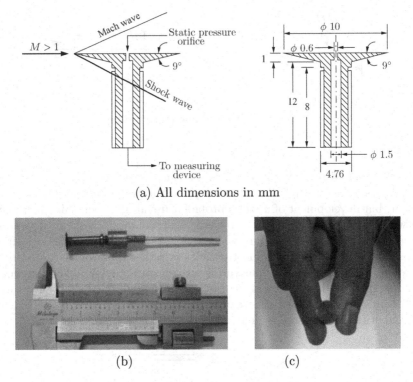

(a) All dimensions in mm

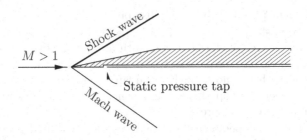

(b)   (c)

Figure 7.33: (a) Schematic of disc-probe; (b) and (c) views of disc-probe

a Mach wave formation at its leading edge, as shown in Figure 7.34. In other words, formation of detached shock should be avoided. This kind of probe does not disturb the flow much and even the mild disturbances created by the probe will propagate only in the downstream direction, since the flow is supersonic. Hence, the freestream flow is totally unaffected by probe presence. Therefore, measured pressure will be the actual static pressure of the stream, provided the probe is properly aligned.

Figure 7.34: Flat-plate static probe

Now we should realize that, even though it is easy to make the orifice of the disc or flat plate probes to sense near exact static pressure of the supersonic stream, these probes, shown in Figure 7.33 and Figure 7.34, can be placed in the stream only with supports. Therefore, care must be taken to ensure that the supporting structures also do not disturb the flow much. For instance, supports should not have blunt edges, since such edges will result in the formation of detached shocks in supersonic flows. Considering all disturbances due to the complete setup of the probe, it must be ensured that the Mach wave at the leading edge of the probe is the only wave upstream of the static hole. If this is achieved, the static pressure measured by such a probe can be regarded as the actual static pressure of the stream.

Example 7.3

A pitot probe has to be designed for measurement in a Mach 2 air stream. (a) If the probe blockage should not cause more than 1 percent error to the flow Mach number, what should be the ratio of the flow field area to the probe projected area? (b) What will be the error in the flow Mach number measured, due to this probe blockage, in Mach 3, 1.5, and 1.2 flows of the same flow cross-sectional area as the Mach 2 flow field? (c) From these error estimates will it be possible to identify the probe blockage which will cause only acceptable error to the Mach number over the entire range of supersonic flow beginning from Mach 1.5?

Solution

(a) Given, $M = 2$. Let A be the flow field area and A_b be the net area of the flow field, after accounting for the probe blockage. In a supersonic flow, decrease of area would cause decrease of flow Mach number. Let M_b be the Mach number of the flow when the probe is kept in the flow. The decrease of flow Mach number should not be more than 1 percent of the actual Mach number M.

For $M = 2$,

$$M_b = 0.99 \times M$$

$$= 0.99 \times 2$$

$$= 1.98$$

By Equation (2.83),

$$\text{for } M = 2, \quad \frac{A}{A^*} = 1.688$$

$$\text{for } M_b = 1.98, \quad \frac{A_b}{A^*} = 1.66$$

Thus the projected area of the probe is

$$\frac{A}{A^*} - \frac{A_b}{A^*} = 1.688 - 1.660$$

$$= 0.028$$

Therefore,

$$\frac{A_{\text{flow}}}{A_{\text{probe}}} = \frac{1.688}{0.028}$$

$$= \boxed{60.29}$$

(b) For $M = 3$,

$$M_b = 0.99 \times M$$

$$= 0.99 \times 3$$

$$= 2.97$$

The area ratios for these Mach numbers are

$$\text{for } M = 3, \ \frac{A}{A^*} = 4.235$$

$$\text{for } M_b = 2.97, \ \frac{A_b}{A^*} = 4.115$$

Therefore,

$$\frac{A_{\text{flow}}}{A_{\text{probe}}} = \frac{4.235}{4.235 - 4.115}$$

$$= \boxed{35.3}$$

If the error caused by the probe blockage to the flow Mach number, M, has to be less than 1 percent, the area ratio has to be

$$\frac{A_{\text{flow}}}{A_{\text{probe}}} = 35.3$$

When $\frac{A_{\text{flow}}}{A_{\text{probe}}} = 60$, the area ratio becomes

$$\frac{A_b}{A^*} = 4.235 - \frac{4.235}{60}$$

$$= 4.164$$

For $\dfrac{A_b}{A^*} = 4.164$,

$$M_b = 2.982$$

Therefore, the error in M is

$$\text{Error} \;=\; \frac{3 - 2.982}{3} \times 100$$

$$=\; \boxed{0.6 \,\text{percent}}$$

For $M = 1.5$,

$$M_b \;=\; 0.99 \times M$$

$$=\; 0.99 \times 1.5$$

$$=\; 1.485$$

For these Mach numbers, the corresponding area ratios are

$$\text{for } M \;=\; 1.5, \; \frac{A}{A^*} = 1.176$$

$$\text{for } M_b \;=\; 1.485, \; \frac{A_b}{A^*} = 1.166$$

Therefore,

$$\frac{A_{\text{flow}}}{A_{\text{probe}}} \;=\; \frac{1.176}{1.176 - 1.166}$$

$$=\; \boxed{117.6}$$

For error in M caused by probe blockage to be less that 1 percent, the area ratio is

$$\frac{A_{\text{flow}}}{A_{\text{probe}}} = 117.6$$

For $\frac{A_{\text{flow}}}{A_{\text{probe}}} = 60$, the area ratio becomes

$$\frac{A_b}{A^*} \;=\; 1.176 - \frac{1.176}{60}$$

$$=\; 1.1564$$

For this area ratio,

$$M_b \approx 1.47$$

Therefore, the error in M is

$$\text{Error} \quad = \quad \frac{1.5 - 1.47}{1.5} \times 100$$

$$= \quad \boxed{2.0\,\text{percent}}$$

For $M = 1.2$,

$$M_b \quad = \quad 0.99 \times M$$

$$= \quad 0.99 \times 1.2$$

$$= \quad 1.188$$

Area ratio for these Mach numbers are

$$\text{for} \quad M \quad = \quad 1.2, \quad \frac{A}{A^*} = 1.030$$

$$\text{for} \quad M_b \quad = \quad 1.188, \quad \frac{A_b}{A^*} = 1.02524$$

Therefore,

$$\frac{A_{\text{flow}}}{A_{\text{probe}}} \quad = \quad \frac{1.030}{1.030 - 1.02524}$$

$$= \quad \boxed{216.39}$$

When $\frac{A_{\text{flow}}}{A_{\text{probe}}} = 60$, the area ratio becomes

$$\frac{A_b}{A^*} \quad = \quad 1.030 - \frac{1.030}{60}$$

$$= \quad 1.0128$$

For this area ratio,

$$M_b = 1.13$$

Therefore, the error in M is

$$\text{Error} \quad = \quad \frac{1.2 - 1.13}{1.2} \times 100$$

$$= \quad \boxed{5.83\,\text{percent}}$$

(c) For the range of Mach number from 1.5 to 3, $\frac{A_{flow}}{A_{probe}} = 60$ causes error in the range of 0.6 to 1.3 percent of the actual Mach member. Therefore, the probe blockage which can be used to measure Mach number in flows of supersonic Mach number 1.5 and above is

$$\text{Blockage} = \frac{1}{60} \times 100 = \boxed{1.67 \text{ percent}}$$

Note that, this blockage causes 5.83 percent error at Mach 1.2. Thus, in the transonic range of Mach number, the flow blockage should be less than 1.67 percent for proper measurement. But for supersonic flow of Mach number above 2, blockage ratio above 60 will result in less than 1 percent error in the Mach number measured.

Thus it may be stated that the ratio of the flow field area to probe projected area, $\frac{A_{flow}}{A_{probe}} \geq 64$, i.e. $\frac{A_{flow}}{A_{probe}} \geq 8^2$ will cause only less than 1 percent error in the flow Mach number measured for flow fields of Mach numbers above 2.

7.14   Determination of Flow Direction

Determination of flow direction becomes essential in many problems. For instance, in measurement of wind speed with a pitot-static probe the probe axis has to be aligned in line with the local flow direction, within a few degrees to achieve the required accuracy in the measurement. The instruments used for determining the flow direction are termed *yaw probes*. These yaw probes are generally used for determining the flow direction as well as its magnitude. An experimenter, depending on her/his imagination can think of any number of yaw probe designs. Let us see some of the popular yaw probes.

7.14.1   Yaw Sphere

A yaw sphere, like the one shown in Figure 7.35, can be used to determine the flow direction. For a two-dimensional flow, just two holes are sufficient to measure the flow angularity. But, for three-dimensional flow four holes are necessary since both yaw and pitch of the flow have to be measured simultaneously to determine the flow direction. The measuring procedure for yaw head of a flow, say in a two-dimensional wind tunnel test-section, is the following.

- Place a yaw sphere having orifices 90 degrees apart on the forward part, as shown in Figure 7.35, in the flow field.

- Adjust the instrument axis so that both the orifices A and B read the same pressure ($p_A = p_B$). Measure the angle between the tunnel axis and instrument axis as the flow angularity.

It is difficult to measure small angles by this method. An alternate procedure for yaw measurement is to align the instrument axis with respect to tunnel axis

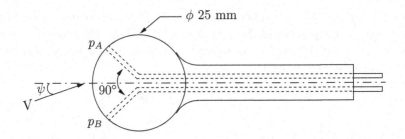

Figure 7.35: Yaw sphere

and note the pressure difference $(p_A - p_B)$. The instrument may be calibrated by standard experiments for yaw head, defined as $(\Delta K/\Delta \psi)$, where ψ is the yaw angle and $\Delta K = \Delta p/q$, Δp is the pressure difference and q is dynamic pressure of the flow. A calibration curve as shown in Figure 7.36 may be plotted, to be used as a standard for the instrument.

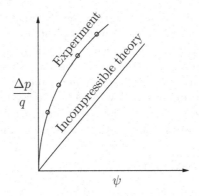

Figure 7.36: Yaw sphere calibration curve

Usually the yaw head varies from 0.04 to 0.07 per degree of yaw. Therefore, for each instrument the yaw head should be determined experimentally.

7.14.2   Claw Yaw Probe

The claw-type yaw meter, as shown in Figure 7.37, works on the same principle as that of yaw sphere. It is used to measure the rotation and direction of flow near a model at any point, because of its slender nature and low blockage.

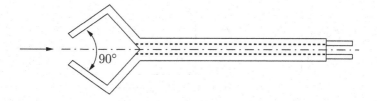

Figure 7.37: Claw yaw probe

7.14.3 Three–Hole and Five–Hole Yaw Probes

The three-hole and five-hole probes, shown in Figure 7.38, are used for measuring the flow direction in two-dimensional and three-dimensional flows, respectively.

The three-hole yaw probe is provided with a stagnation pressure hole C between the two-dimensional holes A and B, located at 45 degrees on either side of hole C. When the pressures measured by the holes A and B are equal, the tube is considered to be aligned to the direction of the flow. A properly made probe of this type is capable of measuring yaw angles up to an accuracy of 0.5 degree. The *cylindrical portion of the probe should extend at least three diameters downstream of the pressure taps*, in order to eliminate the three-dimensional effects due to the blunt nose of the probe. The pressures p_A and p_B can be calibrated for velocity at null position. That is, we may express

$$p_C = p_A + \frac{1}{2}\rho V^2 K$$

where V is the flow velocity and K has to be determined experimentally.

The measuring principle of a five-hole probe is similar to a three-hole probe except that it is *capable of measuring yaw as well as pitch of the flow*. The tube should have provision to rotate it in two directions, i.e., in yaw and pitch. By using a single calibration curve, as shown in Figure 7.38, the flow direction can be determined.

7.14.4 Cobra Probe

The three- and five-hole probes described above cannot be used to get flow direction near any solid surface, because of their large size. Therefore, based on the same principle, probes which are small in size can be fabricated for measurements near the wall region of any model. The cobra probe, shown in Figure 7.39, is one such special probe. This is a two-dimensional probe similar to three-hole yaw probe. Tubes A, B, C are just soldered together and then A and B are ground or chamfered to 45 degrees.

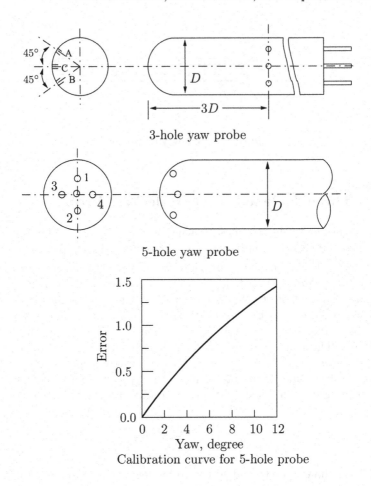

3-hole yaw probe

5-hole yaw probe

Calibration curve for 5-hole probe

Figure 7.38: Three- and five-hole yaw probes

7.15   Low–Pressure Measurement

Low-pressure gauges are usually referred to as *vacuum gauges*. Measurement of low-pressures requires considerable care on the part of the experimentalist. Here in this section we will see some of the popular types of vacuum instruments and the physical principle upon which they operate. For more specialized information about low-pressure measurements one may refer to books like the monograph by Dushman and Lafferty (1962).

For moderate vacuum measurements the manometers, dial gauges, and diaphragm gauges may be employed with reasonable accuracy. However, our aim here is to discuss the measurement of low pressures which cannot be measured by the conventional gauges. In other words, we are primarily concerned with the measurement of absolute pressures below 1 torr (1 mm of Hg or 133 Pa).

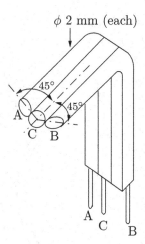

ϕ 2 mm (each)

Figure 7.39: Cobra probe

The *McLeod gauge, Pirani gauge, Knudsen gauge, ionization gauge,* and *alphatron gauge* are some of the commonly used pressure gauges for low-pressure measurements.

7.15.1 The McLeod Gauge

The McLeod gauge is used as a standard for measuring low vacuum pressures. It is basically a mercury manometer. A typical McLeod gauge is shown schematically in Figure 7.40. The measuring procedure of a McLeod gauge is as follows. The mercury reservoir is lowered until the mercury column drops below the opening O. Now the bulb B and the capillary C are at the same pressure as the vacuum space p. The reservoir is raised until the mercury fills the bulb and rises in the capillary up to a level at which the mercury in the reference capillary is at zero level marked on it. Let the volume of the capillary per unit length be "a."

The volume of the gas in the capillary, $\mathbb{V}_C$, becomes

$$\mathbb{V}_C = ay \tag{7.4}$$

where y is the gas column length in the capillary.

Let the volume of capillary, bulb, and the tube down to the opening be $\mathbb{V}_B$. Assuming the compression of the gas in the capillary to be isothermal, we can write

$$p_C = p\mathbb{V}_B/\mathbb{V}_C \tag{7.5}$$

where p_C is pressure of the gas trapped inside the capillary volume $\mathbb{V}_C$. The pressure indicated by the capillary is

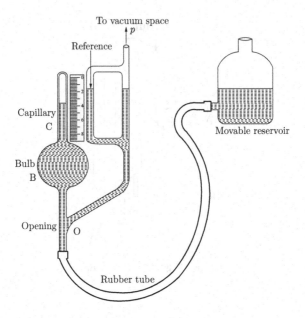

Figure 7.40: The McLeod gauge

$$p_C - p = y \qquad (7.6)$$

Now, it is essential to note that, in Equation (7.6), *the pressure is expressed in terms of mercury column height.* That is, y on the right-hand side of Equation (7.6) is in pressure units. Combination of Equations (7.4) and (7.6) results in

$$p = \frac{ay^2}{\mathbb{V}_B - ay} = \frac{y\mathbb{V}_C}{\mathbb{V}_B - ay} \qquad (7.7)$$

Assuming $ay \ll \mathbb{V}_B$, which is the case mostly, we have

$$\boxed{p = \frac{ay^2}{\mathbb{V}_B}} \qquad (7.8)$$

Usually McLeod gauges have the capillary calibrated in terms of micrometers. The gauge can be used for measuring low pressures in the range from 10^{-2} to 10^2 μm of mercury, i.e., from 0.00133 to 13.3 Pa, for dry gases. The McLeod gauge is very sensitive to vapors which may be present in the sample, since vapor can condense upon compression. For such cases Equation (7.7) will not be valid.

Example 7.4

A McLeod gauge with $\mathbb{V}_B = 100$ cm^3 and capillary diameter 1 mm measures the pressure of a vacuum chamber as 10 mm of mercury. Calculate the chamber pressure in Pa.

Solution

$$\mathbb{V}_C = \frac{\pi 1^2}{4} \times 10 = 7.85 \text{ mm}^3$$

$$\mathbb{V}_B = 10^5 \text{ mm}^3$$

By Equation (7.7), we have

$$p = \frac{10 \times 7.85}{10^5 - 7.85} = 0.000785 \text{ torr}$$

$$= 0.785 \,\mu \text{ m}$$

$$= 0.785 \times 10^{-3} \text{ mm of Hg}$$

But 1 mm of Hg is $(1/760) \times 101325$ Pa, thus,

$$p = \frac{0.785 \times 10^{-3}}{760} \times 101325$$

$$= \boxed{0.105 \text{ Pa}}$$

7.15.2 Pirani Gauge

The Pirani gauge is a device that measures pressure through *the change in thermal conductance of the gas.* The fact that the effective thermal conductivity of a gas decreases at low pressures is made use of in the measurement of pressures by Pirani gauge. The Pirani gauge consists of an electrically heated filament placed inside a vacuum space, as shown in Figure 7.41.

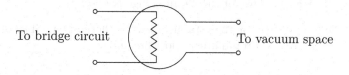

To bridge circuit   To vacuum space

Figure 7.41: Schematic of Pirani gauge

The heat loss from the filament depends on the thermal conductivity of the gas in the vacuum space and the temperature of the filament. The thermal conductivity of the filament decreases with decrease in pressure and hence for a given electric energy the filament temperature increases with decrease of pressure. The filament temperature is measured from the variation of the resistance of the filament material. Usually tungsten or platinum is used as the filament. The heat loss from the filament is influenced by the ambient temperature also. Therefore, to compensate for the loss of heat due to ambient temperature variations, two gauges are connected in series, as shown in Figure 7.42. The measurement gauge is evacuated, and both the measuring and the compensation gauge are exposed to the same environment conditions. The bridge is adjusted to null position, using the resistance R_2. Now the gauge is compensated for the ambient temperature and is ready for measurement.

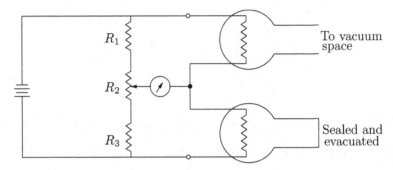

Figure 7.42: Pirani gauge circuit with compensation for ambient temperature variation

Pirani gauges are suitable for measurement of pressures in the range from 1 μm to 1 torr. For pressures higher than 1 torr the thermal conductance of the filament changes very little with pressure and hence the sensitivity of the instrument becomes poor. Even though simple to construct, the Pirani gauge has the following disadvantages.

- It requires empirical calibration and is not generally suitable for measuring pressures below 1 μm.

- The transient response of the Pirani gauge is poor.

- The time necessary for the establishment of thermal equilibrium may be of the order of several minutes at low pressures.

7.15.3 Knudsen Gauge

The Knudsen gauge is capable of measuring low pressures in the range between 10^{-5} m and 10 μm of Hg. This is used as a calibration device for other gauges meant for pressure measurement in this range. A typical Knudsen gauge has an arrangement as shown in Figure 7.43.

As shown in the figure, two vanes V along with mirror M are mounted on a thin filament suspension. Two heated plates P, both maintained at the same

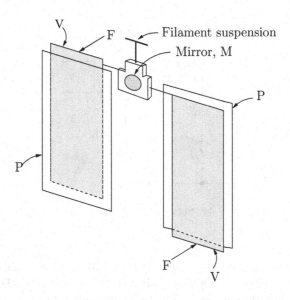

Figure 7.43: Schematic of Knudsen gauge

temperature T, are kept near the vanes. The gap between the plates and the vanes is less than the mean free path of the surrounding gas. The temperature of the plates is always kept higher than the surrounding gas temperature T_g. The vanes are at the temperature of the gas. Because of this difference in temperature between the plates and vanes, the gas molecules striking the vanes from the hot plates have a higher velocity than those leaving the vanes. Thus, there is a net momentum imparted to the vanes which may be measured from the angular displacement of the mirror. The total momentum exchange with the vanes is a function of molecular density, which, in turn is related to the pressure and temperature of the gas. Thus, an expression for the gas pressure may be derived in terms of the temperatures and measured force. For small temperature difference $(T-T_g)$, it may be shown that this relation is (Dushman and Lafferty, 1962)

$$p = 4F \frac{T_g}{T - T_g}$$

where the pressure is in dynes per square centimeter when the force is in dynes(1 dyne $= 10^{-5}$ newton) and the temperature is in kelvin.

7.15.4   Ionization Gauge

The ionization gauges are capable of measuring pressures as low as 10^{-12} torr. This device has an arrangement as shown in Figure 7.44.

The heated cathode emits electrons, which are accelerated by the positively charged grid. As the electrons move towards the grid, they produce ionization

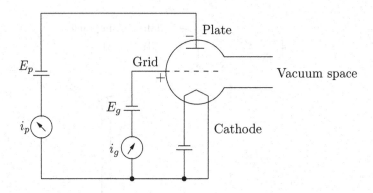

Figure 7.44: Schematic of ionization gauge

of the gas molecules through collisions. The plate is maintained at a negative potential so that the positive ions are collected there, producing the plate current i_p. The grid generates current i_g, due to the collision of electrons and negative ions. The pressure of the gas is found to be proportional to the ratio of plate current to grid current. That is,

$$p = \frac{1}{S} \frac{i_p}{i_g}$$

where the proportionality constant S is called the *sensitivity* of the gauge. A typical value of S for nitrogen is approximately 20 torr^{-1}, but the exact value must be determined by calibration of the particular gauge. The value of S is a function of the type of gas and the tube geometry. While using an ionization gauge, it must be ensured that the gauge is not exposed to gases at pressures higher than 1 μm, since at higher pressures there is a danger of burning out the cathode.

In this section on low-pressure measurements, only a few of the pressure measuring devices which are popularly employed have been described. For further details about this kind of instrument one may refer to books specializing in this topic, such as *Scientific Foundations of Vacuum Techniques* by Dushman and Lafferty (1962).

7.16   Preston/Stanton Tubes

The Preston and Stanton tubes are essentially small impact (pitot) tubes meant for the indirect measurement of skin friction, using some well-established mean velocity correlations to infer mean shear stress at the wall.

As we know, normal and tangential stresses are exerted on a wall by a flowing fluid. The measurement of normal stress, mainly pressure, was discussed in detail in different sections of this chapter. In this section let us see some of the popularly employed methods for measuring the tangential stress. Determination of the wall shear stress (skin friction) in laminar, transitional, or turbulent boundary layers is important both from practical and fundamental points of view. The local, time-averaged value of the wall shear could be integrated spatially to obtain the skin friction drag of a body moving in a fluid. In transitional and turbulent flows, both instantaneous pressure and skin fiction are the causes for momentum transfer to the wall. Therefore, statistical analysis of the pressure and wall shear stress data will be of high value to understand this momentum transfer process.

7.17   Sound Measurements

7.17.1   Introduction

Like any science, the discipline of sound has a unique vocabulary associated with it. Precise definition of the basic concepts forms the foundation for the development of a science and prevents possible misunderstandings. Therefore, let us first acquaint ourselves with the unit systems that will be used in sound measurement and the basic concepts of sound propagation, before getting into the details of sound measurement.

7.17.2   Sound and Noise

Sound may be defined as *any pressure variation* (in air, water, or other medium) *that the human ear can detect. Unpleasant or unwanted sound is termed noise.* The most familiar instrument for measuring pressure variations in air is the barometer. However, the pressure variations which occur with changing weather conditions are much too slow for the human ear to detect and hence do not meet our definition of sound. But if variations in atmospheric pressure occur more rapidly (at least 20 times per second) they can be heard and hence called sound. Barometers cannot respond quickly enough to this kind of rapid change and therefore cannot be used to measure sound.

The *frequency* of sound is defined as the number of pressure variations per second. The frequency is measured in *hertz* (Hz). The frequency of a sound produces its distinctive *tone*. For example, the rumble of distant thunder has a low frequency, while a whistle has a high frequency. The normal range of hearing for a healthy person extends from approximately 20 Hz up to 20,000 Hz. The range from the lowest to highest note of a piano is from 27.5 Hz to 4186 Hz. These limits are subjective and will vary slightly from person to person. An average human ear is not able to hear sound if the frequency is outside this range. Electronic detectors can detect waves of lower and higher frequencies as well. A dog can hear sound of frequency up to about 50 kHz and a bat up to about 100 kHz. The waves with frequency below human audible range are called *infrasonic waves* and those with frequency above the audible range are called *ultrasonic.*

Example 7.5

A sound wave of wavelength 0.5 cm is produced in air. If it travels at a speed of 340 m/s, will it be audible?

Solution

The velocity of sound travel is given by

$$V = f\lambda$$

where f is the frequency and λ is the wavelength. Thus,

$$f = \frac{V}{\lambda} = \frac{340}{0.005} = 68,000 \, \text{Hz}$$

This is well above the audible upper limit of 20,000 Hz. It is an ultrasonic wave and will not be audible.

These pressure variations travel through any elastic medium (such as air) from the sound source to the listener's ears. We know that, at room temperature the sound travels at a speed of about 340 m/s. The *wavelength*, defined as the distance from one wave top or pressure peak to the next, as shown in Figure 7.45, can be calculated if the speed and the frequency of a sound are known.

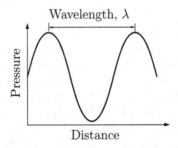

Figure 7.45: Sound pressure propagation

$$\text{Wavelength} \, (\lambda) = \frac{\text{Speed of sound}}{\text{Frequency}}$$

From this relation, the wavelength at different frequencies of propagation of sound can be obtained. For example, at 20 Hz frequency one wavelength of sound is (340/20) 17 m, while at 20 kHz, it is just 1.7 mm, at room temperature. Thus, high-frequency sounds have short wavelengths and low-frequency sounds have long wavelengths, as shown in Figure 7.46.

7.17.3   Pure Tone

When a sound has only one frequency it is referred to as a *pure tone*. However, in practice pure tone never exists. Most sounds are made up of different frequencies.

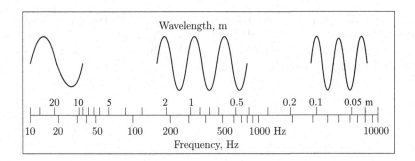

Figure 7.46: Wavelength at different frequencies

For instance, even a single note on a piano has a complex wave form.

7.17.4 Broadband and White Noise

Broadband noise is a noise consisting of a mixture of a wide range of frequencies. Most industrial noise is broadband noise.

White noise is a noise having frequencies evenly distributed throughout the audible range. White noise will sound like rushing water.

7.17.5 Sound Units

One of the main characteristics used to describe a sound is the size or *amplitude* of the pressure fluctuations. The weakest sound a healthy human ear can detect has an amplitude of $20\,\mu$Pa (i.e., millionth of a pascal). Note that $20\,\mu$Pa is about 5×10^9 times less than normal atmospheric pressure. A pressure change of $20\,\mu$Pa is so small that it causes the eardrum to detect a wavelength less than the diameter of a single hydrogen molecule. Surprisingly the ear can tolerate sound pressures more than a million times higher than the hearing threshold. Thus, if sound is measured in Pa it gives some quite large, unmanageable numbers. To overcome this situation, another scale, namely the *decibel*, dB, scale is used.

The *decibel* is a ratio between a measured quantity and an agreed reference level. Thus, decibel is not an absolute unit of measurement. The reference level used is the hearing threshold of 20 μPa. The dB scale is *logarithmic* and the reference level $20\,\mu$Pa is defined as 0 dB.

$$\text{Sound pressure level (dB)} = 20\log_{10}\left(\frac{p}{p_0}\right)$$

where p is any sound pressure in Pa and p_0 is the reference sound pressure (20 μPa). This reference level implies that *multiplication* of the sound pressure in Pa by 10 is equivalent to *addition* of 20 dB to the dB level. For example, with

20 μPa as 0 dB, 200 μPa corresponds to 20 dB, 2000 μPa corresponds to 40 dB, 20,000 μPa corresponds to 60 dB, and so on. Thus, the dB scale compresses a range of a million into a range of only 120 dB. The *sound pressure levels* (SPL) in dB and Pa of various familiar sounds are shown in Figure 7.47.

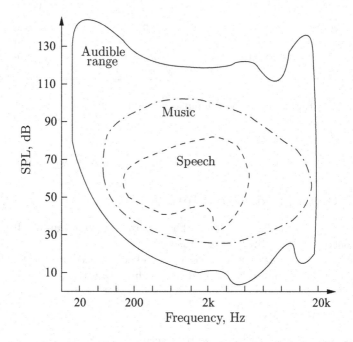

Figure 7.47: Levels of some familiar sounds

It is interesting to note here that, the human ear reacts to a logarithmic change in sound level, which corresponds to the decibel scale where 1 dB is the same relative change everywhere on the scale.

7.17.6 Human Hearing Limits

We defined sound as any pressure variation which can be heard by a human ear. This implies a range of frequencies from 20 Hz to 20 kHz for a healthy human ear. In terms of sound pressure level, audible sound ranges from the *threshold of hearing* at 0 dB to the *threshold of pain* which can be over 130 dB. The sound pressure levels and frequencies of speech, music, and audible range are as shown in Figure 7.48.

The subjective or *perceived* loudness of a sound is determined by several complex factors. One such factor is that the human ear is not equally sensitive at all frequencies. It is most sensitive to sound between 2 kHz and 5 kHz, and less sensitive at higher and lower frequencies. Further, this difference in sensitivity to different frequencies is more pronounced at low *sound pressure level*, SPL

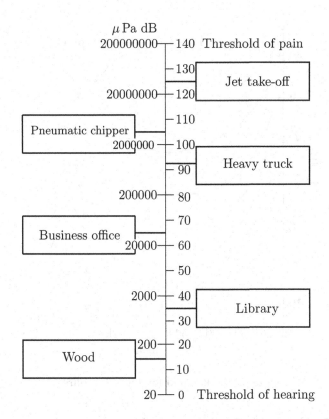

Figure 7.48: SPL-frequency zones of human hearing

than at high SPL. This can be seen in Figure 7.49 which shows a family of *equal loudness contours*.

From Figure 7.49 it is evident that a 50 Hz tone must be 15 dB higher than a 1 kHz tone at a level of 70 dB in order to give the same subjective loudness. The loudness level is a relative measure of sound strength judged by an average human listener. The unit of loudness level is *phon*.

7.17.7   Impulse Sound

An *impulse* or *impulsive* sound is a short duration sound of less than 1 second duration. For example, typewriter and hammering noises are impulse sounds. Because of the short duration of such sounds the ear is less sensitive in perceiving its loudness. It is shown by researchers that the perceived loudness of sounds shorter than 70 milliseconds is less than that of sounds of longer durations having the same level.

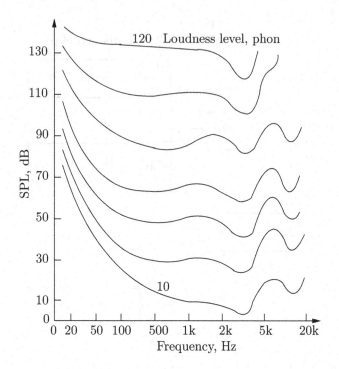

Figure 7.49: Equal loudness contours

7.18   Dynamic Pressure Gauges

So far in our discussions on pressure measurements, we were focusing on the measurement of mean or time-averaged pressure. Measurement of steady or slowly varying pressure is more or less straightforward. If the pressure is rapidly changing, for example as in the case of sound transmission or turbulent flow, it is often necessary to measure the fluctuating pressure and compute its mean as well as its higher-order statistics such as variance, skewness, flatness, correlation coefficient, spectral density, probability distribution, etc. In that case, the temporal and spatial responses of pressure gauge becomes important. However, computing even the time-averaged pressure becomes problematic if the gauge output is not linearly related to the pressure input. Further, for nonlinear transducers, the mean output is not simply related to the mean pressure and errors might be introduced in their measurements.

Transducers measuring turbulence pressure fluctuations, or any other field quantity in a turbulent flow, must resolve a broad spectrum of eddy sizes. These might range from the order of 1 m to 0.01 mm at Reynolds number typical of laboratory experiments, and even a broader range of scales is found in industrial and geophysical flows. There is no probe at the moment which is capable of covering the entire range even for a modest Reynolds number flow. But traversing a sufficiently small gauge through the flow region could provide the spatial variations of the turbulence statistics, in a piecemeal fashion, i.e., not simultaneously. However, the temporal resolution required for such measurements ranges from the order of 1 minute down to a millisecond or microsecond.

The determination and control of sound and noise intensity are technologically important applications of dynamic pressure measurements. Sound transmitted in a fluid manifests itself as pressure fluctuations (called acoustic waves) in that elastic medium. Sound may be generated from a variety of sources such as musical instruments or flow-induced noise. Here too the pressure is time dependent and the signal is often stochastic.

Dynamic pressure gauges are called *microphones* or, for liquid application, *hydrophones*. These are typically elastic transducers having sufficiently fast response. Obviously, liquid manometers and Bourdon tubes are not suitable for dynamic pressure measurements. A fluctuating pressure causes an elastic membrane to oscillate, and this motion is transduced to an electrical signal preferably linearly proportional to the sensed pressure. Common microphones are classified in five categories, namely, piezoelectric transducers, variable capacitance, resistance or reluctance transducers, and linear-variable differential transformers (LVDT).

Piezoelectric transducers utilize crystal quartz, Rochelle salt (potassium sodium tartrate), barium titanate, or lead zircorate titanate, all generate a potential difference between two surfaces when subjected to a force along certain directions. These crystals are shaped as cylinders or bars, and are loaded in normal or shearing modes for measuring pressure or shear stress, respectively.

Capacitance transducers, also called *condenser microphones*, are the most commonly used transducers for dynamic pressure measurements. They measure the change in capacitance in a small air gap between two electrically charged metallic surfaces. One of these surfaces is rigid and the other is a deformable diaphragm or membrane subjected to the source pressure fluctuations.

Variable resistance transducers are essentially strain gauges bonded to appropriate elastic links. While measuring resistance, these transducers also measure the modified magnetic flux linkage of two electromagnets due to the motion of a metallic diaphragm. The latter kind has extremely high sensitivity but limited frequency response. The LVDT gauges sense the displacement of a magnetic core attached to an elastic element subjected to the applied pressure. The motion of the core causes a voltage imbalance between two symmetrically placed secondary coils of a transformer.

These dynamic pressure gauges typically produce 1 volt for 1 μbar of pressure change. This means that, for most applications, signal amplification is required and the signal-to-noise ratio has to be carefully considered. Microphones have typical frequency response ranging from a few hertz to a few kilohertz, and come in sizes ranging from less than 1 mm to a few centimeters. The upper limit of microphone frequency is roughly doubled for each halving of the sensor diameter.

In aerodynamic and hydrodynamic measurements, dynamic gauges are occasionally used to measure both the mean and fluctuating pressures. The frequency response in such measurements ranges from direct current (DC) to some high-frequency cutoff that generally depends on the Reynolds number. These dynamic gauges are not usually called microphones. The terminology "microphone" is used for devices that measure only the fluctuating part of the pressure signal. Sound measuring devices, on the other hand, are intentionally designed so as not to respond to constant or slowly varying pressure (having frequency less than the lower limit of the gauge frequency response). This is because the slow changes in the atmospheric pressure are much larger than the typical sound-pressure fluctuations to which the microphone must respond. The high-pass filtering is accomplished simply by connecting the two sides of the sensor elastic membrane using a capillary tube. The resulting slow leak protects the diaphragm from bursting if the ambient pressure changes significantly.

7.19   Summary

Pressure measuring devices may broadly be grouped into *manometers* and *pressure transducers*. The U-tube manometers, multitube manometers, and Betz-type manometers are some of the popular liquid manometers used for fluid flow pressure measurements. The pressure transducers used for fluid flow experimentation may be classified as *electrical type*, *mechanical type*, and *optical type*.

Measuring principle of the liquid manometer is that *the pressure applied is balanced by the weight of a liquid column.* Some of the commonly used manometric liquids are water, alcohol, and mercury. A mercury barometer is quite satisfactory for measuring atmospheric pressure. In international standard atmosphere, the atmospheric pressure is 101325 Pa (1.01325 bar), which corresponds to 760 mm of mercury.

The multitube manometer operates on the same principle as the U-tube manometer. Inclined manometers are generally used for measuring very small pressure differences. *Micro manometer* is basically a precision U-tube manometer. It is a rugged instrument capable of measuring liquid column height with an accuracy of ± 0.1 mm. *Betz manometer* is yet another U-tube manometer in a general sense. The fluid column carries a float and a graduated scale is attached to the float. The scale is transparent and the graduation markings could be projected on a screen. Thus, the manometer readings can be read directly. Measurement range of this instrument is usually about 0 to 300 mm and its sensitivity is 0.1 mm.

Dial-type pressure gauges generally operate on the principle of a bellows or Bourdon tube which is usually an elliptical cross-sectional tube having a "C"-shape configuration. When pressure is applied to the inside of the tube, an elastic deformation results, which is proportional to the pressure applied.

Pressure transducers are *electromechanical* devices that convert pressure to electrical signal which can be recorded with a data acquisition system such as that used for recording strain gauge signals.

Linear differential transformer works on the induction principle, like any other transformer, but with a movable core fixed to the diaphragm. This instrument is capable of detecting displacements of the *order of a few microns*.

In *capacitance pickup*, the capacitance between two metallic plates with a dielectric in between is varied. One of the plates is fixed and the other one is connected to a pressure capsule. The change in capacitance is a measure of the pressure.

In the *optical-type* pressure transducer, the sensing element is a thin diaphragm which is highly polished on both sides.

Pitot-static tube is used for measuring the total and static pressures simultaneously. *Total pressure* is that which results when a flow is decelerated to rest isentropically. *Static pressure* is the pressure which acts normal to the flow direction. For incompressible flows, the total, static, and dynamic pressures are linked together by the Bernoulli equation as

$$p_0 - p = \frac{1}{2}\rho V^2$$

In subsonic flows, a pitot probe facing the flow always measures the stagnation pressure. But at supersonic speeds there will be a detached shock formed and positioned in front of the pitot probe nose. It implies that the tube does not measure the actual stagnation pressure but it only measures the stagnation pressure behind a normal shock. This new value is called *pitot pressure* and in modern terminology it refers to a supersonic stream. This indicated pitot pressure p_{02}, may be used for calculating the stream Mach number, as follows. For a supersonic stream with Mach number M_1 and p_1 and p_{01} as the static and stagnation pressures, respectively, by isentropic and normal shock relations, we have

$$\frac{p_{01}}{p_1} = \left(1 + \frac{\gamma-1}{2}M_1^2\right)^{\frac{\gamma}{\gamma-1}}$$

$$\frac{p_{02}}{p_{01}} = \left(1 + \frac{2\gamma}{\gamma+1}(M_1^2 - 1)\right)^{\frac{-1}{\gamma-1}}\left[\frac{(\gamma+1)M_1^2}{(\gamma-1)M_1^2+2}\right]^{\frac{\gamma}{\gamma-1}}$$

Multiplying these two equations, we obtain,

$$\frac{p_1}{p_{02}} = \frac{\left[\frac{2\gamma}{\gamma+1}M_1^2 - \frac{\gamma-1}{\gamma+1}\right]^{\frac{1}{\gamma-1}}}{\left(\frac{\gamma+1}{2}M_1^2\right)^{\frac{\gamma}{\gamma-1}}}$$

This relation is known as the *Rayleigh supersonic pitot formula*. Once the static pressure p_1 and pitot pressure p_{02} are known, M_1 can be calculated using this formula.

For a compressible flow the pressures and the velocity are related by the equation

$$\frac{\gamma}{\gamma-1}\frac{p}{\rho} + \frac{1}{2}V^2 = \frac{\gamma}{\gamma-1}\frac{p_0}{\rho_0}$$

This is called the *compressible Bernoulli equation.*

Static pressure measurements are very sensitive to the inclination of the tube to the flow direction. Inclinations beyond 5° result in large errors in the measured pressures. For measurement of static pressures in supersonic flow, the static pressure hole should be located beyond 18D from the nose and the velocity change due to the probe should be kept to a minimum. For this the nose should be very sharp with an included angle less than 10°. Also, it is an usual practice to employ a straight probe (without vertical stem) for static pressure measurements in supersonic wind tunnels. The tip of the probe should be located at the subsonic zone of the flow field (for instance upstream of the throat in a Laval nozzle) and hence there is no formation of shock because of the probe presence.

The dynamic pressure of a flow can be measured directly by pitot-static probe, which is the combination of pitot and static probes. For compressible flows, the dynamic pressure can be expressed in terms of Mach number M, as

$$q = \frac{\gamma p}{2} M^2$$

But in practice, because of the nature of static pressure, it is seldom possible to find a location for static orifice, which will measure the correct static pressure. In most of the cases, the fabricated probe will indicate a pressure difference Δp given by

$$\Delta p = F \frac{1}{2} \rho V^2$$

where F is the correction factor which has to be determined from experimental calibration. A proper pitot-static tube is then that with $F \approx 1$.

The major factors influencing pitot-static tube measurements are turbulence, velocity gradient, vibration, viscosity, wall proximity, and misalignment. Even though the turbulent fluctuations of a flow have a definite effect on pitot-static probe readings, errors introduced by this effect are generally small. A small correction is essential for accuracy when there is a velocity gradient across the mouth of the pitot tube.

For boundary layer studies, because the region in which measurements have to be made is thin, usually flattened pitot-tubes are employed.

At low Reynolds numbers pitot-tube measurements are severely influenced by viscosity. This error can be avoided by choosing bigger size tubes for pitot probes, thereby keeping the tube Reynolds number above the limiting value at which the viscosity effects begin to be significant.

The effect of vibration on pitot-static tubes may sometimes be considerable to introduce significant errors in their measurements. These vibrations can be due to aerodynamic or mechanical reasons. Vibrations transmitted to the probe from *mechanical sources are likely to cause more problems than those due to vortex shedding*, especially when they make the probe head vibrate along the wind direction.

The pressure measured by a pitot or static probe will be in error if it is not aligned with the flow direction. But for small angles of misalignment the error is generally small, especially for pitot probes, which are far less sensitive to this effect than static probes. Pitot tube is insensitive up to about 20° yaw.

The appreciable insensitivity of the pitot probe reading to misalignment in incompressible flow continues in compressible flow also, right up to supersonic regimes. The yaw insensitivity is usually defined as *the angular misalignment at which the pitot reading has changed by 1 percent of the difference between the zero yaw pitot and static pressures of the undisturbed flow. In incompressible flow this is 1 percent of the dynamic pressure of undisturbed flow*. With the above definition the value of yaw insensitivity for any pitot probe can be comfortably taken as ± 11 degrees for compressible flow. In general, static probes with measuring orifice located sufficiently far downstream is suitable for measurement of static pressure in supersonic flows. A typical probe measures static pressure within ± 0.5 percent at Mach 1.6, if the pressure tapings are located more than 10 tube diameters downstream of the shoulder. This distance increases with increase of Mach number, in the supersonic regime.

The instrument used for determining the flow direction is termed *yaw probes*. Three-hole and five-hole probes are used in two-dimensional and three-dimensional flows, respectively. Five-hole probe is similar to three-hole probe except that it is *capable of measuring yaw as well as pitch of the flow*. The tube should have provision to rotate in two directions, i.e., in yaw and pitch.

Low-pressure gauges are usually referred to as *vacuum gauges*. Measurement of low pressures requires considerable care on the part of the experimentalist. For moderate vacuum measurements manometers, dial gauges, and diaphragm gauges may be employed with reasonable accuracy. But for measurement of absolute pressures below 1 torr (1 mm of Hg or 133 Pa), the *McLeod gauge, Pirani gauge, Knudsen gauge, ionization gauge*, and *alphatron gauge* are some of the commonly used pressure gauges. The McLeod gauge is used as a standard for measuring low vacuum pressures. It can be used for measuring low pressures in the range from 10^{-2} to $10^2 \, \mu m$ of mercury, i.e., from 0.00133 to 13.3 Pa, for dry gases. Pirani gauge is a device that measures pressure through *the change in thermal conductance of the gas*. Pirani gauges are suitable for measurement of pressures in the range from 1 μm to 1 torr. The Knudsen gauge is capable of measuring low pressures in the range between 10^{-5} m and 10 μm of Hg. The ionization gauges are capable of measuring pressures as low as 10^{-12} torr.

The *Preston* and *Stanton tubes* are essentially small impact (pitot) tubes meant for the indirect measurement of skin friction, using some well-established mean velocity correlations to infer mean shear stress at the wall.

Sound may be defined as *any pressure variation* (in air, water, or other medium) *that the human ear can detect*. If variations in atmospheric pressure occur more rapidly (at least 20 times per second) they can be heard and hence called sound. A barometer cannot respond quickly enough and therefore cannot be used to measure sound. The *frequency* of sound is defined as the number of pressure variations per second. The frequency is measured in *hertz (Hz)*. The waves with frequency below human audible range are called *infrasonic waves* and those with frequency above the audible range are called *ultrasonic*.

The *wavelength* is defined as the distance from one wave top or pressure peak to the next.

$$\text{Wavelength} (\lambda) = \frac{\text{Speed of sound}}{\text{Frequency}}$$

When a sound has only one frequency it is referred to as a *pure tone*. *Broadband noise* is a noise consisting of a wide mixture of frequencies. Most industrial noise is broadband noise. *White noise* is a noise having frequencies evenly distributed throughout the audible range. White noise will sound like a rushing water.

The *decibel* is a ratio between a measured quantity and an agreed reference level.

$$\boxed{\text{Sound pressure level (dB)} = 20\log_{10}\left(\frac{p}{p_0}\right)}$$

where p is any sound pressure in Pa and p_0 is the reference sound pressure (20 μPa).

An *impulse* or *impulsive* sound is a short duration sound of less than 1-second duration.

Dynamic pressure gauges are called microphones or, for liquid application, hydrophones. These are typically elastic transducers having sufficiently fast response.

Exercise Problems

7.1 A pitot tube has to be designed to measure the settling chamber pressure of a hypersonic tunnel. The maximum speed in the chamber is expected to be 2 m/s. If the pressure and temperature in the chamber are 25 atm and 800°C, determine the limiting minimum diameter for the pitot tube up to which the pitot tube will continue to measure the correct pressure.

[Answer: 1.32 mm]

7.2 Determine the misalignment insensitivity of a pitot probe, (a) when the measured pressure should be within 5 percent of the correct pressure. (b) Will this misalignment limit vary with flow speed?

[Answer: (a) 18.2°, (b) 18.2°]

7.3 If 5 percent error due to angularity is the acceptable upper limit for the pressure measured by a static pressure probe, determine the misalignment angle up to which the probe will measure the correct pressure when the flow speed is 30 m/s and the stagnation pressure and temperature are 1 atm and 20°C, respectively.

[Answer: 2.87°]

7.4 A symmetrical wedge of semi-vertex angle 7° is tested in a Mach 1.6 tunnel. Determine the pressure coefficient C_p at a point just downstream of the model nose and assess its sensitivity to the expansion level at the wind tunnel nozzle exit, when the tunnel is run with settling chamber pressure of (a) 4 atm and (b) 6 atm.

[Answer: (a) 0.23, (b) 0.23]

7.5 A pitot-static probe is used to measure the dynamic pressure of a one-dimensional incompressible flow. If the flow is with 6 percent isotropic turbulence, determine the error in the measured value, caused by the turbulence.

[Answer: − 0.72 percent]

7.6 A liquid manometer with measuring scale resolution of 1 mm is used for measuring air flow speed. (a) Determine the minimum speed that can be measured accurately if the liquid used is mercury, and (b) find the minimum speed that can be measured if the liquid used is water.

[Answer: (a) 14.75 m/s, (b) 4 m/s]

7.7 A pitot-static probe used in a flow vibrates, causing an error of 2.56 percent to the dynamic pressure measured by the probe. If the frequency of vibration is 150 hertz, find the amplitude of the vibration when the flow speed is 25 m/s at sea level condition.

[Answer: 6 mm]

7.8 A three-hole probe is used to measure the flow angularity of a two-dimensional incompressible flow. If the probe constant shows 10 percent error, determine the flow angularity.

[Answer: −2.87°]

7.9 An induction type supersonic tunnel draws air from standard atmosphere. If the test-section temperature is −60°C, determine the dynamic pressure of the flow.

[Answer: 43.679 kPa]

7.10 Two oblique shocks of equal strength but opposite family, causing a flow turning of 5°, in a Mach 2 air stream of stagnation pressure of 5 atm cross each other, as shown in Figure P7.10. (a) Determine the pressures a pitot probe will measure at locations 3, 3u, and 3l. (b) Check whether the strength of the shock cross-over point is equal to the combined strength of these shocks. (c) Can the shock cross-over point be taken as identical to a normal shock?

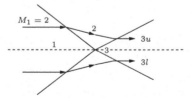

Figure P7.10: Flow past two oblique shocks of equal strength but opposite family, crossing each other

[Answer: (a) The pitot tube will measure the pressure at region 3u as 4.982 atm. At region 3l also, the pitot probe will measure 4.982 atm, because the two fields are identical. (b) The shock strength at the cross-over point is approximately equal to the sum of the strengths of the individual oblique shocks. (c) Yes.]

7.11 A pitot tube of diameter 0.6 mm is to be used for measurement in a fluid stream of velocity 10 m/s. If the pressure and temperature of the flow are 101 kPa and 20°C, will the measurement be accurate?

[Answer: No]

7.12 A pitot-static probe measuring in a subsonic flow is influenced by turbulence. If the error in the dynamic pressure measured, caused by the turbulence, is 1.25 percent, determine the turbulence level, assuming the turbulence as isotropic.

[Answer: 7.9 percent]

7.13 A McLeod gauge, of bulb volume 95 cm^3 and capillary diameter 1 mm, is connected to a low-pressure chamber at 0.09 Pa. Determine the mercury column indicated by the capillary tube.

[Answer: 9.04 mm]

7.14 Determine the $C_{p_{\min}}$ over a NACA0012 aerofoil, at zero angle of attack in an air stream at 30 m/s, 1 atm, and 300 K.

[Answer: −0.131]

7.15 A pitot probe is to be designed for measurement in a Mach 2 uniform air stream of diameter 10 mm. What should be the limiting maximum diameter of the probe to restrict the error in the Mach number measured to be less than 5 percent?

[Answer: 2.81 mm]

Chapter 8

Velocity Measurements

8.1 Introduction

Flow velocity and its nondimensional form known as the Mach number are some of the most important parameters in fluid flow analysis. Flow velocity can be measured by many techniques. Some of the popularly used techniques are velocity measurement through pressure measurements, through optical properties like Doppler shift using laser Doppler anemometer, through measurement of the vortex shedding frequency, and through the heat transfer principle; using hot-wire anemometer. Also, we know that for incompressible flows the speed parameter usually used is the flow velocity and for compressible flows the parameter commonly used is the Mach number. In this chapter let us see how we can obtain the velocity and Mach number from the quantities measured by the techniques mentioned above.

8.2 Velocity and Mach Number from Pressure Measurements

For an incompressible flow, velocity V can be calculated from the measured total and static pressures, using the Bernoulli equation

$$\boxed{p_0 - p = \frac{1}{2} \rho V^2}$$ (8.1)

For compressible flows, Mach number M is one of the most important parameters. From the measured pressures, the flow Mach number may be obtained using the pressure Mach number relations which are well established.

If the flow at the measuring point experiences only isentropic changes, its stagnation pressure may be assumed to be the same as the reservoir pressure p_0. A measurement of the static pressure p alone is necessary to compute the

Mach number using the isentropic relation,

$$\frac{p_0}{p} = \left(1 + \frac{\gamma - 1}{2}M^2\right)^{\frac{\gamma}{\gamma-1}} \tag{8.2}$$

This relation is applicable to both subsonic and supersonic flows, and is used for obtaining the Mach number distribution along an aerodynamic surface, using surface static pressure holes, and for flow-field Mach number, using a static probe.

In an isentropic supersonic flow, the local Mach number can be computed using the pitot pressure measured by a pitot probe. But we know that a detached shock will be formed at the nose of the probe when it is placed in a supersonic stream. Therefore, what the probe measures is the total pressure behind the shock and not the actual total pressure of the stream. Also, we know that the portion of the detached shock (also called bow shock) ahead of the probe nose can be approximated as a normal shock. Therefore, the flow Mach number can be calculated, using the normal shock relation,

$$\frac{p_{01}}{p_{02}} = \left(\frac{2\gamma}{\gamma + 1}M^2 - \frac{\gamma - 1}{\gamma + 1}\right)^{\frac{1}{\gamma-1}} \left(\frac{1 + \frac{\gamma - 1}{2}M^2}{\frac{\gamma + 1}{2}M^2}\right)^{\frac{\gamma}{\gamma-1}} \tag{8.3}$$

where p_{01} is the reservoir pressure which is the same as the total pressure of the stream in an isentropic flow, and p_{02} is the pressure measured by the pitot probe. In flow fields where the reservoir conditions are not known, as in the case of measurement from an aircraft, it is not possible to use either of the above two equations (Equations (8.2) and (8.3)) to obtain the Mach number, since it is necessary to know both the static and total pressures for using them. In subsonic flow, the pitot pressure measured is the actual total pressure, and the Mach number can be obtained from Equation (8.2), once the static pressure is known. But in supersonic flows, the measured pitot pressure p_{02} is the total pressure behind the normal shock at the probe nose, and not the actual total pressure of the flow. Therefore, to eliminate the actual total pressure p_0 (or p_{01}), Equation (8.2) may be divided by Equation (8.3), resulting in

$$\frac{p}{p_{02}} = \frac{\left(\frac{2\gamma}{\gamma + 1}M_1{}^2 - \frac{\gamma - 1}{\gamma + 1}\right)^{\frac{1}{\gamma-1}}}{\left(\frac{\gamma + 1}{2}M_1{}^2\right)^{\frac{\gamma}{\gamma-1}}} \tag{8.4}$$

This equation is popularly known as the *Rayleigh supersonic pitot formula*. In this equation both the pressures involved are measurable in supersonic streams.

Note that, among the isentropic relation, normal shock relation and Rayleigh pitot formula, Rayleigh formula is the most accurate for determining the flow

Mach number through pressure measurement. This is because, in the isentropic relation given by Equation (8.2), the total pressure p_0 is assumed to be constant throughout, which is not true in actual flows. In the normal shock relation, the total pressure, p_{01} upstream of the shock, is taken as the reservoir pressure, which again neglects the losses from the reservoir to the test location. But in Rayleigh formula the pressures p and p_{02} used are the actual static and total pressures measured at the location where the Mach number is to be determined.

8.3 Laser Doppler Anemometer

The novel method of measuring velocities in fluid flows using laser Doppler technique has been subjected to a rapid development in the past few decades. The usefulness and reliability of this instrument in the measurement of low as well as high speed flows have been successfully established. This technique is commonly referred to as *Laser Doppler Anemometry* or *Laser Doppler Velocimetry*. In short, it is termed as LDA or LDV in literature. This instrument, though somewhat expensive, has many advantages over other conventional instruments such as pitot-static probe, hot-wire anemometer, and so on. The specific advantages of LDA are the following.

- The measurement with LDA is absolute, linear with velocity, and requires no precalibration.

- It is applicable to a wide range of flow velocities, say from 10^{-4} m/s to 10^3 m/s.

- It has negligible probe interference.

- It has high frequency response.

- It is insensitive to temperature, hence can be used in cold flows as well as in hot flows like rocket exhaust or plasma flows.

- It has high resolution since probe volume as small as 10^{-6} cubic centimeter could be obtained.

- It is capable of measuring a single component of velocity as well as its temporal variations. It is possible to measure even reversals of flow direction, with an appropriate optical configuration.

8.3.1 LDA Principle

The principle underlying laser Doppler anemometer is that *a moving particle illuminated by a light beam scatters light at a frequency different from that of the original incident beam*. This difference in frequency is known as *Doppler shift*, and it is proportional to the velocity of the particle. In fluid flows where the velocities are much lower than the speed of light the Doppler shift is small and orders of magnitude less than the frequency of the incident beam. The detection of it is possible only if the light source is highly stable. *LASER* is the ideal light source for this purpose. The word *LASER* stands for *Light Amplification by Stimulated Emission of Radiation*. For instance, the Doppler shift encountered in high-speed flows is of the order of a few MHz and the laser radiation for a helium–neon laser is about 5×10^{14} Hz. Even the highest quality monochrometer using a conventional arc-lamp has a band of 10^8 Hz, while a laser has fundamental frequencies which are stable within a few *hertz*. The laser generally used is of the helium–neon type radiating at 6328 Å, with a power output between 5 and 50 milliwatts. However, for improved detector output, high-power argon lasers (1 to 5 watts) are used, wherever necessary.

The Doppler shift is detected by a device called *photomultiplier* which gives an electrical output whose frequency is proportional to the velocity of the scattering particle. This frequency is converted into voltage using electronic devices, thus, the final voltage output is a measure of the flow velocity.

For measurement with LDA, the fluid which is in motion should contain scattering particles. Often artificial scattering centers are introduced into the flow to increase the intensity of the scattering. These are tiny particles with a diameter of a few microns and added in very small quantities, without introducing change to the gross properties of the fluid as well as to the flow.

The size of the particles being of the order of the wavelength of light, the scattering is of the *Mie* type and the scattered light intensity is orders of magnitude larger than Rayleigh scattering. Different materials are used as scattering particles depending on the fluid flow to be studied. The following are the commonly used scattering substances.

- Micro polythene spheres.

- Diluted milk droplets.

- Diluted smoke particles.

- Aerosol.

- Fine alumina powder.

- Water glycerin mixture droplets.

Very often naturally occurring particles in city tap water and dust in the atmosphere are sufficient to act as scattering centers.

8.3.2 Doppler Shift Equation

The Doppler shift in frequency occurs when a particle moves with a component of its velocity in the direction of the light beam. Let us consider a light source radiating at a frequency ν_L and illuminating a particle moving with a velocity $\overline{V}$ away from it, as illustrated in Figure 8.1. This particle due to its movement encounters fewer wave fronts than those emitted by the source. Thus, the particle is illuminated by a wave frequency ν_p, which is different from ν_L, given by

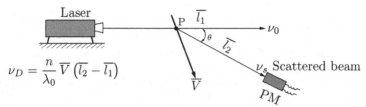

$$\nu_D = \frac{n}{\lambda_0} \overline{V} \left(\overline{l}_2 - \overline{l}_1\right)$$

(a) Principle of laser Doppler anemometer

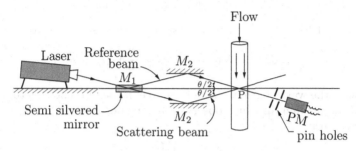

(b) Reference beam method

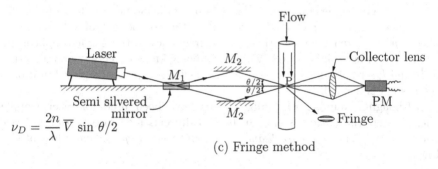

$$\nu_D = \frac{2n}{\lambda} \overline{V} \sin \theta/2$$

(c) Fringe method

Figure 8.1: Laser Doppler anemometer

$$\nu_p = \frac{1}{\lambda_L} \left(C - \overline{V}\,\overline{l}_1\right) = \nu_L \left(1 - \frac{\overline{V}\,\overline{l}_1}{C}\right) \qquad (8.5)$$

where C is velocity of the light in vacuum, l_1 is unit vector in the direction of propagation of the light beam, and λ_L is wavelength of the light.

The scattered light in the direction of the photodetector (unit vector $\overline{l}_2$) will have an additional frequency shift ν_S due to the component of velocity in the direction of scattered light. The frequency of the scattered light ν_S is given by

$$\nu_p = \nu_S \left(1 - \frac{\overline{V}\,\overline{l}_2}{C}\right) \qquad (8.6)$$

The resultant shift in frequency ν_D is then given by

$$\nu_D = \nu_L - \nu_S \tag{8.7}$$

Substituting for ν_S from Equation (8.6) into Equation (8.7), we get

$$\nu_D = \nu_L \left[1 - \frac{\left(1 - \dfrac{\overline{V}\,\overline{l_1}}{C} \right)}{\left(1 - \dfrac{\overline{V}\,\overline{l_2}}{C} \right)} \right] \tag{8.8}$$

We know that $|\overline{V}| \ll C$, therefore, Equation (8.8) simplifies to

$$\nu_D = \nu_L \overline{V} \left(\overline{l_1} - \overline{l_2} \right) / C = \frac{n}{\lambda_0} \overline{V} \left(\overline{l_1} - \overline{l_2} \right)$$

where λ_0 is the wavelength of the light source in vacuum and n is the refractive index of the medium in which observation is made.

The equation

$$\boxed{\nu_D = \frac{n}{\lambda_0} \overline{V} \left(\overline{l_1} - \overline{l_2} \right)} \tag{8.9}$$

is the basic governing equation for laser Doppler anemometry. The Doppler shift in frequency ν_D is directly proportional to the component of velocity in the direction of reference vector $(\overline{l_1} - \overline{l_2})$ and frequency of the incident light beam. In the case of laser light source, value of λ_0 being known precisely and directions of $\overline{l_1}$ and $\overline{l_2}$ are obtained from the experimental setup, hence no calibration of the system is needed.

Although a number of configurations have been used for laser Doppler anemometer, the basic system can be broadly classified into two major categories, namely the *reference beam system* and the *fringe system*, illustrated in Figure 8.1.

8.3.3   Reference Beam System

In a reference beam system the scattered beam of light is optically mixed with the original beam and the difference is obtained as the Doppler shift in frequency. This technique is known as *optical hetrodyning* and it is a characteristic of the photomultiplier. Detector cathodes with square law property are essential for this purpose. Adjustments of the two beams to obtain optical hetrodyning requires high skill.

8.3.4   Fringe System

In this system the fringes obtained during interference of the two beams are made use of for the measurement of flow velocities. Particles passing through the fringes scatter light which is modulated at a frequency ν_M depending on the fringe spacing and velocity of the scattering particle. The scattered light is focused on a photomultiplier. Mathematically it can be shown that ν_M obtained from the fringe technique has the same magnitude as ν_D. That is, both systems yield the same result.

Referring to Figure 8.1, it can be seen that,

$$\nu_D = \nu_M = \frac{2n\overline{V}}{\lambda_0} \sin \frac{\theta}{2} \qquad (8.10)$$

A variety of physical configurations are used for the laser Doppler anemometer. Readers are highly encouraged to consult books specializing in this topic for further details. The development of laser Doppler anemometer has been rapid during the past few decades. The present status of the technique is that it can be regarded as a technique which is complementary to the pitot-static probe and hot-wire anemometer. Its distinct advantage is that the method is totally free from probe disturbances, since all the instrumentation is external to the flow. Further, it is capable of measuring a single component, as well as temporal variations. This method may also be used as a calibration standard. However, it does have many disadvantages also. The major disadvantage is that the flow ceases to be a single phase flow when liquid droplets or solid particles are introduced into it as scattering particles. Further,

- flow has to be strictly two-dimensional for the light beam to pass through, giving the properties at its path. Otherwise, the passing light beam carries the integrated effect in its path. In most practical situations the flow is three-dimensional, thereby rendering the LDA unsuitable for measurements.

- Laser source and measuring equipment are far more complex than a simple pressure probe.

- Eye protection against direct or reflected laser beam is essential.

- LDA system is very very expensive compared to any other measuring device of its kind.

Example 8.1

Determine the Doppler shift caused by an air jet of average velocity 30 m/s, intercepting a helium-neon laser beam of wavelength 6328 angstrom, and at an angle of 2° with reference to the laser beam.

Solution

Given, $\overline{V} = 30$ m/s, $\lambda_0 = 6328 \times 10^{-10}$ m and $\theta = 2°$.

For air the refractive index n is approximately 1.0.

By Equation (8.10), the Doppler shift is

$$\nu_D \ = \ \frac{2\,n\overline{V}}{\lambda_0}\,\sin\frac{\theta}{2}$$

$$= \ \frac{2 \times 1 \times 30}{6328 \times 10^{-10}} \times \sin(1°)$$

$$= \ \boxed{1.65 \times 10^6\,\text{Hz}}$$

8.4 Measurement of Velocity by Hot-Wire Anemometer

In Chapter 5 we saw that hot-wire anemometer can be used for measuring flow velocity for fluids like air. For liquids a *hot-film* anemometer can be employed. Due to the small size of the sensing element used in hot-wire systems the frequency response of the system is quite high, of the order of a few kilohertz. Hence, it is used extensively for the measurement of turbulent fluctuations. As illustrated in Chapter 5, a precision Wheatstone bridge is sufficient to make mean velocity measurements with the hot-wire anemometer.

8.5 Measurement of Velocity Using Vortex Shedding Technique

Measurement of low velocities, below 2 m/s, are difficult to do with instruments like a pitot-static probe in conjunction with a manometer. For measuring such low velocities, vortex shedding techniques could be used. It is well known that a cylinder immersed in a moving fluid sheds vortices alternately from the two sides. At low Reynolds numbers between 40 and 150, these vortices are shed in stable, well-defined patterns, which persist for long distances downstream and this is termed the *Karman vortex street*. Periodic vortex shedding continues up to Reynolds numbers of about 10^5 or more, but the vortices then quickly lose their identity in a generally turbulent wake. However, just downstream of the cylinder it is possible to detect the frequency even at these Reynolds numbers.

The frequency of the vortex shedding from a cylinder can be related to flow Reynolds number in the following way (Ower and Pankhurst, 1977)

$$S = 0.212\left(1 - \frac{c}{Re}\right)$$

where Re is the Reynolds number based on cylinder diameter, c is a constant having values 21.2 and 12.7 for Reynolds number range 50–150 and 300–2000, respectively, S is called Strouhal number, defined nd/U, where n is the frequency of vortex shedding, d is the diameter of the cylinder, and U is the velocity of flow upstream of the cylinder. The Reynolds number Re is given by Ud/ν, where ν is the kinematic viscosity of the fluid. The above equation can be rewritten as

$$U = 4.72\,nd + \frac{c\nu}{d} \qquad (8.11)$$

This equation is used to measure velocity of air. A cylinder of suitable diameter is inserted into the air stream and the vortex-shedding frequency n is determined. This may be done accurately using a hot-wire system with an electronic circuit containing an oscilloscope. This was the method used by Roshko (Ower and Pankhurst, 1977). The hot-wire should not be placed more than 5 or 6 diameters behind the cylinder, since Roshko's results show that, although for Re up to 150 satisfactory observations can be made even at some 50 diameters downstream, this is not true for Re above 200.

The vortex-shedding frequency n may also be determined by simple methods like smoke trails used in conjunction with a stroboscope, however, this kind of technique has not become popular for determining n.

It is essential that Re should be within one of the ranges 50–150 or 300–2000. Specifically, Re should not be in the range from 150 to 300, since the observations in that range showed a much wider scatter than elsewhere. Therefore, it is important to choose an appropriate cylinder ensuring that the Re lies within the desired limits. Table 8.1 (Ower and Pankhurst, 1977) connects the air velocity with Re, n, and cylinder diameter d for some cases. The kinematic viscosity ν is taken as 1.461×10^{-5} m^2/s in the table data.

Table 8.1 Relation between vortex shedding frequency and cylinder diameter (valid for the range of Re 50–150 and 300–2000)

V m/s	$d = 10$ mm		$d = 5$ mm		$d = 2$ mm		$d = 1$ mm	
	Re	n Hz	Re	n Hz	Re	n Hz	Re	n Hz
0.5	342	10	–	–	68	37	–	–
1.0	684	21	342	41	137	90	68	146
1.5	1026	31	513	62	–	–	103	252
2.0	1368	42	684	83	–	–	137	358
2.5	1710	53	855	104	342	255	–	–
5.0	–	–	1710	210	684	520	342	1021
10.0	–	–	–	–	1368	1050	684	2080
15.0	–	–	–	–	–	–	1026	3142

The major advantage of this method is that it is capable of measuring low speeds of air which cannot be measured accurately with a "conventional" manometer.

8.6   Fluid-Jet Anemometer

The difficulty in the measurements of small differential pressure with sufficient accuracy imposes limitation on pitot-static probes for measurement of flow velocities of the order of 2 m/s. This problem may be sorted out by using the device referred to as fluid-jet anemometer, shown in Figure 8.2.

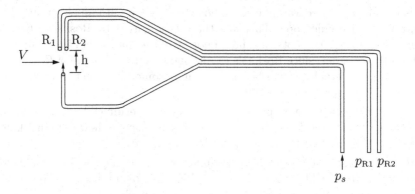

Figure 8.2: Fluid-jet anemometer

As seen from Figure 8.2, a jet is directed across the flow and is deflected to an extent. The deflection of the jet strongly depends on the ratio of flow velocity V to jet velocity. The pressure difference Δp measured with the tubes R_1 and R_2 is made use of for determining V. Design of appropriate geometry of the probe, together with suitable choice of jet supply pressure p_s, creates pressure differences Δp which are many times the dynamic pressure $\frac{1}{2}\rho V^2$. This kind of instrument has been used successfully in various industrial processing plants.

8.7   Summary

Flow velocity and its nondimensional form known as the Mach number are some of the most important parameters in fluid flow analysis. Some of the popularly used techniques are velocity measurement through pressure measurements, through optical properties like Doppler shift using laser Doppler anemometer, through measurement of the vortex shedding frequency and through heat transfer principle, using hot-wire anemometer.

For an incompressible flow, velocity can be calculated using the Bernoulli equation

$$p_0 - p = \frac{1}{2}\rho V^2$$

For compressible flows, the flow Mach number may be obtained using the pressure Mach number relation,

$$\frac{p_0}{p} = \left(1 + \frac{\gamma - 1}{2}M^2\right)^{\frac{\gamma}{\gamma - 1}}$$

In an isentropic supersonic flow, the local Mach number can be measured by a pitot probe. But we know that a detached shock will be formed at the nose of the probe when it is placed in a supersonic stream. Therefore, the flow Mach number can be calculated, using the normal shock relation,

$$\frac{p_{01}}{p_{02}} = \left(\frac{2\gamma}{\gamma + 1}M^2 - \frac{\gamma - 1}{\gamma + 1}\right)^{\frac{1}{\gamma - 1}} \left(\frac{1 + \frac{\gamma - 1}{2}M^2}{\frac{\gamma + 1}{2}M^2}\right)^{\frac{\gamma}{\gamma - 1}}$$

where p_{01} is the reservoir pressure which is the same as the total pressure of the stream in an isentropic flow, and p_{02} is the pressure measured by the pitot probe. In flow fields where the reservoir conditions are not known, as in the case of measurement from an aircraft, it is not possible to use any of the above two equations to obtain Mach number,

$$\frac{p}{p_{02}} = \frac{\left(\frac{2\gamma}{\gamma + 1}M_1^2 - \frac{\gamma - 1}{\gamma + 1}\right)^{\frac{1}{\gamma - 1}}}{\left(\frac{\gamma + 1}{2}M_1^2\right)^{\frac{\gamma}{\gamma - 1}}}$$

This equation is popularly known as the *Rayleigh supersonic pitot formula*.

Laser Doppler Anemometry is a method of measuring velocities in fluid flows using laser Doppler technique. It is termed *LDA* or *LDV* in literature. The specific advantages of LDA are the following.

- The measurement with LDA is absolute, linear with velocity, and requires no precalibration.

- It is applicable to a wide range of flow velocities, say from 10^{-4} m/s to 10^3 m/s.

- It has negligible probe interference.

- It has high frequency response.

- It is insensitive to temperature, hence can be used in cold flows as well as in hot flows like rocket exhaust or plasma flows.

- It has high resolution since probe volume as small as 10^{-6} cubic centimeter could be obtained.

- It is capable of measuring a single component of velocity as well as its temporal variations. It is possible to measure even reversals of flow direction, with an appropriate optical configuration.

Hot-wire anemometer can be used for measuring flow velocity for fluids like air. For liquids a hot-film anemometer can be employed.

Measurement of low velocities below 2 m/s, are difficult to do with instruments like a pitot-static probe in conjunction with a manometer. For measuring such low velocities, vortex shedding techniques could be used.

The difficulty in the measurements of small differential pressure with sufficient accuracy imposes limitation on pitot-static probes for measurement of flow velocities of the order of 2 m/s. This problem may be sorted out by using a fluid-jet anemometer.

Exercise Problems

8.1 A slender cylindrical wire of diameter 1 mm is placed in an air stream of velocity 25 m/s. If the pressure and temperature are standard sea-level values, determine the frequency of the vortices shed by the wire.

[Answer: 5257 hertz]

8.2 If the Doppler shift caused by an air stream intercepting a helium-neon laser beam of wavelength 6300 angstrom, at an angle of 5° with reference to the laser beam is 2 million hertz, determine the average velocity of air stream.

[Answer: 43.33 m/s]

Chapter 9

Temperature Measurement

9.1 Introduction

Although it is well known that temperature is a measure of the intensity of "hotness" or "coldness," it is not easy to give an exact definition for it. Based on physiological sensations, we express the level of temperature qualitatively with words, such as freezing cold, cold, warm, hot, red-hot, and so on. However, we cannot assign numerical values to temperature based on our sensations alone. Furthermore, our senses may be misleading. Fortunately, several properties of materials change with temperature in a repeatable and predictable manner, and this forms the basis for the accurate measurement of temperature.

We know from experience that, when a hot body is brought into contact with a cold body, heat is transferred from the hot body to the cold body until both the bodies attain the same temperature. At that point, the heat flow stops, and the two bodies are said to have reached *thermal equilibrium*. The equality of temperature is the only requirement for thermal equilibrium. The *zeroth law of thermodynamics* states that, "*if two bodies are in thermal equilibrium with a third body, they are in thermal equilibrium with each other.*" This zeroth law serves as a basis for the validity of temperature measurement. By replacing the third body with a thermometer, the zeroth law can be restated as "*two bodies are in thermal equilibrium if both have the same temperature reading even if they are not in contact.*"

9.2 Temperature Scales

Temperature scales aim at using a common basis for temperature measurements. All temperature scales are based on some easily reproducible states, such as the freezing and boiling points of water, also called the *ice point* and the *steam point*.

The temperature scales in *SI* and in English systems are the *Celsius scale* (formerly called the centigrade scale) and the *Fahrenheit scale*. On the Celsius scale, the ice and steam points are assigned the values 0°C and 100°C, respectively. The corresponding values on the Fahrenheit scale are 32°F and 212°F. These are often referred to as *two-point scales*, since temperature values are assigned at two different points. A more useful scale is the *absolute temperature scale*. As the name implies, there are no negative temperatures on an absolute temperature scale, and the lowest attainable temperature is absolute zero. The absolute temperature scale in the *SI* system is the *Kelvin scale*. The relation between the Celsius scale and Kelvin scale is

$$\boxed{T(\text{K}) = T(^\circ\text{C}) + 273.15}$$

where T is the temperature.

In the English system, the absolute temperature scale is the *Rankine scale*. It is related to the Fahrenheit scale by

$$\boxed{T(\text{R}) = T(^\circ\text{F}) + 459.67}$$

The temperature scales in the above two systems are related by

$$T(\text{R}) = 1.8T(\text{K})$$

and
$$T(^\circ\text{F}) = 1.8T(^\circ\text{C}) + 32$$

In the two-point temperature scale, a fixed scale of temperature between the melting point of pure ice and the boiling point of pure water at standard atmospheric pressure has been laid down. But in practical situations, temperatures much lower than the melting point of pure ice and much higher than the boiling point of pure water are commonly encountered. Therefore, the fixed point scale becomes inadequate for the measurement of low or high temperatures.

An extension of the thermometric scale is therefore made by introducing further fixed points, using substances other than water. The fixed points are laid down in what is known as the *International Practical Temperature Scale* (IPTS). The IPTS is regularly reviewed by the international committee of weights and measures. The latest recommendations of the committee is that of the International Temperature Scale of 1968. The recommendations are published fully in the booklet entitled, *The Practical Temperature Scale of 1968*, sponsored by the National Physical Laboratory (NPL) and the Ministry of Technology, now the department of Trade and Industry, and published by Her Majesty's Stationery Office.

Quoting from the booklet (Joel, 1991): "The International Practical Temperature Scale gives the practical realization of thermodynamic temperatures and is recognized as the scale on which all national standardizing laboratories and others should base their measurements."

9.2.1   The International Practical Temperature Scale

In the International Practical Temperature Scale of 1968, the practical Kelvin temperature is given the symbol T_{68}. The international practical Celsius temperature is given the symbol t_{68}. The relationship between them is

$$t_{68} = T_{68} - 273.15\,\mathrm{K}$$

The scale of fixed points given in IPTS–68 is listed in Table 9.1.

The standard instrument used for measurement of temperature from 13.8 K to 630.74°C is the platinum resistance thermometer. The standard instrument used for measurement from 630.74°C to 1064.43°C is the platinum –10 percent rhodium/platinum thermocouple.

Above 1337.58 K the International Practical Temperature of 1968 is defined by the Planck's law of radiation with 1337.58 K as the reference temperature. It should be noted that the list of reference points given above is only partial. The international practical temperature scale also gives secondary reference points for a range of other substances such as nitrogen, mercury, cadmium, nickel, tungsten, etc.

9.3 Temperature Measurement

A substance used for the measurement of temperature must experience some recognizable change when its temperature is changed. Further, this change must admit a constant repetition without deterioration.

The changes which have been commonly used for temperature measurement are the following.

- Increase or decrease in size, that is, expansion or contraction.

- Increase in pressure if enclosed.

- Change of color.

- Change of state.

- Change of electrical resistance.

- Generation of electromotive force.

- Change in degree of surface radiation.

The choice of material and effect used for measurement depends upon the temperature range considered, degree of accuracy required, type of installation, and the cost.

Now let us see the details of the thermometric devices commonly used for temperature measurements.

9.3.1 Fluid Thermometers

Thermometer is a general name given to thermometric measuring devices. Some of the commonly used thermometers are the *mercury-in-glass thermometer*, *Beckmann thermometer*, and *gas thermometer*.

Table 9.1 International practical temperatures

Equilibrium state	Assigned value of International Practical Temperature	
	T_{68}	t_{68}
Equilibrium between solid, liquid, and vapor phases of equilibrium hydrogen (triple point of equilibrium hydrogen)	13.81 K	− 259.34 °C
Equilibrium between the liquid and vapor phases of equilibrium hydrogen at a pressure of 33330.9 N/m² (25/76 standard atmosphere)	17.042 K	− 256.108 °C
Equilibrium between the liquid and vapor phases of equilibrium hydrogen (boiling point of equilibrium hydrogen)	20.28 K	− 252.87 °C
Equilibrium between the liquid and vapor phases of neon (boiling point of neon)	27.102 K	− 246.048 °C
Equilibrium between the solid, liquid, and vapor phases of oxygen (triple point of oxygen)	54.361 K	− 218.789 °C
Equilibrium between the liquid and vapor phases of oxygen (boiling point of oxygen)	90.188 K	− 182.962 °C
Equilibrium between the solid, liquid, and vapor phases of water (triple point of water)	273.15 K	0.01 °C
Equilibrium between the liquid and vapor phase of water (freezing point of water)	373.15 K	100 °C
Equilibrium between the solid and liquid phases of zinc (freezing point of zinc)	692.73 K	419.58 °C
Equilibrium between the solid and liquid phases of silver (freezing point of silver)	1235.08 K	961.93 °C
Equilibrium between the solid and liquid phases of gold (freezing point of gold)	1337.58 K	1064.43 °C

9.3.1.1 Mercury-in-Glass Thermometer

Mercury-in-glass thermometer the most common type of thermometer. A typical form of mercury-in-glass thermometer is illustrated in Figure 9.1.

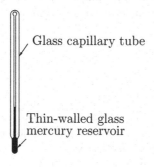

Figure 9.1: Mercury-in-glass thermometer

It consists of a fine-bore glass tube, called the capillary tube, on the bottom of which is fused a thin-wall glass tube which is generally cylindrical but can sometimes be spherical. The tube and bulb are then completely filled with mercury by repeated heating and cooling and finally, in the filled condition, it is brought to a temperature higher than its intended operational range. It is then sealed. Upon cooling the mercury will contract and the mercury level in the capillary tube will fall to some level depending upon the prevailing ambient temperature. The scale is then etched onto, or fitted to, the glass tube. To avoid error due to the shrinkage of the glass, the scale is not added until the thermometer has been left to age for some time. The glass stem of thermometer requires a long time to contract after heating. Because of this when the thermometer is cooled after heating, the glass will tend to remain enlarged due to expansion; hence a depression to zero is experienced. This depression for some glasses may be as much as 0.11°C after heating to 100°C. An accurate thermometer is often given a calibration to offset this.

Mercury is well suited for a thermometer since it does not wet the glass, and it has a reasonable expansion. It has a freezing point of −38.86°C and a boiling point of 356.7°C. These temperatures therefore limit the range of the mercury-in-glass thermometer. The upper temperature limit can be increased by the introduction of a gas, such as nitrogen, into the thermometer. However, at too high temperatures, the glass itself becomes the limiting factor, since the glass will soften and eventually melt at such high temperatures.

For temperatures lower than −38.86°C, liquids other than mercury are sometimes used. Alcohol, sometimes dyed red, is commonly used. This has a freezing point of −113°C.

9.3.2   Beckmann Thermometer

The Beckmann thermometer is used for the accurate determination of small temperature change, such as that encountered while using the bomb calorimeter. This is meant only for the determination of temperature change and not the temperature itself. The maximum temperature change that can be measured is usually 6°C. The construction details of a Beckmann thermometer are illustrated in Figure 9.2.

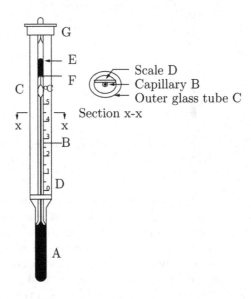

Figure 9.2: Beckmann thermometer

For accurate measurements, the thermometer should have a large bulb, about 50 mm long and about 10 mm in diameter. The overall length of the thermometer is about 0.6 m, and the 6°C scale has a length of about 180 mm. Each degree on the scale is divided into 100 divisions, resulting in an accuracy of 1/100 of a degree Celsius. The mercury bulb A is fitted with a long capillary tube B. The capillary tube is surrounded by an outer glass tube C into which is inserted the scale D. The section of the arrangement is shown in Figure 9.2. A small mercury reservoir E is kept at the top of the capillary tube. The capillary tube feeds into the reservoir via the drop point F. The thermometer is surrounded by cap G.

Before use, the thermometer must be set. For measurement of a small increase in temperature, the mercury level is set low on the scale. For measuring a small decrease in temperature the mercury level is set high on the scale. If the mercury level in the thermometer is too low then the bulb is heated until the mercury rises up through the drop point to join the reserve mercury in the reservoir. The thermometer is then cooled and some mercury from the reservoir will be pulled down into the capillary. A light tap on the thermometer at the drop point will separate the mercury again and, when settled, the mercury level in the capillary will be higher.

If the mercury level is too high in the capillary then the bulb is heated until the mercury level rises up through the drop point to join the mercury in the reservoir. Heating is continued and in this way mercury from the capillary is driven into the reservoir. A light tap at the drop point will again separate the mercury and the level in the capillary will drop to a lower level. Thus, the mercury level can be set to any part of the scale.

9.3.3 Gas Thermometer

It was originally used as the fundamental temperature measuring device to establish the various temperatures given as the International Temperature Scale in relationship to the originally conceived fixed points of the freezing point and boiling point of pure water. The parts of a simple gas thermometer are shown in Figure 9.3. It consists of a bulb A onto which a bent glass tube B is fixed.

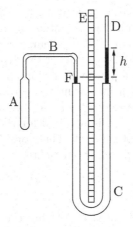

Figure 9.3: A gas thermometer

The bent tube B is connected to a vertical glass tube D, using a rubber or plastic tube C, as shown in Figure 9.3. A scale E is fixed between the glass tubes B and D. It is filled with mercury and the mercury level in the bent tube is always adjusted to a fixed level, marked F. Thus, at all times, there is a fixed volume of gas in the tube.

The pressure of the gas in the bulb A can be determined by measuring the difference in mercury levels h. It is assumed here that *equal changes of pressure of the gas in the bulb are produced by equal changes of temperature of the gas.* Thus, the pressure–temperature relation for the gas in the bulb is linear. Hydrogen gas is normally used in this apparatus.

The thermometer has to be calibrated before use. To do this, the bulb is immersed first in pure melting ice and it was recommended by the international committee on *weights and measures* that the pressure should be adjusted to 100 cm of mercury column height. This pressure (p_0) will then correspond to 0°C. The bulb is then immersed in the stream of pure boiling water. The gas will expand but its volume is brought back to the original volume by raising the vertical glass tube. Hence, the pressure of the gas would have been raised. Let this pressure be denoted by p_{100}, which corresponds to the pressure at 100°C. Since the pressure rise per degree is constant, we have

$$\text{Pressure rise/°C} = (p_{100} - p_0)/100 \text{ cm Hg} = \alpha$$

Hence, for some unknown temperature $T°C$ and the corresponding pressure p_T, we have the relation

$$\alpha T = p_T - p_0$$

or

$$T = \boxed{\frac{p_T - p_0}{\alpha} \text{ °C}}$$

This thermometer can be used for measuring a wide range of temperatures. At very low temperatures, helium is substituted for hydrogen, since helium has a lower condensation temperature. For high temperatures nitrogen is used instead of hydrogen. For use at temperatures higher than the melting point of glass, the bulb is made of platinum or a platinum–iridium alloy.

The bulb will expand with increasing temperature and also will elastically expand or contract as the pressure increases or decreases. Therefore, for accurate measurements, correction is required for these effects. Handling of this thermometer is comparatively difficult. Therefore, it is usually used as a standard for calibrating other types of thermometer.

Example 9.1

A gas thermometer was calibrated by placing the bulb in melting ice at 0°C and the difference in height of the mercury column was 820 mm. The bulb was then placed in steam at 100°C and the mercury column was adjusted to be 1300 mm. The bulb was then placed in a fluid of unknown temperature and, after adjustment, the difference in height of the mercury column read was 97 cm. Determine the temperature of the fluid.

Solution

Let the atmospheric pressure be 76 cm of Hg. At 0°C,

$$p_0 = 82 + 76 = 158 \text{ cm Hg}$$

At 100°C,

$$p_{100} = 130 + 76 = 206 \text{ cm Hg}$$

Therefore,

$$\alpha = \frac{p_{100} - p_0}{100} = \frac{206 - 158}{100} = 0.48$$

The pressure at unknown temperature T is

$$p_T = 97 + 76 = 173 \text{ cm Hg}$$

Therefore, the unknown temperature becomes

$$T = \frac{p_T - p_0}{\alpha} = \frac{173 - 158}{0.48} = \boxed{31.25°C}$$

Note: The problem can also be solved without converting the pressures to absolute values, since at all places only the difference in pressure is used. However, it is better to make it a practice to convert the pressure and temperature to absolute units before using in any calculation.

Example 9.2

Determine the response time (i.e., the time constant) of the mercury-in-glass thermometer of 3 mm bulb diameter, in an environment with heat transfer coefficient 100 W/(m² K). Take the specific heat coefficient of mercury as 0.14 kJ/(kg K).

Solution

Given, $d = 0.003$ m, $h = 100$ W/(m² K), $c = 140$ J/(kg K).

For mercury, $\rho = 13600$ kg/m³. The mass of the mercury in the bulb is

$$
\begin{aligned}
m &= \rho \mathbb{V} \\[2mm]
&= \rho \times \frac{4}{3}\pi r^3 \\[2mm]
&= 13600 \times \frac{4}{3}\pi \times 0.0015^3 \\[2mm]
&= 1.923 \times 10^{-4} \text{ kg}
\end{aligned}
$$

The surface area of the bulb is

$$
\begin{aligned}
A &= 4\pi r^2 \\[2mm]
&= 4\pi \times 0.0015^2 \\[2mm]
&= 2.827 \times 10^{-5} \text{ m}^2
\end{aligned}
$$

The response time, by Equation (9.30), is

$$\tau \;=\; \frac{mc}{hA}$$

$$=\; \frac{(1.923 \times 10^{-4}) \times 140}{100 \times (2.827 \times 10^{-5})}$$

$$=\; \boxed{9.52\,\text{s}}$$

9.3.4   Temperature Gauges Using Fluids

Heating of a fluid in a confined space results in its pressure rise. The fluid can fill the space completely or partially. The change of pressure as a result of the change of temperature of the fluid is used for temperature measurement. A typical temperature gauge using fluid is illustrated in Figure 9.4.

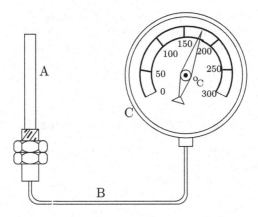

Figure 9.4: Temperature gauge using fluid

This gauge has a temperature pick-up unit A which is fitted into a suitable pocket at the point where the temperature is to be measured. The pick-up unit is filled with the working fluid, such as mercury. The unit is coupled by means of a long metal capillary tube B to a Bourdon pressure gauge C which is calibrated to measure temperature, instead of pressure, which it measures usually. In operation, as the temperature changes so the pressure of the fluid system changes and hence the Bourdon gauge pointer moves on over the scale. If the gauge is not required for distance reading then the pick-up unit is mounted directly to the gauge. This is a robust type of thermometer which has wide industrial application where extreme accuracy of temperature measurement is not stringent.

9.4 Temperature Measurement by Thermal Expansion

The phenomenon of thermal expansion is made use of for temperature measurement in a number of temperature sensing devices. The expansion of solids is employed mainly in bimetallic elements by utilizing the differential expansion of bonded strips of two metals. Liquid expansion is used in the common liquid-in-glass thermometers. Restrained expansion of liquids, gases, or vapor results in a pressure rise, which is the basis of pressure thermometers.

9.4.1 Bimetallic Thermometers

When two metallic strips A and B with different thermal expansion coefficients α_A and α_B but at the same temperature are firmly bonded together, as shown in Figure 9.5, any temperature change will cause a differential expansion and the bonded strip, if unrestrained, will deflect into an uniform circular arc.

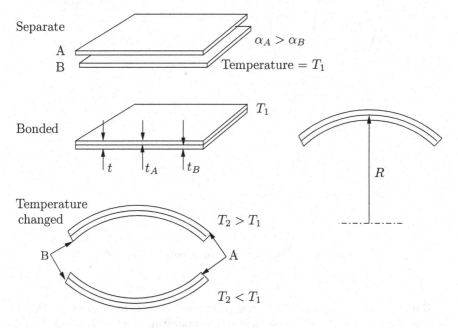

Figure 9.5: Bimetallic sensors

In most practical cases, the radius of curvature R of the deflection is given by

$$R \approx \frac{2t}{3(\alpha_A - \alpha_B)(T_1 - T_2)}$$

where t is the thickness of the bounded strip and $(T_1 - T_2)$ is the temperature rise. Combination of this equation with associated strength of material relations allows calibration of the deflections of various types of elements in practical use.

Since there are no practically usable metals with negative thermal expansion, the element B is generally made of Invar, a nickel steel with a nearly zero expansion coefficient. While brass was employed originally, a variety of alloys are used now for the high-expansion strip A, depending on the mechanical and electrical characteristics required. Details of materials and bonding processes are, in some cases, considered trade secrets. Some of the popular configurations developed to meet application requirements are shown in Figure 9.6.

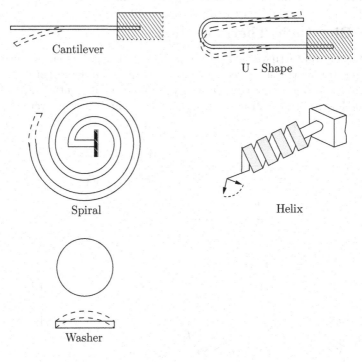

Cantilever

U - Shape

Spiral

Helix

Washer

Figure 9.6: Bimetallic sensor configurations

Bimetallic devices are used for temperature measurements and vary widely as combined sensing and control elements in temperature-control systems, mainly of the on–off type. Also, they are used as overload cut-out switches in electric apparatus by allowing the current to flow through the bimetal, heating and expanding it, and causing a switch to open when excessive current flows. The accuracy of bimetallic elements varies largely, depending on the application requirements. Since majority of control applications are not extremely critical, requirements can be satisfied with a rather low-cost device. The working temperature range is about −35 to 550°C. Errors of the order of 0.5 to 1 percent of scale range may be expected in bimetal thermometers of high quality.

9.5 Temperature Measurements by Electrical Effects

Temperature measurements based on electrical effects such as change in resistance with temperature have become popular because they are very convenient techniques and they furnish a signal that is easily detected, amplified, or used for control purposes. Further, they are generally quite accurate when proper calibration and compensation are done. Some of the popular temperature measuring devices based on electrical effects are

- Thermocouples.

- Resistance temperature detectors, RTD.

- Thermistors.

Among these, thermocouples are the most popularly used temperature measuring devices.

9.5.1 Thermocouples

Thermocouples are widely used for temperature measurements in fluid streams. Thermocouples are devices which operate on the principle that "*a flow of current in a metal accompanies a flow of heat.*" This principle is popularly known as the *Seebeck effect*. In some metals, such as copper, platinum, iron, and chromal, the flow of current is in the direction of heat flow. In some other metals, such as constantan, alumel, and rhodium, the flow of current is in the direction opposite to that of the heat flow. These two groups are called *dissimilar metals*. Thermocouples consist of two dissimilar metals joined together at two points, one point being the place where the temperature is to be measured and the other point being a place where the temperature is known, termed the *reference junction*.

Now, let us see the thermocouples in some detail. When two dissimilar metal wires are joined at both ends and one of the ends is heated, as shown in Figure 9.7, there is a continuous flow of current in the thermocouple circuit, in accordance with *Seebeck effect*.

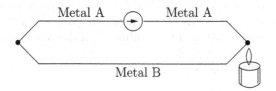

Figure 9.7: The Seebeck effect

If this circuit is broken at the center, the net open circuit voltage, also known as Seebeck voltage becomes a *function of the junction temperature* and the *composition of the two metals*. All dissimilar metals exhibit this effect. Some of the popular dissimilar metals are *copper–constantan, iron–constantan, nickel–constantan, chromal–constantan, chromal–alumal, and platinum–rhodium.* Copper–constantan is widely used in gas dynamic studies at moderate temperatures, since it has a temperature measuring range of $-270°C$ to $400°C$. For small changes in temperature the Seebeck voltage is linearly proportional to the temperature.

9.5.2   Measurement of Thermocouple Voltage

It is not possible to measure the Seebeck voltage directly, since we must connect a voltmeter to the thermocouple, and the leads of the voltmeter themselves create a new thermoelectric circuit. Let us connect a voltmeter across a copper–constantan thermocouple, as shown in Figure 9.8.

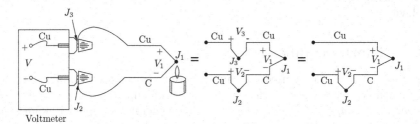

Figure 9.8: Measurement of junction voltage

We are interested in measuring only the voltage V_1 due to junction J_1. However, by connecting the voltmeter to measure the output of junction J_1, we have two more metallic junctions; J_2 and J_3. But J_3 is a copper-to-copper junction and therefore, it creates no *thermal electromotive force* (EMF) and hence $V_3 = 0$. The junction J_2 is copper-to-constantan and hence it will add an EMF (V_2) in opposition to V_1. The resultant voltmeter reading V will be proportional to the temperature difference between J_1 and J_2. This implies that, for measuring temperature at J_1, the temperature of J_2 must be known. One way to obtain the temperature of J_2 is to physically place the junction into an ice bath, forcing its temperature to be $0°C$ and establishing J_2 as the *reference junction*, as shown in Figure 9.9. Since both voltmeter terminal junctions are now copper–copper, they create no thermal EMF and the reading V on the voltmeter is proportional to the difference between J_1 and J_2.

Now, the voltmeter reading is given by

$$V = (V_1 - V_2) \approx \alpha \left(T_{j1} - T_{j2} \right)$$

where α is a constant of proportionality called *Seebeck coefficient* and T_{j1} and

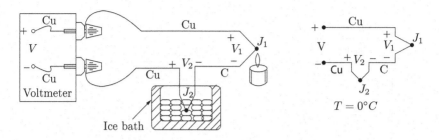

Figure 9.9: Reference junction

J_{j2} are the absolute temperatures. Therefore,

$$\boxed{V = \alpha T_{j1} - 0 = \alpha T_{j1}}$$

By adding the voltage of the ice point reference junction, we have now referenced the reading V to $0°C$. This method is very accurate because the ice point temperature can be precisely controlled. The ice point is used by the National Bureau of Standards (NBS) as the fundamental reference point for their thermocouple tables, so we can now look at the NBS tables and directly convert the measured voltage V to temperature T_{j1}. Even though the ice point proves to be a good reference junction, it is not a convenient one in practice, since care must be taken to keep the ice in proper enclosure and the molten water has to be removed continuously, in order to keep the temperature of the bath at $0°C$. Therefore, in laboratories it is a common practice to use baths like oil baths kept at constant temperature as the reference junction. In this case the temperature of the bath has to be added to the measured temperature to get the resultant temperature of the junction T_{j1}.

In the copper–constantan thermocouple, shown in Figure 9.9, the copper wire and the voltage terminal are made of the same material. But in other thermocouples like the iron–constantan thermocouple, the iron wire increases the number of dissimilar metal junctions in the circuit, as illustrated in Figure 9.10, as both voltmeter terminals become C–Fe thermocouple junctions.

If the junctions J_3 and J_4 are not at the same temperature, there will be an error introduced in the measurement. For more precise measurement, the copper leads of the voltmeter should be extended so that the copper-to-iron junctions are made on an isothermal block, as shown in Figure 9.11.

The isothermal block is an electrical insulator but a good heat conductor. It serves to keep the junctions J_3 and J_4 at the same temperature. The absolute temperature of the block is of no consequence in the thermocouple operation, since the two copper–iron junctions act in opposition. The relation

$$V = \alpha \left(T_1 - T_{\text{ref}} \right)$$

is still valid.

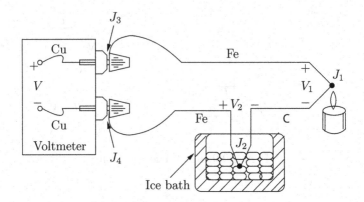

Figure 9.10: Iron–constantan thermocouple

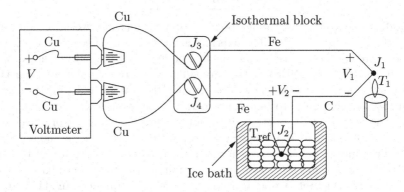

Figure 9.11: Iron–constantan couple with an isothermal block

Now, let us go a step further and replace the ice bath with another isothermal block, as shown in Figure 9.12.

The new block is at a reference temperature T_{ref}, and since J_3 and J_4 are still at the same temperature, it can be shown that,

$$V = \alpha \left(T_1 - T_{ref} \right)$$

Even this circuit is not compact since two thermocouples are connected in it. Let us eliminate the extra Fe wire in the negative (LO) lead containing the Cu–Fe junction (J_4) and the Fe–C junction (J_{ref}). This can be done by just joining the two isothermal blocks, as shown in Figure 9.12(b). The output voltage is still

$$V = \alpha(T_{J1} - T_{Jref})$$

The *law of intermediate metals* states that *a third metal (here iron) inserted between the two dissimilar metals of a thermocouple junction will have no effect*

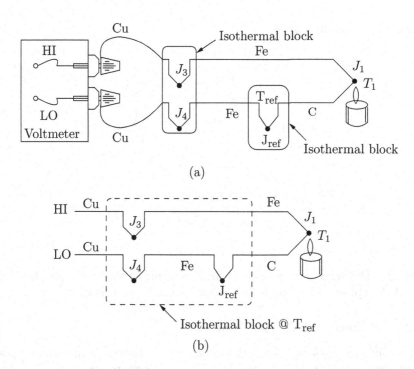

Figure 9.12: (a) Ice bath elimination; (b) unification of isothermal blocks

on the output voltage as long as the two junctions formed by the additional metal are at the same temperature. The law of intermediate metals may be illustrated schematically as shown in Figure 9.13.

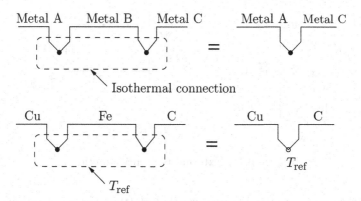

Figure 9.13: Law of intermediate metals

With this law, the equivalent circuit of iron–constantan thermocouple becomes that shown in Figure 9.14.

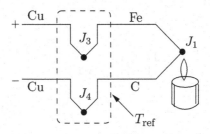

Figure 9.14: Equivalent circuit

Here again,

$$V = \alpha(T_{J1} - T_{\text{ref}})$$

where α is the Seebeck coefficient for iron–constantan thermocouple. Junctions J_3 and J_4 replace the ice bath and become the reference junction.

9.5.2.1 External Reference Junction

It is possible to replace the ice bath with an external temperature reference, like a thermistor or RTD. The thermistor, which is a *temperature sensitive resistor*, and the RTD, which is a *resistance temperature detector*, will be seen in detail in the following sections. Examine the thermocouple circuit illustrated in Figure 9.15.

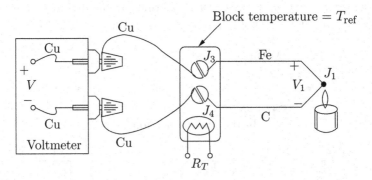

Figure 9.15: External reference junction

Here the temperature of the isothermal block (the reference junction) is measured directly and this measured temperature is used to determine the unknown temperature T_{J1}. A thermistor, whose resistance R_T is a function of temperature is made use of for measuring the absolute temperature of the reference junction. The isothermal block is designed in such a way to keep the junctions

J_3 and J_4 as well as the thermistor at the same temperature. Using a computer controlled digital multimeter we can measure

- The thermistor resistance R_T to find T_{ref} and convert T_{ref} to its equivalent junction voltage, V_{ref}.
- From the voltage V subtract V_{ref} to find V_1, and convert V_1 to temperature T_{J1}.

The above procedure which depends on the software of a computer to compensate for the effect of the reference junction is known as *software compensation.* The isothermal terminal block temperature sensor can be any device which has a characteristic proportional to absolute temperature: a thermistor, an RTD, or an integrated circuit sensor.

At this stage, it is natural to ask the question, what is the need for a thermocouple which requires a reference junction compensation, when we have devices like thermistor and RTD that will measure the absolute temperature directly? The answer to this question is that the thermistor, the RTD, and the integrated circuit transducers are useful only over a certain range of temperatures. The thermocouples are much more rugged than thermistors. They can be made on the spot, either by soldering or by welding. In short, thermocouples are the most versatile temperature transducers among the available ones. Further, since the measurement system performs the entire task of reference compensation and software voltage-to-temperature conversion, using a thermocouple becomes as easy as connecting a pair of wires.

In situations where it is necessary to measure more than one temperature, thermocouple measurement becomes convenient. This is accomplished with an isothermal reference junction for more than one thermocouple element, as illustrated in Figure 9.16.

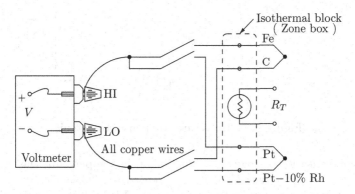

Figure 9.16: Zone box switching

As shown in Figure 9.16, a reed relay scanner connects the voltmeter to different thermocouples in sequence. All of the voltmeter and scanner wires are copper, irrespective of the type of thermocouple chosen. As long as we know the type of each thermocouple, we can mix thermocouple types on the same isothermal junction block, also called a *zone box*, and make the appropriate modifications in software. The junction block temperature sensor R_T is located at the center of the block to minimize errors due to thermal gradients.

9.5.2.2   Software Compensation

The most versatile technique available for thermocouple measurements is the software compensation. This technique is independent of the types of thermocouples chosen from a set of thermocouples connected on the same block, using copper leads throughout the scanner. When using a data acquisition system with a built-in zone box, the thermocouple is just connected like a pair of test leads. All of the conversions are performed by the computer. Even though software compensation is simple and versatile, it is slow since the computer requires an additional amount of time to calculate the reference junction temperature. To increase the speed of measurement, hardware compensation may be used.

9.5.2.3   Hardware Compensation

Instead of measuring the temperature of the reference junction and computing its equivalent voltage, as is done with software compensation, a battery could be inserted to cancel the offset voltage of the reference junction. The combination of the *hardware compensation* voltage and the reference junction voltage is equal to that of a 0°C junction.

A typical hardware compensation circuit is illustrated in Figure 9.17.

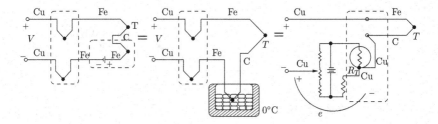

Figure 9.17: Hardware compensation circuit

The compensation voltage e is a function of the temperature sensing resistor resistance R_T. The voltage V is now referenced to 0°C, and may be directly read and converted to temperature by using the NBS tables. This circuit is also called *electronic ice point reference*. These circuits are commercially available for use with any voltmeter and with a wide variety of thermocouples. The major disadvantage with this circuit is that a unique ice point reference circuit is usually needed for each individual thermocouple type.

A practical ice point reference circuit that can be used in conjunction with a reed relay scanner to compensate an entire block of thermocouple inputs is shown in Figure 9.18.

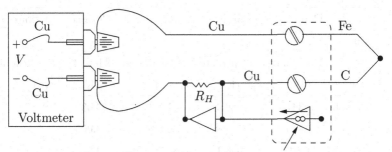

Integrated temperature sensor

Figure 9.18: Practical hardware compensation

All the thermocouples in the block must be of the same type, but each block input can accommodate a different thermocouple type by simply changing gain resistors. The advantage of the hardware compensation circuit is that it eliminates the need to compute the reference temperature. This saves two computation steps and makes a hardware compensation temperature measurement comparatively faster than a software compensation measurement. However, the hardware compensation is restricted to one thermocouple type per card, whereas the software compensation is versatile and accepts any thermocouple.

9.5.2.4 Voltage-to-Temperature Conversion

It is a common practice to express the thermoelectric *emf* in terms of the potential generated with a reference junction at 0°C. We have used hardware and software compensations to synthesize an ice-point reference. Now, all we have to do is to read the digital voltmeter and convert the voltage reading to temperature. But the job is not that simple due to the nonlinear relationship between the voltage and the temperature of a thermocouple. Output voltages for more common thermocouples are plotted as a function of temperature in Figure 9.19.

The slope of the curves (the Seebeck coefficient) in Figure 9.19 is plotted with temperature in Figure 9.20.

It is obvious from this figure that the thermocouple is a nonlinear device. A summary of the output characteristics of the most common thermocouple combinations is given in Table 9.2.

The output voltage V of a simple thermocouple circuit is usually expressed as

$$V = AT + \frac{1}{2}BT^2 + \frac{1}{3}CT^3 \tag{9.1}$$

where V is the voltage based on a reference junction temperature of 0°C, T is

Table 9.2 Thermal emf in absolute millivolts for commonly used thermocouple combinations (reference junction at 0°C)

Temperature °F	°C	Copper vs. Constantan (T)	Chromel vs. Constantan (E)	Iron vs. Constantan (J)	Chromel vs. Alumel (K)	Platinum vs. Platinum 10% Rhodium (S)
−300	−184.4	−5.341	−8.404	−7.519	−5.632	
−250		−4.745	−7.438	−6.637	−5.005	
−200	−128.9	−4.419	−6.471	−5.760	−4.381	
−150		−3.365	−5.223	−4.623	−3.538	
−100	−73.3	−2.581	−3.976	−3.492	−2.699	
−50	−1.626	−2.501	−2.186	−1.693		
0	−17.8	−0.674	−1.026	−0.885	−0.692	−0.092
50		0.422	0.626	0.526	0.412	0.064
100	37.8	1.518	2.281	1.942	1.520	0.221
150		2.743	4.075	3.423	2.667	0.408
200	93.3	3.967	5.869	4.906	3.819	0.597
250		5.307	7.788	6.425	4.952	0.807
300	148.9	6.647	9.708	7.947	6.092	1.020
350		8.085	11.728	9.483	7.200	1.247
400	204.4	9.523	13.748	11.023	8.314	1.478
450		11.046	15.844	12.564	9.435	1.718
500	260.0	12.572	17.942	14.108	10.560	1.962
600		15.834	22.287	17.178	12.865	2.472
700	371.1	19.095	26.637	20.253	15.178	2.985
800			31.108	23.338	17.532	3.524
1000	537.8		40.056	29.515	22.251	4.609
1200		48.927		26.911	5.769	
1500	815.6		62.240		33.913	7.514
1700					38.287	8.776
2000	1093.3				44.856	10.675
2500	1371.1				54.845	14.018
3000	1648.9					17.347

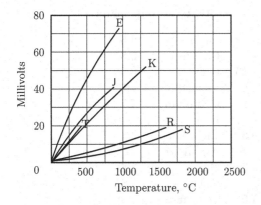

Figure 9.19: Thermocouple temperature versus voltage

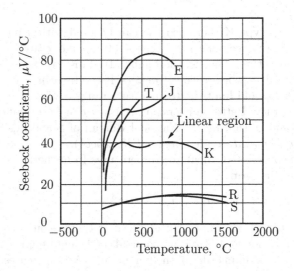

Figure 9.20: Seebeck coefficient versus temperature

the temperature in degree Celsius, and A, B, and C are constants dependent on the thermocouple material.

The thermoelectric power or sensitivity of a thermocouple is given by

$$S = \frac{dV}{dT} = A + BT + CT^2 \tag{9.2}$$

Appropriate values of the sensitivity of various thermocouple materials relative to platinum at 0°C is given in Table 9.3 (Holman, 1989).

Now examine Figure 9.20. A horizontal line in the figure would indicate that the thermocouple device is linear. It is seen that the shape of the type K thermocouple approaches constant over a temperature range from 0°C to 1000°C.

Table 9.3 Themoelectric sensitivity of some materials

Bismuth	−72	Silver	6.5
Constantan	−35	Copper	6.5
Nickel	−15	Gold	6.5
Potassium	−9	Tungsten	7.5
Sodium	−2	Cadmium	7.5
Platinum	0	Iron	18.5
Mercury	0.6	Nichrome	25
Carbon	3	Antimony	47
Aluminum	3.5	Germanium	300
Lead	4	Silicon	440
Tantalium	4.5	Tellurium	500
Rhodium	6	Selenium	900

Consequently, the type K thermocouple can be used with a multiplying voltmeter and an external ice point reference to obtain a moderately accurate direct readout of temperature. That is, the temperature display involves only a scale factor. This procedure works with voltmeters. It can be seen that use of one constant scale factor would limit the temperature range of the system and restrict the system accuracy. Better conversion accuracy can be obtained by reading the voltmeter and consulting the National Bureau of Standards thermocouple tables (*Omega Temperature Measurement Handbook*).

To develop software to calculate temperatures from thermocouple voltages, a polynomial in the form

$$T = a_0 + a_1 x + a_2 x^2 + a_3 x^3 + \ldots\ldots + a_n x^n \tag{9.3}$$

can be used. where T is temperature (°C), x is thermocouple voltage (volts), with reference junction at 0°C, a is polynomial coefficient unique to each thermocouple, and n is maximum order of the polynomial. As n increases, the accuracy of the polynomial improves. A representative number is $n = 9$, for ±1°C accuracy. Lower-order polynomials may be used over a narrow temperature range to obtain higher system speed.

The accuracy with which each polynomial fits the NBS tables is indicated in Table 9.4, which is an example of the polynomials used for the packages for a data acquisition system. Rather than directly calculating the exponentials, the computer is programmed to use the nested polynomial form to save execution time.

The polynomial fit rapidly degrades outside the temperature range shown in Table 9.3 and should not be extrapolated outside those limits. As mentioned above, the nested polynomial form minimizes the execution time. For example, the fifth-order polynomial for iron-constantan thermocouple would be written as

$$T = a_0 + x(a_1 + x(a_2 + x(a_3 + x(a_4 + a_5 x)))) \tag{9.4}$$

Table 9.4 NBS polynomial coefficients

	Type E	Type J	Type K
	Nickle-10 % Chromium(+) versus Constantan(−)	Iron(+) versus Constantan(−)	Nickel-10 % Chromium(+) versus Nickel-5 % (Aluminum Silicon)
	−100°C to 1000°C ±0.5° 9th order	0°C to 760°C ±0.1° 5th order	0°C to 1370°C ±0.7° 8th order
a_0	0.104967248	-0.048868252	0.226584602
a_1	17189.45282	19873.14503	24152.10900
a_2	−282639.0850	−218614.5353	67233.4248
a_3	12695339.5	11569199.78	2210340.682
a_4	−448703084.6	−264917531.4	−860963914.9
$a_5.$	$1.10866E+10$	2018441314	$4.83506E+10$
a_6	$−1.76807E+11$		$−1.18452E+12$
a_7	$1.71842E+12$		$1.38690E+13$
a_8	$−9.19278E+12$		$−6.33708E+13$
a_9	$2.06132E+13$		

	Type R	Type S	Type T
	Platinum-13 % Rhodium(+) versus Platinum(−)	Platinum-10 % Rhodium(+) versus Platinum(−)	Copper(+) versus Constantan(−)
	0°C to 1000°C ±0.5° 8th order	0°C to 1750°C ±1° 9th order	−160°C to 400°C ±0.5° 7th order
a_0	0.263632917	0.927763167	0.100860910
a_1	179075.491	169526.5150	25727.94369
a_2	−48840341.37	−31568363.94	−767345.8295
a_3	$1.90002E+10$	8990730663	78025595.81
a_4	$−4.82704E+12$	$−1.63565E+12$	−9247486589
$a_5.$	$7.62091E+14$	$1.88027E+14$	$6.97688E+11$
a_6	$−7.20026E+16$	$−1.37241E+16$	$−2.66192E+13$
a_7	$3.71496E+18$	$6.17501E+17$	$3.94078E+14$
a_8	$−8.03104E+19$	$−1.56105E+19$	
a_9			$1.69535E+20$

Table 9.5 Digital voltmeter sensitivity required for the thermocouple voltage measurements

Thermocouple type	Seebeck coefficient $\mu V/°C$ @ 20°C	DVM sensitivity for 0.1°C μV
E	62	6.2
J	51	5.1
K	40	4.0
R	7	0.7
S	7	0.7
T	40	4.0

Equation (9.4) is a fifth-order polynomial.

The calculation of higher-order polynomial is a time-consuming task for a computer. As mentioned above, we can save time by using a lower-order polynomial for a smaller temperature range. In the software for one data acquisition system, the thermocouple characteristic curve is divided into 8 sectors, as shown in Figure 9.21, and each sector is approximated by a third-order polynomial. All the foregoing procedures assume that the thermocouple voltage can be mea-

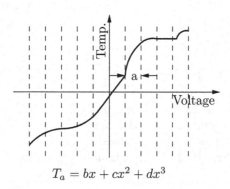

$$T_a = bx + cx^2 + dx^3$$

Figure 9.21: Curve divided into sectors

sured easily and accurately. However, in reality it is not so due to the fact that thermocouple output voltages are very small. To get an idea of the sensitivity of the digital voltmeter required for the measurement of the thermocouple voltage, examine Table 9.5.

Even for the common type K thermocouple, the voltmeter must be able to resolve 4 μV to detect a 0.1°C change. The magnitude of this signal is susceptible for noise creep into any system. To overcome this problem instrument designers utilize several fundamental noise rejection techniques, including tree switching, normal mode filtering, integration, and guarding.

9.5.2.5 Noise Rejection

The noise rejection methods used in thermocouple measurements include trees switching, normal mode filtering, integration, and guarding.

9.5.2.6 Tree Switching

It is a method of organizing the channels of a scanner into groups, each with its own main switch. Without tree switching, every channel can contribute noise directly through its stray capacitance. With tree switching, groups of parallel channel capacitance are in series with a single tree switch capacitance. It results in considerable reduction of crosstalk in a large data acquisition system, due to the reduced inter channel capacitance.

9.5.2.7 Analogue Filter

A filter may be directly used at the input of a voltmeter to reduce noise. It reduces interference dramatically. However, this causes the voltmeter to respond more slowly to step inputs.

9.5.2.8 Integration

It is an analogue to digital, A/D, technique which essentially averages noise over a full line cycle, thus power-line-related noise and its harmonics are totally eliminated. If the integration period is chosen to be less than an integer line cycle, its noise rejection properties are essentially negated. Since thermocouple circuits that cover long distances are especially susceptible to power-line-related noise, it is advisable to use an integrating analogue to digital converter to measure thermocouple voltage.

9.5.2.9 Guarding

The guard, physically a metal box surrounding the entire voltmeter circuit, is connected to a shield surrounding the thermocouple wire, and serves to shunt the integrating current. The digital voltmeter (dvm) guard is especially useful in eliminating noise voltages created when the thermocouple junction comes into contact with a common mode noise source.

9.6   Practical Thermocouple Measurements

So far, we have discussed different aspects about the thermocouple device including the concepts of the reference junction, a polynomial fit to process temperature data, methods to reduce noise and so on. Now, let us have a close look at thermocouple wire itself. This is essential since the polynomial curve fit relies on the thermocouple wire being perfect. In other words, if the wire maintains its perfectness, the initial calibration is valid throughout the life of the wire. Otherwise, it will become decalibrated while employing it for temperature measurement. Let us see the causes which will affect the perfectness of a thermocouple wire, introducing error in the temperature measurements. Most of the measurement errors may be due to one of the following sources.

- Poor junction connection.

- Decalibration of thermocouple wire.

- Shunt impedance and galvanic action.

- Thermal shunting.

- Noise and leakage currents.

- Thermocouple specifications.

- Documentation.

9.6.1   Poor Junction Connections

Some of the well-established methods to connect two thermocouple wires are soldering, silver-soldering, and welding. When the thermocouple wires are soldered together, a third metal is introduced into the thermocouple circuit. But the solder will not introduce any error as long as the temperatures of both sides of the thermocouple are the same. However, the solder imposes a limit on the maximum temperature to which the junction can be subjected. To reach a higher measurement temperature, the joint must be welded. But in welding the thermocouple wire utmost care must be exercised, since over heating can degrade the wire, and the welding gas and atmosphere in which the wire is welded can both diffuse into the thermocouple metal, changing its characteristics. The difficulty is compounded by the very different nature of the two metals being joined. Commercial thermocouples are welded on expensive machinery using a capacitive–discharge technique to ensure uniformity. A poor weld can result in an open connection, which can be detected by performing an open thermocouple check. This is a common test function available with data loggers. The open thermocouple is the easiest malfunction to detect, but it is not necessarily the most common mode of failure.

9.6.2   Decalibration

The decalibration is a process which alters the physical makeup of the thermocouple wire making it not conform with the NBS polynomial within the specified limit. Decalibration may be caused by diffusion of atmospheric particles into the metal. It may also be caused by high temperature annealing or by cold-working the metal.

Thermocouple wires are bound to have some imperfection due to manufacturing limitations. These defects will result in output voltage errors. These inhomogeneities can be especially disruptive if they occur in a region of the steep temperature gradients. Since it is not possible to have any idea about the location of imperfection within a wire, the best thing one can do is to avoid creating a steep gradient. Gradients can be reduced by using metallic sleeving or by careful placement of the thermocouple wire.

9.6.3 Shunt Impedance

High temperature can spoil the thermocouple wire insulation. The resistance of the insulation decreases exponentially with increasing temperature, even to the extent that it creates a virtual junction. Consider a completely open thermocouple operating at a high temperature, as shown in Figure 9.22.

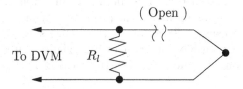

Figure 9.22: Leakage resistance

The leakage resistance R_l can be sufficiently low to complete the circuit path and give us an improper voltage reading. Now, let us assume that the thermocouple is not open, but a very long section of small diameter wire is used, as shown in Figure 9.23.

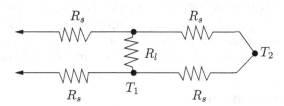

Figure 9.23: Virtual junction

When the thermocouple wire is small, its series resistance R_s will be quite high and under extreme condition, $R_l \ll R_s$. It means that the thermocouple junction will appear to be at R_l and the output will be proportional to T_1 and not T_2.

High temperature can also have other detrimental effects on thermocouple wires. The impurities and chemicals within the insulation can actually diffuse into the thermocouple metal, causing the temperature–voltage dependence to deviate from the published values. The insulation should be carefully chosen for using thermocouple at high temperatures. Atmospheric effects can be minimized by choosing proper protective metallic or ceramic sheath.

9.6.4　Galvanic Action

The dyes used in some thermocouple insulation will form an electrolyte in the presence of water. This creates a galvanic action, with a resultant output hundreds of times more than the Seebeck effect. Precautions should be taken to shield the thermocouple wires from all harsh atmospheres and liquids.

9.6.5　Thermal Shunting

It is a problem associated with the mass of the thermocouple. Since some energy is used to heat the mass of the thermocouple, the thermocouple will slightly alter the temperature it was meant to measure. If the temperature to be measured is small, the thermocouple must naturally be small. But a thermocouple made with a small wire is far more susceptible to the problems of contamination, annealing, strain, and shunt impedance. To minimize these effects, thermocouple extension wire can be used. Extension wire is a commercially available wire primarily intended to cover long distances between the measuring thermocouple and the voltmeter.

Extension wire is made of metals having a Seebeck coefficient very similar to a particular thermocouple type. Generally it is large in size so that its series resistance does not become a factor when traversing long distances.

9.6.6　Wire Calibration

Thermocouple wire is manufactured to a certain specification, signifying its conformance with the NBS tables. The specifications can sometimes be enhanced by calibrating (testing it at known temperatures) the wire. If the wire is calibrated intending to confirm its specifications, it becomes more imperative that all of the aforementioned conditions be heeded in order to avoid decalibration.

9.6.7　Documentation

If might look silly to speak of documentation as being a source of voltage measurement error. But the fact is that thermocouple systems, by their very ease of use, invite a large number of data points. The sheer magnitude of the data can become quite large. When a large amount of data is taken, there is an increased probability of error due to mistaking of lines, using the wrong NBS curve, etc.

9.6.8　Diagnostics

The diagnostic system for detecting a faulty thermocouple and data channels consists of three components. They are the *event record*, the *zone box test*, and the *thermocouple resistance history*. Now let us see the details of these components.

9.6.8.1 Event Record

Even though it is not a test by itself, a recording of all pertinent events will be of great value as a diagnostic tool. For instance, let us examine the following event record.

- September 21 event record.

- 10.00 power failure.

- System power returned.

- Changed TC07 to type K thermocouple.

- New data acquisition program.

- TC07 appears to be incorrect reading.

Scrutiny of the program listing reveals that number TC07 uses a type J thermocouple, thus interpreting it as a type J. But from the event record, apparently thermocouple TC07 was not entered into the program. Even though not all anomalies can be discovered this easily, the event record can provide valuable insight into the reason for an unexplained change in a system measurement.

9.6.8.2 Zone Box Test

As we know, the zone box is the isothermal terminal block of known temperature used in place of an ice bath reference. If the thermocouple is short circuited temporarily at the zone box, as shown in Figure 9.24, the system should read a temperature very close to that of the zone box.

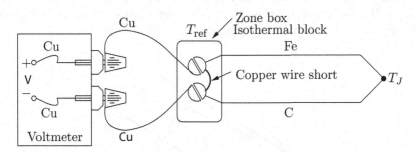

Figure 9.24: Shorting the thermocouple at the terminals

If the thermocouple lead resistance is much larger than the shunting resistance, the copper wire shunt makes the voltage V become zero. In the normal unsorted case, we want to measure T_j, and the system reads

$$V \approx \alpha \left(T_j - T_{\mathrm{ref}} \right)$$

But for the functional test, we have shorted the terminals so that $V = 0$. The indicated temperature T_j' is thus,

$$0 = \alpha \left(T_j' - T_{\text{ref}} \right)$$

i.e.,

$$T_j' = T_{\text{ref}}$$

Thus, for a DVM reading of $V = 0$, the system will indicate the zone box temperature. First we read T_j (different from T_{ref}), then we short the thermocouple with a copper wire and make sure that the system indicates the zone box temperature instead of T_j. This test ensures that the controller, scanner, voltmeter, and zone box compensation are all operating properly.

9.6.9 Thermocouple Resistance

An abrupt change in the resistance of a thermocouple circuit can act as a warning indicator. If the resistance versus time for each set of thermocouple wires is plotted, then any sudden change in resistance can be spotted, which could be an indication of an open wire, a wire shorted due to insulation failure, change due to vibration fatigue, or due to one of many failure mechanisms.

9.7   The Resistance Temperature Detector

Temperature can also be measured using *resistance temperature detector* (RTD) devices. They work on the principle that metals have a marked temperature dependence. The classical resistance temperature detector construction using platinum was proposed by C.H. Meyers in 1932 (Omega temperature measuring devices). He wound a helical coil of platinum on a crossed mica bed, as shown in Figure 9.25, and mounted the assembly inside a glass tube.

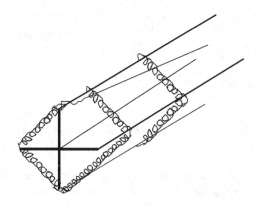

Figure 9.25: RTD construction by Meyer

This construction resulted in minimized strain on the wire while maximizing resistance. However, the thermal contact between the platinum and the measured point becomes poor due to this construction. This results in poor thermal response of the device. Further, fragility of the structure limits its use primarily to that of a laboratory standard.

Evans and Burns proposed a modification to Meyers' design, which is called the *bird-cage element*. The platinum element remains largely unsupported, which allows it to move freely when expanded or contracted due to temperature variations. Strain-induced resistance changes with time and temperature are thus minimized, and the bird-cage becomes the ultimate laboratory standard. Because of its unsupported structure and subsequent susceptibility to vibration, this configuration is still a bit too fragile for an industrial environment.

To make the probe rugged, the platinum wire is bifilar wound on a glass or ceramic bobbin. The winding reduces the effective enclosed area of the coil to minimize the magnetic pickup and its related noise. After winding the wire onto the bobbin, the assembly is sealed with a coating of molten glass. The sealing process assures that the RTD will maintain its integrity under extreme vibrations. But it limits the expansion of the platinum metal at high temperatures. Unless the coefficients of expansion of the platinum and the bobbin match perfectly, stress will be placed on the wire as the temperature changes. This will result in a strain-induced resistance change, which may result in a permanent charge in the resistance of the wire.

There are partially supported RTD probes which offer a compromise between the bird-cage approach and the sealed helix. One such approach uses a platinum helix threaded through a ceramic cylinder and affixed via glass-frit. These devices will maintain stability in rugged vibrational applications.

The RTD devices measure temperature quite accurately. As seen, RTD is some type of resistance element which is exposed to the temperature to be measured. The temperature is indicated as a change in the resistance of the element.

A list of materials which may be used as resistance elements and their characteristics are given in Table 9.6.

The linear temperature coefficient of resistance α is defined as

$$\alpha = \frac{R_2 - R_1}{R_1(T_2 - T_1)} \tag{9.5}$$

where R_1 and R_2 are the resistances of the material at temperatures T_1 and T_2, respectively. This relationship is valid only over a narrow range of temperature, ensuring that the variation of resistance with temperature is approximately linear. For a wide range of temperatures the resistance of the material is usually expressed by a quadratic relation

$$R = R_0 \left(1 + a\,T + b\,T^2\right) \tag{9.6}$$

where R is the material resistance at temperature T, R_0 is the material resistance at a reference temperature T_0, and a, b are constants to be determined experimentally.

Table 9.6 Resistivity and resistance–temperature coefficients at 20°C

Substance	$\alpha\,(°C^{-1})$	$\rho\,(\mu\Omega.cm)$
Nickel	0.0067	6.85
Iron (alloy)	0.002 to 0.006	10
Tungsten	0.0048	5.65
Aluminum	0.0045	2.65
Copper	0.0043	1.67
Lead	0.0042	20.6
Silver	0.0041	1.59
Gold	0.004	2.35
Platinum	0.00392	10.5
Mercury	0.00099	98.4
Manganin	$\pm$ 0.00002	44
Carbon	$-$ 0.0007	1400
Electrolytes	$-$ 0.02 to $-$ 0.09	variable
Semiconductor (thermistors)	$-$ 0.0068 to $+$ 0.14	10^9

The RTD devices are more suitable for measurements of the temperature of a block of metal than for measurements in gases like air. This is due to the fact that in a block of metal the temperature communication between the metal and the RTD wire is good but the communication between the wire and air is poor.

Even though the RTD is considered to be a quite accurate device for temperature measurement, it does have some disadvantages. They are the lead error and relatively bulky size which sometimes give rise to poor transient response and conduction error. Further, since a current must be fed to the RTD for bridge measurement, there is the possibility of self-heating which may alter the temperature of the element.

9.7.1 Metal Film RTDs

In modern construction technique, a platinum or metal–glass slurry film is deposited as a screen onto a small ceramic substrate, etched with a laser-trimming system, and sealed. The film RTD offers substantial reduction in assembly time and has the further advantage of increased resistance for a given size. Due to the manufacturing technology, the device size itself is small, which makes it respond quickly to step changes in temperature. Film RTDs are less stable than their hardware counterparts, but they are becoming more popular because of their advantages in size and production cost.

9.7.1.1 Resistance Measurements

The resistance for a platinum RTD range from 10 ohms for the bird-cage model to several thousand ohms for the film RTD. The single most common value is

100 ohms at 0°C. The DIN 43760 standard temperature coefficient of platinum wire is $\alpha = 0.00385$ ohms/°C. For a 100 ohms wire this corresponds to +0.385 ohms/°C at 0°C. This value for α is actually the average slope from 0°C to 100°C. The more chemically pure platinum wire used in platinum resistance standards has an α of +0.00392 ohms/ohm/°C. Both the slope and the absolute value are small numbers, especially when we consider the fact that the measurement wires leading to the sensor may be several ohms or even tens of ohms. Therefore, even a small lead impedance can constitute a significant error to our temperature measurement. For example, a ten ohm lead impedance implies $10/0.385 \approx 26$°C error in our measurement. The classical method of avoiding this problem is to use a bridge, as shown in Figure 9.26.

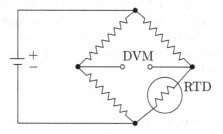

Figure 9.26: Wheatstone bridge

The bridge output voltage is an indirect indication of the RTD resistance. The bridge requires four connection wires, an external power source, and three resistors that have a zero temperature coefficient. To avoid subjecting the three bridge-completion resistors to the same temperature as the RTD, the RTD is separated from the bridge by a pair·of extension wires, as shown in Figure 9.27.

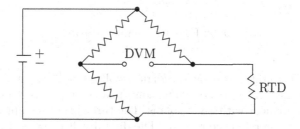

Figure 9.27: Bridge with separated RTD

These extension wires have their impedance and this can affect the temperature reading. To minimize the problem due to the impedance of the extension wires, a *three-wire bridge* configuration shown in Figure 9.28, can be used.

If the lengths of wire A and B are perfectly matched, their impedance effects will cancel each other because each is in an opposite leg of the bridge. The

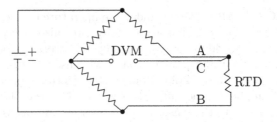

Figure 9.28: Three-wire bridge

third wire C acts as a sense lead and carries no current. However, the Wheatstone bridge shown in Figure 9.28 creates a nonlinear relationship between the resistance change and the bridge output voltage change. This compounds the already nonlinear temperature-resistance characteristics of the RTD by requiring an additional equation to convert the bridge output voltage to equivalent RTD impedance.

9.7.1.2　Four-Wire Ohms

Many problems associated with the RTD bridge can be eliminated by using a current source along with a remotely sensed digital voltmeter, as illustrated in Figure 9.29.

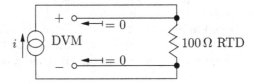

Figure 9.29: Four-wire ohms measurement

The output voltage read by the digital voltmeter is directly proportional to RTD resistance; therefore, only one conversion equation is required for temperature measurement with 4-wire ohms. The three bridge-completion resistors are replaced by one reference resistor. The digital voltmeter measures only the voltage drop across the RTD and is insensitive to the length of the lead wires. However, the 4-wire ohms reads an additional extension wire than the 3-wire bridge. This may be called a disadvantage, but it is a small price to pay for the accuracy of the temperature measurement.

9.7.2　Measurement Errors with Three-Wire Bridge

Consider the 3-wire bridge shown in Figure 9.30.

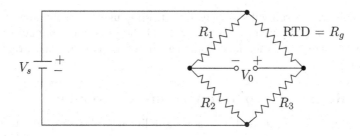

Figure 9.30: Three-wire bridge

If the voltages V_s and V_0 are measured, we can determine the RTD resistance R_g and then solve for temperature. The unbalance voltage V_0 of a bridge built with $R_1 = R_2$ is given by

$$V_0 = V_s \left(\frac{R_3}{R_3 + R_g} \right) - V_s \left(\frac{1}{2} \right) \tag{9.7}$$

If $R_g = R_3$, then $V_0 = 0$ and the bridge is balanced. This can be done manually, but if manual bridge balance is undesirable, we can just solve for R_g in terms V_0, to result in

$$R_g = R_3 \left(\frac{V_3 - 2V_0}{V_3 + 2V_0} \right) \tag{9.8}$$

This expression assumes the lead resistance to be zero. If R_g is located at some distance from the bridge in a 3-wire configuration, as shown in Figure 9.31, the lead resistance R_l will appear in series with both R_g and R_3.

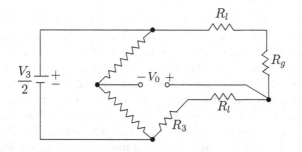

Figure 9.31: RTD in a 3-wire bridge

The resistance R_g will be given by

$$R_g = R_3 \left(\frac{V_3 - 2V_0}{V_3 + 2V_0} \right) - R_l \left(\frac{4V_0}{V_s + 2V_0} \right) \tag{9.9}$$

This circuit works well with devices like strain gauges, which change resistance value by only a few percent, but for an RTD the resistance changes drastically with temperature and therefore, some care must be paid in measurement with this circuit.

9.7.3   Resistance to Temperature Conversion

The resistance temperature device is a more linear probe compared to the thermocouple. But for measurement with reasonable accuracy, the RTD still requires curve–fitting. The Callender–Van Dusen equation (*Omega Temperature Measurement Handbook*) has been used for years to approximate the RTD curve to result in

$$R_T = R_0 + R_0\, \alpha \left[T - \delta \left(\frac{T}{100} - 1 \right) \left(\frac{T}{100} \right) - \beta \left(\frac{T}{100} - 1 \right) \left(\frac{T^3}{100} \right) \right] \quad (9.10)$$

where R_T is the resistance of RTD at temperature T, R_0 is the RTD resistance at $T = 0°C$, α is the temperature coefficient at $T = 0°C$ (typically + 0.00392 $\Omega/\Omega/°C$), $\delta = 1.49$ (typical value for 0.00392 platinum) and $\beta = 0$ for $T > 0$ and 0.11 for $T < 0$. The exact values of the coefficients α, β, and δ are determined by testing the RTD at four temperatures and solving the resultant equations. It is essential to note that attention must be paid to account for the following problems associated with RTD.

9.7.3.1   Protection

The RTD is somewhat more fragile than the thermocouple, due to its construction. In measurements it has to be gently heated to protect it.

9.7.3.2   Self-Heating

In the measurement process, a current must be passed through the RTD device to provide a voltage that can be measured. The current causes Joule heating (I^2R) within the RTD, changing its temperature. This *self-heating* results in a measurement error. Therefore, attention must be paid to account for the error in the measured temperature due to the magnitude of the current supplied. A typical value for self-heating error is 0.5°C per milliwatt in free air. Obviously, an RTD placed in a thermally conductive medium will distribute its Joule heat to the medium and the error due to self-heating will become smaller.

To minimize the self-heating error, minimum ohms measurement current that will give the required resolution should be used, in conjunction with the largest possible size of RTD that can still give good response time.

9.7.3.3   Thermal Shunting

An act of altering the measurement temperature by inserting a measurement transducer is termed *thermal shunting*. The problem of thermal shunting is comparatively more for RTDs than the thermocouples, since the physical size of an RTD is larger than the thermocouple.

9.7.3.4 Thermal EMF

The platinum-to-copper connection that is made when the RTD is used for temperature measurement can cause a thermal offset voltage. The offset-compensated *ohms* technique can be used to eliminate this effect.

9.7.4 Thermistors

The thermistor is a semiconductor device that has a negative temperature co-efficient of resistance, in contrast to the positive coefficient displayed by most metals. Like the RTD, the thermistor is also a temperature-sensitive resistor. While the thermocouple is the most versatile temperature transducer and RTD is the most stable temperature transducer, the best description of the thermistor is that it is a *most sensitive* temperature transducer. Of the above three temperature sensors, the thermistor exhibits the largest parameter change with temperature.

Thermistors are generally composed of semiconductor materials. Most thermistors have a negative temperature coefficient; that is, their resistance decreases with increasing temperature. The negative temperature coefficient can be as large as several percent per degree Celsius, allowing the thermistor to detect minute changes in temperature which could not be observed with a thermocouple or RTD. But the increased sensitivity results in loss of linearity. The thermistor is an extremely nonlinear device which is highly dependent upon process parameters. Consequently, the thermistor curves are not standardized to the extent that RTD and thermocouple curves have been standardized. A comparison of the temperature versus resistance curves for the above three devices are shown in Figure 9.32.

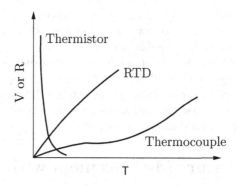

Figure 9.32: Temperature-resistance curves

An individual thermistor curve can be very closely approximated through the equation (*Omega Temperature Measurement Handbook*)

$$\frac{1}{T} = A + B \ln R + C \left(\ln R\right)^3 \tag{9.11}$$

where T is temperature in kelvin, R is the resistance of the thermistor, A, B, C are the curve-fitting constants. Equation (9.11) is called the *Steinhart–Hart equation*. The constants A, B, and C are found by selecting three data points on the published data curve and solving the three simultaneous equations. When the data points are chosen to span more than 100°C within the nominal centers of the thermistor temperature range, Equation (9.11) approaches a ±0.02°C curve fit.

Somewhat faster computer execution time is achieved through a simpler equation

$$T = \frac{B}{(\ln R)(A)} - C \tag{9.12}$$

where A, B, and C are again found at selecting three data points and solving for three resultant simultaneous equations. Equation (9.12) must be applied over a narrower temperature range in order to approach the accuracy of Equation (9.11).

9.7.4.1 Measurement with Thermistor

The high resistivity of a thermistor is an advantage in temperature measurement. The four-wire resistance measurement like RTD is necessary for thermistors. For example, a common thermistor has a resistance of 5000 Ω at 25°C. With a typical thermal coefficient of 4 percent per °C, a measurement lead resistance of 10 Ω produces an error of only 0.05°C. This error is a factor of 500 times less than the equivalent RTD error.

The major disadvantage with thermistors is that they are more susceptible to permanent decalibration at high temperatures. Further, the use of thermistors is generally limited to a few hundred degrees Celsius and manufactures warn that extended exposures even below maximum operating limits will cause the thermistor to drift out of its specified tolerance.

Thermistors can be made very small implying that they will respond quickly to temperature changes. The small size of the device also implies that their small thermal mass makes them especially susceptible to self-heating errors. Compared to RTDs and thermocouples, the thermistors are more fragile and they must be carefully mounted to avoid damage to the device.

9.8   Temperature Measurement with Pyrometers

9.8.1   Optical Pyrometer

Some solids, such as metals, when they become hot, begin to emit light with a very dull red color and as the temperature increases the color becomes a brighter red, orange, yellow, and so on. The particular color emission occurs at a particular temperature and hence can be used as a means of measuring the temperature if suitable calibration is adopted. It is this principle which is used in the *optical pyrometer*.

The optical pyrometer has some form of a telescope through which the high temperature object is observed. Usually the telescope will have a light filter installed in order to cut down the light intensity at the higher temperature so as not to damage the eye. The telescope is focused on the object whose temperature is required, such as a furnace, re-entry object, and so on. Color filters can be introduced by the side of the image of the object and the color of the filter is matched with the color of the object. There is a calibrated scale on the side of the instrument which records the temperature.

The most common optical pyrometer, however, is the disappearing filament pyrometer. If the filament of a lamp has an electric current passed through it, the filament will glow. The color of the filament will depend upon the current passing though it. In the disappearing filament pyrometer a small lamp is inserted in the field of vision of the telescope. After focusing the telescope on the object, the current through the filament is adjusted such that the color of the filament merges with the color of the object and hence the filament optically disappears. A meter in the filament electric circuit is calibrated to read temperature, and hence the temperature of the object is determined.

From the above discussions it is seen that the optical pyrometer is used for high temperature measurements. Any light-absorbing gases between the pyrometer and the object or through any other cause which restricts the entry of light into the pyrometer will introduce significant errors in the measurements with pyrometer. Also, in pyrometer using color comparison, error may occur due to color interpretation. The different colors and the approximate temperature corresponding to them are given in Table 9.7.

Table 9.7 Color and the corresponding temperature

Color	Temperature (°C)
Dazzling	1500
White	1300
Yellow	1100
Orange	1000
Bright red	900
Cherry red	800
Dark cherry red	700
Dark red	600
Red just visible	500

9.8.2 Radiation Pyrometer

In this type of pyrometer, arrangement is made to focus the radiant energy on the hot junction of a thermocouple. The cold junctions of the thermocouple are shielded from the radiation and are coupled externally to a galvanometer, as shown in Figure 9.33.

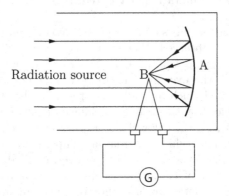

Figure 9.33: A radiation pyrometer

As seen from the figure, the radiant energy from the radiation source is received by a concave mirror A in the apparatus. By means of a focusing adjustment the radiant energy is focused directly on the hot junction of thermocouple B, after reflection from the mirror surface. The galvanometer G, connected to the thermocouple is calibrated directly to read temperature. While using, care must be taken to ensure that the reflecting power of the mirror is not impaired and hence it must be kept clean. Also, no high-energy absorbing media must come between the radiation source and the pyrometer.

9.8.3   Infrared Thermography

It is a technique, based on radiation, to assess energy losses by radiation. An important energy radiation level is that of the *infrared* waveband. This occurs at wavelength between the appropriate limits of 10^{-3} to 10^{-6} meters. This is below the visual limit of the red color. Because it is not visible, in the normal sense by eye detection, special detectors are required to record its presence. This has led to the development of infrared thermography.

In infrared thermography a detector is aimed at a surface. The infrared radiant energy is received and through subsequent apparatus, it is translated into a visual screening (bright for high temperature, through to dark for low temperature), and thus energy loss by radiation can be assessed and calibrated.

In the case of photography, infrared energy received from an object can be translated into a photographic image, and, thus, such cameras can be used at night since no direct visual wavelengths are required from the energy transmitting object.

9.8.4   Fusion Pyrometers

The variation of melting point of certain chemical substances is utilized for temperature measurements in fusion pyrometers. Combining certain clays, metals, and salts, it is possible to obtain mixtures of these substances which soften and melt at definite temperatures. These mixtures are made generally in the form of slender cones and pyramids, as shown in Figure 9.34.

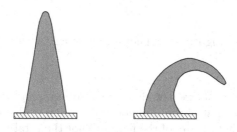

Figure 9.34: Fusion pyrometer cones

They are used for measuring the temperature of a furnace, kilns, etc. When the required temperature is reached the cone signifies this fact by beginning to become soft and curls over, as shown in Figure 9.34. The disadvantage of this technique is that these cones can be used only once.

9.8.5   Thermal Paints

These are special paints which are applied on surfaces whose temperature distribution is required. When the temperature of the surface increases, establishing a distribution over the area, the color of the paint, being a property dictated by the temperature level, changes. Thus, temperature distribution is determined by color calibration against known surface temperatures.

9.9   Temperature Measurement in Fluid Flows

In fluid flow analysis, we are interested in measuring the static and the total temperatures of the flow at any specified location. In principle, *for direct measurement of the static temperature, the measuring device should travel at the velocity of the flow without disturbing the flow.* But it is impossible to make such an infinitely thin device. Therefore, alternatively, locate a temperature measuring probe, for instance a thermocouple, at the wall surface over which fluid flow is taking place, as shown in Figure 9.35.

It is seen from this figure that the thermocouple is located inside the boundary layer. At a wall surface the flow velocity is equal to zero and we would expect the thermocouple to measure the wall temperature, which may be closer to the total temperature $T_{t\infty}$ of the flow than its static temperature T_{∞}. It will be interesting to compare the present situation to that of pressure measurement.

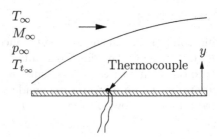

Figure 9.35: Thermocouple at a wall

In pressure measurements we saw that a pressure tapping situated on a wall at the same location as that of the thermocouple shown in Figure 9.35 is expected to measure the static pressure of the flow and not the total pressure. That is, in the case of static pressure measurement, the fact that the static pressure does not vary through the boundary layer in the y direction (transverse direction) is made use of. In the case of temperature, the very fact that the thermocouple in Figure 9.35 measures only the total temperature $T_{t\infty}$, rules out the possibility of measuring the static temperature, as in the pressure measurement.

A temperature distribution in a compressible boundary layer is as shown in Figure 9.36.

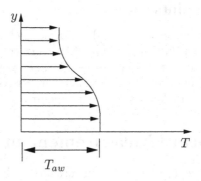

Figure 9.36: Temperature variation in a boundary layer

Assume that the wall is perfectly insulated. Therefore,

$$\frac{\partial T}{\partial y} = 0, \quad \text{at } y = 0$$

and the temperature is called adiabatic wall temperature T_{aw}. There is heat flow in the positive y direction, due to conduction. Due to high velocity gas flow near a surface, there can be appreciable frictional heating of the fluid. In other words, the faster moving layers in the boundary layer do work on the slowly moving layers.

If the heat loss due to conduction and energy gain from viscous heating cancel each other, then the flow can be considered to be brought to rest adiabatically in the boundary layer, resulting in $T_{aw} = T_{t\infty}$. That is the adiabatic wall temperature T_{aw} is identically equal to the freestream total temperature $T_{t\infty}$. A measure of the relative importance of heat conduction and viscous heating is given by the Prandtl number Pr,

$$Pr = \frac{\mu C_p}{K}$$

where μ is the viscosity coefficient, C_p is the specific heat at constant pressure and K is the conduction coefficient. For a fluid with $Pr = 1$, the adiabatic wall temperature is equal to the freestream stagnation temperature. If $Pr < 1$, then $T_{aw} < T_{t\infty}$. This can be summarized by defining a *recovery factor R*, where

$$R = \frac{T_{aw} - T_\infty}{T_{t\infty} - T_\infty} \tag{9.13}$$

Since

$$T_{t\infty} = T_\infty \left(1 + \frac{\gamma - 1}{2} M_\infty^2\right)$$

From isentropic relations, it follows that,

$$T_{aw} = T_\infty \left(1 + R\frac{\gamma - 1}{2} M_\infty^2\right) \tag{9.14}$$

It can be shown that, for laminar compressible boundary layer, $R = \sqrt{Pr}$, whereas, for a turbulent boundary layer, $R \approx Pr^{1/3}$. For air, up to moderately high temperature, $Pr = 0.72$, so that $R \approx 1$ for a turbulent boundary layer of air.

9.9.1 Static Temperature Determination

From the above discussions it is evident that direct measurement of static temperature is not possible. A direct measurement of adiabatic wall temperature can be used to determine $T_{t\infty}$, then by measuring $P_{t\infty}$ and P_∞, the Mach number M_∞ and the static temperature can be calculated.

9.9.2 Total Temperature Measurement

Measurement of total or stagnation temperature T_t is simple in principle. The temperature inside a pitot tube, where the flow is brought to rest, should be the stagnation temperature, in both subsonic and supersonic flows, and could be measured by a thermometer placed inside the tube. The difficulty in this measurement is that the thermal equilibrium does not actually exist, since heat is conducted away from the thermometer placed inside the tube walls. This reduces the temperature of the stagnant fluid to a value lower than $T_{t\infty}$.

In the absence of a wall, as in Figure 9.35, the stagnation temperature probe such as that shown in Figure 9.37 can be used to obtain the freestream total temperature $T_{t\infty}$.

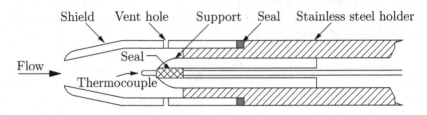

Figure 9.37: Total temperature probe

This total temperature probe uses a thermocouple as the sensing element. The shield and support have to be designed to keep the rate of heat loss by conduction and radiation to a minimum. To replenish some of the lost energy, a small amount of flow through the probe is permitted by the vents provided.

When the flow is supersonic, there will be a detached shock standing in front of the probe. However, the measurement of $T_{t\infty}$ is unaffected by the presence of the shock, since the flow across a shock is adiabatic. In such streams the temperature indicated by the probe is from slightly below to considerably below the true stagnation temperature. The performance of such a probe is usually defined by a *recovery factor K*, defined as

$$K = \frac{T_{ti} - T_\infty}{T_t - T_\infty} \tag{9.15}$$

where T_{ti} is the indicated or measured temperature in kelvin, T_t is the actual total temperature in kelvin, and T_∞ is the static temperature in kelvin. By suitable design, K can be made very close to unity, for air. In any case, such a probe must be calibrated to define K as a function of Reynolds number Re, Prandtl number Pr, and Mach number M.

Example 9.3

A total temperature probe measures the temperature of a Mach 2 air stream as 600 K. If the probe recovery factor is 0.98, determine the stream static temperature.

Solution

From isentropic table, for $M_\infty = 2.0$, we have $\dfrac{T_\infty}{T_{t\infty}} = 0.5556$.

The recovery factor becomes

$$K = \frac{T_{ti} - T_\infty}{T_{t\infty} - T_\infty} = 0.98 = \frac{600 - T_\infty}{\left(\frac{1}{0.5556} - 1\right) T_\infty}$$

This gives the stream static temperature as

$$T_\infty = \boxed{336.36 \text{ K}}$$

9.10 Temperature-Measuring Problems in Fluid Flows

In the measurement of temperature in flowing fluids, certain types of problems are encountered irrespective of the nature of the device being used. These problems are mainly due to the heat transfer between the probe and its environment. Let us see the error due to the heat transfer effect one by one.

9.10.1 Conduction Error

Examine the probe kept in a flow system, shown in Figure 9.38.

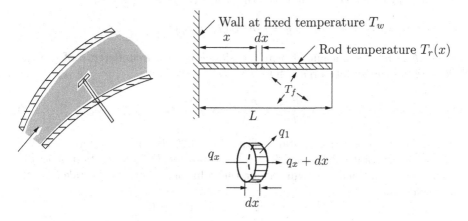

Figure 9.38: Temperature probe in a flow field

Usually the wall will be hotter or colder than the flowing fluid. Therefore, there will be heat transfer, and this leads to a probe temperature which is different from that of the fluid. To analyze the heat transfer effect, assume the probe support to be a slender rod, as shown in Figure 9.38. Let the rod temperature T_r be a function of x alone; it does not vary with time or over the rod cross-section, at a given x. A fluid of constant and uniform temperature T_f completely surrounds the rod and exchanges heat with it by convection. For a steady-state condition,

Heat flow in at x = Heat out at $(x + dx)$ + Heat loss at the surface

i.e.,

$$q_x = q_{x+dx} + q_l$$

By Fourier law for one-dimensional heat transfer, we have

$$q_x = -kA\frac{dT_r}{dx} \tag{9.16}$$

where k is the thermal conductivity of the rod and A is its cross-sectional area. Therefore,

$$q_{x+dx} = q_x + \frac{d}{dx}(q_x)dx = -kA\frac{dT_r}{dr} + \frac{d}{dr}\left(-kA\frac{dT_r}{dr}\right)dx \tag{9.17}$$

Assuming k and A to be constants, we have

$$q_{x+dx} = -kA\frac{dT_r}{dx} - kA\frac{d^2T_r}{dx^2}dx \tag{9.18}$$

The heat loss by conduction at the surface of the rod is given by

$$q_l = h(c\,dx)(T_r - T_f) \tag{9.19}$$

where h is film coefficient of heat transfer, c is the circumference of the rod, thus $c\,dx$ is the surface area. Thus, we have

$$\frac{d^2T_r}{dx^2} - \frac{hc}{kA}T_r = -\frac{hc}{kA}T_f \tag{9.20}$$

If h and c are treated as constants, Equation (9.20) is a linear differential equation with constant coefficients. It can be solved for T_r as a function of x. Two boundary conditions are required for the solution. They can be taken as the following.

B.C 1 : $T_r = T_w$ at $x = 0$

B.C 2 : $\dfrac{dT_r}{dx} = 0$ at $x = L$,

treating the rod end $x = L$ as insulated. With these boundary conditions, Equation (9.20) gives

$$T_r = T_f - (T_f - T_w)\left[\left(1 - \frac{e^{mL}}{2\cosh mL}\right)e^{mx} + \frac{e^{mL}}{2\cosh mL}e^{-mx}\right] \tag{9.21}$$

where

$$m = \sqrt{\frac{hc}{kA}} \tag{9.22}$$

Generally the temperature measuring device (thermocouple bead, thermistor, etc.) is located at $x = L$. Therefore, we can evaluate Equation (9.21) to obtain

$$\text{Temperature error} = T_r - T_f = \frac{T_w - T_f}{\cosh mL} \qquad (9.23)$$

It is seen from Equation (9.22) that the error becomes very small when T_w is close to T_f. Insulating the support or actively controlling the wall temperature, this can be achieved. The term $\cosh mL$ will be large if m and L are large. Thus the probe should be located as far from the wall as possible.

If the boundary condition at $x = L$ is changed to a realistic one in which there is convective heat transfer with a film coefficient h_e at the end, then at $x = L$,

$$T_r - T_f = \frac{T_w - T_f}{\cosh mL + [h_e/(mL)]\, \sinh mL} \qquad (9.24)$$

Note that, the error predicted with Equation (9.24) is less than that predicted with Equation (9.23).

9.10.2 Radiation Error

The radiant heat-exchange between the probe and its surroundings causes additional error. Even though this occurs simultaneously with the conduction loss, it is treated separately here for simplicity. Neglecting conduction loss we may write, for the probe shown in Figure 9.39, under steady-state conditions,

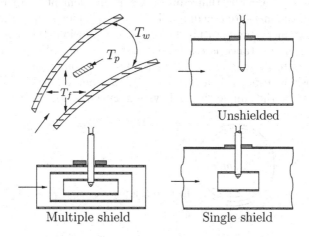

Figure 9.39: Probe with and without radiation shield

$$\text{Heat convected to probe} = \text{Net heat radiated to wall} \qquad (9.25)$$

$$hA_s\,(T_f - T_p) = 0.174\,\epsilon_p\,A_s \left[\left(\frac{T_p}{100} \right)^4 - \left(\frac{T_w}{100} \right)^4 \right] \qquad (9.26)$$

where h is film coefficient at probe surface, A_s is probe surface area, ϵ_p is the emittance of probe surface, T_p is the absolute temperature of the probe and T_w is the absolute temperature of the wall. Equation (9.26) assumes the radiation configuration described as "a small body completely enclosed by a larger one." The error due to radiation is given by

$$\text{Error} = T_p - T_f = \frac{0.174\,\epsilon_p}{h}\,\frac{T_w^4 - T_p^4}{10^8} \tag{9.27}$$

The error can be reduced by making the difference between T_w and T_p as small as possible. This can be achieved by insulating the wall or controlling its temperature. A probe surface of low emittance ϵ_p (a shiny surface) will reduce the error further. Also, high value of h will reduce the error.

A significant error reduction can be achieved by more shields, as shown in Figure 9.39. However, for multiple shields the spacing between shields must be sufficiently large to prevent excessive conduction heat transfer between the shields and to allow sufficient flow velocity for good convection from the gas to the shields.

9.11 Dynamic Response of Temperature Sensors

Dynamic characteristics of temperature sensors are related to the heat-transfer and the storage parameters that cause the sensing-element temperature to lag behind that of the measured medium. This poses problems since the conversion from sensing-element temperature to thermal expansion, thermoelectric voltage, or electrical resistance is essentially instantaneous. When a temperature measurement is to be made under non-steady-state conditions, it is important that the transient response characteristics of the thermal system be taken into account. Examine the simple thermal system shown in Figure 9.40.

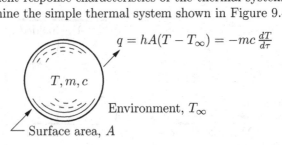

Figure 9.40: First-order sensor model

The thermometer is represented by the mass m and specific heat c and is suddenly exposed to a fluid environment at temperature T_∞. Let h be the convective heat transfer coefficient between the thermometer and the fluid. Assume the radiation heat transfer to be negligible and the thermometer is uniform in temperature at any instant of time. In other words, the thermal conductivity of the thermometer material is sufficiently large in comparison with the surface conductance because of the convective heat transfer coefficient. Conservation of energy principle gives

$$h\,A\,(T - T_\infty) = m\,c\,\frac{dT}{d\tau} \qquad (9.28)$$

where h is in watts/(m^2 °C), A in m^2, T in °C, m in kg, c in J/(kg °C) and τ in seconds.

The solution of Equation (9.28) gives the temperature of the thermometer as

$$\frac{T - T_\infty}{T_0 - T_\infty} = e^{(-hA/mc)\tau} \qquad (9.29)$$

where T_0 is the thermometer temperature at $\tau = 0$. The time constant for the system is

$$\tau = \frac{mc}{hA} \qquad (9.30)$$

It is seen from Equation (9.28) that the speed of response may be increased by decreasing m and c and/or increasing h and A. When h is sufficiently large, there may be substantial temperature gradients within the thermometer itself and a different type of analysis must be employed to determine the temperature variation with time. For such analysis one may consult books specializing in heat transfer.

9.12 Summary

Temperature is a measure of "hotness" or "coldness." Several properties of materials change with temperature, and this forms the basis for temperature measurement.

The *zeroth law of thermodynamics* which states that "*if two bodies are in thermal equilibrium with a third body, they are in thermal equilibrium with each other,*" serves as a basis for the validity of temperature measurement.

All temperature scales are based on some easily reproducible states, such as the freezing and boiling points of water. On the Celsius scale, the ice and steam points are assigned the values 0°C and 100°C, respectively. The absolute temperature scale in *SI* system is the *Kelvin scale*. The relation between the Celsius scale and Kelvin scale is

$$\boxed{T(\text{K}) = T(°\text{C}) + 273.15}$$

In the English system, the absolute temperature scale is the *Rankine scale*. It is related to the Fahrenheit scale by

$$\boxed{T(\text{R}) = T(°\text{F}) + 459.67}$$

The temperature scales in the above two systems are related by

$$T(\text{R}) = 1.8T(\text{K})$$

and

$$T(°F) = 1.8T(°C) + 32$$

In practical situations, temperatures much lower than the melting point of pure ice and much higher that the boiling point of pure water are commonly encountered. Therefore, the fixed point scale becomes inadequate for the measurement of low or high temperatures. An extension of the thermometric scale is therefore made by introducing further fixed points, using substances other than water. The fixed points are laid down in what is known as the *International Practical Temperature Scale* (IPTS).

In the International Practical Temperature Scale of 1968, the practical kelvin temperature is given the symbol T_{68}. The international practical Celsius temperature is given the symbol t_{68}. The relationship between them is

$$t_{68} = T_{68} - 273.15\,K$$

The standard instrument used for measurement of temperature from 13.8 K to 630.74°C is the platinum resistance thermometer. The standard instrument used for measurement from 630.74°C to 1064.43°C is the platinum −10 percent rhodium/platinum thermocouple.

Above 1337.58 K, the International Practical Temperature of 1968 is defined by the Planck's law of radiation with 1337.58 K as the reference temperature.

A substance used for the measurement of temperature must experience some recognizable change when its temperature is changed.

The changes which have been commonly used for temperature measurement are the following.

- Increase or decrease in size, that is, expansion or contraction.

- Increase in pressure if enclosed.

- Change of color.

- Change of state.

- Change of electrical resistance.

- Generation of electromotive force.

- Change in degree of surface radiation.

Thermometer is a general name given to thermometric measuring devices. Some of the commonly used thermometers are the *mercury-in-glass thermometer*, *Beckmann thermometer*, and *gas thermometer*.

Mercury-in-glass thermometer is the most common type of thermometer. Mercury is well suited for a thermometer since it does not wet the glass, and it has a reasonable expansion. It has a freezing point of −38.86°C and a boiling point of 356.7°C. The upper temperature limit can be increased by the introduction of a gas, such as nitrogen, into the thermometer. However, at too high temperatures, the glass itself becomes the limiting factor, since the glass will soften and eventually melt at such high temperatures. For temperatures lower than $-38.86°C$, liquids other than mercury are sometimes used. Alcohol, sometimes dyed red, is commonly used. This has a freezing point of $-113°C$.

Beckmann thermometer is used for the accurate determination of small temperature change. This is meant only for the determination of temperature change and not the temperature itself. The maximum temperature change that can be measured is usually $6°C$.

Gas thermometer was originally used as the fundamental temperature measuring device to establish the various temperatures given as the International Temperature Scale in relationship to the originally conceived fixed points of the freezing point and boiling point of pure water.

Heating of a fluid in a confined space results in its pressure rise. The fluid can fill the space completely or partially. The change of pressure as a result of the change of temperature of the fluid is made use of for temperature measurement.

The phenomenon of thermal expansion is made use of for temperature measurement in a number of temperature sensing devices. The expansion of solid is employed mainly in bimetallic elements by utilizing the differential expansion of bonded strips of two metals. Liquid expansion is used in the common liquid-in-glass thermometers. Restrained expansion of liquids, gases, or vapor results in a pressure rise, which is the basis of pressure thermometers.

Bimetallic devices are used for temperature measurements and vary widely as combined sensing and control elements in temperature-control systems, mainly of the on–off type.

Temperature measurements based on electrical effects have become popular because they are very convenient techniques and they furnish a signal that is easily detected, amplified, or used for control purposes. Some of the popular temperature measuring devices based on electrical effects are

- Thermocouple.

- Resistance temperature detector, RTD.

- Thermistor.

Thermocouples are devices which operate on the principle that "*a flow of current in a metal accompanies a flow of heat.*" This principle is popularly known as the *Seebeck effect.* Thermocouples consist of two dissimilar metals joined together at two points, one point being the place where the temperature is to be measured and the other point being a place where the temperature is known, termed the *reference junction.*

At this stage, one may ask what is the need for a thermocouple which requires a reference junction compensation, when we have devices like thermistor and RTD that will measure the absolute temperature directly? The answer to this question is that the thermistor, the RTD, and the integrated circuit transducers are useful only over a certain range of temperature. The thermocouples are much more rugged than thermistors. They can be made on the spot, either by soldering or by welding. In short, thermocouples are the most versatile temperature transducers among the available ones.

In situations where it is necessary to measure more than one temperature, thermocouple measurement becomes convenient. This is accomplished with an isothermal reference junction for more than one thermocouple element.

The most versatile technique available for thermocouple measurements is the software compensation. This technique is independent of the types of thermocouples chosen from a set of thermocouples connected on the same block, using copper leads throughout the scanner.

Instead of measuring the temperature of the reference junction and computing its equivalent voltage, as is done with software compensation, a battery could be inserted to cancel the offset voltage of the reference junction. The combination of the *hardware compensation* voltage and the reference junction voltage is equal to that of a 0°C junction.

It is a common practice to express the thermoelectric *emf* in terms of the potential generated with a reference junction at 0°C. We have used hardware and software compensations to synthesize an ice-point reference. Now all we have to do is read the digital voltmeter and convert the voltage reading to temperature. But the job is not that simple due to the nonlinear relationship between the voltage and the temperature of a thermocouple.

The output voltage V of a simple thermocouple circuit is usually expressed as

$$V = AT + \frac{1}{2}BT^2 + \frac{1}{3}CT^3$$

where V is the voltage based on a reference junction temperature of 0°C, T is the temperature in degrees Celsius and A, B, and C are constants dependent on the thermocouple material.

The thermoelectric power or sensitivity of a thermocouple is given by

$$S = \frac{dV}{dT} = A + BT + CT^2$$

The noise rejection methods used in thermocouple measurements include trees switching, normal mode filtering, integration, and guarding.

Thermocouple measurement errors may be due to

- Poor junction connection.

- Decalibration of thermocouple wire.

- Shunt impedance and galvanic action.

- Thermal shunting.

- Noise and leakage currents.

- Thermocouple specifications.

- Documentation.

The dyes used in some thermocouple insulation will form an electrolyte in the presence of water. This creates a galvanic action, with a resultant output hundreds of times more than the Seebeck effect. Precautions should be taken to shield the thermocouple wires from all harsh atmospheres and liquids.

Thermal shunting is a problem associated with the mass of the thermocouple. Since some energy is used to heat the mass of the thermocouple, the thermocouple will slightly alter the temperature it was meant to measure. If the temperature to be measured is small, the thermocouple must naturally be small.

Thermocouple wire is manufactured to a certain specification, signifying its conformance with the NBS tables. The specifications can sometimes be enhanced by calibrating (testing it at known temperatures) the wire. If the wire is calibrated intending to confirm its specifications, it becomes more imperative that all of the aforementioned conditions be heeded in order to avoid decalibration.

The diagnostic system for detecting a faulty thermocouple and data channels consists of three components. They are the *event record*, the *zone box test*, and the *thermocouple resistance history*. Now let us see the details of these components.

An abrupt change in the resistance of a thermocouple circuit can act as a warning indicator.

Temperature can also be measured using *resistance temperature detector* (RTD) devices. They work on the principle that metals possess a marked temperature dependence. The RTD devices are more suitable for measurements of the temperature of a block of metal than for measurements in gases like air. This is due to the fact that in a block of metal the temperature communication between the metal and the RTD wire is good but the communication between the wire and air is poor.

Even though the RTD is considered to be a quite accurate device for temperature measurement, it does have some disadvantages. They are the lead error and relatively bulky size which sometimes give rise to poor transient response and conduction error. Further, since a current must be fed to the RTD for bridge measurement, there is the possibility of self-heating which may alter the temperature of the element. The resistance temperature device is a more linear probe compared to the thermocouple. But for measurement with reasonable accuracy, the RTD still requires curve-fitting. It is somewhat more fragile than the thermocouple, due to its construction. In measurements it has to be gently heated to protect it.

Thermistor is a semiconductor device that has a negative temperature coefficient of resistance, in contrast to the positive coefficient displayed by most metals. Like the RTD, the thermistor is also a temperature-sensitive resistor. While the thermocouple is the most versatile temperature transducer and RTD is the most stable temperature transducer, the best description of the thermistor is that it is a *most sensitive* temperature transducer. Of the above three temperature sensors, the thermistor exhibits the largest parameter change with temperature.

Some solids, such as metals, when they become hot, begin to emit light with a very dull red color and as the temperature increases the color becomes a brighter red, orange, yellow, and so on. The particular color emission occurs at a particular temperature and hence can be used as a means of measuring the temperature if suitable calibration is adopted. It is this principle which is used in the *optical pyrometer*.

Infrared thermography is a technique, based on radiation, to assess energy losses by radiation. An important energy radiation level is that of the *infrared* waveband. This occurs at wavelength between the appropriate limits of 10^{-3} to 10^{-6} meters. This is below the visual limit of the red color. Because it is not visible, in the normal sense by eye detection, special detectors are required to record its presence. This has led to the development of infrared thermography.

The variation of melting point of certain chemical substances is utilized for temperature measurements in fusion pyrometers. Combining certain clays, metals, and salts, it is possible to obtain mixtures of these substances which soften and melt at definite temperatures. They are used for measuring the temperature of a furnace, kilns, etc.

Thermal paints are special paints which are applied on surfaces whose temperature distribution is required. When the temperature of the surface increases, establishing a distribution over the area, the color of the paint, being a property dictated by the temperature level, changes. Thus, temperature distribution is determined by color calibration against known surface temperatures.

In fluid flow analysis, we are interested in measuring the static and the total temperatures of the flow at any specified location. In principle, *for direct measurement of the static temperature, the measuring device should travel at the velocity of the flow without disturbing the flow.*

Direct measurement of static temperature is not possible. But measurement of adiabatic wall temperature can be used to determine $T_{t\infty}$, then by measuring $p_{t\infty}$ and p_∞, the Mach number M_∞ and the static temperature can be calculated.

Measurement of total or stagnation temperature T_t is simple in principle. The temperature inside a pitot tube, where the flow is brought to rest, should be the stagnation temperature, in both subsonic and supersonic flows, and could be measured by a thermometer placed inside the tube.

When the flow is supersonic, there will be a detached shock standing in front of the probe. However, the measurement of $T_{t\infty}$ is unaffected by the presence of the shock, since the flow across the shock is adiabatic. In such streams the temperature indicated by the probe is from slightly below to considerably below the true stagnation temperature. The performance of such a probe is usually defined by a *recovery factor K*, defined as

$$K = \frac{T_{ti} - T_\infty}{T_t - T_\infty}$$

Dynamic characteristics of temperature sensors are related to the heat-transfer and the storage parameters that cause the sensing-element temperature to lag behind that of the measured medium. This poses problems since the conversion from sensing-element temperature to thermal expansion, thermoelectric voltage, or electrical resistance is essentially instantaneous. When a temperature measurement is to be made under non-steady-state conditions, it is important that the transient response characteristics of the thermal system be taken into account.

Exercise Problems

9.1 Air at 230 m/s and 30°C flows over a flat plate. If the total pressure of the flow is 1 atm, what will be the error in the temperature measured by a thermocouple located at 0.5 m from the leading edge of the plate? Take the Prandtl number as 0.72.

[Answer: 0.909%]

9.2 Determine the radiation error for a temperature probe of emittance 0.1, which measures the temperature of a stream of actual temperature 323 K and film coefficient as 20 W/(m² K), as 320 K.

[Answer: 0.347%]

9.3 If the Prandtl number for an air stream at 33°C is 0.8, determine the thermal conductivity of air.

[Answer: 0.02354 W/(m °C)]

9.4 A metallic transducer of diameter 5 mm and length 60 mm is used to measure the temperature of fluid in a chamber at 405 K. If the wall temperature of the transducer is 400 K and its thermal conductivity and film transfer coefficients are 300 W/(m °C) and 150 W/(m² °C), respectively, determine the error in the temperature measured, due to conduction.

[Answer: 2.76 unit]

9.5 A thermometer with time constant 12 seconds, initially at 300 K is exposed to an environment at 350 K. What will be the temperature indicated by the thermometer after 5 seconds from the exposure?

[Answer: 317.05 K]

9.6 Helium gas flows over a thin flat plate, with freestream Mach number and temperature of 1.6 and 300 K, respectively. If a thermocouple of recovery factor 0.98 is located at the wall surface, what will be the temperature recorded by the thermocouple?

[Answer: 552 K]

9.7 If a thermometer, initially at 300 K, exposed to an environment at 375 K indicates the temperature as 360 K, at the end of 7 seconds from the instant of exposure, determine its time constant.

[Answer: 4.35 s]

Chapter 10

Measurement of Wall Shear Stress

10.1 Introduction

The force due to wall shear stress is a quantity of primary importance in aerodynamic problems. In engineering applications, its overall value is usually estimated from the measurement of the gross drag force. This involves an estimate of the pressure drag, which must be subtracted from the total drag to get the drag due to shear stress. Drag due to the shear stress is also called *skin friction drag*. Flow over any solid surface gives rise to shear stress τ_w. At low speeds, this is generally determined from a measurement of velocity profile near the surface and an application of Newton's friction law,

$$\boxed{\tau_w = \mu_w \left(\frac{du}{dy}\right)_w}$$

where μ_w and $(du/dy)_w$ are the coefficient of viscosity and the velocity gradient at the wall, respectively. The measurement of velocity profiles at high speeds becomes very difficult, especially in turbulent boundary layers, in which the measurement must be made very close to the surface in order to get the velocity profile. But it is just for turbulent boundary layers this measurement is most needed, since theory is not available. In this region, the physical size of the pitot tube gives rise to displacement errors. Similarly, a hot-wire even though smaller in size yields erroneous results near the wall, due to conduction effects. Even though some corrections could be applied to these readings, they are not reliable enough to estimate the shear stress accurately. Hence, it calls for proper instrumentation for a reasonably accurate estimate of it.

10.2   Measurement Methods

Even though there are various direct and indirect methods available for determining the wall shear stress, each technique has got its own limitations and could be applied only under certain specified circumstances. Some of the well-known methods are

- Floating element method.

- Momentum integral method.

- Preston tube technique.

- Force technique.

- Heat transfer gauge.

Let us examine the measurement principles of these techniques in some detail in this chapter.

10.2.1   Floating Element Method

The floating element method is a direct method for measuring the wall shear stress. This method is used for measuring the skin friction directly on a flat plate. A small segment of the flat plate surface is separated from the remaining portion of the plate by a very small gap on either side but kept flush by suspending the element from a set of leaf springs forming a parallelogram linkage, as shown in Figure 10.1.

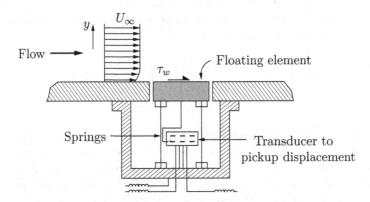

Figure 10.1: Shear stress balance

When there is a flow over the entire plate, the element is subjected to a shear flow which causes a displacement. The vertical faces of the element being subjected to the same normal pressure all around, the forces acting on these faces get canceled. The movement of the element is an indication of the shear stress and by precalibration of the complete floating system, the actual shear stress can be determined. The leaf springs are made of steel and the spring constants are so chosen that the total movement is of the order of few microns and it is measured using a transducer such as a linear differential transformer. The gap between the main plate and the floating element is kept very small, of the order of 0.2 mm, to prevent flow through the gap. The surface area of the floating element can be as small as 2 to 3 square centimeters.

Though the floating element method is a direct method for measuring the shear force, it can be used only for zero pressure gradient flows. If there is any pressure gradient, the pressure difference acting in the direction normal to the vertical faces becomes appreciable compared to the shear force. Further, the pressure difference can induce a flow inside the instrument exerting forces on the supporting system. The principle of the shear stress balance described above was due to Satish Dhawan (Liepmann and Roshko, 1956). There are various adaptations and modifications of the skin friction meter developed by Dhawan. The instrument illustrated in Figure 10.1 is one such modified version.

10.2.2 Momentum Integral Method

The momentum integral method in literature uses the famous relation of Karman (White, 1991), which is of the form

$$\frac{d\theta}{dx} + (2+H)\frac{\theta}{U}\frac{dU}{dx} = \frac{C_f}{2} = \frac{\tau_w}{\rho U^2} \tag{10.1}$$

The momentum thickness θ and shape factor H (= displacement thickness δ^*/momentum thickness θ) can be easily determined from the measured mean velocity profiles at flow stations. In principle, $d\theta/dx$ and dU/dx could also be obtained once θ and U distributions are known. In the case of a flat plate, that is, zero pressure gradient flow, the above equation reduces to

$$C_f = 2\frac{d\theta}{dx} \tag{10.2}$$

In this case, it is easy to estimate C_f (τ_w) if θ distribution is known. Measure the mean velocity profile at few stations and calculate θ. The advantage of this method is that θ is very sensitive to the velocity profile in the region very close to the wall.

However, in the case of a pressure gradient flow τ_w cannot be determined accurately since the first and second terms of Equation (10.1) are large and opposite in sign. The value of τ_w can be appreciably different if the errors in determining the first and second term of Equation (10.1) are even as low as 1 to 2 percent.

10.2.3   Preston Tube

Preston tube used with success a hypodermic needle tube to determine the
local *wall shear stress* from the dynamic pressure indicated by the tube (Hinze,
1977). To this end a circular tube with 1-mm outer diameter and inner to outer
diameter ratio of 0.6 was placed parallel to the main stream and against the
wall, as shown in Figure 10.2.

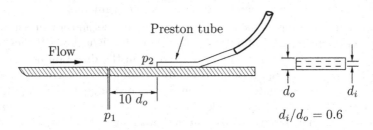

Figure 10.2: Preston tube

The diameter of the tube is still large enough to ensure that the main part
of the tube diameter is in the fully turbulent zone of the flow along the wall.
The dynamic pressure indicated by the tube depends on the local velocity distribution close to the wall. The Preston tube has been studied in detail, encom-
passing different geometrical parameters like the inner to outer diameter ratio
of the tube, the cross-sectional shape of the mouth of the tube and so on by
many investigators. From the above details, it is evident that the Preston tube
is a small pitot tube placed on the wall with its open end facing the flow direction. It works on the principle of similarity and it is used mainly in turbulent
flows. This instrument needs precalibration in a flow where the wall shear stress
is already known and similarity exists. It can be shown that, in a turbulent
boundary layer, the flow near the wall region follows the law (White, 1991)

$$\frac{U}{U^*} = f\left(\frac{y\,U^*}{\nu}\right) \tag{10.3}$$

called the dimensionless inner law, where the variable

$$U^* = \left(\frac{\tau_w}{\rho}\right)^{1/2} \tag{10.4}$$

has the units of velocity and is called the *wall-friction velocity*. U is the local
velocity, ν is kinematic viscosity, and y is coordinate normal to the wall.

In a nondimensional form we can write $\Delta p/(\frac{1}{2}\rho U^{*2})$ or $\Delta p/\tau_w$ as a function
of U^*d/ν, where Δp is the difference between the pressure measured by the
Preston tube and the undisturbed static pressure. When the ratio of inner to
outer diameter of the Preston tube is kept close to 0.6, dimensional analysis

shows that, for the universal velocity distribution, also called the *law of the wall*,

$$\boxed{\frac{\tau_w d^2}{\rho \nu^2} = f\left(\frac{\Delta p\, d^2}{\rho \nu^2}\right)} \qquad (10.5)$$

This functional relation should be the same for all flows near the wall region. That is, the Preston tube should be capable of measuring the shear stress, in a pipe flow, channel flow, or a boundary layer flow with a single calibration curve. For the sake of convenience, the instrument is calibrated in a two-dimensional channel or a pipe flow where the skin friction could easily be determined from the pressure gradient.

10.2.4 Fence Technique

The fence technique is an indirect method for measuring wall shear stress. A fence is just a small narrow strip of material placed on the wall, as shown in Figure 10.3.

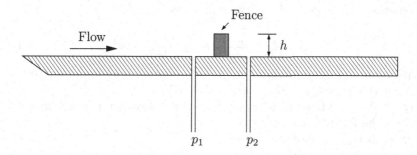

Figure 10.3: Fence on a wall

The pressure difference across the two faces of the fence is a function of wall shear. The height of the fence should be very small such that it is fully submerged in the sublayer.

$$p_1 - p_2 = a\,\tau_w^b \qquad (10.6)$$

where p_1 and p_2 are the pressures just upstream and downstream of the fence and a and b are constants. The constant a is considered to be a function of the height h of the face and b depends on $U^* h/\nu$.

Experiments indicate that, in the range $4 < U^* h/\nu < 8$, b is approximately 1.5. Therefore, for this case

$$p_1 - p_2 = a\,\tau_w^{1.5} \qquad (10.7)$$

The value a has to be obtained by precalibration using a channel or pipe flow. The above relation is valid for zero pressure gradient only. For use in pressure gradients some corrections are needed. Let $p_1 - p_2 = \Delta p$.

$$\underset{\text{pressure gradient}}{\Delta p} = \underset{\text{zero gradient}}{\Delta p} \left[1 + \frac{1}{2}k\left(\frac{U^*h}{\nu}\right)\right] \qquad (10.8)$$

where k is the correction factor to be determined from experiments. The advantage of the fence technique is that Δp is reasonably large so that it can be easily measured with a micro manometer. A fence can be easily made by just sticking a small piece of metal to the surface. The use of the fence is still in the research stage and it is yet to become a popular technique for the measurement of skin friction.

10.2.5   Heat Transfer Gauge

The heat transfer gauge can be used to measure wall shear over a very small area in laminar or turbulent flows with or without pressure gradient. This instrument operates on the principle that the heat transfer from the wall to the flow is a function of wall shear stress. An approximate relation between heat transfer and skin friction can be derived and the expression is given by

$$\tau_w = k_1 L \frac{dp}{dx}\frac{1}{Nu} + k_2 \frac{\mu^2}{\rho Pr}\frac{1}{L^2}Nu^3 \qquad (10.9)$$

where

Nu is the Nusselt number $= Q_w/k(T_w - T_1)$.

Q_w is the total heat transfer rate from the heated film per unit width.

Pr is the Prandtl number $= \mu C_p/K$.

dp/dx is the pressure gradient.

μ = viscosity coefficient.

T_w = hot gauge temperature.

T_1 = freestream temperature.

L = length of the heated film.

K = thermal conductivity of the flow medium.

k_1 and k_2 are constants.

When the pressure gradient is zero, the heat transfer is proportional to $\sqrt{3}\,\tau_w$.

An electrically heated thin film embedded in the surface of the wall is used to measure the heat transfer rate. The heat transfer gauge is made by placing a small length of platinum wire or by coating a thin film of platinum on the surface of an insulated plug. The plug can be of ceramic or glass. If a wire is used the diameter of the wire could be around 0.2 mm. The ends of the wire are connected to a constant temperature hot-wire anemometer system so that the wire temperature is maintained constant, that is $(T_w - T_1)$ is held constant. The constructional details of the heat transfer gauge and a typical calibration curve are shown in Figure 10.4.

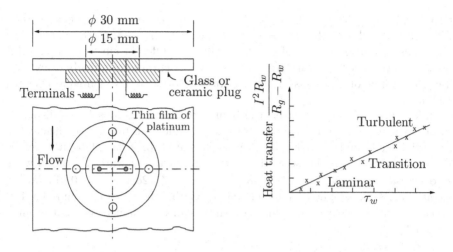

Figure 10.4: Heat transfer gauge and its calibration plot

The gauge is calibrated in a given flow and the value of current or the corresponding voltage is measured. It is necessary to keep the temperature of the heated element below a certain value so that the thermal boundary layer thickness is small compared to the velocity or momentum boundary layer thickness. This is one of the major assumptions in the derivation of the basic equation. The value of $(T_w - T_1)$ is kept between 10°C and 20°C. Care must be taken in measuring T_1 to a minimum accuracy of ± 0.1°C. Since the heat transfer rate is proportional to the cube root of the wall shear stress the instrument cannot be a sensitive one. However, this is the only method available for the measurement of the wall shear stress under various conditions with a single calibration. Further, this instrument is a special kind in the sense that it can be used for measurement in laminar, transitional, and turbulent boundary layers with or without pressure gradient.

10.2.6   Law of the Wall

The wall shear stress in a turbulent boundary layer can also be estimated using the law of the wall. Even though in principle the wall shear stress can be obtained from the velocity distribution in the wall region, that is, $\tau_w = \mu(\partial U/\partial y)_{y=0}$, in practice it is difficult to measure the velocity near the surface on account of the pitot tube size and its displacement effect. To circumvent this difficulty, the law of the wall may be used to determine the velocity U. It is sufficient if the velocity distribution is known in the logarithmic region, say between $yU^*/\nu = 10$ and 100, where U^* is the characteristic velocity $\sqrt{\tau_w/\rho}$. To evaluate the coefficient of skin friction C_f, plot the mean velocity profile in the form $U/U^* = A \ln (yU^*/\nu) + B$, using various values for U^*, where A and B are constants. When the correct U^* is used, the slope of the straight line will be equal to 5.6 and the intercept will be equal to 5.4. This process is somewhat tedious. Alternately, a plot of the mean velocity profile in the form U/U_∞ vs $y\,U_\infty/\nu$ can be used. For each value of C_f there will be a different curve. Compare the present plotting with a standard chart where the curves are already drawn for different values of C_f. This method is known as Clauser's technique (White, 1991).

Like laminar boundary layer, the boundary layer thickness δ, the displacement thickness δ^*, the momentum thickness θ, and skin friction coefficient C_f for turbulent boundary layer also can be estimated for a flat plate in terms of Reynolds number Re. These relations are empirical in nature. For flows with zero pressure gradient $\delta = 0.376\ Re_x^{-1/5}$; $\quad \theta = 7/12\delta$; $\quad C_f = 0.074 Re_x^{-1/5}$ where Re_x is the Reynolds number, based on the outer velocity and the distance x from the transition point. Transition can be assumed to occur around Re_x of 10^5.

10.3   Summary

Wall shear stress is a quantity of primary importance in aerodynamic problems. Drag due to the shear stress is also called *skin friction drag*. At low speeds, this is generally determined from Newton's friction law,

$$\tau_w = \mu_w \left(\frac{du}{dy}\right)_w$$

There are various direct and indirect methods available for determining the wall shear stress. Some of the well-known methods are

- Floating element method.

- Momentum integral method.

- Preston tube technique.

- Force technique.

- Heat transfer gauge.

Floating element method is a direct method for measuring the wall shear stress. This method is used for measuring skin friction directly on a flat plate. Though the floating element method is a direct method for measuring the shear force, it can be used only for zero pressure gradient flows. If there is any pressure gradient, the pressure differences acting in the direction normal to the vertical faces becomes appreciable compared to the shear force.

Momentum integral method makes use of the relation

$$\frac{d\theta}{dx} + (2+H)\frac{\theta}{U}\frac{dU}{dx} = \frac{C_f}{2} = \frac{\tau_w}{\rho U^2}$$

to obtain the shear force. In the case of a flat plate, that is, zero pressure gradient flow the above equation reduces to

$$C_f = 2\frac{d\theta}{dx}$$

In this case, it is easy to estimate C_f (τ_w) if θ distribution is known. The advantage of this method is that, θ is very sensitive to the velocity profile in the region very close to the wall. However, in the case of a pressure gradient flow τ_w cannot be determined accurately since the first and second terms of Equation (10.1) are large and opposite in sign.

Preston tube makes use of a hypodermic needle tube to determine the local *wall shear stress* from the dynamic pressure indicated by the tube. The dynamic pressure indicated by the tube depends on the local velocity distribution close to the wall. It is a small pitot tube placed on the wall with its open end facing the flow direction. It works on the principle of similarity and it is used mainly in turbulent flows. This instrument needs precalibration in a flow where the wall shear stress is already known and similarity exists.

Fence technique is an indirect method for measuring wall shear stress. A fence is just a small narrow strip of material placed on the wall. The pressure difference across the two faces of the fence is a function of wall shear. The height of the fence should be very small such that it is fully submerged in the sublayer.

The heat transfer gauge can be used to measure wall shear over a very small area in laminar or turbulent flows with or without pressure gradient. This instrument operates on the principle that the heat transfer from the wall to the flow is a function of wall shear stress. An electrically heated thin film embedded in the surface of the wall is used to measure the heat transfer rate. The heat transfer gauge is made by placing a small length of platinum wire or by coating a thin film of platinum on the surface of an insulated plug.

The wall shear stress in a turbulent boundary layer can also be estimated using the *law of the wall*.

Chapter 11

Mass and Volume Flow Measurements

11.1 Introduction

Mass and volume flow measurements become essential in many applications of practical interest. The principle involved in flow measurement may cover a wide variety of areas incorporating steady and unsteady flow of liquids, gases, two-phase media, and laminar and turbulent flows. Basically the following two approaches are followed for measuring flow rate.

Direct Quantity Measurements

* Weighing or volume tanks
* Positive displacement meters

Indirect, or Rate, Measurements

* Obstruction devices
* Velocity probes
* Special methods

Flow rate measurement devices often require accurate measurement of pressure and temperature in order to calculate the output of the instrument. In fact, accuracy of many flow measurement devices depends on the accuracy of pressure and temperature measurements. Flow rate is expressed in both volume and mass units. Some commonly used terms in flow measurement are liter and gallon.

1 liter = 1000 cubic centimeters = 0.26417 gallons
1 gallon per minute (gpm) = 63.09 cm^3/s

It should be noted that the commercial gas flow meters specify the flow rate as volume flow rate at standard conditions of 1 atm and 20°C. The units employed are standard cubic feet per minute (scfm) and standard cubic centimeters per minute (sccm). In this chapter we will be discussing some of the popularly employed techniques and devices for flow measurements.

11.2   Direct Methods

11.2.1   Tanks

Tanks are considered to be the basic reference device for steady-flow measurements. Usually a tank is a known volume which is filled, and the time of filling noted. For liquids of known density, weight measurements may replace volume. The measurement accuracy of these devices can be quite high when precise tanks and scales are used. The direct weighing technique is frequently employed for calibration of water and other liquid flow meters. Thus, it may be taken as a standard calibration technique. Obviously, it is not suitable for transient flow measurements.

11.2.2   Displacement Meters

Most of the displacement meters are basically positive-displacement meters, used for volume flow measurements. Positive-displacement flow meters are generally used in places where consistently high accuracy is desired under steady flow conditions. The home water meter shown in Figure 11.1 is a typical positive displacement meter.

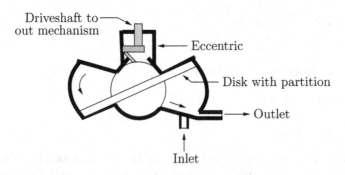

Figure 11.1: A nutating-disk meter

This meter works on the principle of nutating disk. The top and bottom of the disk remain in contact with the mounting chamber, thus allowing the disk to wobble or nutate about the vertical axis. Water enters the left side of the meter and strikes the disk which is eccentrically mounted. The disk wobbles and allows the fluid to move through the meter. A partition which separates the inlet and outlet chambers of the disk gives direct indication of the volume of liquid which has passed through the meter as it nutates. The nutation of the disk is recorded as volume flow by connecting a gear and register arrangement to the nutating disk. The nutating disk meter can be used with an accuracy of within ±1 percent.

The rotating-vane meter shown in Figure 11.2 is another kind of positive-displacement meter.

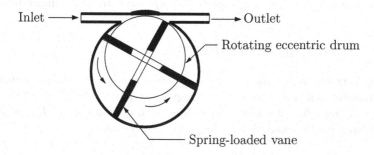

Figure 11.2: A rotating-vane flow meter

The vanes maintain continuous contact with the casing of the meter through spring loading. A fixed quantity of fluid is trapped in each section as the eccentric drum rotates, and this fluid ultimately flows out of the exit. An appropriate device to record the volume of the displaced fluid is connected to the shaft of the eccentric drum.

The lobed-impeller meter, shown in Figure 11.3, is yet another positive-displacement meter meant for mass flow measurement of gas and liquid flows.

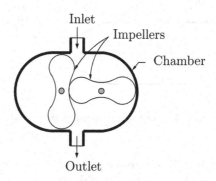

Figure 11.3: Lobed-impeller flow meter

Accurate fit of the casing and impeller are ensured in these meters. In this way, the incoming fluid is always trapped between the rotors and is conveyed to the outlet by the rotation of the rotors. Number of revolutions per unit time of the rotors is a measure of the volumetric flow rate.

The accuracy of the displacement meters can be very good even at the low end of the mass/volume range. Since these are devices with moving parts, wear of the parts may alter the measurement accuracy. Measurement of liquids with entrained vapor may also pose problems unless a good vapor trap is used. Most of these devices are totalizers, and do not measure instantaneous flow rates. Remote sensing of all the positive displacement meters may be accomplished with rotational transducers or sensors, and appropriate electronic counters.

11.3 Indirect Methods

Indirect methods determine the flow rate from the measurement of quantities other than the flow rate. Some of the popularly used devices for indirect measurement of mass flow are the *variable-head meter, drag-body meter, pitot tube, variable-area meter, mass flow meter,* and *hot-wire anemometer.*

11.3.1 Variable-Head Meters

All meters which measure the pressure drop across a restriction in a pipeline are *variable-head meters.* Several types of flow meters fall under the category of variable-head meters. Such devices are also called *flow-obstruction meters* or *differential-pressure meters.* The principle of measurement in these devices is the following.

Consider the flow through a duct shown in Figure 11.4.

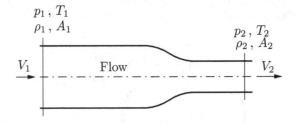

Figure 11.4: One-dimensional flow system

Let us assume the flow to be one-dimensional and incompressible. By continuity equation, we have the mass flow rate $\dot{m}$ as

$$\dot{m} = \rho_1 A_1 V_1 = \rho_2 A_2 V_2 \tag{11.1}$$

where A is cross-sectional area, V is velocity, and ρ is density. Also, $\rho_1 = \rho_2$ since the flow is incompressible. Let us treat the flow to be isentropic. By the

Bernoulli equation (for incompressible flows), we have

$$p_1 + \frac{1}{2}\rho V_1^2 = p_2 + \frac{1}{2}\rho V_2^2 \tag{11.2}$$

The pressure drag $(p_1 - p_2)$ may be written, using Equations (11.1) and (11.2), as

$$p_1 - p_2 = \frac{V_2^2 \rho}{2}\left[1 - \left(\frac{A_2}{A_1}\right)^2\right] \tag{11.3}$$

The volumetric flow rate $\dot{Q}$ may be written as

$$\dot{Q} = A_2 V_2 = \frac{A_2}{\sqrt{1 - (A_2/A_1)^2}} \times \sqrt{\frac{2}{\rho}(p_1 - p_2)} \tag{11.4}$$

It is evident from Equation (11.4) that a flow duct or channel like that shown in Figure 11.4 could be used for flow measurement by simply measuring the pressure drop $(p_1 - p_2)$ and calculating the flow rate from Equation (11.4). However, the assumption of isentropic flow based on which the equation derived is never true, since no duct can be without friction, and some losses are always incurred in the flow. The volumetric flow rate given by Equation (11.4) is therefore only for an *ideal value*, and it is usually related to the actual flow rate through an empirical *discharge coefficient* C, given by the following relation.

$$\frac{\dot{Q}_{\text{actual}}}{\dot{Q}_{\text{ideal}}} = C \tag{11.5}$$

The discharge coefficient C is not a constant and depends upon the duct geometry and the Reynolds number of the flow.

If the flow through the channel is a gas which can be treated as a perfect gas, then by perfect gas state equation, we have

$$p = \rho RT \tag{11.6}$$

where T is absolute temperature and R is the gas constant for the particular gas, which can be expressed in terms of universal gas constant R_u and the molecular weight M as

$$R = \frac{R_u}{M} \tag{11.7}$$

where $R_u = 8314$ J/(kmol.K). For isentropic flow, by energy equation, we have

$$h_1 + \frac{V_1^2}{2} = h_2 + \frac{V_2^2}{2} \tag{11.8}$$

where h is the enthalpy. Expressing enthalpy h as $C_p T$, where C_p is the specific heat at constant pressure, we have

$$C_p T_1 + \frac{V_1^2}{2} = C_p T_2 + \frac{V_2^2}{2} \tag{11.9}$$

Combining Equations (11.1), (11.6), and (11.9), we obtain

$$\dot{m}^2 = 2A_2^2 \left(\frac{\gamma}{\gamma - 1}\right) \left(\frac{p_1^2}{RT_1}\right) \left[\left(\frac{p_2}{p_1}\right)^{\frac{2}{\gamma}} - \left(\frac{p_2}{p_1}\right)^{\frac{\gamma+1}{\gamma}}\right] \tag{11.10}$$

where γ is the ratio of specific heats (C_p/C_v) and also is called the *isentropic index*. When V_1 is small compared to V_2, Equation (11.10) gets simplified to

$$\dot{m} = \sqrt{\frac{2}{RT_1}} A_2 \left[p_2 \Delta p - \left(\frac{1.5}{\gamma} - 1\right)(\Delta p)^2 + ...\right]^{\frac{1}{2}} \tag{11.11}$$

where $\Delta p = p_1 - p_2$. Equation(11.11) is valid for $\Delta p < p_1/4$. For $\Delta p < p_1/10$, the equation gets further simplified to result in

$$\dot{m} = A_2 \sqrt{\frac{2p_2}{RT_1}(p_1 - p_2)} \tag{11.12}$$

Thus, as seen from Equations (11.4) and (11.12), measurement of only p_1 and p_2 is necessary for determining the flow rate.

The three types of obstruction or constriction meters which are widely used are the venturi, flow nozzles, and orifice. Typical shapes of these constriction meters are shown schematically in Figure 11.5.

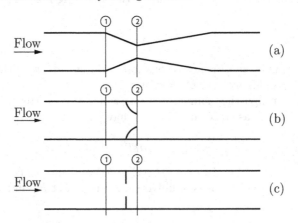

Figure 11.5: Schematic of (a) venturi, (b) flow nozzle, (c) orifice

The advantages and disadvantages of the above constriction meters are the following.

- The venturi offers the best accuracy, least head loss, and best resistance to abrasion and wear due to dirt particles in the fluids. But it is expensive and occupies considerable space.

- The nozzle type meters offer all advantages of the venturi to a slightly lesser extent, but require only a small space. However, proper installation of the nozzle is difficult.

- The orifice is less expensive, easy to install, and occupies a very small space. However, it suffers from considerable head loss and is very sensitive to abrasion and other damage.

Flow rate calculations of these devices are made on the basis of Equation (11.4), with appropriate empirical constants, namely the velocity approach factor M_v, the flow coefficient K, the diameter ratio β, and the expansion factor Y, defined as follows.

$$M_v = \frac{1}{\sqrt{1 - (A_2/A_1)^2}} \qquad (11.13)$$

$$K = CM_v \qquad (11.14)$$

$$\beta = \frac{d}{D} = \sqrt{\frac{A_2}{A_1}} \qquad (11.15)$$

The empirical constant Y, which is meant only for compressible flows is defined for venturi, nozzles, and orifices separately. For venturi and nozzle this factor is given by (Holeman and Gajda, 1989),

$$Y_a = \left[\left(\frac{p_2}{p_1} \right)^{\frac{2}{\gamma}} \frac{\gamma}{\gamma - 1} \frac{1 - (p_2/p_1)^{\frac{\gamma-1}{\gamma}}}{1 - (p_2/p_1)} \frac{1 - (A_2/A_1)^2}{1 - (A_2/A_1)^2 (p_2/p_1)^{\frac{2}{\gamma}}} \right]^{1/2} \qquad (11.16)$$

For orifice with either flange taps or vena contracta taps,

$$Y_b = 1 - \left[0.41 + 0.35 (A_2/A_1)^2 \right] \frac{p_1 - p_2}{\gamma p_1} \qquad (11.17)$$

For orifices with pipe taps the expansion factor Y is given by

$$Y_c = 1 - \left[0.333 + 1.145 \left(\beta^2 + 0.75 \beta^5 + 12 \beta^{13} \right) \right] \frac{p_1 - p_2}{\gamma p_1} \qquad (11.18)$$

The empirical expansion factor given by Equations (11.17) and (11.18) are accurate within $\pm$ 0.5 percent for $0.8 < p_2/p_1 < 1.0$. The expansion factors Y_a and Y_c, in Equations (11.16) and (11.18) are shown graphically in Figures 11.6 and 11.7, respectively.

With the above-mentioned empirical constants, the following semi-empirical equations may be written.

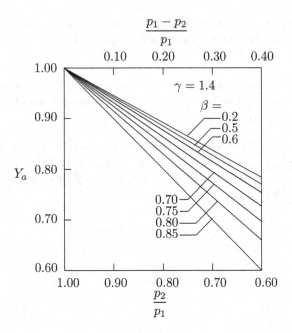

Figure 11.6: Adiabatic expansion factors for venturi and nozzles, given by Equation (11.16)

For Venturis in Incompressible Flows

$$Q_{\text{actual}} = CM_v A_2 \sqrt{\frac{2}{\rho}} \sqrt{p_1 - p_2} \qquad (11.19)$$

For Nozzles and Orifices in Incompressible Flows

$$Q_{\text{actual}} = K A_2 \sqrt{\frac{2}{\rho}} \sqrt{p_1 - p_2} \qquad (11.20)$$

For compressible flows, the above equations get modified as follows.

For Venturis in Compressible Flows

$$\dot{m}_{\text{actual}} = Y_a C M_v A_2 \sqrt{2\rho_1 (p_1 - p_2)} \qquad (11.21)$$

11.3.1.1   For Nozzles and Orifices in Compressible Flows

$$\dot{m}_{\text{actual}} = Y_b K A_2 \sqrt{2\rho_1 (p_1 - p_2)} \qquad (11.22)$$

The coefficients described above have been tabulated in *Fluid Meters, Their Theory and Applications*, 6th ed., ASME, New York, 1971.

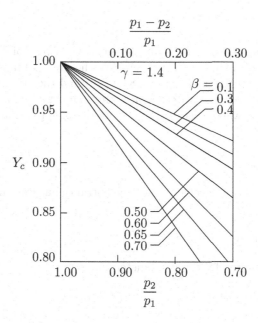

Figure 11.7: Expansion factor for square-edged orifices with pipe taps given by Equation (11.18)

Even though all these devices can be used for flow measurements without calibration if made to standard dimensions, they suffer from a limited useful range because of the square root relationship between pressure drop and flow rate. The ASME power test code (1959) gives extensive data and specifications for these three types of meters.

These devices are used for measurements in compressible fluids in the subsonic regime of flow speeds by applying a simple correction for expansion of the fluid, as discussed earlier. However, they are not too useful for measuring flow at small Reynolds numbers ($Re < 15,000$), when accuracy of measurement is of prime importance, the discharge coefficient being strongly influenced by the Reynolds number. Viscous effects become dominant at these Reynolds numbers. For highly viscous fluids or very small flow rates, capillary-type meters are used to overcome this problem. The Hagen–Poiseuille law governs this type of viscous flow and the flow rate $\dot{Q}$ is given by

$$\dot{Q} = \frac{\pi r^4}{8\mu L} \Delta p \tag{11.23}$$

where r is the tube radius, L is its length, Δp is the pressure drop, and μ is the dynamic viscosity coefficient.

Example 11.1

A venturi of throat diameter 125 mm is inside a water pipe of diameter 200 mm. If a U-tube mercury manometer connected to the pressure taps at the entrance and throat of the venturi measures 87.8 cm, (a) find the velocity at the venturi throat, (b) the volume flow rate of water through the pipe, and (c) the power associated with the water jet discharged from the pipe to atmospheric pressure.

Solution

(a) Let subscripts 1 and 2 refer to the pipe entrance and venturi throat, respectively.

Given, $d_1 = 0.2$ m, $d_2 = 0.125$ m.

The pressure measured by the manometer is

$$p_1 - p_2 = \frac{87.8}{76} \times 101325$$

$$= 117057 \, \text{Pa}$$

By Equation (11.3),

$$p_1 - p_2 = \frac{\rho V_2^2}{2}\left[1 - \left(\frac{A_2}{A_1}\right)^2\right]$$

Therefore,

$$V_2^2 = \frac{2(p_1 - p_2)}{\rho\left[1 - \left(\frac{A_2}{A_1}\right)^2\right]}$$

$$= \frac{2 \times 117057}{10^3 \times \left[1 - \left(\frac{0.125}{0.2}\right)^4\right]}$$

$$= \frac{2 \times 117057}{10^3 \times 0.8474}$$

$$= 276.27$$

$$V_2 = \sqrt{276.27}$$

$$= \boxed{16.62 \, \text{m/s}}$$

(b) The volume flow rate through the pipe is

$$\dot{Q} = A_2 V_2$$

$$= \left(\frac{\pi \times 0.125^2}{4}\right) \times 16.62$$

$$= \boxed{0.204\,\text{m}^3/\text{s}}$$

(c) The power of the water jet is

$$\text{Power} = \dot{m}V_e$$

$$= \rho\dot{Q}V_e$$

$$= 10^3 \times 0.204 \times 16.62$$

$$= \boxed{3.39\,\text{kW}}$$

11.3.2 Some Practical Details of Obstruction Meters

The American Society of Mechanical Engineers, ASME, has standardized the geometry of the obstruction or variable head meters for flow measurements under various flow conditions of practical importance (Holman and Gajda, 1989). The ASME standard venturi meter is shown schematically in Figure 11.8.

It is seen that the pressures at the entry to venturi and at the throat are measured by connecting these pressure taps to manifolds. These manifolds receive a sampling of the pressure all around these sections so that these pressures are the actual representative pressures at these locations. The discharge coefficients for such venturi meters are shown in Figure 11.9. Generally, the discharge coefficients are smaller for pipes with diameter less than 50 mm. The approximate behavior of discharge coefficient variation with Reynolds number for different pipe diameters is shown in Figure 11.10.

Appropriate discharge coefficient for a venturi may be obtained precisely by direct calibration. It is possible to obtain values accurate up to ±5 percent by direct calibration. The dimensions prescribed by ASME for flow nozzles are shown in Figure 11.11.

The discharge coefficients for the above ASME standard nozzles are given graphically in Figure 11.12.

The measurement accuracy of the flow rates depends strongly on the proper installation of the flow nozzles. The recommended installations for concentric, thin-plate orifices are shown in Figure 11.11. The discharge coefficient as a function of Reynolds number for the ASME nozzles shown in Figure 11.11 is shown in Figure 11.12.

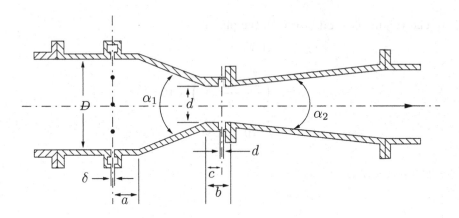

Figure 11.8: Details of a venturi meter D = pipe diameter, d = venturi throat diameter α = 0.25 D to 0.75 D for 100 mm < D < 150 mm, 0.25 D to 0.5 D for 150 < D < 800 mm $b = d$, $c = d/2$, δ = 1.5 to 12 mm according to D. Annular pressure chamber with at least four piezometer vents, r_2 = 3.5 d to 3.75 d, r_1 = 0 to 1.375 D, $\alpha_1 = 21° \pm 2°$, $\alpha_2 = 5°$ to 15°

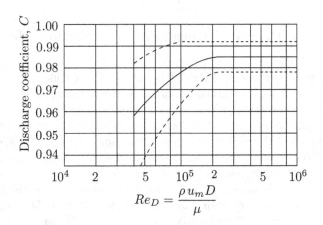

Figure 11.9: Discharge for the venturi meter shown in Figure 11.8, valid for 0.25 < β < 0.75 and D > 50 mm (the dashed lines are the tolerance limits)

It is seen from Figure 11.13 that pressure taps for measuring pressure are provided at three locations. They are the stations at inlet to orifice, at outlet of orifice, and at stations just upstream and just downstream of orifice.

- The pressure taps adjacent to the orifice plate are made on the flanges themselves.

- The inlet pressure tap is located at one diameter distance upstream, and the outlet pressure tap at one and a half diameter location downstream of

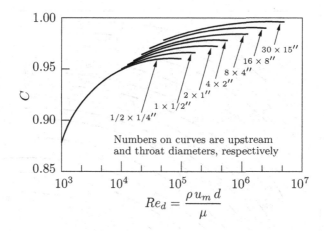

Figure 11.10: Venturi discharge coefficient for various throat diameters (the numbers on the curves are upstream and throat diameters of the venturi)

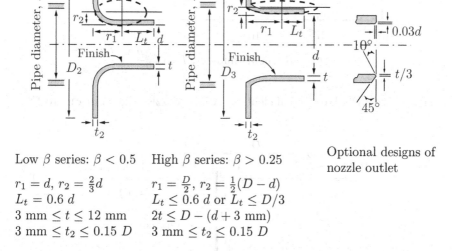

Low β series: $\beta < 0.5$ High β series: $\beta > 0.25$ Optional designs of nozzle outlet

$r_1 = d$, $r_2 = \frac{2}{3}d$	$r_1 = \frac{D}{2}$, $r_2 = \frac{1}{2}(D - d)$
$L_t = 0.6\,d$	$L_t \leq 0.6\,d$ or $L_t \leq D/3$
$3\text{ mm} \leq t \leq 12\text{ mm}$	$2t \leq D - (d + 3\text{ mm})$
$3\text{ mm} \leq t_2 \leq 0.15\,D$	$3\text{ mm} \leq t_2 \leq 0.15\,D$

Figure 11.11: ASME long-radius flow nozzles (Holman and Gajda, 1989)

the orifice, measured from the upstream face of the orifice.

The outlet pressure tap is located at the vena contracta of the orifice. The vena contracta location variation with diameter ratio β is given in Figure 11.14, for concentric orifices.

The values of the orifice flow coefficient M_c as a function of the ratio of orifice diameter to pipe diameter, β, for pipes with diameters in the range 30 mm to

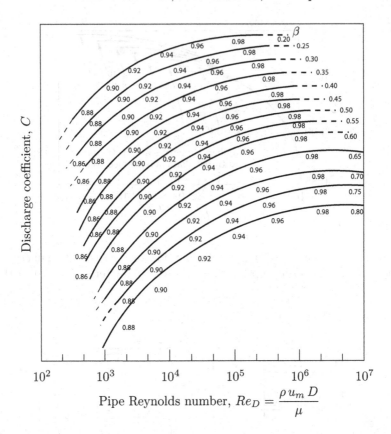

Figure 11.12: Discharge coefficient for ASME nozzles shown in Figure 11.11

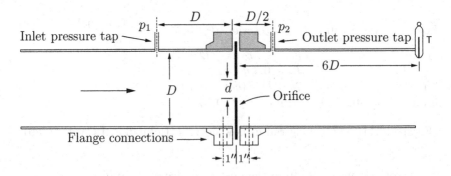

Figure 11.13: Location of pressure taps for use with concentric, thin-plate, square-edged orifice

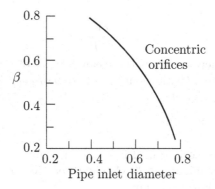

Figure 11.14: Location of outlet pressure tap for orifices (Holeman and Gajda, 1989)

75 mm with pressure taps located as per the above prescriptions are given in Figure 11.15.

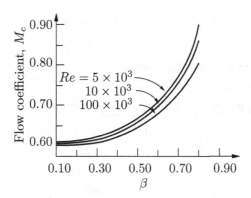

Figure 11.15: Flow coefficient for concentric orifices in pipes

The various flow coefficients in our present discussions are plotted as a function of Reynolds number, defined as

$$R_e = \frac{\rho V_m d}{\mu} \tag{11.24}$$

where ρ is the fluid density, V_m is the mean velocity of flow, μ is the dynamic viscosity coefficient, and d is the diameter of the *particular section* for which the *Reynolds number is specified.*

Example 11.2

Determine the size of the venturi if it has to measure a maximum flow rate of water of 100 liters per minute at 20°C. The throat Reynolds number has to be at least 10^4 at these flow conditions.

Solution

$$\text{Volume flow rate} = 100 \text{ l/min}$$

$$= 100 \times 10^3 \text{ cm}^3/\text{min}$$

$$= 0.00167 \text{ m}^3/\text{s}$$

The maximum mass flow rate becomes
$$\dot{m} = 1.67 \text{ kg/s}$$

since $\rho_w = 998 \text{ kg/m}^3$ at 20°C. The Reynolds number based on venturi throat diameter d is

$$R_{ed} = \frac{\rho V_m d}{\mu} = \frac{\rho A V_m d}{\mu A}$$

where A is the cross-sectional area of venturi throat.

$$R_{ed} = \frac{\dot{m} d}{(\pi d^2/4)\,\mu} = \frac{4\dot{m}}{\pi d \mu} = 10^4$$

The viscosity of water at 20°C is $\mu = 1.003 \times 10^{-3} \text{ N.s/m}^2$
Thus,

$$d_{\max} = \frac{4(1.67)}{\pi \times 1.003 \times 10^{-3} \times 10^4}$$

$$= \boxed{2.12 \times 10^{-4} \text{ m}}$$

Example 11.3

Find the time required to empty a square water tank of side height 6 m and depth 2 m, through a 20 cm diameter faired orifice, which is fitted to a pipe of length 1 m, below the tank bottom.

Solution

By Equation (11.20), the flow rate is

$$Q = KA_{\text{orifice}} \sqrt{\frac{2}{\rho} \Delta p}$$

For the faired orifice, the flow coefficient $K = 1$, therefore,

$$Q = A_{\text{orifice}} \sqrt{\frac{2}{\rho} \Delta p}$$

Also, $\Delta p = \rho g h$. Thus

$$Q = A_{\text{orifice}} \sqrt{2gh}$$

But $Q = A_{\text{tank}} V$ and $V = \frac{dh}{dt}$, therefore,

$$A_{\text{tank}} \frac{dh}{dt} = A_{\text{orifice}} \sqrt{2gh}$$

This gives

$$\frac{dh}{\sqrt{h}} = \frac{A_{\text{orifice}}}{A_{\text{tank}}} \sqrt{2g}\, dt$$

Integrating between time $t = 0$, corresponding to $h = 2$ and $t = t$, corresponding to $h = 0$, we get

$$\int_0^2 \frac{dh}{\sqrt{h}} = \frac{A_{\text{orifice}}}{A_{\text{tank}}} \sqrt{2g} \int_0^t dt$$

$$2\sqrt{h} = \frac{A_{\text{orifice}}}{A_{\text{tank}}} \sqrt{2g}\, t$$

$$2\sqrt{2} = \frac{A_{\text{orifice}}}{A_{\text{tank}}} \sqrt{2g}\, t$$

Given,

$$A_{\text{orifice}} = \frac{\pi \times 0.2^2}{4}$$

$$= 0.0314\,\text{m}^2$$

$$A_{\text{tank}} = 6 \times 6$$

$$= 36\,\text{m}^2$$

Substituting these, the time to drain the water tank of depth $h = 2$ m, becomes

$$2\sqrt{2} = \frac{0.0314}{36} \times \sqrt{2 \times 9.81} \times t$$

$$2\sqrt{2} = 3.863 \times 10^{-3} \times t$$

$$t = \frac{2\sqrt{2}}{3.863 \times 10^{-3}}$$

$$= 732.18 \text{ seconds}$$

$$= \boxed{12.203 \text{ minutes}}$$

11.3.3   Sonic Nozzle

In all the obstruction meters discussed above, when the flow rate is sufficiently high, the pressure differential becomes very large, and ultimately flow may choke attaining sonic velocity at the minimum area location. Under these conditions the flow rate becomes the maximum for the given inlet conditions and the flow is called *choked flow*. For a perfect gas, as we know, the pressure ratio for choked flow is

$$\left(\frac{p_{\text{th}}}{p_{01}}\right)_{\text{critical}} = \left(\frac{2}{\gamma+1}\right)^{\frac{\gamma}{\gamma-1}} \tag{11.25}$$

where pressure p_{01} is the stagnation pressure at the inlet and p_{th} is the static pressure at the throat. This pressure ratio is termed the *critical pressure ratio*. Substitution of Equation (11.25) into Equation (11.10) gives the mass flow rate as

$$\dot{m} = A_{\text{th}}\, p_{01} \sqrt{\frac{2}{RT_{01}} \left[\frac{\gamma}{\gamma+1}\left(\frac{2}{\gamma+1}\right)^{\frac{2}{\gamma-1}}\right]^{\frac{1}{2}}} \tag{11.26}$$

where T_{01} is the stagnation temperature at the inlet and A_{th} is the throat area. Equation (11.26) is commonly used for nozzles operating under choked conditions, that is, when p_{th}/p_{01} is less than the critical value given by Equation (11.25). From Equation (11.26) it is evident that under choked condition the mass flow rate of an ideal (perfect) fluid flow depends only on the stagnation pressure p_{01} and stagnation temperature T_{01} at the inlet.

For air with $\gamma = 1.4$ and $R = 287$ J/(kg K), Equation (11.26) reduces to

$$\dot{m} = \frac{0.6847\, p_{01}}{\sqrt{RT_{01}}} A_{\text{th}} \tag{11.27}$$

The stagnation pressure and temperatures can be measured easily, therefore the sonic nozzle is considered to be a convenient device for measuring gas flow rates.

It should be noted that the above equations for sonic nozzles are based on ideal gas assumptions. They must be corrected by an appropriate discharge coefficient which is a function of the geometry of the nozzle and other flow parameters. This discharge coefficient for sonic nozzles is usually about 0.97.

Example 11.4

A sonic nozzle is fitted in a 100-mm-diameter pipe for measuring the mass flow rate. Air at a stagnation condition of 1.5 MPa and 303 K is flowing through the pipe. If the flow rate is 0.55 kg/s, calculate the throat diameter of the nozzle required to run at just choked condition.

Solution

By Equation (11.27), we have the choked mass flow rate as

$$\dot{m} = \frac{0.6847 p_0}{\sqrt{RT_0}} A_{th}$$

Given that, $\dot{m} = 0.55$ kg/s, $p_0 = 1.5 \times 10^6$ Pa and $T_0 = 303$ K. For air the gas constant $R = 287$ J/(kg K). Therefore,

$$0.55 = \frac{0.6847 \times 1.5 \times 10^6}{\sqrt{287 \times 303}} A_{th}$$

$$A_{th} = \frac{162.19}{1.027} \times 10^{-6} \ \text{m}^2$$

$$= 157.926 \times 10^{-6} \ \text{m}^2$$

Therefore,

$$d_{th} = \sqrt{\frac{157.926 \times 4}{\pi}} \times 10^{-3} \ \text{m}$$

$$= 14.18 \times 10^{-3}$$

$$= \boxed{1.418 \, \text{cm}}$$

11.3.4   Pitot Tubes

The conventional pitot tube or the pitot-static tube can also be used for measurement of flow rate by calibrating them for volume flow rate by locating them at a specific point in a duct. But such single-point flow rate measurements become inaccurate when velocity profiles differ from calibration conditions. The error due to the above effect can be minimized by dividing the total flow area into several equal-area annuli and measuring the total velocity at the center of the area of each annulus by moving the pitot tube to such points, and summing these individual flow rates to obtain the total flow rate. This method is alright for laboratory applications but impractical for online monitoring of flow rates.

11.3.5 Rotameters

A *rotameter* is a direct-reading meter, usually employed for flow rate measurements in passages with small diameters. Basically it is a *constant-pressure-drop, variable-area meter*. It consists of a tube with tapered bore in which a *float* takes a vertical position corresponding to each flow rate through the tube, as shown in Figure 11.16.

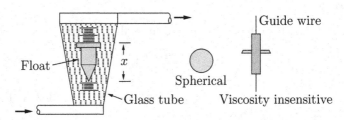

Figure 11.16: Rotameter

For a given flow rate, the float remains stationary and the vertical component of the forces of differential pressure, viscosity, buoyancy, and gravity are balanced. This force balance is self-maintaining since the flow area of the rotameter, which is the annular area between the float and the tube, varies continuously with the vertical displacement of the float; thus the device may be thought of as an orifice of adjustable area. The downward force, which is the gravity minus the buoyancy, is constant, therefore, the upward force, which is due to pressure drop, must also be constant. Since the float area is constant, the pressure drop should be constant. For a fixed flow area, the pressure drop Δp varies with the square of the flow rate, and so to keep the Δp constant for different flow rates, the area must vary. The tapered tube of the rotameter provides this variable area. The position of the float is a measure of volume flow rate, and the position of the float can be made linear with flow rate by making the tube area vary linearly with the vertical distance.

Assuming incompressible flow, the volume flow rate $\dot{Q}$ can be expressed as

$$\dot{Q} = C_d \frac{(A_t - A_f)}{\sqrt{1 - \left[(A_t - A_f)/A_t\right]^2}} \sqrt{2g\mathbb{V}\frac{w_f - w_{ff}}{A_f w_{ff}}} \tag{11.28}$$

where C_d is discharge coefficient, A_t is area of tube, A_f is area of float, g is the gravitational acceleration, $\mathbb{V}$ is volume of float, w_f is specific weight of float, and w_{ff} is specific weight of flowing fluid.

If the variation of C_d with float position is small and if $[(A_t - A_f)/A_t]^2$ is very small compared to 1, then Equation (11.28) can be shown to be

$$\dot{Q} = K\,(A_t - A_f) \tag{11.29}$$

and if the tube is shaped in such a manner that A_t varies linearly with float position x, then $\dot{Q} = K_1 + K_2\,x$, is a linear relation, where $K_1 = (-\,KA_f)$ is a constant, $K_2 = \text{constant} \times K$ and $K = C_d\,\sqrt{2g\mathbb{V}\frac{w_f - w_{ff}}{A_f w_{ff}}}$.

The rotameter floats may be made of various materials to obtain the desired density difference $(w_f - w_{ff})$ for metering a particular fluid. Some float shapes, such as spheres, require no guiding in the tube; others are kept central by guide lines or by internal ribs in the tube. Floats shaped to induce turbulence can give viscosity insensitivity over a 100:1 range.

The tube usually made of high-strength glass allows direct observation of the float position. For higher strength, metal tubes can be used and the float position detected magnetically through the metal wall. When an electrical or pneumatic signal related to the flow state is wanted, the float motion can be measured with a suitable displacement transducer. Accuracy of a rotameter is typically ±2 percent full scale and repeatability is about 0.25 percent of the reading.

11.3.6   Drag-Body Meters

Insertion of an appropriately shaped body into a flow stream can serve as a flow meter. The drag experienced by the body becomes a measure of flow rate after suitable calibration. A body immersed in a flow stream is subjected to a drag force F_D, given by

$$F_D = \frac{1}{2}\,\rho V^2\,A\,C_D \tag{11.30}$$

where ρ is fluid density, V is fluid velocity, A is cross-sectional area of the body facing the flow, and C_D is the drag coefficient. For sufficiently high Reynolds number and a properly shaped body, the drag coefficient is almost a constant. Therefore, for a given density, F_D is proportional to V^2 and thus to the square of volume flow rate.

The drag force can be measured by attaching the drag producing body to a strain gauge force measuring transducer. A typical drag force flow meter using a cantilever beam with bonded strain gauges is shown in Figure 11.17.

A hollow tube arrangement with the strain gauges fixed on the out surface, as shown in Figure 11.17, serves to isolate the strain gauges from the flowing fluid. If the drag force is made symmetric, reversed flows can also be measured. The main advantage of this type of flow meter is the high dynamic response. The type of gauge described above is basically second order with a natural frequency

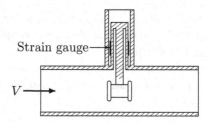

Figure 11.17: Drag-body flow meter

of 70 to 200 Hz. However, this type of gauge suffers from small damping and thus sharp transients may cause difficulty.

11.3.7   Ultrasonic Flow Meters

The fact that small-magnitude pressure disturbances travel through a fluid medium at a definite velocity, the speed of sound, relative to the fluid is made use of in ultrasonic flow meters. The word ultrasonic refers to the fact that in reality the pressure disturbances generally are short bursts of sine waves whose frequency is above the range audible to human hearing (about 20 kHz). A typical frequency may be about 10 MHz.

Ultrasonic flow meters make use of transmitters and receivers of acoustic energy in their measurements. A common approach is to utilize piezoelectric crystal transducers for both the functions. Electrical energy in the form of a short burst of high-frequency voltage is applied to a crystal, in a transmitter, causing it to vibrate. If the crystal is in contact with a fluid, the vibration will be transmitted to the fluid and propagated through it. The receiver crystal when exposed to these pressure fluctuations, responds to this vibration. The vibrating motion of the crystal produces an electrical signal proportional to the vibrational excitation, in accordance with the usual action of piezoelectric displacement transducers.

Application of these principles is illustrated in Figure 11.18.

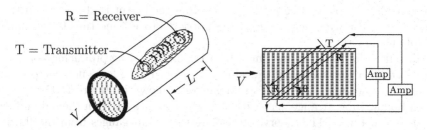

Figure 11.18: Schematic of ultrasonic flow meter

When the flow velocity V is zero, the transit time t_0 of pulse from the transmitter to the receiver is given by

$$t_0 = \frac{L}{a} \qquad (11.31)$$

where L is the distance between transmitter and receiver and a is the acoustic velocity in fluid. For instance, in air at $15°C$, $a = 340\,m/s$, and so if $L = 1\,m$, $t_0 = 0.00294$ s. If the fluid is moving with velocity V, the transit time becomes

$$t = \frac{L}{a + V} = L\left(\frac{1}{a} - \frac{V}{a^2} + \frac{V^2}{a^3} - \ldots\right) \approx \frac{L}{a}\left(1 - \frac{V}{a}\right) \qquad (11.32)$$

Let $\Delta t = t_0 - t$, then

$$\Delta t \approx \frac{LV}{a^2} \qquad (11.33)$$

Thus, knowing a and L, V can be calculated once Δt is measured. The length parameter L can be taken as constant but a varies with temperature. In fact, a is a strong function of temperature. Also, Δt is quite small since in low-speed streams, flow velocity V is small compared to speed of sound a. For instance, if $V = 5$ m/s, $L = 0.2$ m, and $a = 340$ m/s, then $\Delta t = 8.6$ μs, a very short increment of time for accurate measurement. Since there is no provision for the direct measurement of t_0 in this arrangement, the device has been modified like that shown in Figure 11.19 to measure t_0 directly.

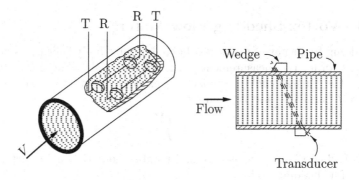

Figure 11.19: Modified ultrasonic flow meter

Taking t_1 and t_2 to be the transit times with the flow and against the flow, we can write

$$\Delta t \approx t_2 - t_1 = \frac{2LV}{a^2 - V^2} \approx \frac{2LV}{a^2} \qquad (11.34)$$

This Δt is two times the Δt in Equation (11.33). This is a physical time increment that may be measured directly.

Two self-excited oscillations are created by using the received pulses to trigger the transmitted pulses in a feedback arrangement, as shown in Figure 11.18.

The pulse repetition frequency in the forward propagating loop is $1/t_1$, while that in the reverse loop is $1/t_2$. The frequency difference is

$$\Delta f \approx \frac{1}{t_1} - \frac{1}{t_2}$$

But $t_1 = L/(a + V \cos \theta)$ and $t_2 = L/(a - V \cos \theta)$. Substituting these into the above equation, we obtain

$$\boxed{\Delta f = \frac{2V \cos \theta}{L}} \tag{11.35}$$

Equation (11.35) is independent of the speed of sound a and thus not subject to the errors due to changes in a. This technique is practical and forms the basis of many commercially available ultrasonic flow meters. Usually two methods are used for reading the frequency difference Δf. They are called the *sing-around scheme* and *up-down counter scheme*.

In the sing-around scheme, two signals of different frequencies are multiplied, giving an output with sum and difference frequencies. The difference frequency is extracted by filtering. This approach responds to the flow changes very quickly, but it requires pulses with well-defined leading edges. But such pulses are not always available.

The up-down counter scheme accumulates the two frequencies separately for 5 to 20 seconds and then subtracts them giving an average effect which may reduce the noise problem.

11.3.8 Vortex-Shedding Flow Meters

The shedding frequency of the vortex behind a blunt body kept in a steady flow is made use of for the measurement of flow in vortex-shedding flow meters.

When the pipe Reynolds number Re_d exceeds about 10^4, vortex shedding frequency f is reliable and is given by

$$f = \frac{St\, V}{D} \tag{11.36}$$

where V is flow velocity, St is Strouhal number, and D is a characteristic dimension of a shedding body.

The Strouhal number can be kept nearly constant, by properly designing the vortex shedding body shape over a wide range of Re. That is, f can be made proportional to V, thus giving a digital flow metering principle based on counting the vortex shedding rate. Some typical vortex shedding shapes used in flow meters and flow meter principle are shown in Figure 11.20.

The vortices shed causes alternative forces or local pressures on the shedder. Strain gauge or piezoelectric methods can be employed to detect these forces. Hot film thermal anemometer sensors embedded in the shoulder of the body can detect the periodic flow-velocity fluctuations. The interruption of ultrasonic beams by the passing vortices may be used for counting the vortices. The differential pressures induced by the vortices may be detected by making use of

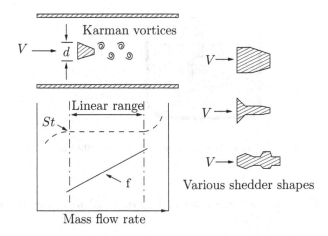

Figure 11.20: Vortex shedding body and flow meter principle

oscillations created by the pressure in a small caged bellow hose motion, with a magnetic proximity pickup.

11.3.9 Measurement of Gross Mass Flow Rate

The measurement of mass flow rate rather than volume flow rate becomes essential in many practical applications. For instance, the range and endurance of an aircraft are determined by the mass flow rate of the fuel, and not by its volume flow rate. Therefore, mass flow meters used for such applications should indicate the mass not the volume.

Generally two approaches are employed for mass flow rate measurement. The first approach makes use of

- Some kind of volume flow meter.

- Some means of density measurement.

Both these methods make use of a computer to calculate the mass flow rate.

The second approach is to find flow metering concepts that are inherently sensitive to mass flow rate.

11.4 Volume Flow Meter

Let us see some basic methods of fluid density measurements here. Examine the float and tank arrangement shown in Figure 11.21.

A portion of the flowing liquid is bypassed through the still well. The buoyant force on the float is directly related to the density and may be computed by measuring the buoyant force.

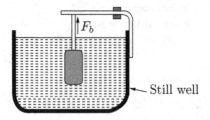

Figure 11.21: Density measurement by buoyant force

The U-tube arrangement shown in Figure 11.22 measures the density as follows:

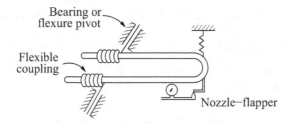

Figure 11.22: Density measurement by volume trap

A definite volume of flowing liquid contained within the U-tube is continuously weighed by a spring and pneumatic displacement transducer. The flexible couplings, shown in Figure 11.22, is meant to isolate external forces from the U-tube.

The buoyant force and volume trap methods are suitable for measurements of liquid density. For measurements of gas density, the arrangement shown in Figure 11.23 may be employed.

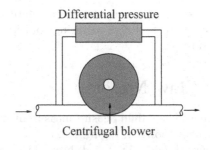

Figure 11.23: Device for gas density measurement

A small centrifugal blower, running at a constant speed, pumps continuously a sample of gas flow. The pressure drop across such a blower is proportional to density and may be measured with a suitable differential pickup.

Density may also be measured by making use of the fact that acoustic impedance depends on the product of the density and speed of sound. The crystal transducer shown in Figure 11.24 serves as an acoustic-impedance detector.

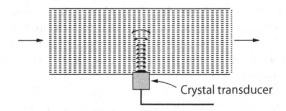

Figure 11.24: Density measurement by acoustic-impedance

The signal output from the transducer will be proportional to the speed of sound available from the flow stream. Division of this signal into the acoustic-impedance signal gives a density signal (Doebelin, 1986).

The arrangement in Figure 11.25 measures density by making use of the fact that the attenuation of radiation from a radioisotope source depends on the density of the material through which the radiation passes.

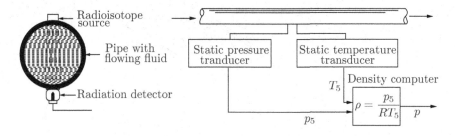

Figure 11.25: Density measurement based on radiation attenuation

The output current of radiation detector is nearly linear with density, for a given fluid, over a limited range of density. For the densities normally encountered the range is adequate.

Even though devices such as those described above are available for measurement of density directly, they are mostly used for measurement of liquid densities only. For obtaining the gas densities, it is preferred to make use of the pressure and temperature measurement to compute the densities. This may be because the transducers for temperatures and pressures are available for any value encountered in practice. Further, these transducers measure the pressures and temperatures with a very high degree of accuracy.

In all the measurements described above in this section, mass flow rate has to be computed from the measured volume flow and density, using appropriate relations.

11.4.1　Direct Mass Flow Meters

Even though it is possible to measure the mass flow rate to a desired level of accuracy, using the techniques described above, they are all only indirect methods. Flow metering concepts can be devised to measure the mass flow rate directly. In certain applications, direct measurement of mass flow proves to be advantageous over the indirect methods. Let us see some of the commonly used direct mass flow meters in this section.

From fluid mechanic concepts, it can be shown that for one-dimensional, incompressible, frictionless flow through a turbine or an impeller wheel, the torque τ exerted by an impeller wheel on the fluid (with a negative sign) or on a turbine wheel by the fluid (with a positive sign) is given by

$$\tau = -\dot{m}\left(V_{ti}\,r_i - V_{to}\,r_o\right) \tag{11.37}$$

where $\dot{m}$ is the mass flow rate through the wheel, V_{ti} and V_{to} are the tangential velocity at the inlet and outlet, respectively, r_i and r_o are the radius at the inlet and outlet respectively, and τ is the torque.

Examine the system shown in Figure 11.26.

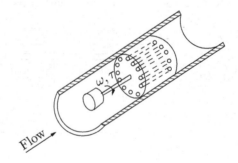

Figure 11.26: An impeller in a duct

The flow is directed through an impeller wheel which is motor driven at constant speed. If the incoming flow has no rotational component of velocity (i.e., $V_{ti} = 0$), and if the axial length of the impeller is enough to make $V_{to} = r\omega$, the driving torque necessary on the impeller is

$$\tau = r^2\,\omega\,\dot{m} \tag{11.38}$$

where r and ω, respectively, are the impeller radius and its rotational rpm; also they are constants. Thus, the torque τ is a direct and linear measure of mass flow rate $\dot{m}$. It is essential to note here that, for $\dot{m} = 0$ the torque τ will not be zero because of frictional effects. Further, the viscosity changes also would cause this zero flow torque to vary. The torque change due to these causes can be avoided by driving the impeller at constant torque, with some kind of slip clutch. Then, impeller speed becomes a measure of mass flow rate according to

$$\omega = \frac{\tau/r^2}{\dot{m}} \tag{11.39}$$

The direct mass flow meter described above gives an idea about the direct measurement. There are a number of devices developed for the direct measurement of mass flow rate. For specific details about such devices one may refer to books specializing in this topic.

11.5 Summary

Basically the following two approaches are followed for measuring flow rate.

Direct Quantity Measurements

    * Weighing or volume tanks
    * Positive displacement meters

Indirect, or Rate, Measurements

    * Obstruction devices
    * Velocity probes
    * Special methods

The accuracy of many flow measurement devices depends on the accuracy of pressure and temperature measurements. Flow rate is expressed in both volume and mass units. Some commonly used terms in flow measurement are liter and gallon.

    1 liter = 1000 cubic centimeters = 0.26417 gallons
    1 gallon per minute (gpm) = 63.09 cm^3/s

Tanks are considered to be the basic reference device for steady-flow measurements. The measurement accuracy of these devices can be quite high when precise tanks and scales are used. The direct weighing technique is frequently employed for calibration of water and other liquid flow meters. Thus, it may be taken as a standard calibration technique.

Displacement meters are basically positive-displacement meters, used for volume flow measurements. Positive-displacement flow meters are generally used in places where consistently high accuracy is desired under steady flow conditions. The home water meter is a typical positive displacement meter. This meter works on the principle of a nutating disk. The top and bottom of the disk remain in contact with the mounting chamber, thus allowing the disk to wobble or nutate about the vertical axis. Water enters the left side of the meter and strikes the disk, which is eccentrically mounted. The disk wobbles and allows the fluid to move through the meter. A partition which separates the inlet and outlet chambers of the disk gives direct indication of the volume of liquid which has passed through the meter as it nutates. The nutation of the disk is recorded as volume flow by connecting a gear and register arrangement to the nutating disk. The nutating disk meter can be used with an accuracy of within ±1 percent.

The rotating-vane meter is another kind of positive-displacement meter. The vanes maintain continuous contact with the casing of the meter through spring loading. A fixed quantity of fluid is trapped in each section as the eccentric drum rotates, and this fluid ultimately flows out of the exit. An appropriate device to record the volume of the displaced fluid is connected to the shaft of the eccentric drum.

The lobed-impeller meter is yet another positive-displacement meter meant for mass flow measurement of gas and liquid flows. The incoming fluid is always trapped between the rotors and is conveyed to the outlet by the rotation of the rotors. Number of revolutions per unit time of the rotors is a measure of the volumetric flow rate.

Some of the popularly used devices for indirect measurement of mass flow are the *variable-head meter, drag-body meter, pitot tube, variable-area meter, mass flow meter, and hot-wire anemometer.*

All meters which measure the pressure drop across a restriction in a pipeline are *variable-head meters.* The three types of obstruction or constriction meters which are widely used are the venturi, flow nozzle, and orifice.

The advantages and disadvantages of the above constriction meters are the following.

- The venturi offers the best accuracy, least head loss, and best resistance to abrasion and wear due to dirt particles in the fluids. But it is expensive and occupies considerable space.

- The nozzle type meters offer all advantages of the venturi to a slightly lesser extent, but require only a small space. However, proper installation of the nozzle is difficult.

- The orifice is less expensive, easy to install, and occupies a very small space. However, it suffers from considerable head loss and is very sensitive to abrasion and other damage.

For venturis in incompressible flows

$$Q_{\text{actual}} = CM_v A_2 \sqrt{\frac{2}{\rho}} \sqrt{p_1 - p_2}$$

For nozzles and orifices in incompressible flows

$$Q_{\text{actual}} = K A_2 \sqrt{\frac{2}{\rho}} \sqrt{p_1 - p_2}$$

For venturis in compressible flows

$$\dot{m}_{actual} = Y_a C M_v A_2 \sqrt{2\rho_1 (p_1 - p_2)}$$

For nozzles and orifices in compressible flows

$$\dot{m}_{\text{actual}} = Y_b K A_2 \sqrt{2\rho_1 (p_1 - p_2)}$$

In all the obstruction meters, when the flow rate is sufficiently high, the pressure differential becomes very large, and ultimately flow may choke attaining sonic velocity at the minimum area location. Under these conditions the flow rate becomes the maximum for the given inlet conditions and the flow is called *choked flow*. For a perfect gas, as we know, the pressure ratio for choked flow is

$$\left(\frac{p_2}{p_{01}} \right)_{\text{critical}} = \left(\frac{2}{\gamma + 1} \right)^{\frac{\gamma}{\gamma - 1}}$$

where pressure p_{01} is the stagnation pressure at the inlet and p_2 is the static pressure at the throat. This pressure ratio is termed the *critical pressure ratio*. The mass flow rate is given by

$$\dot{m} = A_{\text{th}}\, p_{01} \sqrt{\frac{2}{RT_{01}} \left[\frac{\gamma}{\gamma + 1} \left(\frac{2}{\gamma + 1} \right)^{\frac{2}{\gamma - 1}} \right]^{\frac{1}{2}}}$$

where T_{01} is the stagnation temperature at the inlet and A_{th} is the throat area. This relation is commonly used for nozzles operating under choked conditions. For air with $\gamma = 1.4$ and $R = 287\,\text{J/(kg\,K)}$, the choked mass flow rate reduces to

$$\dot{m} = \frac{0.6847\, p_{01}}{\sqrt{RT_{01}}} A_{\text{th}}$$

The stagnation pressure and temperatures can be measured easily, therefore the sonic nozzle is considered to be a convenient device for measuring gas flow rates.

The conventional pitot tube or the pitot-static tube can also be used for measurement of flow rate by calibrating them for volume flow rate by locating them at a specific point in a duct. But such single-point flow rate measurements become inaccurate when velocity profiles differ from calibration conditions.

A *rotameter* is a direct-reading meter, usually employed for flow rate measurements in passages with small diameters. Basically it is a *constant-pressure-drop, variable-area meter*.

Insertion of an appropriately shaped body into a flow stream can serve as a flow meter. The drag experienced by the body becomes a measure of flow rate after suitable calibration. A body immersed in a flow stream is subjected to a drag force F_D given by

$$F_D = \frac{1}{2} \rho V^2 A\, C_D$$

where ρ is fluid density, V is fluid velocity, A is cross-sectional area of the body facing the flow, and C_D is the drag coefficient. For sufficiently high Reynolds number and a properly shaped body, the drag coefficient is almost a constant. Therefore, for a given density, F_D is proportional to V^2 and thus to the square of volume flow rate.

The fact that small-magnitude pressure disturbances travel through a fluid medium at a definite velocity; the speed of sound, relative to the fluid is made use of in ultrasonic flow meters. Ultrasonic flow meters make use of transmitters and receivers of acoustic energy in their measurements. A common approach is to utilize piezoelectric crystal transducers for both functions. Electrical energy in the form of a short burst of high-frequency voltage is applied to a crystal in a transmitter causing it to vibrate. If the crystal is in contact with a fluid, the vibration will be transmitted to the fluid and propagated through it. The receiver crystal when exposed to these pressure fluctuations responds to this vibration. The vibrating motion of the crystal produces an electrical signal proportional to the vibrational excitation, in accordance with the usual action of piezoelectric displacement transducers.

The shedding frequency of the vortex behind a blunt body kept in a steady flow is made use of for the measurement of flow in *vortex-shedding flow meters*.

The measurement of mass flow rate rather than volume flow rate becomes essential in many practical applications. Therefore, mass flow meters used for such applications should indicate the mass not the volume. Generally two approaches are employed for mass flow rate measurement. The first approach makes use of

- Some kind of volume flow meter.

- Some means of density measurement.

Both these methods make use of a computer to calculate the mass flow rate.

The second approach is to find flow metering concepts that are inherently sensitive to mass flow rate.

In certain applications, direct measurement of mass flow proves to be advantageous over the indirect methods. Let us see some of the commonly used direct mass flow meters in this section.

From fluid mechanic concepts, it can be shown that for one-dimensional, incompressible, frictionless flow through a turbine or an impeller wheel, the torque τ exerted by an impeller wheel on the fluid (with a negative sign) or on a turbine wheel by the fluid (with a positive sign) is given by

$$\tau = -\dot{m}\left(V_{ti}\, r_i - V_{to}\, r_o\right)$$

where $\dot{m}$ is the mass flow rate through the wheel, V_{ti} and V_{to} are the tangential velocity at the inlet and outlet, respectively, r_i and r_o are the radius at the inlet and outlet, respectively, and τ is the torque. The flow is directed through an impeller wheel which is motor driven at constant speed. If the incoming flow has no rotational component of velocity (i.e., $V_{ti} = 0$), and if the axial length of the impeller is enough to make $V_{to} = r\omega$, the driving torque necessary on the impeller is

$$\tau = r^2\, \omega\, \dot{m}$$

where r and ω, respectively, are the impeller radius and its rotational rpm; also they are constants. Thus the torque τ is a direct and linear measure of mass flow rate *dotm*. It is essential to note here that for $\dot{m} = 0$, the torque τ will not be zero because of frictional effects. Further, the viscosity changes also would cause this zero flow torque to vary. The torque change due to these causes can be avoided by driving the impeller at constant torque, with some kind of slip clutch. Then, impeller speed becomes a measure of mass flow rate according to

$$\omega = \frac{\tau/r^2}{\dot{m}}$$

The direct mass flow meter described above gives an idea about the direct measurement.

Exercise Problems

11.1 Oxygen gas, at 10 atm and 300 K, enters a variable head meter and leaves at 9.5 atm. (a) If the exit diameter is 100 mm, determine the mass flow rate through the meter. (b) What will be the mass flow rate through the meter, if the pressure at the exit is 8 atm?

[Answer: (a) 8.786 kg/s, (b) 15.98 kg/s]

Chapter 12

Special Flows

12.1　Introduction

It is the intention here to deal with certain flow fields which are special in the sense that the experimentation in such fields requires a slightly different approach than those employed in other flow fields. Even though it is possible to identify a large number of special kinds of flows, we will see only a selected few, just to demonstrate the kind of approach required to study such fields and to have an idea about such fields. When dealing with special flows like *rotational flows* even the coordinate system employed has to be different from that used for studying general flows like pipe flow, boundary layer flows, and so on. In fluid flow analysis, a majority of the problems are dealt with in a fixed or inertial coordinate system. But for flow over rotating turbine blades and the geophysical boundary layers on the rotating earth, it is advantageous to use *non-inertial coordinates* (an accelerated reference frame and if the fluid is rotating uniformly, the analysis must consider the centrifugal force and the Coriolis force) moving with the accelerating system. In this case, Newton's law $F = ma$ which is valid only when a is the absolute acceleration of the particle relative to inertial coordinates (an inertial frame of reference is an unaccelerated frame of reference, that is, a frame in which Newton's laws of motion are applicable) must be modified.

12.2　Geophysical Flows

Rotational fluid flows occur in a variety of technical contexts. In atmospheric and ocean currents rotation plays an important role. Flow patterns are appreciably changed by rotation and in this chapter an attempt is made to bring out certain important features of such flows. Many peculiar flow patterns could be seen in this class of flow. The formation of hurricanes and tornadoes is highly influenced by rotation. Being a very vast field, we will confine our attention only to low rotational speeds where the motion of the fluid does not vary appreciably from rigid-body rotation. Further, only homogeneous fluids will be considered in the present study. In nature, the atmosphere and ocean are stratified, that is, large density variations exist and in such situations the flow pattern could be quite different.

The forces acting in a rotating fluid flow are: inertia force, viscous force, pressure force, gravitational force, centrifugal force, and Coriolis force. Let us assume that the viscous forces are small on account of slow rotation. Then the basic equation of motion in steady state is

$$U \cdot \nabla U = -\frac{1}{\rho}\nabla p - \nabla \phi_g - \nabla \left(\frac{\Omega^2 r^2}{2}\right) - 2\Omega \times U \qquad (12.1)$$

where U and Ω are the linear and angular velocities, respectively, and $\Omega^2 r^2/2$ is the centrifugal force, $\nabla \phi_g$ is the gravitational force, and $\Omega \times U$ is the Coriolis force. The last two terms are responsible for many of the special flow patterns observed in rotating flows. The mathematical aspect of the rotating flow is highly complex especially when viscosity is to be considered and hence it is beyond the scope of this book to explain all the aspects of the experiments in detail. The experiments here are meant just to show the important aspects of such flows. The nondimensional parameter used to describe the rotating flows is *Rossby number* $Ro = U/L\Omega$, where L is some characteristic length. In the present study we will be dealing with low Rossby number flows only. The Coriolis term can be neglected if Ro is large, which will be true if the motion scale L is small compared to the earth's radius.

12.2.1   Rotating Tank

Rotating flows described above can be established in laboratories with a rotating tank. The details of the rotating tank are shown in Figure 12.1. A transparent plastic tank of 500-mm diameter and 300-mm height is mounted on a rotating table. The table is connected to a $\frac{1}{8}$ hp induction motor with a belt driven reduction pulley. The perspex material is suitable for making the transparent tank. For instance, a thick (say 6 mm) perspex sheet when heated uniformly to a temperature of the order of about 100°C can be rolled into cylindrical form over a mandrel and slowly cooled to room temperature. The rolled cylinder can be closed at the bottom with a suitably sized disc of perspex or any other material to form the tank.

12.3   Experiment on Taylor-Proudman Theorem

In a slowly rotating fluid it can be shown theoretically that the Coriolis force is completely balanced by the pressure gradient induced by the centrifugal force. This is known as *geostrophic motion*. The equations for geostrophic motion in a homogeneous incompressible fluid rotating about a vertical axis with an angular

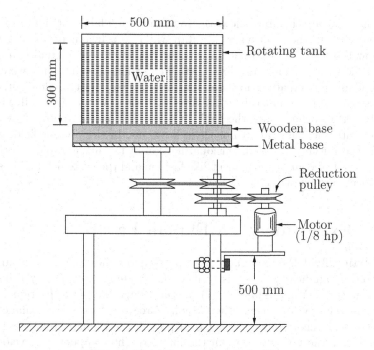

Figure 12.1: Rotating tank facility

velocity Ω are given by

$$-2\Omega v = -\frac{1}{\rho_0}\frac{\partial p}{\partial x} \tag{12.2}$$

$$2\Omega u = -\frac{1}{\rho_0}\frac{\partial p}{\partial y} \tag{12.3}$$

$$\frac{\partial p}{\partial z} = 0 \tag{12.4}$$

By continuity, we have

$$\frac{\partial u}{\partial x} + \frac{\partial v}{\partial y} + \frac{\partial w}{\partial z} = 0 \tag{12.5}$$

Combining these it can be shown that,

$$\boxed{\frac{\partial}{\partial z}(u,v,w) = 0} \tag{12.6}$$

where u, v, and w are the velocity components in the x, y, and z directions, respectively. The above equation implies that *the relative velocity field does not vary in the direction of the rotation axis and that the flow tends to be two-dimensional in planes perpendicular to the rotational axis.* This is known as the *Taylor–Proudman Theorem.*

To demonstrate the above flow, the rotational tank facility (Figure 12.1) can be employed. Fill the tank with clear tap water to a height of about 200 mm and allow the water to settle down without ripples. Inject a small quantity of ink into the water, using an ordinary ink filler. Soon after injecting we can see the ink filaments spreading in a random fashion like any turbulent jet. Drain the tank and fill it again to the same level with clean water. After allowing the water to settle down rotate the tank by switching on the motor. Now inject some ink into the water at two or three locations. In the present field, we can see that the initially three-dimensional motion of the ink is rapidly converted into thin vertical columns in a sheet-like fashion and the system is highly stable. These sheets are known as *Taylor walls.*

12.4   Experiment on Ekman Layer

In a rotating fluid the boundary layer formed on the floor of the tank has some special significance. This is termed the *Ekman boundary layer* and it is spiral in nature (Frank White, 1991). In the Ekman layer the frictional forces reduce the velocity and hence the Coriolis force is no longer in balance with the pressure gradient; hence the net transport in the Ekman boundary layer is at an angle to the net transport in the interior. The component of velocity in the direction of pressure gradient changes the direction of the fluid flow. Ekman layers are divergent and this creates a suction directed towards the interior whose magnitude is proportional to the vorticity of the flow above this layer. This is known as *Ekman suction.*

Fill the rotating tank, shown in Figure 12.1, with water up to a height of nearly 200 mm and allow the water to settle down. Switch on the motor and drop a few crystals of potassium permanganate near the axis of rotation. We will notice that spiral streaks violet in color are formed on the floor which represents the formation of the Ekman layer with an appreciable radial velocity component.

12.5   Experiment on Spin-Up and Spin-Down

In a rotating system, changes in angular velocity with the time affects the flow pattern appreciably. Take the rotating tank, described in Figure 12.1. Fill the tank with clean water up to about 200 mm height and allow it to settle down. Sprinkle some light material powder like cork powder on the water surface and start the motor. In the beginning, we will notice that the particles near the center do not move appreciably compared to those at the periphery of the tank. Gradually, as time increases the particles gather more momentum, but the angular velocity is different at each station and a shear is produced. After a time period, say 10 minutes, the complete fluid mass goes into a spin as though it is a rigid body. The time required to achieve this rigid body condition is called *"spin-up time."* The reverse of it is called *"spin-down time."*

The spin-up occurs in three stages. As soon as the tank starts rotating, a thin shear layer is created near the vertical solid wall, similar to a boundary layer. Within a few revolutions the Ekman layer gets established (time of the order of Ω^{-1}). The fluid particles in the Ekman layer gain angular momentum from the boundaries and are therefore displaced outwards against the pressure gradient due to the excess centrifugal force. This creates a divergence in the Ekman layer which is compensated by a mass flux to the interior. Thus, a suction is created towards the faster rotating region. In the case of spin-down, the mass flux is in the opposite direction. This can be easily visualized in the rotating tank. Mix the water with some fine particles which are slightly heavier than the water. Rotate the tank for few minutes and then stop it, allowing the flow to spin-down. We will find all the soaked particles slowly gathering towards the center. Similar trends could be seen by stirring tea in a cup and hence it is known as the *"tea-cup"* problem in fluid dynamics.

In the rotating tank experiment, dyes like potassium permanganate or ink after dissolving in water, stain the entire mass. This calls for a fresh charge of water for each experiment. This could be avoided by using a dye whose color changes with the pH value of the water. Thymol blue is the most suitable dye for this purpose. A solution of thymol blue is orange in color when acidic and violet when alkaline. Before starting the experiment, add a little quantity (say 10 mg) of thymol blue crystals in water and allow it to dissolve. Few drops of dilute nitric acid or sulphuric acid will turn the entire water orange in color. Whenever the flow patterns have to be observed, inject some liquid ammonia with a fine glass tube. The color of the water will locally change to violet. After the experiment is over, the water can be changed to nearly clear condition by adding a few drops of acid. Thymol blue dissolves easily in alcohol without leaving any sediment. Therefore, it may be dissolved in a small quantity of alcohol before adding it to the rotating tank.

12.6 Transition and Reverse Transition

From our studies on the fundamentals of fluid dynamics in Chapter 2, we know that a laminar flow becomes turbulent through the process of transition. The transition from laminar to turbulent nature of the flow is due to the disturbances experienced by the flow. When a small disturbance is introduced to a laminar boundary layer, the disturbance either gets amplified or damped out depending on the Reynolds number. Experimental results indicate that for a flat plate boundary layer, any disturbance introduced dies down when the Reynolds number, based on the momentum thickness θ, $Re_\theta = \frac{U_\infty \theta}{\nu}$, is less than 350. When the Reynolds number is increased beyond 350 the flow becomes turbulent. The transition process is nonlinear and the physics of the processes is yet to be understood clearly.

It is a well-established fact that the initial disturbances amplify in the early stage of transition in a well-organized manner. At some stage of this amplification these disturbances produce turbulent spots which contain random fluctu-

ations and grow rapidly turning the whole flow fully turbulent. The turbulent
spots can be considered as seeds of turbulence and have certain characteristic
properties. If a hot-wire is placed in the transition region, we can see the bursts
of high-frequency signals on the oscilloscope. A typical trace of the transition
signal is as shown in Figure 12.2.

Figure 12.2: Hot-wire trace of a transition signal

These high-frequency components, marked "x" in Figure 12.2, appear whenever
a turbulent spot crosses the hot-wire. The ratio of the time duration for which
the flow is turbulent to the total duration is known as *intermittency*. For laminar
flow, the intermittency = 0 and it is equal to unity for fully turbulent flow.
During transition, the intermittency varies between 0 to 1.0. A turbulent spot
has a special shape and it spreads at a particular angle. The typical shape of
turbulent spot is illustrated in Figure 12.3.

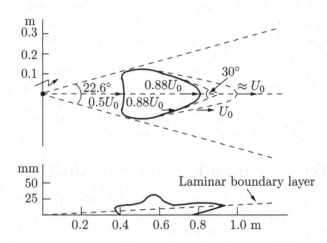

Figure 12.3: Turbulent spot details

It looks like a kidney. Turbulent spots can also be generated artificially by
introducing a small electrical spark and a fine jet of air. By artificially creating
a spot its characteristics can be studied easily. As shown in Figure 12.3, the front
and rear regions of the spot travel at different velocities. Because the velocities
aside the spot are different at different locations, it spreads with time as it is
washed downstream. Measurement of the spot velocities can be made with hot-
wire traces using artificially created spots. These experiments require recording
facilities which are sophisticated. It is not our intention here to go into such

specialized details. For such details the reader is encouraged to refer to books specializing in this topic.

12.6.1 Transition in a Channel Flow – A Visualization

An experimental facility of a small two-dimensional water flow channel to visualize transition from laminar to turbulent flow is shown schematically in Figure 12.4.

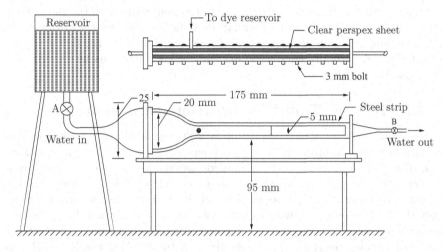

Figure 12.4: Water flow channel

It is fabricated by joining two transparent perspex sheets together with a spacer in between them. Water enters the channel from a large tank through a contraction. The flow is controlled by a valve. Along the centerline of the channel, tiny holes are provided for dye injection, as shown in Figure 12.4. These tiny holes are connected to a small reservoir through rubber tubing with a pinch cock and the dye is gravity fed. Bursts of dye to create artificial spots, which are termed *turbulent plugs* in channel flows, could be injected into the channel by suddenly pinching the rubber tubing.

Experimental Procedure

- Fill the water tank with clear tap water.

- Fill the dye reservoir with any colored liquid, say ink or potassium permanganate solution.

- Allow the water to flow through the channel at a low speed by controlling the valve.

- Allow the dye to flow. The dye will flow like a streak in the channel. The nature of the streak will indicate whether the flow is laminar or turbulent.

- Adjust the water flow velocity until the dye streak indicates that the flow is laminar but just below the critical Reynolds number.

- Now, by pressing the rubber tube between the pinch cock and the injector hole, produce bursts of dye. These bursts behave like turbulent plugs and are adequate for flow visualization of transition.

- Create a single plug and observe its growth along the channel. The turbulence inside the plug can be seen. When the flow Reynolds number is subcritical, the local activity inside the plug reduces as the plug moves downstream. On the other hand, if the velocity is beyond the critical value the plug grows and occupies the whole channel, making the flow fully turbulent. A carefully conducted experiment will enable us to see all these details clearly.

12.6.2 Reverse Transition or Relaminarization

As the name implies, reverse transition is a process in which a turbulent flow changes over to a laminar nature. This phenomenon is also known as relaminarization. The physics of reverse transition has not yet been understood fully. We know that to maintain a flow in turbulent conditions the turbulence generation has to be greater than its dissipation. Reverse transition can take place when the dissipation of turbulence exceeds its generation or when the turbulence production process is completely suppressed. A turbulent boundary layer when subjected to high acceleration reverts back to the laminar condition. Reversion to laminar condition could also be achieved by applying centrifugal force. Buoyancy forces also suppress turbulence under certain conditions. Reduction of Reynolds number below the critical value in a duct flow causes reverse transition. The onset of reverse transition is not as rapid as the regular transition where spots suddenly appear and make the flow turbulent.

For visualizing the phenomenon of reverse transition, consider the two-dimensional channel experiment setup shown in Figure 12.5. As seen from the figure, it is essentially a continuous channel with a divergent section in between. The first stage width of the channel is increased to nearly three times, after the expansion. The channel height is kept constant throughout. For a given mass flow the Reynolds number decreases to one third as the flow proceeds through the divergent section, when the flow rate is adjusted such that the Reynolds number upstream of the contraction is above the critical value. Under this condition the originally turbulent flow will have a Reynolds number much below the critical value. Under this condition the originally turbulent flow will revert back to laminar condition.

Experimental Procedure

- Fill the reservoir with clean water.
- Fill the dye reservoir with ink or potassium permanganate solution.

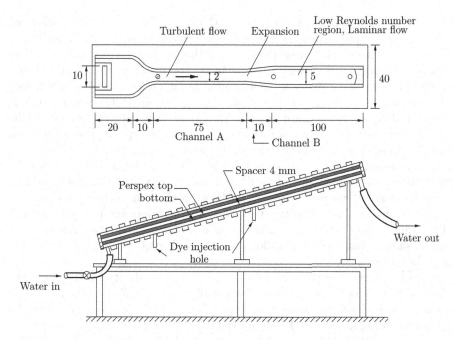

Figure 12.5: Water channel setup for reverse transition.

- Open the valve and allow the water to flow through the channel.

- Allow the dye to flow through the injection hole provided at the beginning of channel A. Also inject dye at the end of channel B.

- Adjust the flow rate of water until the dye injected at A just becomes turbulent. Increase the flow rate a little more. Now we will observe the flow at B is laminar.

- From mass flow measurement, determine the critical Reynolds number. Mass flow rate can be obtained simply by collecting the water at the outlet over a time interval.

- The flow of water through the channel should be completely free from air bubbles. To facilitate this the channel setup is kept at an angle to the ground.

12.7 Measurement in Boundary Layers

From our studies on fluid flow we know that the boundary layer is a thin region adjacent to a solid boundary where the flow velocity increases from zero to 99 percent of freestream value. In general, the boundary layer thickness is of the order of a few centimeters except for situations like atmospheric boundary layers where it could be even many meters in thickness. Even though conventional instruments like pitot-static and hot-wire probes can be used for measurements in boundary layers, the small thickness of the boundary layer makes us pay special attention to any measurements in it. For instance, the mean velocity profiles in a boundary layer can be measured using a pitot tube. In a laboratory the boundary layer being small and of the order of one or two centimeters, it is essential to use a fine pitot tube so that some measurement points can be obtained in the inner region of the boundary layer. A typical pitot probe used for this purpose is shown in Figure 12.6. The measuring tip of the pitot tube is flattened and its lower surface ground so that it can be placed very close to the wall. The flattening should be done in such a way that the gap should be uniform all across its mouth region. With a properly made fine pitot tube, it is possible to measure velocities very close to the boundary up to a distance of 0.3 mm in the direction normal to the boundary. The static pressures can be measured using fine holes drilled into the surface of the body on which the boundary layer is formed.

Measurements near the wall do contain some errors due to the displacement effect. When the probe is very close to the wall, the flow in front of the pitot tube is disturbed and some local separation takes place. This introduces error to the pressure reading measured. To obtain the true velocity, this measured pressure has to be corrected. The error is large very near the wall and becomes negligible as the pitot tube is moved away towards the outer region of the boundary layer. The correction needed for a circular pitot tube, with the ratio of inner diameter to outer diameter of 0.6, is given in Table 12.1.

Table 12.1 Correction for pitot pressure measurement

$\dfrac{y}{d}$	Percentage correction for velocity reading
0.5	1.50
0.6	0.95
0.7	0.68
0.8	0.60
0.9	0.39
1.0	0.30
1.2	0.16
1.4	0.06
1.6	0.02
1.8	0.00

Here, d is the inner diameter of pitot tube and y is the distance from the wall. The movement of the pitot tube inside the boundary layer to obtain the velocity profile has to be made with high precision, especially when measurements have to be made near the wall region. A mechanical traverse with a vernier or micrometer movement may be employed for this purpose and a simple design is

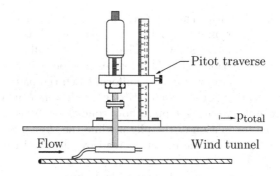

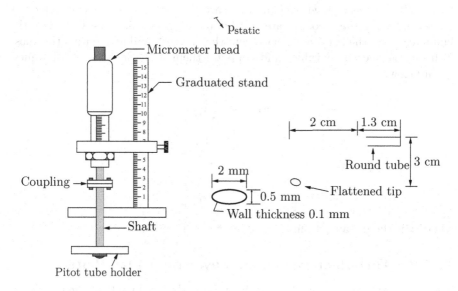

Figure 12.6: Experimental setup for flat plate boundary layer measurement

shown in Figure 12.6. It consists of a micrometer head to which the pitot tube is attached directly.

12.7.1 Laminar Boundary Layer on a Flat Plate

To conduct experiments on a laminar boundary layer of reasonable thickness, it is essential to have a good low-turbulence wind tunnel. A smooth flat plate which is spanning the tunnel test-section and held by the two side-walls is well suited for these measurements. A faired nose of the plate in the form of an ellipse may avoid leading edge separation which can cause early transition which is not desirable in this experiment. Initially the flat plate has to be adjusted for zero pressure gradient. This can be done using the static pressure tapings along the length of the flat plate which are connected to a multiple manometer. On account of the boundary layer on either side of the plate which is positioned along the centerline of the tunnel, as shown in Figure 12.6, as well as the boundary layer growth on the test-section walls, the freestream velocity in the far field on either side, normal to the plate, will not be uniform when the plate is kept horizontal, due to manufacturing errors. Slight tilting of the plate will be necessary, in most cases, to obtain true flat plate flow conditions. This can easily be done by adjusting the inclination of the plate, while the tunnel is running, until the readings of all the wall static pressures read by the multi-tube manometer are the same.

As an exercise, measure the boundary layer profiles at a few locations on the flat plate for different freestream velocities. Plot the mean velocity profiles in the form U/U_∞ verses y/δ, where U and U_∞ are the local and freestream velocities and y and δ are the y-coordinate and boundary layer thickness. Calculate the boundary layer thickness δ, displacement thickness $\delta^\star$, and momentum thickness θ, using the measured values and compare them with the theoretical values, given below.

$$\delta_{0.995} = \frac{5.2}{\sqrt{Re_x}}$$

$$\delta^\star = \frac{1.73}{\sqrt{Re_x}}$$

$$\theta = \frac{0.664}{\sqrt{Re_x}}$$

where x is the distance from the leading edge of the plate.

12.7.2   Turbulent Boundary Layer on a Flat Plate

The boundary layer on the plate can be made turbulent by introducing a tripper downstream of the leading edge. A rough emery paper can serve as a tripper when a small strip of it is pasted across the width of the plate. In a manner similar to that for laminar boundary layer, make measurements at least 50 cm downstream of the trip. Plot the mean velocity profile and obtain the skin friction coefficient C_f, using Clauser's method. Calculate $U^\star = (\tau_w/\rho)^{1/2}$, plot the mean velocity profile $U/U^\star$ versus $yU^\star/\nu$. Determine δ, $\delta^\star$ and θ from the velocity profiles and compare them with the theoretical values;

$$\delta = \frac{0.37x}{(Re_{x_t})^{\frac{1}{5}}}$$

$$\theta = \frac{7}{12}\delta$$

$$C_f = 0.074 \left(Re_{x_t}\right)^{-\frac{1}{5}}$$

$$H \approx 1.4 = \delta^*/\theta$$

Take $x_t = 0$ at the leading edge of the plate.

12.8 Summary

In fluid flow analysis, a majority of the problems are dealt with in a fixed or inertial coordinate system. But for flows over rotating turbine blades and the geophysical boundary layers on the rotating earth, it is advantageous to use *non-inertial coordinates* (an accelerated reference frame and if the fluid is rotating uniformly, the analysis must consider the centrifugal force and the Coriolis force) moving with the accelerating system. In this case, Newton's law $F = ma$ which is valid only when a is the absolute acceleration of the particle relative to inertial coordinates (an inertial frame of reference is an unaccelerated frame of reference, that is, a frame in which Newton's laws of motion are applicable) must be modified.

Rotational flows play an important role in atmospheric and ocean currents. Many peculiar flow patterns could be seen in this class of flow. The formation of hurricanes and tornadoes is highly influenced by rotation. The atmosphere and ocean currents are stratified, that is, large density variations exist and in such situations the flow pattern could be quite different.

The forces acting in a rotating fluid flow are: inertia force, viscous force, pressure force, gravitational force, centrifugal force, and Coriolis force. Let us assume that the viscous forces are small on account of slow rotation. Then the basic equation of motion in steady state is

$$U \cdot \nabla U = -\frac{1}{\rho}\nabla p - \nabla \phi_g - \nabla \left(\frac{\Omega^2 r^2}{2}\right) - 2\Omega \times U$$

The last two terms are responsible for many of the special flow patterns observed in rotating flows. *Rossby number* $Ro = U/L\Omega$ is used to describe the rotating flows. The Coriolis term can be neglected if Ro is large.

In a slowly rotating fluid, it can be shown theoretically that the Coriolis force is completely balanced by the pressure gradient induced by the centrifugal force. This is known as geostrophic motion.

In a rotating fluid, the boundary layer formed on the floor of the tank has some special significance. This is termed the *Ekman boundary layer* and it is spiral in nature. In the Ekman layer the frictional forces reduce the velocity and hence the Coriolis force is no longer in balance with the pressure gradient; hence the net transport in the Ekman boundary layer is at an angle to the net transport in the interior.

In a rotating system, changes in angular velocity with the time affects the flow pattern appreciably.

The spin-up occurs in three stages. As soon as flow starts rotating, a thin shear layer is created near the vertical solid wall something similar to a boundary layer. Within a few revolutions the Ekman layer gets established. The fluid particles in the Ekman layer gain angular momentum from the boundaries and are therefore displaced outwards against the pressure gradient due to the excess centrifugal force. This creates a divergence in the Ekman layer which is compensated by a mass flux to the interior. Thus, a suction is created towards the faster rotating region. In the case of spin-down, the mass flux is in the opposite direction. This can be easily visualized in the rotating tank. Similar trends could be seen by stirring tea in a cup and hence it is known as the *"tea-cup"* problem in fluid dynamics.

For a flat plate boundary layer, any disturbance introduced dies down when the Reynolds number, based on the momentum thickness θ, $Re_\theta = \frac{U_\infty \theta}{\nu}$, is less than 350. When the Reynolds number is increased beyond 350 the flow becomes turbulent. The transition process is nonlinear and the physics of the processes is yet to be understood clearly. The initial disturbances amplify in the early stage of transition in a well-organized manner. At some stage of this amplification, these disturbances produce turbulent spots which contain random fluctuations and grow rapidly turning the whole flow fully turbulent. The turbulent spots can be considered as seeds of turbulence and have certain characteristic properties.

The ratio of the time duration for which the flow is turbulent to the total duration is known as *intermittency*. For laminar flow the intermittency $= 0$, and it is equal to unity for fully turbulent flow. During transition, the intermittency varies between 0 to 1.0.

Reverse transition is a process in which a turbulent flow changes over to laminar nature. This phenomenon is also known as relaminarization. Reverse transition can take place when the dissipation of turbulence exceeds its generation or when the turbulence production process is completely suppressed. A turbulent boundary layer when subjected to high acceleration reverts back to laminar condition. Reversion to laminar condition could also be achieved by applying centrifugal force. Buoyancy forces also suppress turbulence under certain conditions. Reduction of Reynolds number below the critical value in a duct flow causes reverse transition. The onset of reverse transition is not as rapid as the regular transition where spots suddenly appear and make the flow turbulent.

Boundary layer is a thin region adjacent to a solid boundary where the flow velocity increases from zero to 99 percent of freestream value. In general, it is of the order of a few centimeters except for situations like atmospheric boundary layers where it could be even many meters in thickness. In a laboratory the boundary layer being small and of the order of one or two centimeters, it is essential to use a fine pitot tube so that some measurement points could be obtained in the inner region of the boundary layer. With a properly made fine pitot tube, it is possible to measure velocities very close to the boundary up to a distance of 0.3 mm. The static pressures can be measured using fine holes drilled on the surface of the body on which the boundary layer is formed.

To conduct experiments on a laminar boundary layer of reasonable thickness, it is essential to have a good low-turbulence wind tunnel. The boundary layer on the plate can be made turbulent by introducing a tripper downstream of the leading edge. A rough emery paper can serve as a tripper when a small strip of it is pasted across the width of the plate.

Chapter 13

Data Acquisition and Processing

13.1 Introduction

In many experiments of practical interest such as jet acoustics, a large quantity of data has to be measured within a very short period of time. For such measurements, a data acquisition and processing system for continuous and automatic acquisition of data becomes essential. In some experiments, the acquired data need to be processed in real time and appropriate action has to be taken based on the processed result. Such a comprehensive and stringent requirement demands a data acquisition system possessing the necessary characteristics of speed, power, and so on.

Presently there are several systems commercially available for rapidly collecting a large amount of data, processing it, and displaying the desired results in visual and printed forms. But such commercially available units are expensive and most of them are not flexible in their configurations. For laboratory research experiments with a limited number of transducers, low-cost data acquisition units can be configured. Such systems have become popular alternatives to a standalone system. These PC-based or microprocessor-based laboratory systems are highly flexible in their configurations and therefore, a host of options and expansion possibilities are available with them. In this chapter, let us briefly discuss such systems, and the functions of the elements that go together to make up an overall data acquisition and processing installation, with attention focused mainly on the general principles and their applications to PC-based systems.

13.2 Data Acquisition Principle

The essential sequence of operation involved in any data acquisition system, regardless of the size of the system are the following.

- Generation of input signals by transducers.

- Signal conditioning.

- Multiplexing.

- Data conversion from analogue to digital form and vice versa.

- Data storage and display.

- Data processing.

13.2.1 Generation of Signal

The essential element in a modern data acquisition system is the instrument transducer, which furnishes an electrical signal that is indicative of the physical quantity being measured. The manner in which an input signal is generated depends on the type of the transducer or sensor used. *Active transducer* generates an analogue electrical signal which varies in proportion to the quantity being measured. An active transducer is an auxiliary source of power which supplies a major part of the output power while the input signal is only an insignificant portion. Further, there may or may not be a conversion of energy from one form to another. Thermocouple, piezoelectric sensors, and photo diodes are some of the popular active sensors. *Passive sensors*, on the other hand, alter their state when the quantity being measured changes its magnitude. A component whose output energy is supplied entirely or almost entirely by its input signal is usually called a passive transducer. The output and input signals may involve energy of the same form (say, mechanical to electrical) in a passive sensor. Resistance thermometer and hot-wire probes are typical examples of passive sensors. To detect the alternation in the states of passive sensors, additional circuitry or a measuring device is required. Depending on the nature of the measuring device/circuitry, the signal generated may be analogue or digital.

13.2.2 Signal Conditioning

The objective of signal conditioning is to modify the signal received from the sensor to suit the requirements of further processing. The signal conditioning may consist of one or more of the following operations.

- Amplification of the signal from the sensor.

- Filtering out the unwanted frequencies originating from the sensor and their associated circuitry.

- Providing independence matching.

- Compensation for the limitations of the sensor and/or receiving devices (e.g., recorders) so as to extend their frequency range.

- Correction of thermoelectric errors at input functions.

- Performance of arithmetic operations on the outputs of two or more sensors.

The method of signal conditioning is based on the possibility of introducing certain elements ("filters") into the instrument or its circuitry which in the same fashion blocks the spurious signals, so that their effects on the output are removed or reduced. Some filters are composed only of passive elements (e.g., resistors, capacitors, and inductors), while others contain amplifiers. Circuits of the first type are called passive filters, and circuits of the second type are called active filters.

13.2.3 Multiplexing

In many measurements of practical interest, more than one channel of data must be repetitively transduced to voltage signals. For example, in wind tunnel testing, hundreds of test points distributed over the surface of a model may be instrumented for static or stagnation pressure sensing. When the required data rates are sufficiently low, a cost-effective solution time shares a single transducer/amplifier by multiplexing the pressure lines using a scanning device. In other words, by the multiplexing techniques, various systems of a data acquisition can be time-shared by two or more input sources.

One of the popular devices in this field is the scanivalve. It is a multiported rotating valve which sequentially connects each of many pressure lines (up to 64) distributed around its circumference to a single flush-diaphragm transducer. A typical scanivalve is shown schematically in Figure 13.1. Unlimited number

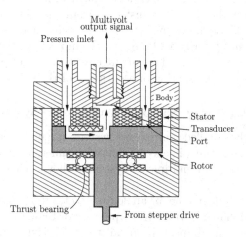

Figure 13.1: Pressure signal multiplexer

of channels can be accommodated by using as many scanivalves as necessary; however, each requires its own pressure transducer.

For very high data rate requirements, systems using a separate analogue transducer for each channel, with electronic multiplexing into a single analogue/digital converter are available. But such devices are very expensive. According to the characteristics of input signals and further processing requirements, multiplexing can be done by one of the following methods.

13.2.3.1 Multichannel Analogue Multiplexed System

In the multichannel analogue multiplexed system, shown in Figure 13.2, the individual analogue signals are fed to the multiplexer directly or after pre-amplification and/or signal conditioning, if required.

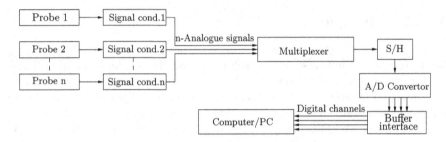

Figure 13.2: Multichannel multiplexed system

The output signal from the multiplexer is temporarily stored in a sample-and-hold (S/H) device, followed by conversion to digital form in an analogue-to-digital (A/D) converter. The S/H device has two modes, namely the sample mode during which the data is transferred from the multiplexer to the S/H device and the hold mode during which A/D conversion takes place. During hold mode, the multiplexer is made to seek the input from the next channel for an efficient utilization of time. This arrangement in a multiplexing makes it a low-cost device due to the sharing of a majority of the subsystems.

13.2.3.2 Simultaneously Sampled Multiplexer System

In a simultaneously sampled multiplexer system, large number of channels are monitored at the same time, in a synchronized manner, at moderate data rates. A simultaneously sampled multiplexer system is shown schematically in Figure 13.3.

In this system, each channel of analogue input, after signal conditioning, is assigned to an individual S/H device. Updating of the data during the sample mode is carried out simultaneously for all the S/H devices with the help of a clock. After the data is received and locked in the hold mode, the multiplexer scans each S/H and feeds the output to the A/D converter.

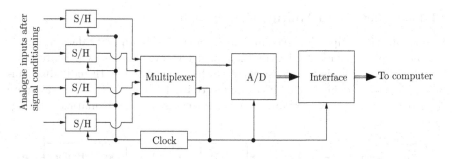

Figure 13.3: Simultaneously sampled multiplexer system

13.2.3.3 Multichannel Digital Multiplexer System

In this system, each analogue input after signal conditioning is assigned to a separate S/H device and an A/D converter and then the digitized outputs are multiplexed with the help of a digital multiplexer. For measurement involving transmission of data over a long distance before it is processed, as in industrial applications, such a configuration will be useful. The conversion of data from analogue to digital form can be done at the source location itself in order to provide immunity to the data against line frequency and other group-loop interferences. For a group of input channels having slowly varying data, a single set of analogue multiplexer, S/H and A/D converter can be used. The digitized outputs from the A/D converters are sometimes passed through additional processors to perform mathematical operations such as calculation of running average or rate of change with respect to time. Schematic of a multichannel digital multiplexer is shown in Figure 13.4.

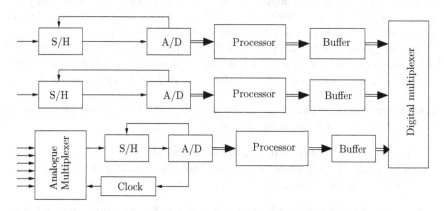

Figure 13.4: Multichannel multiplexer system

13.2.3.4 Low-Level Multiplexing System

In a low-level multiplexing system, usually a single high-quality data amplifier will be employed for handling multichannel low-level inputs, which are weak signals. This system is more cost-effective than the system with individual amplifiers for each channel, when the number of input channels is large (say more than 20). A typical low-level multiplexing system is shown schematically in Figure 13.5.

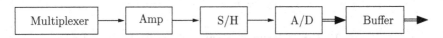

Figure 13.5: Schematic of low-level multiplexing system

Several important factors must be taken into consideration for a successful operation of a low-level multiplexing system. In this system, there is a possibility of signal-to-signal or common-mode signal to differential mode signal cross talk. Guarding may have to be employed for each channel in order to protect signal pick-up. Capacitive balancing will also be necessary, taking into account the contributions from two multiplex switches.

13.3   Data Conversion

The data conversion generally involves conversion of data from analogue to digital or from digital to analogue. In analogue representation, the physical variables of interest are treated as continuous quantities, while in digital representations, the physical quantities are restricted to discrete, non-continuous "chunks." The conversion of data from analogue to digital form and vice versa is necessitated by the fact that the real-world events are continuous with a string of binary digits 0 and 1. Therefore, data has to be converted to digital form before being fed to a computer and the output from the computer may have to be converted back to analogue form in order to control or operate any analogue device. There are many other advantages with such data conversions. They are the following.

- Conversion of signals from analogue to digital form gives noise immunity to the data during its transmission.

- It is easy to determine whether or not a single digital pulse (either a binary 0 or binary 1) is present at a given time, determining the presence of a pulse in analogue form.

- Coding techniques have been developed only for digital signals and thus, to take advantage of the error-detection and error-correction capabilities of these codes, it is essential to correct all the data signals of interest to digital form.

In analogue to digital conversion, the output will be a digital code which may be a straightforward binary representation of the signal, that is, a binary code (BC), or a binary coded decimal (BCD). The binary coded decimal code is generally used because of the fact that the conversion of a decimal number to BCD is much easier. In BCD code a decimal number such as 999 will be represented as 1001 1001 1001. For unipolar signals, signals which do not oscillate between +ve and −ve values, only the magnitude of the signal needs to be converted. But for bipolar signals, signals which oscillate between +ve and −ve values, both the magnitude and sign of the signal must be converted. The leading digit of the digital code is meant for this purpose. The resolution of the conversion itself is specified in terms of Bits (a binary digit of 0 and 1). For example, a 12-Bit resolution with BCD digital code implies approximately 1/1000 accuracy of conversion.

Successive approximation converter is a popular device used in A/D conversion. This uses a digital-analogue converter (DAC) under proper feedback, along with a register. After checking whether the signal is above or below the middle of the value of the maximum allowed signal size, the most significant bit (MSB) of the register is made 1 or 0. Then it tests whether the signal is above or below 75 percent (or 25 percent) of the maximum signal size and sets the next bit as 1 or 0, and so on. This procedure is repeated until all the digits up to the lowest significant bit (LSB) of the register are assigned their values. In this converter, A/D conversion is quite fast and the conversion time is independent of the signal amplitude.

Analogue to digital converters (ADC) are available with 8, 10, 12, 14, and 16 bit resolutions. But 8, 12, and 16 bits ADC are more common. The larger the number of bits, the better the resolution. Digital to analogue conversion (DAC) can also be done in many ways. The popular ones are the weighted register method and the R-2R method. In these procedures, for n-bit conversion, a resistance network with n loops is used with each loop contributing to one-bit of the digital code. In some applications a DAC is required to respond only at selected instants of time to the input digital code, which may be intermittent in nature, and memorize the output until the next update occurs. In such DACs it is essential to incorporate an additional memory register (buffer) at the input, so as to hold the input digital code up to the desired time.

In A/D conversion, high-speed conversion and high resolution require a sample-and-hold device at the input end of the ADC. Without the S/H device substantial nonlinearity errors will occur due to the change in the signal during the conversion process. The S/H device is particularly required for the successive approximation type ADCs which operate at high speed.

13.3.1 Data Storage and Display

The digital data, after conversion from analogue to digital, is usually stored in floppies/winchester drive in the PC-based data acquisition systems and in magnetic tapes in the conventional data acquisition units. Data storage and display generally do not require different apparatus. The relatively simple methods of data display often provide simultaneous storage. Some of the commonly used data storage and display units are the xy plotter, the strip chart recorder, and paper tape output from a computer or a calculator. The data is displayed as well as stored in all these cases. A storage oscilloscope also can be used for this purpose. In a simple data logging systems, the data may be directly sent to a printer or chart recorder or it may also be stored in an intermediate buffer memory which in turn is connected to a printer or a recorder. Some data acquisition systems provide visual displays through digital pond meters.

In situations where the data fluctuates rapidly and the signal bandwidth in the frequency domain is large, storage of analogue data followed by analogue processing may be superior to digital processing. Digital techniques are well suited for limited bandwidth signals, e.g., 1 Hz to 1 kHz, and for applications requiring multiplexing of a large number of signals.

13.3.1.1 Data Processing

The stored data need to be processed to get the results of interest. The processing of digitized data in a computer is done through the use of specific programs developed for this purpose. In computer language, these programs are called *software*. Computation of *means* and *rms* value, auto correlation between two signals, and analysis of the frequency spectrum using fast Fourier transforms (FFT) can be performed by developing suitable software for these operations.

13.3.2 Digital Interfacing

So far we have been discussing the principles and devices involved in the acquisition of data, signal conditioning, analogue to digital and digital to analogue conversion of data, multiplexing of signals, and data processing methods. Now, let us see the procedure involved in the interfacing of these devices with a digital computer. Our discussion will be keeping in mind interfacing of any data acquisition device with a personal computer. However, these concepts will be applicable to any computing system. Before looking into the interfacing details, let us have a look at the personal computer or PC hardware.

13.4   Personal Computer Hardware

A personal computer consist of

- Central Processing Unit, CPU

- Read Only Memory, ROM

- Random Access Memory, RAM

- Digital Clock

- Winchester Drive

- Floppy Drive

- Input/Output Ports

A typical personal computer layout is shown schematically in Figure 13.6.

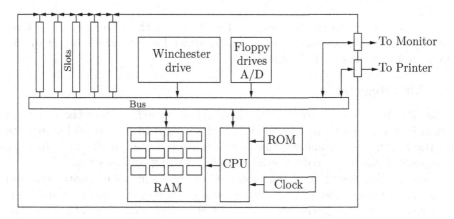

Figure 13.6: Personal computer layout

13.4.1 Central Processing Unit CPU

The central processing unit is a single large-scale semiconductor integrated circuit (LSI) chip and is termed the *microprocessor*. This chip contains the following.

- An instruction register and coder.

- An arithmetic logic unit (ALU).

- A number of registers to store and manipulate data.

- Control and timing circuits.

The microprocessor is connected to other units in the system by means of an address bus, a data bus, and a control bus.

13.4.1.1 Instruction Register and Decoder

The instruction register holds the instructions read from ROM. The decoder sends the control signals approximate to the decoded instruction to the ALU.

13.4.1.2   Arithmetic Logic Unit (ALU)

The ALU has a circuit to perform the basic arithmetic, add and subtract, logical, and/or complement, exclusive operations, register operations, load, move, clear, shift, etc., memory operations, read, store, program sequencing control operations, jump, jump to subroutine, conditional jump, and input/output operations.

In addition, it usually has two registers known as the accumulator and temporary register. It also has a status register which has bits which are set to 1 to indicate the results of ALU operations.

Working Register

The time required to retrieve data from registers and to store the results in them is at least four times faster than that required to retrieve and store data in the RAM. Thus, some working registers are provided in the CPU to store temporarily the intermediate results obtained during computation.

The CPU is the brain of the computer and performs all the arithmetic and logical operations. The ROM contains some of the system programs and basic commands which are permanently stored and are not accessible to the user. The RAM is the memory which is accessed for executing the programs of the user and this memory will be wiped of all its contents when the PC is switched off. The digital clock is a high-frequency clock which aids in the synchronizing of events inside the computer. The winchester drive and the floppy drives are used for off-line storing of programs or data. The input-output (I/O) parts allow the computer to interface with display devices such as printer, plotter, and monitor, or with any other device which is a part of the data acquisition system. All the components of a PC are connected via a communication bus.

Communication Between PC and Devices

We should note that the data requirements, input or output, of the world outside the computer are asynchronous in nature while the internal operations of the computer are perfectly synchronized. Therefore, proper interfacing between these two sets of synchronized and asynchronized events is a must for computer operation. The computer will be able to devote time for input and output operations only at very specific instants of time which are dictated by the program being executed. Thus, the objective of the interfacing operation is to synchronize the asynchronized external devices with the synchronous functioning of the computer. Further, if the external devices are analogue in their operation. Their inputs/outputs have to be digitized using an A to D converter so as to have digital mode communication with the computer.

13.4.2   Input/Output Units

Data to be processed by a computer usually originates as a written document. For example, if the pressure coefficient over a circular cylinder placed in a fluid stream is to be computed, the data will originate as a document which will contain the flow velocity, pressure tap locations over the cylinder, and the measured pressures. The data has to be first converted to a form which can be read by an input unit of the computer. This form is known as the machine readable form. The data in machine readable form is read by an input unit, transformed to appropriate internal code, and stored in the computer's memory.

The processed data stored in the memory is sent to an output unit when commanded by a program. The output unit transforms the internal representation of data into a form which can be read by the user. For example, after processing the measured pressures over a circular cylinder, the output would be the pressure coefficients at different locations on the cylinder, and the geometries of the measurement points. This result would be printed in a format by the output unit. Figure 13.7 illustrates the various states in data transformation.

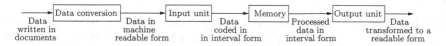

Figure 13.7: Various steps in data input and output

The I/O units of a computer may be classified as either parallel or serial. Parallel units operate with data presented to them in a parallel format, i.e., typically 8 or 16 bits of data presented simultaneously. Serial units, on the other hand, operate with data presented to them one bit at a time. Generally parallel I/O is used for short distance communication between devices and serial I/O is used where the devices to be checked are more widely separated.

Parallel units themselves may be configured into one of the following three ways from the point of view of the microcomputer. It may be an input port, or an output port, or a bidirectional I/O. Various large scale integration (LSI) devices which can be used as either input or output units in a microcomputer system are 8212, 8282, 8286, and 74LS373. Examples of bidirectional I/O are the GFIB interface and digital I/O part of the ADC/DAC interface DT2805. Let us look at the details of interfaces with bidirectional I/O in the section to follow.

It is not cost effective to do parallel data transfers when the communicating devices are far apart. Further, the reliability of data may also deteriorate if parallel transfer is done under such a situation. For such an environment serial I/O is used. With proper convention regarding framing of data and using properly synchronized clocks at the two ends, the serial I/O involves merely sending the data from one end to the other in bit-by-bit form rather than the parallel form. This bit-by-bit transmission is also subdivided into synchronized and asynchronized. In synchronized I/O data transmission must stay on continuously, even when the source runs out of data for some reason. In such events, it sends dummy SYNCH characteristics to satisfy the continuity requirement. In asynchronized I/O, the data words may be sent at arbitrary intervals with idle periods of arbitrary length between them. Examples of synchronous I/O are the IBM's BISYNCH and SDL. The RS232C interface is a very popular asynchronous device.

13.4.2.1 Input/Output Addressing

In most of the microcomputers, devices such as floppy drives and winchester as well as I/O ports are given unique addresses and all of these are connected via "data carrying lines" known as system bus. Such a placement of an I/O port in the address space of the CPU is termed *I/O mapping*. In some microcomputers the I/O port is also treated as a memory device and therefore given specific memory addresses. Such devices are said to be memory mapped.

13.4.2.2 System Buses

The main memory of the computer stores a set of addressed words. Words are retrieved from memory by specifying the addresses. As all the data to be processed by a computer has to be stored in memory, the characteristics of the interconnection paths are determined by the memory structure. For achieving reasonable speed, all the bits in a word are transmitted simultaneously. A set of wires which carries a group of bits in parallel and has an associated control scheme is known as bus. A bus which carries a word to or from memory is known as a data bus. Its width will equal the word length (in bits) of the memory. In order to retrieve a word from memory, it is necessary to specify its address. The address is carried by a memory address bus (MAB) whose width equals the number of bits in the memory address register of the memory. Thus, if a computer's memory has 64 k, 32 bits words, then the data bus (DB) will be 32 bits wide and the address bus 16 bits wide. Besides buses to carry address and data, we also need control signals between the units of a computer. For instance, if the processor has to send READ and WRITE commands to memory, START command to I-O units, etc., such signals are carried by a control bus (CB). Thus, a system bus will consist of a data bus and a memory address bus. Figure 13.8 shows the processor-memory interconnection.

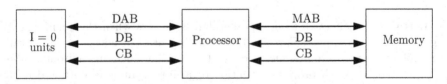

Figure 13.8: Interconnection of computer units via two system bus

One method of connecting I-O units to the computer is to connect them to the processor via bus. This bus will consist of a bus specifying the address of the I-O units to be accessed by the processor, a data bus carrying a word from the addressed input unit to the processor or carrying a word from the processor to the addressed output unit. Besides these two types, a control bus carries commands such as READ, WRITE, START, REWIND TAPE, etc., from the processor to I-O units. It also carries the I-O unit status information to the processor. The I-O to processor system bus will thus encompass I-O device address lines, data lines, and control lines. In Figure 13.8 the I-O to processor bus interconnection is also shown. This method of interconnection will require the processor to completely supervise and participate in the transfer of information from and to I-O units. All information will be first taken to a processor register from the input unit and from there to the memory. Such data transfer is known as *program controlled transfer*. The interconnection of I-O units, processor and memory using two independent system buses is known as the two bus interconnection structure.

It is seen from the above discussion that, this interconnection structure requires the processor to share its resources during I-O transfer. As I-O units are much slower than the processor, this is not desirable. Thus, another interconnection method which minimizes processor participation in I-O transfer is often used. This is also a two bus structure. One bus interconnects the processor to the memory, and the other bus, instead of connecting I-O units to the processor, connects them directly to memory. In this structure, the I-O devices are connected to a special interface logic called direct memory access logic (DMA) or a channel or a peripheral processor unit (PPU). This interconnection structure is illustrated in Figure 13.9.

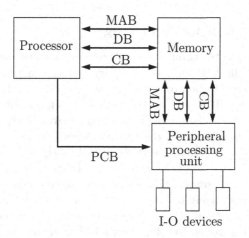

Figure 13.9: Direct PPU-memory connection

The processor issues a READ or WRITE command giving the device address, the address in memory where data read from the input unit is to be stored (or from where data is to be taken to output), and the number of data words to be transferred. This command is accepted by the PPU which now assumes the responsibility of data transfer.

Another popular method of interconnection which is commonly used in small computers is through a single bus, as shown in Figure 13.10.

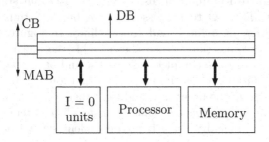

Figure 13.10: A UNIBUS connection of units

This is called a *unibus system*. In this case the bus is shared by the three units and thus the transfer of information can take place at a time only between two of the three units. The main advantage of this system is the addressing of I-O units. These units use the same memory address space. This simplifies programming of I-O units as no special I-O instructions are needed. A READ instruction with an address corresponding to an input device will read information from that device. Besides this advantage, it is easy to add new devices as no new instructions specific to the device would be required.

13.4.3   Input/Output Servicing

The events external to the computer may be asynchronous and hence, data may arrive at or leave the I-O ports at arbitrary instants of time. However, these I-O requirements have to be processed by the computer fairly quickly since the buffer memory space of the I-O ports where data temporarily stays is not very large. The simplest way of servicing is through program controlled I-O in which the CPU monitors the status of the I-O ports periodically, as specified by the program, to see if any of them needs service. If so they are given the required service, e.g., input ports are read and data is latched into output ports. This technique even through simple, is inefficient in time management, since most of the routine checks are likely to go to waste, and requires a significant software overhead. A better method of I-O servicing is to use the intercept units of the CPU in interrupt controlled I-O. In this kind of servicing the I-O ports directly inform the CPU of their service need by means of the interrupt lines. The CPU temporarily suspends all its internal operations and attends to the service needs of I-O ports. After completing the service needs of I-O ports the CPU resumes execution of the interrupt program. This technique is efficient time-wise but it requires additional hardware such as interrupt controllers.

A problem which is common to both the above servicing techniques is that it actually interferes with the normal program execution of the CPU. This is not required when the service required is merely the transfer of data from the

I-O port to the computer memory or vice versa. Such a transfer can be achieved directly by direct memory access (DMA) or hardware controlled I-O. In this case, the system has a separate device called the DMA controller to handle the I-O transaction. When such a transaction is needed, the DMA controller forces the CPU to disconnect itself from the system bus and effects the required transfer of data and returns the control of the system bus back to the CPU at the end of the transaction. Apart from ensuring that the DMA controller has been programmed properly, no other software overhead is required. The execution of program in CPU continues in a normal fashion with DMA transfers taking place in between as required. Typical DMA controllers are available in LSI form. The IBM-PCs use one such device (8239) to provide for the DMA channels needed to transfer data back and forth between memory (RAM) and the floppy/winchester drives. The DMA technique is expensive and hardware intensive. However, it is the fastest method of doing I-O servicing with minimum interference to the functioning of the CPU.

13.5 Data Acquisition Using Personal Computers

For a data acquisition system configured for laboratory applications, the personal computer (PC) may be employed as the controller. A number of measuring instruments, probes, and transducers can be connected to it through appropriate interfaces. In the case of digital instruments, they can be accessed and controlled remotely by a PC. In fact, all the operations which have to be done manually by pressing buttons on the operation panel of any digital instrument can be controlled by a PC. Ready-made standard interfaces are available for controlling digital instruments and also for digitizing analogue signals - A to D cards - and transferring them to the PC. The general purpose interface bus (GPIB) is the most widely used interface for the control of digital measuring equipment. This was developed by the Hewlett-Packard (HP) company in 1972 under the name of HP-IB for the interconnection of instruments. It served as the trend setter for further work in standardizing instrument interfaces and the IEEE-488 standard was finally published in 1975. Presently, GPIB is also called as IEEE-488 bus or HP-IB. For A to D conversion, several ADC/DAC cards are available from various manufacturers, which are IBM-PC compatible. These interface cards can directly be connected to the slots provided in the PC. Most of the cards offer 8 or 16 ADC channels, 1 or 2 DAC channels, and 8 or 16 channels of digital input/output. Additional features such as programmable gain in amplification, sample-and-hold, single ended or differential inputs and DMA are also available with these cards.

Now let us have a close look at the prime features of the GPIB interface and the ADC/DAC card DT 2805 of data translation INC.,USA.

- This GPIB interface can be used with most of the modern measurement devices.

- The ADC/DAC can be connected to probes such as a pressure transducer or thermocouple.

- Each card is capable of acquiring data from 15 devices or probes at a time.

- The GPIB performs digital multiplexing and the DT 2805 does low-level multiplexing of analogue signals.

- It is possible to plug more than one GPIB or DT 2805 ADC/DAC cards in to the same PC, depending on the availability of the extension slots.

13.5.1   The GPIB Interface

The GPIB interface provides a low-cost interconnection for up to 15 digital instruments over a limited distance of about 20 m. It is capable of interfacing instruments with varying data transfer rates from a few hundred bytes/sec to one mega byte/sec. It has been implemented in a modular fashion so that instruments with a wide range of capabilities can be handled and for simple instruments, a subset of the interface would be sufficient. Direct communication between instruments is possible without routing all the data through a control unit. The GPIB interface structure is shown schematically in Figure 13.11.

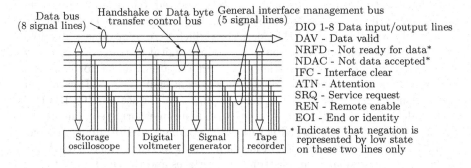

Figure 13.11: GPIB structure

Communication between instruments is possible without needing all the messages/data to be routed through a control unit. The structure of GPIB is shown in Figure 13.11.

The GPIB consist of 16 lines, of which eight are meant for parallel data transfer and eight from the command channel. For regulating and validating the input/output of the instruments, three lines of the command channel are dedicated. A detailed protocol known as the "three-wire handshake" is adapted for this process. The remaining five lines of the command channel have management functions. In the data channel of 8 lines, data flows 8 bits at a time in "bit parallel byte-serial" order.

A maximum of 15 instruments can be simultaneously connected on the GPIB. The bus length is typically 2m/instrument, but not exceeding 20 meters in all. The maximum data rate is about 1 MB/sec, but transfer rates achieved in typical applications consisting of microprocessor-based instruments is about 250 KB/sec. A GPIB cable is terminated at both ends by identical connectors (male-female back to back) which permit piggy-back connections for easy connections/disconnections of an instrument from the bus.

In a configuration of instruments on GPIB, an instrument may be a TALKER, a LISTENER, or the CONTROLLER. There can be only one CONTROLLER at a time which has full control over the bus. Also, there can be only one TALKER which puts data into the 8 data lines (D101 to DI08). There can be many LISTENERS, i.e., instruments which accept data from these 8 data lines. Transfer of each byte of data (1 byte = 8 bits) on the DIO lines is accomplished via the three-wire handshake.

The three lines used for the handshake always operate in the interlocked handshake mode. Each of these may transmit a binary signal 0 (known as low) or 1 (known as high). The lines signify the following messages when they are low: DAV (Data Valid), NRFD (Not Ready For Data), and NDAC (Not Data Accepted). When they are high, the opposite message such as Not Data Valid, etc., is conveyed. Talker drives the DAV line, while all the potential listeners control the NRFD and the NDAC lines. The three-wire handshake is shown in Figure 13.12. The transfer of data occurs following the protocol procedure given

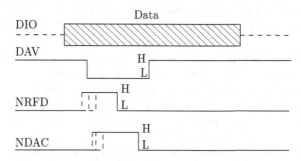

Figure 13.12: Three-wire handshake

below.

- (a) Talker waits for all the potential listeners to signify "ready for data" by pulling the NRFD high.

- (b) Talker asserts "data valid" on DAV (drives it low) after it has stabilized new data on DIO lines.

- (c) Upon receiving data valid the listeners remove "ready for data" by pulling the NRFD low assert "data accepted" on NDAC by driving it high.

- (d) When all listeners have accepted the data, i.e., NDAC is high, the talker removes "data valid" (DAV is pulled high).

- (e) Next data byte is output into the DIO lines by the talker and it waits for step (a).

In this protocol, data is transferred synchronously at the rate of slowest listener participating in the communication. High-speed data transfer between two instruments is not affected by the slow speed of the instruments connected to the bus, provided that they do not participate in the conversation.

To affect a change in talker/listener status of the instrument on GPIB, one of the instruments (which is a microprocessor or a PC) is configured as the controller. The controller may itself be a talker or a listener. It can transmit GPIB commands on the DIO lines and also drive other special lines such as IFC (interface clear) REN (remote enable), ATN (attention), EOI (end or identify). It also senses the service request (SRQ) lines which are used by devices on GPIB to draw the attention of the controller. Driving the IFC line low initializes GPIB of all instruments and lets the controller take charge of the bus. With REN pulled high, the front panel controls of an instrument will be disabled and the instrument will not respond to manual control thereafter. When SRQ is high, the controller interrogates each instrument one by one (called serial poll) or up to eight instruments at a time (called parallel poll) to determine the interrupting instrument. Using the ATN line and the 8 data lines, the controller can configure the instrument which are listener and the instrument which is the talker. The EOI lines is used by the talker to indicate the end of a block of data (see Table 13.1).

The GPIB interface is incorporated in an instrument, not as an integral part but as a separate block. The interface design depends on the instrument. The function of the interface is to convert device-dependent messages into a standard form of data, request or command acceptable to GPIB protocol. In PC-controlled GPIB configurations, each instrument should have its own GPIB interface, including the PC. Some of the devices which are capable of producing only a limited number of messages have a simpler configuration of the interface. For instance, a simple plotter would only have the acceptor handshake and listener function implemented on its GPIB interface. Besides the messages of common GPIB protocol, the interface of device exercises certain controls on it (Figure 13.13) as permitted by the GPIB specifications of the machine (e.g., DEVICE CLEAR, DEVICE TRIGGER etc.).

Several types of modern instruments have been made GPIB compatible by the development of interfaces for them. For instance, measurement devices (e.g., network analyzer, spectrum analyzer, frequency counter, power meter, digital multimeter, oscilloscope), stimulus equipment (e.g., function generator, waveform generator), display equipment (e.g., dot matrix printer, plotter), storage equipment (e.g., analogue instrumentation recorder, magnetic tape drive), and controller equipment (e.g., PC), can all be hooked on a GPIB network.

Table 13.1 The GPIB lines

Line No.	Variable name	Driving	Function
1	DI01	Talker	Data/Command channel - Data
2	DI02		Transmission Between Talker/Listner
3	o	Active	Command transmission to all
4	o	Controller	Instruments
5	o		
6	o		
7	o		
8	DI08		
9	DAV	Talker Active Controller	Data valid - indication of validity of byte on data lines
10	NRFD*	Any Listener	Not read for data - Indication of the readiness of all listeners to accept new byte
11	NDAC*	Any Listener	Not data accepted - Indication of the acceptance of the byte by all the listeners
12	IFC	System Controller	Interface clear - Initialization of GPIB interface. Take over of control of the system
13	ATN	Active Controller	Attention - multiplexing of data / command channel
14	SRQ*	Any Instrument	Service request - Service request to the active controller
15	REN	System Controller	Remote enable - Disabling the front panel controls
16	EOI	Talker System Controller	End or identify - Indication of end of block. Obtaining parallel pol response

* These lines can be pulled low by any of the instruments on the bus.

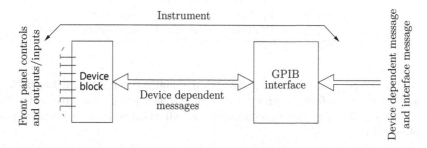

Figure 13.13: Interface device with some controls

For IBM-PC and compatibles, a GPIB interface called GPIB-PC2 (manufactured by National Instruments, USA) is available. This is a half-length card with a GPIB connector. The card can be directly plugged into one of the vacant slots of the PC. It supports all the GPIB interface functions including the controller functions (see Table 13.2). The card is sold along with software, which consists of a MS-DOS device driver called DOS Handler and BASIC interface for the DOS Handler. It is also possible to use this software through user programs written in PASCAL and FORTRAN.

Table 13.2 GPIB interface functions

Neumonic no.	Description name	Function
SH	Source Hand Shake	Handles handshake for the TALKER/ CONTROLLER during transmission (i.e., controls DAV line)
AH	Acceptor Hand Shake	Completes handshake for the LISTENER (Controls NRFD, NDAC lines)
T	Talker	Enables the instrument to send status/DATA bytes when addressed to talk
L	Listener	Enables the instrument to receive data when addressed to listen
SR	Service Request	Sends Service Request to the controller, on SRQ line, on receiving the local Service Request message
RL	Remote Local	Disables/Enables the local controls using REN lines
PP	Parallel Poll	Enables the instrument to present one status bit to the controller
DC	Device Clear	Resets the Device in the instrument
DT	Device Trigger	Triggers the DEVICE in the instrument (have its basic operation started)
C	Controller	Enables the instrument to control the INTERFACE in all the instruments on the bus

For user programs written in BASIC, a manufacturer supplied file BIB.M should be linked before execution. Also, in the user program source code, the file name DECL.BAS which contains the initialization code must be placed at the beginning.

To illustrate the use of function calls through BASIC in a user-written program, let us examine the following code for sending data to a plotter connected on GPIB.

- 100 $BDNAME$ = "PLOTTER"

- 120 CALL IBFIND ($BDNAME$, PLT%)

- 130 WRTS = "PD"

- 140 CALL IBWRT (PLT%, WRTS)

Statement initializes $BDNAME$ as a string variable with the device name. IBFIND is the function call to open the device. It returns an integer PLT% which is used as an identifier of this device later on. WRT is the string variable which is initialized to the command (PD) to the lower pen. IBWRT call sends WRT to the plotter.

The manufacturer supplied package also contains a menu-driven program called IBIC which allows the user to execute function calls interactively from the keyboard one at a time, with the status being displayed on the monitor. With the IBIC menu, the command IBWRT can be used to receive or send data. In order to store the data on the PC, the function call IBRDF (file name) can be used.

Typical commands for sending data from a storage oscilloscope to the PC using the IBIC menu are given below.

- IBWRT "BL = 10 / r / n"

- IBWRT "TRC1 / r / n"

- IBRDF (OUTPUT)

The first command selects the block length of data string to be transferred. The second command brings the oscilloscope in action to get ready to transfer the data. The third command stores the data in PC under the file name OUTPUT. For complete information about IBIC menu and its usage, the manual (which comes with the GPIB for the PC) supplied by the manufacturer can be consulted.

13.5.1.1 DT 2805 ADC/DAC Interface

This is a low-level multiplexer card, manufactured by Data Translation Inc. (USA), which can interface 8 differential channels (or 16 individual channels) of analogue signals with the IBM-PC compatible. It accepts signals as low as 20 mV and amplifies them through a software programmable amplifier (with gains of 1, 10, 100, and 500) and sends them to PC after proper impedance matching. The A/D conversion is 12-bit. The digital input-output (DIO) consists of 16 lines which can be configured as 16 inputs, 16 outputs or as 8 inputs + 8 outputs according to the convenience.

The card can be plugged directly into one of the expansion slots of the PC. It has a termination panel (DT 707) to which all the probes can be connected. Channel 0 of the termination panel has a room temperature sensor which is useful for thermocouple probes. Thus, channel 0 can be taken as the reference junction and it is not to be used for any other input unless the temperature sensor is disconnected. DMA is available on the card and for DMA operation, the selected expansion slot on PC is 1. The DT 2805 occupies two successive addresses (2 EC and 2 ED in Hexa-decimal code) on the I/O bus of the PC (Figure 13.14).

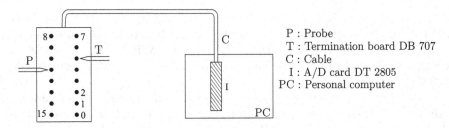

P : Probe
T : Termination board DB 707
C : Cable
I : A/D card DT 2805
PC : Personal computer

Figure 13.14: Hexa-decimal code on the I/O bus of a PC

The card is sold along with user-friendly software such as PC THERM or PCLAB. This software is menu driven and is very easy to operate. The only declarations that the user will have to make are the following.

- Number of channels to be read.

- Rate of data acquisition (up to a maximum of 5000 readings/second).

- Gain of the amplifier (applicable for all signals since only one amplifier is used).

- Whether the signal is unipolar or bipolar.

The software prompts the user to answer a series of queries for the above information. When all the relevant information has been provided, it acquires the data and stores it in a user-specified file.

13.6 Digitization Errors due to A/D Conversion

In most circumstances, the signal from the probe is analogue and it is digitized either in a digital measuring instrument or with the help an ADC/DAC card before being acquired by the PC through appropriate interfaces and software. One of the major sources of errors during data acquisition is the A/D conversion process itself.

Analogue to digital conversion proceeds at a finite rate and during this period of conversion changes that take place in the input voltage level are ignored in S/H circuit. If N is the number of points into which the voltage range 0 - E is discretized by the A/D converter, then the average discretization error in measuring a voltage E(t) in this range is e = E/2N. For a n-bit converter, N = 2. The discretization error spread over the period for which data is collected is equivalent to a noise signal. If the analogue signal has a variance V_{analog} and the noise signal has a variance V_{noise} it is desirable that the signal-to-noise ratio V_{analog}/V_{noise} be as large as possible. This can be achieved by having a suitably high value of n, the number of bits used A/D conversion.

13.7   Summary

In many experiments a large quantity of data has to be measured within a very short time. For such measurements, a data acquisition and processing system for continuous and automatic acquisition of data becomes essential. In some experiments, the acquired data needs to be processed in real time and appropriate action has to be taken based on the processed result. Such a comprehensive and stringent requirement demands a data acquisition system possessing the necessary characteristic of speed, power, and so on.

Presently there are several systems commercially available for rapidly collecting a large amount of data, processing it, and displaying the desired results in visual and printed forms. But such commercially available units are expensive and most of them are not flexible in their configurations. For laboratory research experiments with a limited number of transducers, low-cost data acquisition units can be configured.

The essential sequence of operation involved in any data acquisition system, regardless of the size of the system is the following.

- Generation of input signals by transducers.

- Signal conditioning.

- Multiplexing.

- Data conversion from analogue to digital form and vice versa.

- Data storage and display.

- Data processing.

The essential element in a modern data acquisition system is the instrument transducer, which furnishes an electrical signal that is indicative of the physical quantity being measured.

The objective of signal conditioning is to modify the signal received from the sensor to suit the requirements of further processing. The signal conditioning may consist of one or more of the following operations.

- Amplification of the signal from the sensor.

- Filtering out the unwanted frequencies originating from the sensor and their associated circuitry.

- Providing independence matching.

- Compensation for the limitations of the sensor and/or receiving devices (e.g., recorders) so as to extend their frequency range.

- Correction of thermoelectric errors at input functions.

- Performance of arithmetic operations on the outputs of two or more sensors.

In many measurements of practical interest, more than one channel of data must be repetitively transduced to voltage signals. When the required data rates are sufficiently low, a cost-effective solution time shares a single transducer/amplifier by multiplexing the pressure lines using a scanning device. One of the popular devices in this field is the scanivalve. It is a multiported rotating valve which sequentially connects each of many pressure lines (up to 64) distributed around its circumference to a single flush-diaphragm transducer.

The data conversion generally involves conversion of data from analogue to digital or from digital to analogue. In analogue representation, the physical variables of interest are treated as continuous quantities, while in digital representations, the physical quantities are restricted to discrete, noncontinuous "chunks." The conversion of data from analogue to digital form and vice versa is necessitated by the fact that real-world events are continuous with a string of binary digits 0 and 1. Therefore, data has to be converted to digital form before being fed to the computer and the output from the computer may have to be converted back to analogue form in order to control or operate any analogue device. There are many other advantages with such data conversions. They are the following.

- Conversion of signals from analogue to digital form gives noise immunity to the data during its transmission.

- It is easy to determine whether or not a single digital pulse (either a binary 0 or binary 1) is present at a given time determining the presence of a pulse in analogue form.

- Coding techniques have been developed only for digital signals and thus, to take advantage of the error-detection and error-correction capabilities of these codes, it is essential to correct all the data signals of interest to digital form.

The digital data, after conversion from analogue to digital, is usually stored in floppies/winchester drives in the PC-based data acquisition systems and in magnetic tapes in the conventional data acquisition units.

The stored data needs to be processed to get the results of interest. The processing of digitized data in a computer is done through the use of specific programs developed for this purpose. In computer language, these programs are called *software*. Computation of *means* and *rms* value, auto correlation between two signals and analysis of the frequency spectrum using fast Fourier transforms (FFT) can be performed by developing suitable software for these operations.

A personal computer consists of

- Central Processing Unit, CPU

- Read Only Memory, ROM

- Random Access Memory, RAM

- Digital Clock

- Winchester Drive

- Floppy Drive

- Input/Output Ports

The central processing unit is a single large-scale semiconductor integrated circuit (LSI) chip and is termed the *microprocessor*. This chip contains the following.

- An instruction register and coder.

- An arithmetic logic unit (ALU).

- A number of registers to store and manipulated data.

- Control and timing circuits.

The microprocessor is connected to other units in the system by means of an address bus, a data bus, and a control bus.

The instruction register holds the instructions read from ROM. The decoder sends the control signals approximate to the decoded instruction to the ALU.

The ALU has circuits to perform the basic arithmetic, add and subtract, logical, and, or complement, exclusive operations, register operations, load, move, clear, shift, etc., memory operations, read, store, program sequencing control operations, jump, jump to subroutine, conditional jump, and input/output operations.

In addition, it usually has two registers known as the accumulator and temporary register. It also has a status register which has bits which are set to 1 to indicate the results of ALU operations.

The time required to retrieve data from registers and to store the results in them is at least four times faster than that required to retrieve and store data in the RAM. Thus, some working registers are provided in the CPU to store temporarily the intermediate results obtained during computation.

The CPU is the brain of the computer and performs all the arithmetic and logical operations. The ROM contains some of the system programs and basic commands which are permanently stored and are not accessible to the user. The RAM is the memory which is accessed for executing the programs of the user and this memory will be wiped of all its contents when the PC is switched off. The digital clock is a high-frequency clock which aids in the synchronizing of events inside the computer. The winchester drive and the floppy drives are used for off-line storing of programs or data. The input-output (I/O) parts allow the computer to interface with display devices such as printer, plotter and monitor, or with any other device which is a part of the data acquisition system. All the components of a PC are connected via a communication bus.

We should note that the data requirements, input or output, of the world outside the computer are asynchronous in nature while the internal operations of the computer are perfectly synchronized. The objective of the interfacing operation is to synchronize the asynchronized external devices with the synchronous functioning of the computer.

Data to be processed by a computer usually originates as written documents. The data has to be first converted to a form which can be read by an input unit of computer. This form is known as machine readable form. The data in machine readable form is read by an input unit, transformed to appropriate internal codes, and stored in the computer's memory.

The processed data stored in the memory is sent to an output unit when commanded by a program. The output unit transforms the internal representation of data into a form which can be read by the user.

The I/O units of a computer may be classified as either parallel or serial. Parallel units themselves may be configured into one of the following three ways from the point of view of the microcomputer. It may be an input port, or an output port, or a bidirectional I/O.

In most microcomputers, devices such as floppy drives and winchester as well as I/O ports are given unique addresses and all of these are connected via "data carrying lines" known as the system bus.

One method of connecting I-O units to the computer is to connect them to the processor via bus. This bus will consist of a bus specifying the address of the I-O units to be accessed by the processor, a data bus carrying a word from the addressed input unit to the processor or carrying a word from the processor to the addressed output unit. Besides these two types, a control bus carries commands such as READ, WRITE, START, REWIND TAPE, etc., from the processor to I-O units. It also carries the I-O units status information to the processor. The I-O to processor system bus will thus encompass I-O device address lines, data lines, and control lines.

Another popular method of interconnection which is commonly used in small computers is through a single bus called a unibus system. In this case the bus is shared by the three units and thus the transfer of information can take place at a time only between two of the three units.

For a data acquisition system configured for laboratory applications, the personal computer (PC) may be employed as the controller. A number of measuring instruments, probes, and transducers can be connected to it through appropriate interfaces. In the case of digital instruments, they can be accessed and controlled remotely by a PC. In fact, all the operations which have to be done manually by pressing buttons on the operation panel of any digital instrument can be controlled by a PC.

The prime features of the GPIB interface and the ADC/DAC card (DT 2805 of Data Translation Inc.,USA) are the following.

- This GPIB interface can be used with most modern measurement devices.

- The ADC/DAC can be connected to probes such as a pressure transducer or thermocouple.

- Each card is capable of acquiring data from 15 devices or probes at a time.

- The GPIB performs digital multiplexing and the DT 2805 does low-level multiplexing of analogue signals.

- It is possible to plug more than one GPIB or DT 2805 ADC/DAC card in to the same PC, depending on the availability of the extension slots.

The GPIB interface provides a low-cost interconnection for up to 15 digital instruments over a limited distance of about 20 m. It is capable of interfacing instruments with varying data transfer rates from a few hundred bytes/sec to one mega byte/sec. Communication between instruments is possible without needing all the messages/data to be routed through a control unit.

Chapter 14

Uncertainty Analysis

14.1   Introduction

As we know, the aim of any experiment is to measure some physical quantities as accurately as possible. In a few experiments, we look for only qualitative information, of the type yes or no; e.g., yes, there is a detached shock at the nose of blunt body kept in a supersonic stream. In such cases the discussion of experimental accuracy should be in terms of the probability that the conclusion is correct. But in a vast majority of cases the experiment produces a numerical quantity. In such experiments, the error in the measurement must be estimated to arrive at the conclusion about the accuracy of the results obtained. The error involved may be due to uncertainties of various kinds. The possible imperfection of the theory on which the experiment is based, the imperfection of the instrument used for measurement, and the imperfection on the part of the observer are some of the major causes of uncertainties. In this chapter let us see some of the consequences of these errors on the measured qualities and the methods for analyzing these errors.

14.2   Estimation of Measurement Errors

Let us assume that x is a numerical quantity measured in an experiment. Due to uncertainties of various kinds the measured quantity is not the correct quantity, say, X. Therefore, there is an error e, in the measured quantity which is equal to the difference between the correct and measured quantities, $(X - x)$. The error can be either positive or negative. If the measured value of x is to be put to any use, it is essential that we have some conception about the likely value of error e. We can write a relation involving X, x, and e as

$$X = x \pm e \qquad (14.1)$$

From Equation (14.1) it is clear that, if nothing is known about e, then a measurement of the value x tells nothing about the correct quantity X. An experimenter should cultivate the habit of never making measurements, or calculating any experimental parameter, without having some concept of the error which is likely to be associated with the measurement. In some cases absolute value e of the error is of interest, but normally the relative or fractional error ϵ, namely x/X, is of interest.

The errors that are likely to be associated with the measured quantity x are

- the external estimate ϵ_E, based on the knowledge of experimental work carried out by others.

- the internal estimate ϵ_I, based on the data obtained during the experiment.

When these two estimates agree fairly closely the experiment may be treated as reliable.

14.3   External Estimate of the Error

In measuring a quantity in any experiment there are several steps involved. Let us consider a typical experiment to measure a small pressure with a pressure transducer of diaphragm type. First, the change of position of the diaphragm surface is changed into a change of electrical resistance, then the change of resistance is transformed into a change of voltage, the voltage change is amplified, and finally the amplified voltage is fed into a voltmeter to produce the displacement of a pointer (say). Let us assume that the proportionality errors associated with the various steps are ϵ_a, ϵ_b, ϵ_c, ϵ_d and they are known. If the aim of the experiment is to measure the gain of the system, then it is essential to know the error that would result from the combination of these four errors. Let us write the performance of the four separate stages involved in the measurement in the form

$$\frac{\Delta R}{\Delta \delta} = a\left(1 \pm \epsilon_a\right)$$

$$\frac{\Delta V_{\text{in}}}{\Delta R} = b\left(1 \pm \epsilon_b\right)$$

$$\frac{\Delta V_{\text{out}}}{\Delta V_{\text{in}}} = c\left(1 \pm \epsilon_c\right) \qquad (14.2)$$

$$\frac{\Delta y}{\Delta V_{\text{out}}} = d\left(1 \pm \epsilon_d\right)$$

where R is the resistance, V_{in} and V_{out} are input and output voltages, δ is the diaphragm displacement, and y is the pointer displacement, a, b, c, and d are the outputs associated with each stage of the measurement. Then we have

$$X = \frac{\Delta y}{\Delta \delta} = a\,b\,c\,d\,(1 \pm \epsilon_a)\,(1 \pm \epsilon_b)\,(1 \pm \epsilon_c)\,(1 \pm \epsilon_d) \qquad (14.3)$$

From Equation (14.3) it is evident that the worst possibility is that all four errors are of the same kind, i.e., all are positive or all are negative. In this case the resultant error, assuming that all the errors are very small compared to unity, is

$$\epsilon_E = \epsilon_a + \epsilon_b + \epsilon_c + \epsilon_d \qquad (14.4)$$

where ϵ_E is the *effective* or *resultant* error. When the errors ϵ_a to ϵ_d are due to independent causes, the chance of all errors in the same direction is rare. When the errors are due to independent causes, we may think of a possibility of the error components getting canceled resulting in zero resultant error. Usually this possibility is also unlikely.

The following general procedure is devised (Cook and Rabinowicz, 1963) to treat measurements involving combination of errors, which will reveal what could be the expected resultant error due to the combination of errors. Express Equation (14.3) in its most general form as

$$X = F(a, b, c, d) \qquad (14.5)$$

then

$$dX = \Sigma_{n=a}^{d} \left(\frac{\partial F}{\partial n} \right) dn \qquad (14.6)$$

Let us assume that the variation in X due to errors in the quantities a to d to be e_E. Also, let the errors in a to d to be e_a to e_d. Thus, we have

$$e_E = \Sigma_a^d \left(\frac{\partial F}{\partial n} \right) e_n \qquad (14.7)$$

since the nature of error terms e_a to e_d; whether they are positive or negative or a mixture of these two, is not known, it is not possible to evaluate e_E directly. However, it is possible to overcome this difficulty by evaluating e_E^2 as follows.

$$e_E^2 = \Sigma_{n=a}^{d} \left(\frac{\partial F}{\partial n} \right)^2 e_n^2 + \Sigma_{n=a, m=a}^{d} \left(\frac{\partial F}{\partial n} \right) \left(\frac{\partial F}{\partial m} \right) dndm \qquad (14.8)$$

where $m \neq n$.

When the error terms e_a to e_d are independent and symmetrical in regard to positive and negative values, the cross product terms in Equation (14.8) will tend to zero and hence,

$$e_E^2 = \Sigma_a^d \left(\frac{\partial F}{\partial n} \right)^2 e_n^2 \qquad (14.9)$$

This is the general form of the *resultant error* expression in terms of its components. For the present case under consideration, we have

$$F(a, b, c, d) = abcd = X \tag{14.10}$$

and $\partial F / \partial n$ is in all cases equal to X/n. Hence,

$$e_E^2 = X^2 \Sigma_a^d \frac{e_n^2}{n^2} \tag{14.11}$$

Dividing throughout by X^2 and taking the square root, we get

$$\epsilon_E = \left(\epsilon_a^2 + \epsilon_b^2 + \epsilon_c^2 + \epsilon_d^2 \right)^{1/2} \tag{14.12}$$

The *expected or most probable error* given by Equation (14.12) is significantly smaller than the possibility considered in Equation (14.4). This can be easily verified by the case when

$$\epsilon_a = \epsilon_b = \epsilon_c = \epsilon_d \tag{14.13}$$

For this case with all the error terms equal, Equation (14.4) gives an error of $4\epsilon_a$, while the most probable error, given by Equation (14.12) is only ϵ_a.

14.3.1   Dependence and Independence of Errors

In the derivation of Equation (14.9) it was assumed that the different error components involved in the experimental measurement were independent. In other words, if ϵ_a is assumed, for instance, to be positive, there was no way that this fact could influence ϵ_b, ϵ_c, or ϵ_d. This situation is called the *independence of errors*. To illustrate the *dependence of errors* on each other, let us consider an experiment to determine the density of a gas stream. To estimate the density, it is necessary to measure the pressure and the temperature of the flow. Once the pressure and the temperature are known, it follows from the perfect gas state equation that the density ρ of the stream is

$$\rho = \frac{p}{RT} \tag{14.14}$$

where p and T are the static pressure and static temperature of the stream and R is the gas constant of the flowing gas. The error in the density will be determined by the error in the pressure measurement e_p and error in the temperature measurement e_T. If these errors are independent, applying Equation (14.9), we have

$$\rho = F(p, R, T) = \frac{p}{RT} \tag{14.15}$$

where

$$\frac{\partial F}{\partial p} = \frac{1}{RT}, \quad \frac{\partial F}{\partial R} = 0, \quad \frac{\partial F}{\partial T} = -\frac{p}{RT^2} \tag{14.16}$$

Hence, using Equation (14.16), we get from Equation (14.9)

$$e_E^2 = \frac{e_p^2}{R^2 T^2} + \frac{e_T^2}{R^2 T^4} \tag{14.17}$$

Dividing each term by ρ^2 or $p^2/R^2 T^2$, and taking square root, we obtain,

$$\epsilon_E = \left(\epsilon_p^2 + \epsilon_T^2\right)^{1/2} \tag{14.18}$$

Equation (14.18) has the same form as Equation (14.12), and it may be realized that this must be the case, since we could have transformed Equation (14.3) into the form of Equation (14.14) simply by redefining the constants b, c, and d in Equation (14.2).

There is a simple way (Cook and Rabinowicz, 1963) of visualizing equations such as (14.18). Consider an experiment of a missile aimed at a target. Let the first error-producing term ϵ_1 cause the missile to miss the target and proceed in an arbitrary direction, as shown in the Figure 14.1.

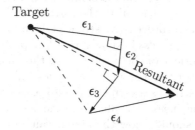

Figure 14.1: Effect of independent errors

If the second error term ϵ_2 is positively related to the first, it will be in the same direction, resulting in a total error of $\epsilon_1 + \epsilon_2$. If ϵ_2 is independent it could be in any direction with respect to ϵ_1. The most random direction is at right angles; this produces a total error of $(\epsilon_1^2 + \epsilon_2^2)^{1/2}$. Further, random errors tend to be at right angles to the resultant of ϵ_1 and ϵ_2, satisfying Equation (14.18). Any correlated error is added up in one straight line, as illustrated in Figure 14.2.

The error combination model described above is analogous to the random-walk problem of particle motion. For example, after n random excitation of comparable magnitude, a particle is likely to be away from its initial position by a distance proportional to $n^{1/2}$.

From the discussion of dependent and independent errors we see that:

- The total error in a measurement will be greater when the errors are dependent than when errors are independent.

- It is preferable, from an accuracy point of view, to design experiments involving independent errors.

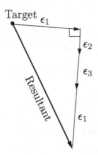

Figure 14.2: Correlated errors

14.3.2   Estimation of External Error

In order to estimate the external error associated with a measurement the error in each step must be considered and combined together as shown in Equation (14.9). However, there is no general method available for estimating the error of a step in a measurement process. Mostly it is done by intuition. Thus, if the limiting resolution of the process is known, then the probable relative error will be no less than the quotient of the resolution to the quantity being measured. In cases where the theory of the process applies only approximately to the experiment, the likely error as a result of this disagreement may be estimated from the deviation of the measured value from the actual or standard value. For this, the experimenter should know the actual value or standard value of the quantity being measured.

In commercial experiments, a commonly used approach for external error estimation is to consult the literature supplied by the manufacturer. In most of the measurements it is not possible to estimate the probable error of all stages of the measurement process. In such situations it is helpful to know that, from Equation (14.12), if one step has a much smaller error than some of the others, it may be ignored. In the same way, if one step has a much larger error than the others, only it needs to be considered.

14.4   Internal Estimate of the Error

In any experiment, it is advisable to check the measured value for repeatability, by carrying out repeated measurements. In other words, it is advisable, when carrying out an experiment, not to rely on a single measurement of a quantity. It is a matter of common experience that these repeat measurements will not give identical values of the quantity being measured, and the variation among the measurements may be used to make an estimate of the accuracy of the measurement as a whole. This is known as the *internal estimate of error* ϵ_I. The reader is encouraged to consult Cook and Rabinowicz (1963) for procedure to estimate ϵ_I.

14.5 Uncertainty Analysis

As we know, no measurement is perfectly accurate. Hence, it is essential to estimate the inaccuracies associated with the measured values to assess the validity of the results. It is now generally agreed that, the appropriate concept for expressing inaccuracies is *uncertainty* and that the value should be provided by an uncertainty analysis (Kline, 1985). An uncertainty is not the same as an error. An error in a measurement is just the difference between the correct value and the recorded value. An error is a fixed number and not a statistical variable. *An uncertainty is a possible value that the error might take on in a given measurement.* Since the uncertainty can take on various values over a range, it is basically a statistical variable. Uncertainty can be considered as a histogram of values.

In a broad sense, a measured value describes the central tendency, usually the mean, whereas the uncertainty describes the dispersion usually in terms of a measure associated with a stated probability level such as the standard deviation. Ideally, this measure of uncertainty is calculated from repeated trials, but it may need to be taken from estimates in whole or part in many experiments.

14.5.1 Uses of Uncertainty Analysis

The primary uses of uncertainty analysis are (Kline, 1985)

- It forces a complete examination of the experimental procedure, including the potential sources of error.

- It is useful to identify the need for employing improved instruments and/or improved procedure to obtain the desired output accuracy.

- To minimize instrument cost for a given output accuracy.

- To identify the instrument and/or procedure which controls accuracy.

- To assess the feasibility of an experiment to yield the results with desired accuracy.

- It serves as an appropriate basis for deciding whether (a) computations agree with data or lie outside acceptable limits, (b) data sets on one phenomenon or situation from two or more laboratories agree or disagree, and (c) testing the performance of a given hardware in separate facilities is necessary.

- To decide on when more accurate experiments must be provided to further "calibrate" approximate theory, e.g., in turbulence modeling.

- To provide the basis for guarantees of accuracy in commercial tests of large equipment, such as power plants.

- Allows design of probes for minimum uncertainty.

The uses of uncertainty analysis, listed above, are some of the primary ones which we think of in any experimental scheme. In addition to the above list there may be many more uses which may also be of high value. Some popularly used terms in uncertainty analysis and their older nomenclatures are given in Table 14.1.

Table 14.1 Some popular terms used in uncertainty analysis

Current usage	Earlier version
Precision	Repeatability
	Random error
	Random component of uncertainty
Bias	Fixed component of uncertainty
	Fixed error
	Systematic error

The first group of elements in Table 14.1 can be sampled (Kline, 1985) with the available procedures and apparatus, and should be based on statistical estimates from samples whenever possible. The second group of elements in this table cannot be sampled (via replication) within available procedures and/or apparatus and therefore must be estimated if required.

14.6 Uncertainty Estimation

The general procedure for estimating the uncertainties in the calculated quantities using measured data is described in this section. The derived general expression (Equation 14.9) has been employed to demonstrate the estimation of uncertainties associated with flow Mach number calculated using the measured values of total pressure and the ambient pressure.

14.7 General Procedure

Let x_1, x_2, x_3, $...x_i$,.. be the independent parameters (variables) in the experimental measurement, and u_1, u_2, u_3, $...u_i$,.. be the relative uncertainties of x_1, x_2, x_3, $..x_i$,... Let R be the experimental result calculated from the measured data.

The first step in the procedure is to analyze how errors in the x_i propagate into the calculation of R from the measured values. The quantity R can be expressed as

$$R = R(x_1, x_2, x_3, ..., x_i, ...x_n) \qquad (14.19)$$

The effect of error in measuring individual x_i on R may be estimated by analogy to a derivative of a function.

A variation δx_i in x_i would cause R to vary according to

$$\delta R_i = \frac{\partial R}{\partial x_i} \delta x_i \tag{14.20}$$

For applications, it is convenient to normalize the above equation by dividing throughout by R to obtain

$$\frac{\delta R_i}{R} = \frac{1}{R} \frac{\partial R}{\partial x_i} \delta x_i = \frac{x_i}{R} \frac{\partial R}{\partial x_i} \frac{\delta x_i}{x_i} \tag{14.21}$$

Equation(14.19) might be used to estimate the uncertainty interval in the result R, due to variation in x_i. To do this, substitute the uncertainty interval for x_i, namely

$$u_{R_i} = \frac{x_i}{R} \frac{\partial R}{\partial x_i} u_{x_i} \tag{14.22}$$

Uncertainty in R due to the combined effect of uncertainty intervals in x_i may be obtained by considering the following.

- The random error in each variable as a range of values within the uncertainty interval.

- The fact that it is unlikely that all errors will add to the uncertainty at the same time.

- It can be shown that the best representation for the uncertainty interval of the result is

$$u_R = \pm \left[\left(\frac{x_1}{R} \frac{\partial R}{\partial x_1} u_1 \right)^2 + \left(\frac{x_2}{R} \frac{\partial R}{\partial x_2} u_2 \right)^2 + \dots + \left(\frac{x_n}{R} \frac{\partial R}{\partial x_n} u_n \right)^2 \right]^{1/2} \tag{14.23}$$

This equation is the general expression for estimating the uncertainties in any calculated value from measured data. However, this expression has to be cast in the appropriate form before using it to estimate the uncertainty.

14.7.1 Uncertainty in Flow Mach Number

In this section, a procedure to estimate the uncertainty in flow Mach number M which is calculated from the measured total pressure and ambient pressure is given. The steps involved are as follows:

Obtain an expression for the uncertainty in determining the Mach number of a flow from measurements of total pressure p_t and the ambient pressure p_a. The Mach number in terms of p_t and p_a is

$$M = \left\{ \left[\left(\frac{p_t}{p_a} \right)^{\frac{\gamma-1}{\gamma}} - 1 \right] \frac{2}{\gamma - 1} \right\}^{1/2} = \chi^{1/2} \text{ (say)}$$

Differentiating, we get

$$
\begin{aligned}
dM &= \frac{\partial M}{\partial p_t}dp_t + \frac{\partial M}{\partial p_a}dp_a \\
&= \frac{1}{\sqrt{\chi}}\left[\frac{1}{\gamma p_a}\left(\frac{p_t}{p_a}\right)^{-\frac{1}{\gamma}}dp_t - \frac{1}{\gamma p_a}\left(\frac{p_t}{p_a}\right)^{\frac{\gamma-1}{\gamma}}dp_a\right]
\end{aligned}
$$

where

$$
\frac{\partial M}{\partial p_t} = \frac{1}{\sqrt{\chi}}\left[\frac{1}{\gamma p_a}\left(\frac{p_t}{p_a}\right)^{-\frac{1}{\gamma}}\right]
$$

$$
\frac{\partial M}{\partial p_a} = -\frac{1}{\sqrt{\chi}}\left[\frac{1}{\gamma p_a}\left(\frac{p_t}{p_a}\right)^{\frac{\gamma-1}{\gamma}}\right]
$$

The uncertainty in M can be obtained from

$$
u_M = \pm\left[\left(\frac{p_t}{M}\frac{\partial M}{\partial p_t}u_1\right)^2 + \left(\frac{p_a}{M}\frac{\partial M}{\partial p_a}u_2\right)^2\right]^{1/2}
$$

where u_1 and u_2 are the relative uncertainties of p_t and p_a.

High fluctuations in pitot (total) pressure were observed during measurements in the underexpanded flow field. By repeated observations, it was estimated that the maximum possible error in the measurement of total pressures (corresponding to the maximum stagnation pressure of 2050 mm (gauge), for $M_j = 1.52$) would be around 60 mm.

Hence, the relative uncertainty in total pressure p_t is

$$
u_{p_t} = u_1 = \pm\frac{\text{expected error in measured } p_t}{p_t \text{ measured}}
$$

$$
= \pm\frac{60 \text{ mm}}{2050 \text{ mm}} \quad \text{(for example)}
$$

$$
= \pm0.03 = 3\%
$$

The relative uncertainty in barometric height (ambient pressure) is

$$
u_{p_a} = u_2 = \pm\frac{0.5 \text{ mm}}{730 \text{ mm}} = 0.000685
$$

$$\chi = 2.329$$

$$\sqrt{\chi} = 1.526$$

$$\frac{\partial M}{\partial p_t} = \frac{1}{1.526}\left[\frac{1}{1.4 \times 730}\left(\frac{2780}{730}\right)^{-\frac{1}{1.4}}\right]$$

$$= 0.0002467$$

$$\frac{\partial M}{\partial p_a} = -\frac{1}{1.526}\left[\frac{1}{1.4 \times 730}\left(\frac{2780}{730}\right)^{\frac{1.4-1}{1.4}}\right]$$

$$= -0.0009398$$

$$\left(\frac{p_t}{M}\frac{\partial M}{\partial p_t}u_1\right)^2 = \left[\frac{2780}{1.52}(0.0002467)(0.03)\right]^2$$

$$= 0.0001832$$

$$\left(\frac{p_a}{M}\frac{\partial M}{\partial p_a}u_2\right)^2 = \left[\frac{730}{1.52}(-0.0009398)(0.000685)\right]^2$$

$$= 0.00000095$$

$$u_M = \pm(0.0001832 + 0.00000095)^{1/2}$$

$$= \pm 0.0135$$

$$= \boxed{\pm 1.35\%}$$

14.8   Uncertainty Calculation

Uncertainty analysis is the procedure used to quantify data validity and accuracy. Uncertainty analysis also will be useful in identifying the potential sources of unacceptable errors and suggesting improved measurement methods. Errors are always present in any experimental measurement methods. Usually experimental errors will be of the following two types.

- *Fixed (or systematic) error,* which makes repeated measurements to be in error by the same amount for each trial. This error is the same for each reading and can be removed by proper calibration or correction.

- *Random error (non-repeatability).* It is different for every reading and hence cannot be removed. The factors that introduce random error are uncertain by their nature.

The primary objective of uncertainty analysis is to estimate the probable random error in experimental results.

In our analysis here let us assume that the construction and calibration of our equipment and instruments are perfect to estimate fixed error, and only random errors are present in the measurements. Thus, our aim in the uncertainty analysis is to estimate the uncertainty of the experimental measurements and calculated results due to random errors. The procedure adopted usually has the following steps.

- Estimation of uncertainty interval for each measured quantity.

- Statement of the confidence limit on each measurement.

- Analysis of the propagation of uncertainty into results calculated from the experimental data.

Example 14.1

The atmospheric pressure at a place, measured by a barometer is 755 mm. If the least count of the vernier attached to the barometer is 0.1 mm, estimate the uncertainty in the atmospheric pressure measured.

Solution

The observed height of mercury column in the barometer is $h = 755$ mm.

The least count of the vernier scale is 0.1 mm.

Therefore, the probable measurement error may be ± 0.05 mm.

It is important to note that this error is ± 0.05 mm considering the vernier resolution alone. Probably this precise measurement is not possible since the barometer slider and meniscus must be aligned by eye. The slider has a least count of 1 mm. As a conservative estimate, a measurement could be made to the nearest millimeter. The probable value of a single measurement then would be expected as

$$755 \pm 0.5 \text{ mm}$$

The relative uncertainty in barometric height would be stated as

$$u_h = \pm \frac{0.5}{755}$$

$$= \boxed{\pm\,0.0006622\,\text{percent}}$$

14.9 Summary

The aim of any experiment is to measure some physical quantities as accurately as possible. Therefore, the error in the measurement must be estimated to arrive at the conclusion about the correctness of the results obtained. The error involved may be due to the uncertainties of various kinds. The possible imperfection of the theory on which the experiment is based, the imperfection of the instrument used for measurement, and the imperfection on the part of the observer are some of the major causes of uncertainties. An experimenter should cultivate the habit of never making measurements, or calculating any experimental parameter, without having a clear idea about the errors which are likely to be associated with the measurement.

The errors that are likely to be associated with the measured quantity x are

- the external estimate ϵ_E, based on the knowledge of experimental work carried out by others;

- the internal estimate ϵ_I, based on the data obtained during the experiment.

When these two estimates agree fairly closely the experiment may be treated as reliable.

The appropriate concept for expressing inaccuracies is *uncertainty*. An uncertainty is not the same as an error. An error in a measurement is just the difference between the correct value and the recorded value. An error is a fixed number and not a statistical variable. *An uncertainty is a possible value that the error might take on in a given measurement.* Since the uncertainty can take on various values over a range, it is basically a statistical variable. Uncertainty can be considered as a histogram of values.

In a broad sense, a measured value describes the central tendency, usually the mean, whereas the uncertainty describes the dispersion usually in terms of a measure associated with a stated probability level such as the standard deviation.

The primary uses of uncertainty analysis are (Kline, 1985)

- It enforces a complete examination of the experimental procedure, including the potential sources of error.

- It is useful to identify the need for employing improved instruments and/or improved procedure to obtain the desired output accuracy.

- To minimize instrument cost for a given output accuracy.

- To identify the instrument and/or procedure which controls accuracy.

- To assess the feasibility of an experiment to yield the results with desired accuracy.

- It serves as an appropriate basis for deciding whether (a) computations agree with data or lie outside acceptable limits, (b) data sets on one phenomenon or situation from two or more laboratories agree or disagree, and (c) testing the performance of a given hardware in separate facilities is necessary.

- To decide on when more accurate experiments must be provided to further "calibrate" approximate theory, e.g., in turbulence modeling.

- To provide the basis for guarantees of accuracy in commercial tests of large equipment, such as power plants.

- Allows design of probes for minimum uncertainty.

Uncertainty in R due to the combined effect of uncertainty intervals in x_i may be obtained by considering

- the random error in each variable as a range of values within the uncertainty interval;

- the fact that it is unlikely that all errors will add to the uncertainty at the same time;

- it can be shown that the best representation for the uncertainty interval of the result is

$$u_R = \pm \left[\left(\frac{x_1}{R} \frac{\partial R}{\partial x_1} u_1 \right)^2 + \left(\frac{x_2}{R} \frac{\partial R}{\partial x_2} u_2 \right)^2 + ... + \left(\frac{x_n}{R} \frac{\partial R}{\partial x_n} u_n \right)^2 \right]^{1/2}$$

This equation is the general expression for estimating the uncertainties in any calculated value from measured data. However, this expression has to be cast in the appropriate form before using it to estimate the uncertainty.

Uncertainty analysis is the procedure used to quantify data validity and accuracy. Uncertainty analysis also will be useful in identifying the potential sources of unacceptable errors and suggesting improved measurement methods. Usually experimental errors will be of the following two types.

- *Fixed (or systematic) error*, which makes repeated measurements to be in error by the same amount for each trial. This error is the same for each reading and can be removed by proper calibration or correction.

- *Random error (non-repeatability)* is different for every reading and hence cannot be removed. The factors that introduce random error are uncertain by their nature.

Exercise Problems

14.1 The pressure and temperature of an air stream are measured as 650 mm of mercury and 32°C, respectively. If the fluctuation in the pressure is 2 mm of mercury and error in the temperature measured is 0.1°, determine the uncertainty in the density calculated using thermal state equation.

[Answer: 0.31%]

14.2 The pressure and temperature of air in a storage tank, of volume 1 m³, are 3 bar and 30°C, respectively. The accuracy of the volume, pressure, and temperature given are ±0.001 m³, ±100 Pa, and ±0.1°C, respectively. Determine the uncertainty in the mass of air in the tank calculated using these parameters.

[Answer: 0.11 percent]

References

Abram, C., Fond, B., Heyes, A. L., Beyrau, F. (2013) High-speed planar thermometry and velocimetry using thermographic phosphor particles. *Applied Physics B-Lasers and Optics* 111 (2): pp. 155-160, 2013, doi:10.1007/S00340-013-5411-8.

Adhikari, D. and Longmire, E. Infrared tomographic PIV and 3D motion tracking system applied to aquatic predator-prey interaction, *Measurement Science and Technology*, Vol. 24, 024011, 2013, doi:10.1088/0957-0233/24/2/024011.

Barat, M., Influence de la turbulence sur les prises de pression statique, *Comptes Rendus*, Vol. 246, 1958, p. 1156.

Bolton, W. *Instrumentation and process measurements*, Orient Longman Ltd., Hyderabad, 1993.

Bradshaw, P. and Goodman, D. G. *The effect of turbulence on static pressure tubes*, R. & M. 1968, p. 3527.

Chiranjeevi Phanindra, B. and Rathakrishnan, E. Corrugated tabs for supersonic jet control, *AIAA Journal*, Vol. 48, No. 2, February 2010, pp. 453-465, DOI: 10.2514/1.44896.

Cook N. H. and Robinowicz, E. *Physical measurement and analysis*, Addison-Wesley, Inc., 1963.

Coutanceau, M. and Defaye J. R., Circular cylinder wake configurations: a flow visualization, *Journal of Applied Mechanics Review*, No. 6, June 1991, pp. 255-305.

Daum, F. L. and Gyarmathy, G. Condensation of air and nitrogen in hypersonic wind tunnels, *AIAA J.*, 6, pp. 458-465, 1968.

Deobelin, E. O. *Measurement systems - Application and design*, 4th ed., McGraw-Hill Book Company, 1986.

Dushman, S. and Lefferty, J. M. *Scientific foundations of vacuum technique*, John Wiley & Sons, Inc., 1962.

Elsinga, G. E., Scarano, F., Wieneke, B. and van Oudheusden, B. W. Tomographic particle image velocimetry, *Experiments in Fluids*, Vol. 41, pp. 933-947,

2006.

Flow meters computations handbook, ASME, 1961.

Fluid meters, their theory and applications, 6th ed., ASME, 1971.

Fond, B., Abram, C., Heyes, A. L., Kempf, A. M., Beyrau, F. Simultaneous temperature, mixture fraction and velocity imaging in turbulent flows using thermographic phosphor tracer particles, *Optics Express* 20 (20): pp. 22118-22133, 2012, doi:10.1364/oe.20.022118

Heiser, W. H. and Pratt, D. T. *Hypersonic air breathing propulsion*, American Institute for Aeronautics and Astronautics Inc., Washington, D.C., 1994.

Hemant Sharma., Ashish Vashistha, and Rathakrishnan, E. Twin vortex flow physics, Institution of Mechanical Engineers (UK), Part G, *Journal of Aerospce Engineering*, Vol. 222, No. 6, pp. 783-788, 2008.

Hinze, J. O. *Turbulence*, McGraw-Hill Book Company, 1987.

Hoerner, S. F. *Fluid dynamic drag*, New Jersey, 1965.

Holman, J. P. and Gajda, W. J. *Experimental methods for engineers*, McGraw-Hill Book Company, 1989.

Houghton, E. L. and Carruthers, N. B. *Aerodynamics for engineering students*, 3rd ed. 1982, Edward Arnold (Publishers) Ltd., Scotland.

Humble, R. A., Elsinga, G. E., Scarano, F. and van Oudheusden, B.W. Three-dimensional instantaneous structure of a shock wave/turbulent boundary layer interaction, *Journal of Fluid Mechanics*, Vol. 622, pp. 33-62, 2009.

Joel, R. *Basic engineering thermodynamics*, ELBS, Longman Group UK Ltd., 1991.

Kettle, D. J. The design of static and pitot-static tubes for subsonic speeds, *J. Royal Aero. Soc.*, Vol. 58, 1954, p. 835.

Kim, H., Große, S. S., Elsinga, G. E. and Westerweel, J. Full 3D-3C velocity measurement inside a liquid immersion droplet, *Experiments in Fluids*, 51(2), 395-405, 2011.

Kinzie, P. A. *Thermocouple temperature measurements*, John Wiley & Sons, Inc, 1973.

Kline, S. J. The purpose of uncertainty analysis, *Journal of Fluids Engineering*, Vol. 107, June 1985, pp. 153–182.

Koppenwallner, G. *Hypersonic flow simulation in Ludwieg tube wind tunnels*, Proceedings of the Recent Advances in Experimental Fluid Mechanics, December 18-20, 2000, IIT Kanpur, Eds. E. Rathakrishnan and A. Krothapalli, pp. 517-535.

Ladenburg, R. W. (Ed.) *Physical measurements in gas dynamics and combustion-part I*, Princeton University Press, Princeton NJ, 1954.

Liepmann, H. W. and Roshko, A. *Elements of gas dynamics*, John Wiley & Sons, New York, 1957.

Loh, W. H. T. (edited), *Modern developments in gas dynamics*, Plenum Press, 1969.

Lomas, C. G. *Fundamentals of hot-wire anemometry*, Cambridge University Press, 1986.

Morkovin, M. V. Fluctuations and hot-wire anemometry in compressible flows, AGARDograph, No. 24, 1956.

Omega temperature measurement handbook.

Omrane, A., Petersson, P., Alden, M., Linne, M. A. Simultaneous 2D flow velocity and gas temperature measurements using thermographic phosphors, *Applied Physics B-Lasers and optics*, Vol. 92: pp. 99-102, 2008.

Ower, E. and Pankhurst, R. C. *The measurement of air flow*, Pergamon Press, 1977.

Perry, A. E. *Hot-wire anemometry*, Clarendon Press, Oxford, 1982.

Pope, A. and Goin, K. L. *High-speed wind tunnel testing*, John Wiley & Sons, New York, 1965.

Quarmby, A. and Das, H. K. Displacement effect on pitot tubes with rectangular mouths, *Aeronautical Quart.*, Vol. 20, 1969, pp. 129-139.

Rathakrishnan, E. *Fluid mechanics an introduction*, 2nd ed., Prentice Hall of India, 2006.

Rathakrishnan, E. *Fundamentals of engineering thermodynamics*, Second edition, Prentice Hall of India, 2005.

Rathakrishnan, E. *Gas dynamics*, Prentice Hall of India, 1995.

Rathakrishnan, E. *Applied gas dynamics*, John Wiley, 2010.

Rathakrishnan, E. *Gas tables*, 3rd ed. 2012, Universities Press, Hyderabad, India.

Rathakrishnan, E. Visualization of the flow field around a flat plate, *IEEE Instrumentation and Measurement Magazine*, Vol. 15, Number 6, December 2012, pp. 8-12, DOI: 10.1109/MIM.2012.6365535.

Scarano, F. and Poelma, C. Three-dimensional vorticity patterns of cylinder wakes, *Experiments in Fluids*, Vol. 47, pp. 69-83, 2009.

Takama, Y., Suzuki, K. and Rathakrishnan, E. Visualization and size measurement of vortex shed by flat and arc plates in an uniform flow, *International*

Review of Aerospace Engineering (IREASE), Vol. 1, No. 1, pp. 55-60, February 2008.

The temperature handbook, Vol. 28, Omega Engineering, Stamford, CT, USA, 1992.

Tuve, G. L. *Mechanical engineering experimentation*, McGraw-Hill Book Company, 1961.

Violato, D., Moore, P. and Scarano, F. Lagrangian and Eulerian pressure field evaluation of rod-airfoil flow from time-resolved tomographic PIV, *Experiments in Fluids*, 50(4), pp- 1057-1070, 2010, DOI: 10.1007/soo348-010-1011-0.

White, F. M. *Fluid mechanics*, 2nd ed., McGraw-Hill Book Company, 1986.

White, F. M. *Viscous fluid flow*, 2nd ed., McGraw-Hill Book Company, 1991.

Index

Printed in the United States
by Baker & Taylor Publisher Services

Printed in the United States
by Baker & Taylor Publisher Services